An
Introduction
to
Quantum
Physics

A. P. French

PROFESSOR OF PHYSICS, THE MASSACHUSETTS INSTITUTE OF TECHNOLOGY

Edwin F. Taylor

SENIOR RESEARCH SCIENTIST, THE MASSACHUSETTS INSTITUTE OF TECHNOLOGY

An Introduction to Quantum Physics

THE M.I.T. INTRODUCTORY PHYSICS SERIES

W·W·NORTON & COMPANY · INC · NEW YORK

Copyright © 1978 by the Massachusetts Institute of Technology

Library of Congress Cataloging in Publication Data
French, Anthony Philip.
An Introduction to Quantum physics.
   (The M.I.T. introductory physics series)
   Bibliography: p.
   Includes index.
   1.   Quantum theory. I. Taylor, Edwin F., co-
author.   II. Title.   III. Series: Massachusetts Institute of Technology.
Education Research Center. M.I.T. introductory physics series.
QC174.12.F73   530.1'2   78-4853
ISBN 0-393-09015-9
ISBN 0-393-09106-0 pbk.

Printed in the United States of America

All Rights Reserved

First Edition

1 2 3 4 5 6 7 8 9 0

Contents

7   Quantum amplitudes and state vectors   279

8   The time dependence of quantum states   315

9   Particle scattering and barrier penetration   367

Contents

Preface

Quantum physics concerns the behavior of the smallest things we know. These smallest things are very small indeed. Hold up two fingers together in front of you. The diameter of an atom is approximately the same fraction of the width of two fingers as the width of two fingers is of the diameter of the earth (10^{-8} cm is to 3 cm as 3 cm is to 10^9 cm). Our expectations about how things behave are shaped by experiences with objects large enough to see and handle; no wonder these expectations are sometimes wrong when applied to objects as small as an atom. In the same way the "classical" laws of physics, particularly Newtonian mechanics, devised to describe the behavior of objects of visible size, must be modified—and in some respects completely replaced—in order to account for the behavior of atoms and subatomic particles.

Although the world of the very small is remote from our senses, it shapes everyday experience. Almost everything we touch and see (together with nerve impulses and light, the messengers of touch and sight) owes its character to the subtle architecture of atoms and molecules, an architecture whose building code is quantum mechanics. And when we come to large-scale phenomena that depend in a direct way on the details of atomic processes—for example lasers, superconductors, and solid-state electronics—then the explicit use of quantum physics is essential.

In contrast to the generally well-established order of presenting the subject matter of classical physics, there are many differing views concerning the strategy and order of presentation appropriate to quantum physics. The sequence that we have chosen for this book begins with a description of some of the "facts of life" in the atomic world (Chapters 1 and 2), and then moves as quickly as possible to the consequences of wave-particle duality in physical systems (Chapters 3 through 5). After reaching this plateau, we use polarized photons to examine more carefully the form and meaning of the wave-mechanical description (Chapters 6 and 7). The remainder of the book (Chapters 8 through 14) then extends the basic ideas to situations and systems of increasing complexity, culminating in the structure of many-electron atoms and a discussion of radiation.

Text and exercises alone provide a pedagogically rather limited introduction to the radically different world of the very small. Happily a growing intuition for quantum physics can also be encouraged by a wide range of learning aids: filmed demonstrations, take-home experiments, and computer-generated films that play out the predictions of quantum physics. Some of these have been developed at the MIT Education Research Center as part of the effort that produced this text. Following this introduction is a brief description of some available learning aids.

This book is not a general survey of "modern physics," with separate sections on such fields as nuclear and solid-state physics. Rather, the emphasis throughout is on the experimental and theoretical underpinnings of quantum mechanics, with examples from various areas of physics selected to illuminate how the theory works in practice. We have chosen to concentrate on a development of the Schrödinger method, but have introduced the matrix notation and indicated some relations between the two descriptions.

This book is one in the MIT Introductory Physics Series, prepared under the auspices of the MIT Education Research Center, which came to the end of its active existence in 1973. (The volumes previously published are *Newtonian Mechanics*, *Vibrations and Waves*, and *Special Relativity*.) This part of the center's work was sponsored mainly by grants from the U.S. National Science Foundation and enabled us, as authors, to benefit from the advice of many people. Physicists from MIT and elsewhere discussed at length how this presentation should be organized and provided criticism and guidance during its many trials. David Park and Walter Knight offered helpful comments on early drafts of the text, and Arthur K. Kerman and Leo Sartori collaborated with one of us (EFT) in writing them. William H. Ingham made helpful suggestions about the text and devised many exercises. Charles P. Friedman also read and commented on the material and (with guidance from Felix Villars) contributed to part of the argument of Chapter 10. David Root and James Rothstein developed solutions to many exercises. Robert I. Hulsizer provided support and encouragement during his time as director of the center. Jerrold R. Zacharias and the late Francis L. Friedman, who founded the center, have influenced our work at every stage through their insistence on a "clean story line" and on the illumination of theory by experiment. Finally, we are deeply grateful to several generations of students, at MIT and elsewhere, whose insights and perplexities have helped us to chart a course through this fascinating subject.

A. P. FRENCH
EDWIN F. TAYLOR

Cambridge, Massachusetts
January, 1978

Learning aids for quantum physics

The fundamental ideas of quantum physics are simple but subtle. A growing intuition for these subtleties can sometimes be encouraged by seeing them in different embodiments: demonstration experiments, take-home experiments, and computer-generated films that play out the predictions of the theory. Listed below, in order of presentation of the corresponding subject matter in the book, are a few such learning aids found useful in teaching from preliminary versions of this text.

We have quoted sources for these materials in the United States. In most cases alternative sources exist, both in the U.S.A. and elsewhere. Distributors change faster than, say, book publishers, so this list may be out of date by the time you read it. We apologize for this but feel that the effort to establish a source and to order these materials in advance of class use can pay off handsomely in increased student interest and understanding.

Sale and rental sources for films in the list are coded as follows: EDC—Education Development Center (39 Chapel Street, Newton, Massachusetts 02160); MLA—Modern Learning Aids (*Sales*, Modern Learning Aids, P.O. Box 1712, Rochester, New York 14603; *Rental*, Modern Film Rentals, 2323 New Hyde Park Avenue, New Hyde Park, New York 11040). Unless otherwise noted, all films are 16 mm.

Polarizing materials (linear, circular, and quarter-wave plate) and plastic diffraction grating are available from scientific supply houses such as Edmund Scientific Co. (Barrington, New Jersey 08007). Large sheets of polarizing materials can be obtained through the Polaroid Corporation (Polarizer Division, Cambridge, Massachusetts 02139).

Probability and Uncertainty: the Quantum-Mechanical View of Nature, Richard P. Feynman, British Broadcasting Corporation, 60 minutes, black and white, sound. Lecture number six in the series The Character of Physical Law. Source: EDC. The best one-hour introduction to quantum physics we have seen. Feynman rings all the changes on the two-slit interference experiment with particles, revealing many of the essential paradoxes and delights of quantum physics. He does not mention that such experiments have actually been done with electrons (page 76 of the text). The film is too long for a fifty-minute lecture hour; we recommend a relaxed evening showing followed by discussion and refreshments. The script is available with those of other lectures in the series in printed form (BBC Publications and MIT Press), but no one should pass up a chance to see Richard Feynman in action. The film could be used at any time during a course, but the earlier the better.

Diffraction grating. Carry a small square plastic diffraction grating everywhere you go and view the spectra of city lights. Neon signs are a favorite. Blue mercury and yellow sodium vapor street lights show characteristic spectra for these elements. They are being replaced with the more efficient high-pressure lamps, in which collision broadening smears out the spectral lines. Chapter 1.

Franck-Hertz Experiment, Byron Youtz, PSSC Physics, 30 minutes, black and white, sound (English or Spanish). Source: MLA. A classic demonstration that electrons require a minimum kinetic energy to raise mercury atoms from their ground state to their first excited state. In a delightful epilogue, James Franck reminisces about the original experiment by himself and Gustav Hertz, and comments on their ignorance of the Bohr theory at the time they performed it. Chapter 1.

Matter Waves, Alan Holden and Lester Germer, PSSC Physics, 28 minutes, black and white, sound (English or Spanish). Source: MLA. A beautifully done film in the style of the fifties that examines the evidence for wavelike properties

Learning aids for quantum physics

of particles. Lester Germer adds personal immediacy to the presentation by describing the original Davisson-Germer experiment and demonstrating a modern application. Chapter 2.

Quantum Mechanical Harmonic Oscillator, Alfred Bork, 4 minutes, black and white, silent, 16 mm and Super-8 cartridge. Source: EDC. Computer-animated film that uses the condition that the wave function approaches zero at infinite distance to generate wave functions and energy eigenvalues for the simple harmonic oscillator. Chapter 4.

Energy Eigenvalues in Quantum Mechanics, Harry Schey, MIT Education Research Center, 3 minutes, black and white, silent. Source: EDC. This computer-animated film shows how the same energy eigenvalues arise for a bound particle when each of four boundary or continuity conditions in turn is temporarily suspended and trial energies are varied until this condition is also satisfied. Chapter 4.

Polarized light kit. Small squares of linear polarizing sheet, circular polarizing sheet, and quarter-wave plate cost but a few cents apiece. The exercises to Chapter 6 begin with some suggested experiments with linearly polarized light. Cleaved calcite rhombs ($\frac{3}{4}$-inch size is sufficient) add to the interest of such take-home experiments, but they probably cost about two dollars apiece and we know of no current source for them.

Polarization of Single Photons, Stephan Berko, MIT Education Research Center, 15 minutes, color, sound. Source: EDC. Produced as part of the effort that resulted in the present text, this film describes the cosine-squared law of transmission through two linear polarizers when the transmission axes are rotated with respect to one another, and carries the experiment down to very low intensities at which the statistical character of quantum events becomes apparent. Chapter 6.

Interference in Photon Polarization, James L. Burkhardt, MIT Education Research Center, 4 minutes, color, silent. Source: EDC. Demonstrates the paradox of the recombined beams presented in Chapter 7, an interference effect that can be used to support the conclusion that amplitudes, not just probabilities, are required to describe quantum events. Chapter 7.

Quantum Physics Series, Judah L. Schwartz, Harry Schey, Abraham Goldberg, MIT Education Research Center, 2 to 4 minutes, black and white, silent, 16 mm and Super-8

Learning aids for quantum physics

cartridge. Sources: 16 mm—EDC; Super-8 cartridge—BFA Educational Media (2211 Michigan Avenue, Santa Monica, California 90404). A series of computer-generated films that show probability packets for various one-dimensional phenomena: free particles, scattering from wells and barriers, bound particles (and three titles not listed below demonstrating propagation in crystals). Momentum representation is shown, too. One film (QP-10, below—available at present only in 16 mm) makes the important connection between probability packet and individual observations.

QP-1   Barriers: Scattering in One Dimension
QP-2   Wells: Scattering in One Dimension
QP-3   Edge Effects: Scattering in One Dimension
QP-4   Momentum Space: Scattering in One Dimension
QP-5   Free Wave Packets
QP-10 Individual Events in One-Dimensional Scattering
QP-15 Particle in a Box

Several figures in Chapters 8 and 9 are stills from these films.

The Stern-Gerlach Experiment, Jerrold Zacharias, MIT Education Research Center, 27 minutes, black and white, sound. Source: MLA. A careful experimental demonstration that a nonuniform magnetic field splits a beam of cesium atoms into two distinct beams corresponding to different angular momentum quantum states. Several figures in Chapter 10 are taken from this film.

An
Introduction
to
Quantum
Physics

It is difficult for a young physicist . . . to realize the state of our science in the early 1920's . . . It was not just that the old theories of light and mechanics had failed. On the contrary. You could say that they had succeeded in regions to which they could hardly have been expected to apply, but they succeeded erratically . . . And over the whole subject brooded the mysterious figure of h.

G. P. THOMSON, *Symposium on the History of Modern Physics* (1961)

It could be that I've perhaps found out a little bit about the structure of atoms. You must not tell anyone anything about it . . .

NIELS BOHR, *letter to his brother* (1912)

1
Simple models of the atom

We know that classical physics, as represented by New-tonian mechanics and Maxwell's laws of electromagnetism, works marvelously well for the analysis of the behavior of macroscopic objects in terms of empirically determined laws of force. But as soon as we enter the world of the atom, we find that new phenomena appear, requiring new concepts for their analysis and description. The whole realm of phenomena at the atomic or subatomic level is the special province of quantum theory. However, because the behavior of matter in bulk ul-timately results from the properties of its constituent atoms, our deeper insights into physical phenomena on the macro-scopic scale will often also depend on quantum theory. For ex-ample: We can do a vast amount of useful analysis of the mechanical behavior of solids using measured values of their elastic constants, tensile strengths, etc. But if we want to ac-count for these measured values in terms of more fundamental processes, we must invoke quantum theory. It is at the root of our whole understanding of the structure of matter.

The properties of atoms—and even the fact of their exis-tence—pose a series of questions unanswerable by classical physics:

Atoms are typically a few angstroms in diameter ($1\text{Å} = 10^{-8}$ cm) with remarkably little difference in size between the lightest and the heaviest (see Figure 1-1).

1

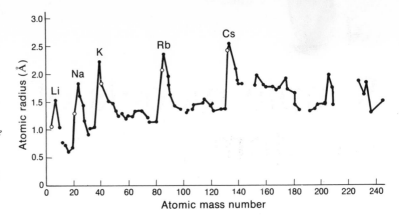

Fig. 1-1 Atomic radii. There is very little increase in atomic size with increasing atomic mass number. Note the periodic variation in radius, with the maximum radii being those of the alkali atoms.

Why this size rather than some other? And why not a wide range of sizes?

When isolated from radiation and other atoms, most atoms remain stable indefinitely: they neither collapse nor explode. Why do not the negatively charged electrons collapse into the positively charged nucleus, thereby destroying the atom to the accompaniment of a burst of radiation?

When atoms are excited electrically or by collisions or otherwise, they emit radiation of discrete wavelengths characteristic of the kind of atoms excited (see Figure 1-2). Why discrete wavelengths rather than a continuous spectrum? And how can a particular spectrum be accounted for, as well as differences between spectra of different kinds of atoms?

These questions are only a beginning. Why are some kinds of atoms more reactive chemically than other kinds? Why are some substances harder, denser, more transparent, more elastic, more electrically conductive, more thermally conductive, more digestible than other substances? All such questions can be related to the properties of atoms, and we can understand them only if we possess the facts and concepts embodied in quantum mechanics.

In this book we will turn our attention again and again to the atom, each time from a different point of view or level of sophistication. In the present chapter we discuss a few of the simplest models of the atom, all of them basically classical in nature, with one or two additional assumptions to help the classical models behave more like the observed quantum sys-

Simple models of the atom

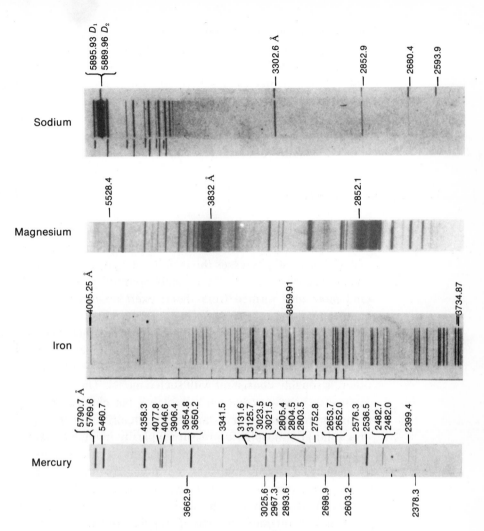

Fig. 1-2 *Emission spectra of various vapors. The pattern of emission lines in each spectrum is characteristic of the particular chemical element. (Spectra reproduced from G. Herzberg, Atomic Spectra and Atomic Structure, Dover Publications, Inc., New York, 1944. Reprinted through the permission of the publisher.)*

tems. Despite their crude nature these models can be used to correlate, even if they cannot be said to explain, a wide range of observations. The ultimate failures of these models force us to look more deeply and to return repeatedly to the atom with models of increasing sophistication.

Why start with crude models? Why return again and again to them when the "real" answers are already known?

3   1-1   Introduction

Why not tell the quantum-mechanical truth straight out and then stop? Some readers may feel equipped to go straight to the now-accepted answers, and they can begin their study with a later chapter of this text. But for most people crude atomic models provide a gradual conceptual transition from classical descriptions to the "true" quantum statements about atoms, statements that seem strange and awkward at first but later on become comfortable, simple, and natural.

1-2 THE CLASSICAL ATOM[1]

The simplest model of the atom is a hard, tiny, electrically neutral sphere—just the smallest possible fragment of the bulk material that still possesses the identity of a given chemical element. According to this conceptually primitive picture, atoms (and molecules formed from them) exert no forces on one another until they are brought in contact, and then they offer infinite resistance to being forced any closer together. The dramatic difference in behavior between a substance in vapor form, on the one hand, and in its solid or liquid phase, on the other, is roughly consistent with such a model. This major difference in behavior does not involve a big change in interatomic or intermolecular distances. The vaporization of a liquid or a solid at atmospheric pressure typically involves a factor of about 1000 increase in volume, but this is only a factor of 10 in particle separation. Yet the atoms or molecules in a vapor move almost independently of one another, whereas the highly incompressible behavior of the same substance in condensed form indicates strong repulsions between atoms once a densely packed arrangement has been reached. Thus the model of atoms as small spheres with more or less well-defined radii is a reasonable starting point. Of course, the model does not explain the cohesive properties of liquids and solids.

How big are atoms, and what are they made of? The answers to both these questions are well known to us today, but they remained mysteries for a very long time. Although some rough estimates of atomic sizes were deduced from the macroscopic properties of gases and liquids (see the exer-

[1]For a good introduction to early models of the atom, see F. L. Friedman and L. Sartori, *The Classical Atom*, Addison-Wesley, Reading, Mass., 1965.

Simple models of the atom

cises), the main clarifications came with the discovery of the essentially electrical substructure of atoms. In the next section we outline some of the main features of this development for those readers who may not be familiar with it.

1-3   THE ELECTRICAL STRUCTURE OF MATTER

The Laws of Electrolysis

Between 1831 and 1834 Michael Faraday carried out a series of decisive experiments on the electrochemical decomposition (electrolysis) of dissolved substances. Direct current was passed between two electrodes suspended in various solutions and measurement made of the relative amounts of solid or gas liberated at each electrode. For example, the electrolysis of water yields oxygen gas at one electrode and twice as great a volume of hydrogen gas at the other electrode. First Faraday discovered that the amount (mass of solid or volume of gas) of any one product extracted was proportional to the total amount of electricity passed through the solution. Then, comparing a variety of substances, he found that the relative masses of different *elements* liberated in electrolysis were (for the same transfer of charge) the same as the proportions in which these elements combined with one another in chemical reactions. (Before Faraday's experiments, the masses of elements that combine with one another had already been standardized, with hydrogen, the lightest gas, taken as unity. The modern term for this measure is *gram-equivalent*: 1 gram of hydrogen combines, for example, with 8 grams of oxygen to yield 9 grams of water; the *gram-equivalent* of oxygen is 8 grams.)

Faraday interpreted his electrolytic results in terms of the transport of current by positive or negative *ions*—individual atomic or molecular fragments—charged with characteristic amounts of electricity. In his own words, "... if we adopt the atomic theory or phraseology, then the atoms of bodies which are equivalents to each other in their ordinary chemical action, have equal quantities of electricity naturally associated with them."[2]

Faraday's electrolysis experiments provided a value for

[2]M. Faraday, *Experimental Researches in Electricity*, Series VII, January 1834, reprinted by Dover Publications, New York, 1965.

1-3   The electrical structure of matter

the amount of charge needed to release a gram-equivalent of any element. In modern terms, the amount of charge carried by one gram-equivalent of any ionic species is appropriately called the *Faraday*:

$$1 \text{ Faraday (F)} = 96,485 \text{ coulombs} \qquad (1\text{-}1)$$

Molecules, Atoms, and Ions

The full interpretation of Faraday's observations required an understanding of the *molecule* as the basic unit of any chemical substance. A crucial feature was the realization that a molecule might contain more than one atom of a constituent element. As the combining properties of atoms became clearer (about the time of Faraday's death in 1867), it was possible to write down dependable formulas for molecules: N_2, H_2O, etc. Thereafter one could devise a measure of the amounts of pure substances that contain *equal* numbers of molecules. Such a measure is the *gram-molecule*, often called the *mole*. One mole of a pure substance is (approximately) that mass in grams equal to its molecular weight, with atomic hydrogen taken to be unity.[3] Thus 1 mole of water, H_2O, has a mass of 18 grams: 2 for the hydrogen atoms in each molecule and 16 for the single oxygen. (This is, of course, the same proportion, 1:8, expressed by the gram-equivalents.) When applied to an element (such as helium or atomic hydrogen) the mole is called the *gram-atom*. Although there is an equal number N of molecules in one mole of every pure substance, the relative proportions of combining elements cannot by themselves tell us the *value* of this number N, called Avogadro's number. To find Avogadro's number one needs the measure of at least one property of an *individual* atom or molecule. The charge carried by an atomic or molecular ion in electrolysis can fill this role. Thus, for example, in the electrolysis of water we have

$$H_2O \rightarrow 2H^+ + O^{--}$$

Each of the atomic hydrogen ions in this relation carries one basic atomic unit of electric charge, the oxygen carries two (negative) units. The passage of one Faraday of electricity

[3]More exactly, the modern definition of the mole uses as a standard reference *exactly* 12 g of the isotope C^{12} of carbon.

Simple models of the atom

then liberates one gram-equivalent (equal to half a gram-atom) of oxygen. If the basic atomic unit of charge can be found, then the absolute number of atoms in one gram-atom of an element can be deduced. The discovery of the electron paved the way to this crucial step.

The Discovery of the Electron

The electron, as a constituent of all kinds of atoms, was discovered by J. J. Thomson in 1897.[4] This was a culmination of several decades of experiments by many different workers on electrical discharges in gases at low pressure. It had been established that a stream of some kind of rays, called cathode rays, was emitted by any electrode charged to a high negative potential in an evacuated tube. Were these rays particles or electromagnetic radiation? There seemed to be evidence for both interpretations, but then Thomson clarified the situation through his decisive experiments.

Figure 1-3a shows the basic arrangement he used. Cathode rays from C passed through the collimating slits G and H, then between the charged metal plates J and K, and finally struck a fluorescent screen at the end of the evacuated tube. An electromagnet (not shown) supplied a magnetic field parallel to the plates J and K (and perpendicular to the cathode-ray beam) over a distance equal to the length l of the plates. In this way the electric and magnetic forces could be balanced. If the cathode rays consist of particles of mass m and charge e traveling with speed v, we have

$$F_{\text{electric}} = eE \text{ (cgs)} \qquad [= eE \text{ (SI)}]$$

$$F_{\text{magnetic}} = \frac{evB}{c} \text{ (cgs)} \qquad [= evB \text{ (SI)}]$$

where E and B are the electric and magnetic field strengths. (The equations are written first in Gaussian units, with e in esu. The corresponding expression in SI units is enclosed in parentheses.) When the fields are adjusted to provide equal and opposite forces, the beam passes through undeflected, and

$$v = \frac{cE}{B} \text{ (cgs)} \qquad \left[= \frac{E}{B} \text{ (SI)} \right]$$

[4]J. J. Thomson, Phil. Mag. **44**, 293 (1897).

1-3   The electrical structure of matter

Thus the speed of the cathode-ray particles is determined. Then, by measuring the deflection of the beam in the electric or magnetic field alone, the charge-to-mass ratio of the particles could be deduced. If (as Thomson did) we take the electrostatic deflection alone, we have an acceleration eE/m (perpendicular to the incident beam) during the time l/v that the particles are traveling between the deflector plates. Thus the transverse velocity component v_t given to the particles is eEl/mv, and the change of direction of the beam as it emerges from the field region is given (Figure 1-3b) by

$$\tan \theta = \frac{v_t}{v} = \frac{eEl/mv}{v} = \frac{eEl}{mv^2}$$

Substituting for v in terms of the measured quantities E and B, this gives

$$\frac{e}{m} = \frac{c^2 E}{B^2 l} \tan \theta \text{ (cgs)} \qquad \left[= \frac{E}{B^2 l} \tan \theta \text{ (SI)} \right]$$

Thomson found that in his experiments the speed of the particles was extremely high (about one-tenth of the speed of light) and that the value of e/m was about 2000 times greater

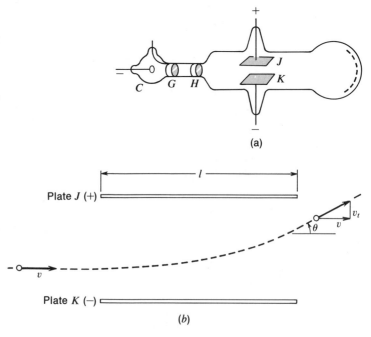

(a)

Plate J (+)

Plate K (−)

(b)

Simple models of the atom

than for the highest charge-to-mass ratio found in electrolysis (for the H^+ ion). The value of e/m for the cathode rays (electrons) according to modern data is given by

$$\frac{e}{m} = 5.2728 \times 10^{17} \text{ esu/g} = 1.7588 \times 10^{11} \text{ coulomb/kg} \qquad (1\text{-}2)$$

Thomson could not feel sure whether this very high value of e/m was due solely to a very small value of m, with the same unit of charge that was instrumental in electrolysis, or whether it was in part due to a large intrinsic charge. But he *was* able to show that this new particle, the electron, with its absolutely characteristic charge-to-mass ratio, was a constituent of every substance that he was able to use as a cathode material.

The Elementary Charge

At the time of Thomson's discoveries concerning free electrons, evidence was being obtained on the size of the charge carried by electrolytic ions. These experiments consisted in measuring the total charge carried by a mist of droplets to which ions had attached themselves. The mean size of the droplets (and hence their number) was inferred from the rate at which the mist settled. Such experiments were the prelude to the famous determination of the fundamental unit of charge e by R. A. Millikan in his "oil-drop" experiments.[5] Millikan demonstrated very directly what he called "the unitary nature of electricity," the fact that electric charge is *quantized* and transferred in integral multiples of e. His first convincing evidence for this was obtained in 1909. However, his precise measurements of e did not come until after Niels Bohr had published his quantum theory of the hydrogen atom in 1913. Bohr appears to have used a value of e inferred much earlier (in 1900) by Max Planck as a by-product of his theoretical analysis of thermal radiation (see Photons, Section 1-6). Planck[6] found

$$e = 4.69 \times 10^{-10} \text{ esu}$$

[5]For a full account of these experiments, as well as Millikan's own work, see R. A. Millikan, *The Electron*, University of Chicago Press, Chicago, 1963.

[6]M. Planck, Verh. Dtsch. Phys. Ges. **2**, 244 (1900).

1-3   The electrical structure of matter

The value of the elementary charge, according to modern determinations, is given by

$$e = 4.80324 \times 10^{-10} \text{ esu} = 1.6022 \times 10^{-19} \text{ Coulomb} \qquad (1\text{-}3)$$

Taking this value of e, together with the charge-to-mass ratio of the electron (Eq. 1-2), we arrive at the electron mass:

$$m = \frac{4.80324 \times 10^{-10}}{5.2727 \times 10^{17}} = 9.109 \times 10^{-28} \text{g}$$

Atomic masses and sizes

We can now determine Avogadro's number, N. Making the reasonable assumption that e is equal both to the magnitude of the charge of the electron and also to the smallest charge carried by any individual ion in electrolysis, we can at once combine Eqs. 1-1 and 1-3 to find the number of atoms or molecules in one mole—that is, Avogadro's number, N:

$$N = \frac{F}{e} = \frac{96,485}{1.6022 \times 10^{-19}} = 6.022 \times 10^{23} \qquad (1\text{-}4)$$

With this crucial constant established, we can then infer the values of individual atomic masses (these are just the gram-atomic weights divided by N) and of atomic volumes and radii. To find these latter, we can use the densities of the elements in solid form. If the density is ρ and the atomic weight is A, the number of atoms per unit volume n is given by

$$n = \frac{\rho N}{A}$$

The volume per atom V_a is then equal to $1/n$. To estimate the atomic radius R, we assume a simple cubic packing of atoms in a solid (Figure 1-4). Then the atomic centers are separated

Fig. 1-4 A simple cubic array of atoms. The center-to-center distance between neighboring atoms is 2R and each atom effectively occupies a cube of edge 2R.

Simple models of the atom

by a distance $2R$, and each atom in effect takes up the volume of a cube of edge $2R$:

$$V_a = (2R)^3 = \frac{1}{n} = \frac{A}{\rho N} \qquad (1\text{-}5)$$

so that

$$R \approx \frac{1}{2} \left(\frac{A}{\rho N} \right)^{1/3}$$

Some results of such calculations for different substances are listed in Table 1-1. Although the analysis is admittedly crude, it does illustrate the point mentioned earlier, that light and heavy atoms have much the same size, corresponding to a radius of the order of 1 Å. This fact is an important datum in any theory of atomic structure. In the next section we shall describe the atomic model that J. J. Thomson based on little else besides this fact and the knowledge that the atom contains individual electrons.

TABLE 1-1   Atomic Radii as Estimated From Density in Solid State

	Density in Solid State (g/cm^3)	Atomic Weight A	Atomic Radius R $(Å)$ From Eq. 1-5
Carbon	3.5	12	0.9
Aluminum	2.7	27	1.3
Iron	7.8	56	1.2
Molybdenum	9.0	96	1.3
Barium	3.8	137	2.9
Platinum	21.4	195	1.2
Uranium	18.7	238	1.4

Compare these values with those in Figure 1-1, which are based on similar but more careful calculations that take account of the particular lattice structure involved in each case.

1-4   THE THOMSON ATOM

After discovering the electron, J. J. Thomson proceeded to develop a theoretical model of the atom as a ball of positively charged material, representing most of the atomic mass, in which the electrons were somehow embedded. The radius of the positively charged ball was to be identified as the radius of the atom itself, that is, $R \approx 10^{-8}$ cm. There was reason to believe that the electrons were almost point charges on this scale of sizes. A calculation based on classical electromagnetic theory indicated that the radius of the electron was given by the relation

$$r_0 = \frac{e^2}{mc^2} = 2.8 \times 10^{-13} \text{ cm} \left[\times \frac{1}{4\pi\epsilon_0} (SI) \right] \qquad (1\text{-}6)$$

This same result can be obtained very directly using special relativity if we assume that the whole mass of the electron is the mass-equivalent of the electrostatic potential energy of the charge e confined with a sphere of radius r_0, so that we can put

$$\text{Energy} = mc^2 \approx \frac{e^2}{r_0} \text{ (cgs)} \qquad \left[\approx \frac{e^2}{4\pi\epsilon_0 r_0} \text{ (SI)} \right]$$

[The above relationship is written with an "approximately equals" sign because the potential self-energy depends on the details of the charge distribution within the radius r_0, but in order of magnitude the potential energy is certainly expressible as e^2/r_0 (cgs).]

Maxwell had long since shown (in 1864) that light was to be understood as electromagnetic radiation, and H. Hertz had demonstrated in 1887 that radiation is emitted by accelerated charges. Thus it was natural to assume that spectral lines were produced by the periodic motions of the electrons inside the atom. Indeed, already in 1896 (a year before Thomson discovered the electron as a free particle) P. Zeeman had studied the splitting of certain spectral lines when the source was placed in a strong magnetic field. The results were interpreted by H. A. Lorentz as showing that the radiation was produced by charged particles with a ratio q/m about 2000 times greater than that of the H^+ ion—that is, by J. J. Thomson's electron![7]

It is easy to calculate the order of magnitude of frequency

[7]For a full account of the development of electromagnetism and early atomic theory, and of the relations between them, see Sir Edmund Whittaker, *A History of the Theories of Aether and Electricity*, Harper & Row, New York, 1960.

Simple models of the atom

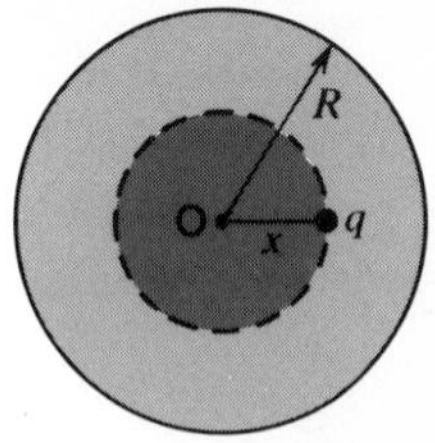

Fig. 1-5 *One-electron Thomson atom. The force on the point charge q, located a distance x from the center O, is due to the charge contained within a sphere of radius x.*

to be expected for the linear oscillations of an electron through the center of a Thomson atom. Assuming for simplicity that the positive charge density is uniform up to the radius R (see Figure 1-5), a point charge q at distance x from the center experiences a force given by

$$F = q \cdot \frac{x^3 Q}{R^3} \cdot \frac{1}{x^2} \text{ (cgs)} \qquad \left[= \frac{q}{4\pi\epsilon_o} \cdot \frac{x^3 Q}{R^3} \cdot \frac{1}{x^2} \text{(SI)} \right]$$

since (by Gauss's law) the force on it is due to the fraction x^3/R^3 of the total charge, acting as though it were concentrated at the center of the sphere.[8] The equation for the linear motion of the electron through the center is thus

$$m \frac{d^2 x}{dt^2} = \frac{qQx}{R^3} \text{ (cgs)} \qquad \left[= \frac{qQx}{4\pi\epsilon_o R^3} \text{ (SI)} \right]$$

Taking the hydrogen atom as the obviously simple example, we put $Q = e$, $q = -e$, so that the above equation becomes

$$\frac{d^2 x}{dt^2} = -\frac{e^2}{mR^3} x \text{ (cgs)}$$

This is the differential equation of simple harmonic motion. The frequency ν is found by substituting into this equation the solution $x = A \sin 2\pi\nu t$, giving

$$4\pi^2\nu^2 = \frac{e^2}{mR^3}$$

$$\nu = \frac{1}{2\pi} \left(\frac{e^2}{mR^3} \right)^{1/2} \text{ (cgs)} \qquad \left[= \frac{1}{2\pi} \left(\frac{e^2}{4\pi\epsilon_o mR^3} \right)^{1/2} \text{ (SI)} \right]$$

[8]Again we are writing the equations of electromagnetism primarily in the Gaussian system.

1-4   The Thomson atom

Putting $e^2 \approx 2 \times 10^{-19}$ (esu)^2, $m \approx 10^{-27}$ g, $R \approx 10^{-8}$ cm, we obtain $\nu \approx 2 \times 10^{15}$ sec^{-1}, thus defining a wavelength $\lambda = c/\nu = 1500$ Å. This is of the correct order of magnitude. [The wavelength of a typical spectral line in the visible region is ≈ 5000 Å (green).] However, with this simple model, we can get only this one frequency. (Any elliptic orbit of the electron, so long as it remains within the uniform sphere of positive charge, has the same frequency as we have calculated for the linear oscillation; see the exercises.)

Thomson worked hard on detailed models for atoms containing several electrons, but with very limited success. His main achievement was to discover various stable (static) arrangements of electrons within the positive charge cloud, but the problem of spectra was essentially intractable on this basis. (Recall the great diversity of atomic spectra, as indicated by Figure 1-2.)

Before turning to the next atomic model—the totally different one proposed by Niels Bohr after Rutherford's discovery of the atomic nucleus—we shall give a brief description of some simple spectra that this new model was able to interpret in a remarkably successful way.

1-5 LINE SPECTRA

It was recognized at an early stage that spectra carry information about the atoms that produce them. This information becomes especially clear-cut if conditions are such that the radiating atoms are almost independent of one another—as in a gas discharge at low pressure, in contrast to conditions in an incandescent solid. We then find the emitted radiation concentrated within narrow wavelength regions (for example, $\Delta\lambda/\lambda \approx 10^{-5}$) forming the familiar sharp spectral lines.

From the standpoint of classical physics, the only reasonable explanation is that the observed wavelengths are an expression of characteristic vibrations within the atom. Furthermore, on the simplest view, each wavelength requires a different periodicity of the atom. However, most spectra are far too complicated to offer any prospect of simple quantitative explanation in terms of the mechanics of atomic vibrations. Historically, the best hope was provided by the regularity of the famous Balmer series of hydrogen (Figure 1-6):

Simple models of the atom

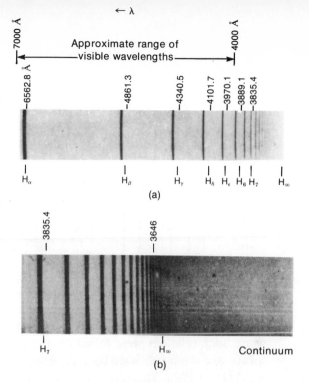

Fig. 1-6 Balmer series. (a) A Balmer spectrum in which the first dozen lines can be distinguished. The position of the Balmer limit is marked. (b) Balmer spectrum employing a higher dispersion, in which individual lines can be seen up to about n = 20. (Spectra reproduced from G. Herzberg, Atomic Spectra and Atomic Structure, *Dover, Publications, Inc., New York, 1944. Reprinted through the permission of the publisher.)*

Line	Color	λ (Å)
H_α	Red	6563
H_β	Turquoise	4861
H_γ	Blue	4341
H_δ	Violet	4102
H_ϵ	Extreme violet	3970

In 1885 J. J. Balmer[9] (a Swiss secondary-school teacher) discovered that these wavelengths were beautifully fitted by the following formula:

$$\lambda_n = 3646 \frac{n^2}{n^2 - 4} \text{ Å} \qquad (n = 3, 4, 5 \ldots)$$

But this relation of course did not supply a key to the actual mechanism of atomic radiation.

A very important clue was provided by the discovery that the reciprocal of λ was a more significant quantity than λ itself. It was found that if spectral lines were characterized by $1/\lambda$ in-

[9]J. J. Balmer, Ann. Phys. **25**, 80 (1885).

1-5   Line spectra

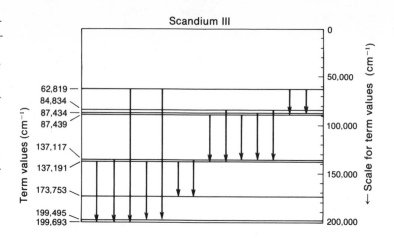

Fig. 1-7 Term diagram for doubly ionized scandium. Each vertical arrow represents an observed spectral line. The length of each arrow is proportional to the wave number of the spectral line. In this particular case, fourteen wave numbers can be found using just nine term values.

stead of λ, then the values of 1/λ for many lines in a given spectrum were equal to the sums or differences of the reciprocal wavelengths of other lines, taken in pairs, from the same spectrum. Such relations were found for the spectra of many elements, and were published by J. R. Rydberg[10] in 1900 and by W. Ritz[11] in 1908. This discovery (known as the *Rydberg-Ritz combination principle*) made it possible to describe a complicated spectrum with the help of a substantially smaller number of basic "terms," as shown in Figure 1-7. Each horizontal line in the diagram represents a particular "term value" expressed in cm⁻¹, and the length of a vertical arrow between two terms is the value of 1/λ of a possible spectral line. (Later we shall see that not all mathematically possible differences show up, but this does not detract from the importance of the basic result.)

In the case of the Balmer series, the use of reciprocal wavelengths leads to the equation

$$\frac{1}{\lambda_n} = R_{\mathrm{H}} \left(\frac{1}{4} - \frac{1}{n^2} \right) \qquad (n = 3, 4, 5 \ldots) \qquad (1\text{-}7)$$

where R_{H} is the experimentally determined "Rydberg constant" for hydrogen, equal to about 109,700 cm⁻¹. Thus the value of 1/λ for any line in the Balmer spectrum is in the form of a difference of two terms, the constant $R_{\mathrm{H}}/4$ and the variable term R_{H}/n^2. As we shall see, this corresponds very directly to

[10]J. R. Rydberg, Report of Intl. Phys. Cong. at Paris, ii, 200 (1900).

[11]W. Ritz, Phys. Z. **9**, 521 (1908); Astrophys. J. **28**, 237 (1908).

Simple models of the atom

the energy difference between one particular energy level in the hydrogen atom and each of a number of other levels.

1-6 PHOTONS

The realization that the reciprocal wavelength, rather than λ itself, is the convenient parameter to describe the lines of a spectrum is profoundly important. It corresponds, as we now know, to the fact that radiant energy is *quantized* in units called *photons*, which have energy E proportional to the optical frequency and hence inversely proportional to wavelength ($\nu = c/\lambda$). This is expressed in the famous equation

$$E = h\nu \tag{1-8}$$

where h is Planck's constant:

$$h = 6.6262 \times 10^{-27} \text{ erg-sec} = 6.6262 \times 10^{-34} \text{ joule-sec} \tag{1-9}$$

If any one quantity can be said to characterize quantum physics, it is Planck's constant. Although Max Planck first introduced the constant (in 1900) to account for the shape of the continuous spectrum of radiation from incandescent objects,[12] it was Einstein[13] (in 1905, the same year in which he published his special relativity theory) who proposed the photon concept and postulated Eq. 1-8. This equation carries far-reaching implications, not only for the description of radiant energy, but also for the energy structure of the atoms from which characteristic spectral radiation is emitted.

The equation $E = h\nu$ is really a very strange one: it is a sort of experiment-based mongrel with parentage in both classical theory ("wave of frequency ν") and quantum theory ("quantum of energy E"). The significance of this relationship is something we shall reexamine more carefully later.

When dealing with radiation of visible wavelengths or shorter, we are not normally able to measure the frequency; it is too high:

[12]M. Planck, Ann. Phys. **1**, 69 (1900) [or see M. Planck, *The Theory of Heat Radiation* (trans. Morton Masius), Dover Publications, New York, 1959].

[13]A. Einstein, Ann. Phys. **17**, 132 (1905) [translated by A. B. Arons and M. B. Peppard, Am. J. Phys. **33**, 367 (1965)].

$$\nu = \frac{c}{\lambda} = \frac{3 \times 10^{10} \text{ cm/sec}}{5000 \times 10^{-8} \text{ cm}} \approx 10^{15}\text{Hz} \qquad \text{(cycles/sec)}$$

But we can measure wavelengths through the use of diffraction gratings, and other devices. Thus it will often be convenient to rewrite Eq. 1-8 in the form

$$E_{\text{photon}} = \frac{hc}{\lambda} \tag{1-10}$$

In atomic physics we most often measure energies in electron-volts:

$$1 \text{ eV} = 1.602 \times 10^{-19} \text{ joule} = 1.602 \times 10^{-12} \text{ erg}$$

It is also standard practice to express the wavelengths λ of atomic spectra in angstroms. It is therefore convenient to express the quantities h and hc in corresponding units. We find

$$h = 4.136 \times 10^{-15} \text{ eV-sec} \tag{1-11}$$

$$hc = 12{,}400 \text{ eV-Å}$$

We can then rewrite Eq. 1-10 numerically as follows:

$$E_{\text{photon}} \text{ (eV)} = \frac{12{,}400}{\lambda(\text{Å})} \tag{1-12}$$

The visible spectrum (4000–7000 Å) thus represents a range of photon energies from about 1.8 to 3.0 eV, indicating the appropriateness of the electron volt as an energy unit for photons in the visible region.

When Einstein first introduced the photon concept, he had very limited evidence with which to support it. The idea has been richly confirmed, however, and we list below three kinds of evidence that indicate the validity of $E = h\nu$ for individual photons.

The Photoelectric Effect

When light falls on a clean metal surface (which means, in practice, a surface cleaned in a vacuum and kept there), electrons are sometimes given off. These are called *photoelectrons*, and their ejection by light is called the *photoelectric effect*. For a given value of ν or λ of the incident light, there is a spread of photoelectron energies down to essentially zero. However, the *maximum* kinetic energy ($K_{\max}$) of a photoelectron is fairly sharply defined and varies linearly with ν as shown in Figure 1-8, which reproduces the first definitive

Simple models of the atom

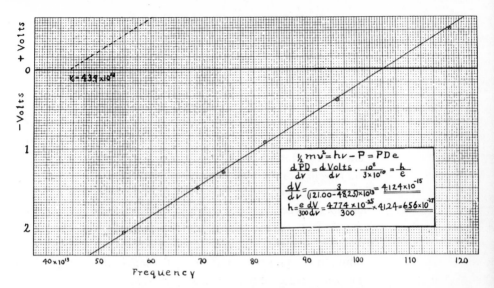

Fig. 1-8 Relation between the maximum kinetic energy of photoelectrons and the frequency of the incident light. This graph is from Millikan's 1916 paper. The ordinate is actually the retarding voltage required to stop the photocurrent. The slope of the straight line directly gives the dependence of K_{max} on v. Note Millikan's computation of h in the lower right corner of the figure. [Figure taken from R. A. Millikan, Phys. Rev. 7, 355 (1916). Reproduced with permission of the University of Chicago Press.]

measurements on the process, obtained by R. A. Millikan in 1916. The energy K_{max} does not depend at all on the intensity of the incident light, only on its frequency. (Experimentally a higher incident intensity, at fixed frequency, yields *more* photoelectrons but not more *energy* for each electron.) Moreover, below a minimum "threshold" frequency v_0 for the incident light, no electrons are ejected at all, irrespective of the intensity of the incident light. The threshold frequency v_0 is characteristic of the metal being used as the photoemitter.

These results are difficult to explain classically. If we suppose electrons in the metal to be ejected as a result of being shaken by the electric field in a classical electromagnetic wave, then the maximum kinetic energy acquired by the photoelectrons should be greater for greater incident light intensity—contrary to the experimental result.

According to Einstein's photon concept, however, the basic process is the absorption of a quantum of energy $E = hv$

by an individual electron. Most such electrons will lose part or all of this energy plowing through the metal. However, some of the electrons situated near the surface will escape with a maximum retained energy. Even these maximum-energy electrons (along with all other escaping electrons) must pay an "energy tax": they must surmount a potential-energy step at the surface of the metal. The height of this step measures the tenacity with which the metal confines its electrons and has different values for different metals. The step height W is called the *work function* of the metal and is typically a few electron-volts. Even the most energetic electrons will not leave the metal if they do not receive an energy in excess of the work function—a fact that explains the existence of a *threshold frequency* $\nu_0 = W/h$ below which no electrons are emitted. According to the photon model, then, the maximum-energy electrons will have an energy given by the expression

$$K_{max} = h\nu - W = h(\nu - \nu_o) \qquad (1\text{-}13)$$

which fits the experimental linear curve of Figure 1-8. The measured slope of this line yields a numerical value for Planck's constant h. In practice one measures the retarding potential V needed to stop the most energetic electrons ($eV = K_{max}$) and plots the function

$$V = \frac{h}{e}\nu - \phi = \frac{hc}{e} \cdot \frac{1}{\lambda} - \phi \qquad (1\text{-}14)$$

where $\phi\ (= W/e)$ is a measure of the work function, expressed in units of electric potential.[14]

Equations 1-13 and 1-14 are consistent with the observation that light of greater intensity yields no greater maximum kinetic energy. According to the photon hypothesis, a more intense beam merely contains a larger number of photons of the same energy, and therefore releases a larger number of photoelectrons, without affecting their energy distribution.

Further support for the photon interpretation is provided by the time delay between the instant the light beam is turned

---

[14]The physical interpretation of the work function has subtleties that we do not need to go into here. See F. Bitter and H. Medicus, *Fields and Particles*, American Elsevier, 1973, p. 149.

Simple models of the atom

on and the onset of a photoelectric current. According to the wave picture, the energy of the light beam is uniformly distributed in both space and time. Knowing the approximate size of an atom, we can estimate the rate at which energy is being deposited on the area that an atom presents to the beam. It seems reasonable to expect that the photocurrent will begin at about the time when this energy reaches a value sufficient to liberate a photoelectron. For a very weak incident beam the expected time delay can be several hours or more (see the exercises). On the other hand, according to the photon hypothesis, the radiant energy arrives as individual quanta, randomly distributed, and there is some chance that the very first photon to arrive will liberate a photoelectron. Experiment confirms the photon picture: Lawrence and Beams showed in 1928 that photoelectrons are sometimes emitted less than 3×10^{-9} sec after initial illumination, even with an incident light beam so weak that the expected time delay according to the wave picture would be much longer.[15]

The photoelectric effect is strong (although not by itself conclusive) evidence for the existence of photons. The fact that the same value of h is obtained from experiments with different metals strengthens this evidence. Further evidence comes from the short-wavelength limit for x rays and from the Compton effect, which we now discuss.

Short-Wavelength-Limit X Rays

When a metal target is bombarded with electrons that have been accelerated through some considerable potential (say 5–50 kV), one obtains what is called *bremsstrahlung* (German for "deceleration radiation"). The spectrum of this radiation is continuous and covers a wide range of wavelengths. Classically, one would try to account for this spectrum by using the equations for the radiation field of an accelerated (or decelerated) charge. But the observed spectra (Figure 1-9a) show a feature which no classical calculation can explain—a sharp cutoff at a certain minimum wavelength (maximum photon energy). This minimum wavelength, λ_m, which is the same for all metals, varies systematically with the accelerating voltage V_o according to a law verified by Duane and Hunt:[16]

[15]E. O. Lawrence and J. W. Beams, Phys. Rev. **32**, 478 (1928).

[16]W. Duane and F. L. Hunt, Phys. Rev. **6**, 166 (1915).

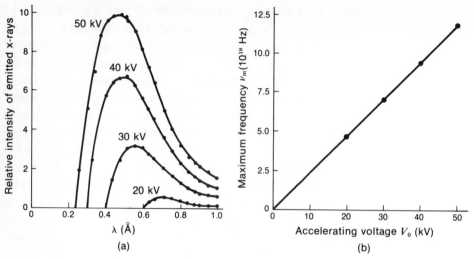

Fig. 1-9 (a) Bremsstrahlung spectra produced by electrons of various energies striking a metal target. Each spectrum exhibits an abrupt cutoff at some minimum wavelength λ_m. (b) The maximum frequency of emission, corresponding to the minimum wavelength λ_m, is proportional to the accelerating voltage V_o. [Data of C. T. Ulrey, Phys. Rev. 11, 401 (1918).]

$$\frac{1}{\lambda_m} = \frac{e}{hc} V_o$$

Then, defining $\nu_m = c/\lambda_m$, we have

$$\left(\begin{array}{c}\text{Maximum}\\ \text{photon energy}\end{array}\right) = h\nu_m = V_o e = \left(\begin{array}{c}\text{Kinetic energy}\\ \text{of incident electron}\end{array}\right) \quad (1\text{-}15)$$

Experimental data on this relationship are shown in Figure 1-9b. The phenomenon resembles the photoelectric effect in reverse. We get the maximum possible photon energy when all the kinetic energy of an incident electron is converted into the energy of a single photon. The work function does not appear in this simple statement of the Duane-Hunt law because it is almost negligible in comparison to the electron and photon energies involved (of the order of 0.1 percent).

The Compton Effect

The Compton effect is a demonstration of the dynamics of individual photons in collision processes. Not only is the pho-

Simple models of the atom

ton a quantum of radiant energy E, it also carries a corresponding amount of linear momentum p; collisions between photons and electrons (or photons and other particles) can be analyzed using the energy and momentum conservation laws of (relativistic) particle dynamics. All observations on such collision events are consistent with the relation $p = E/c$ for an individual photon of any energy. This is the same relation between momentum and energy that holds classically for electromagnetic radiation.

The behavior of individual photons as particles in this dynamical sense was first demonstrated by A. H. Compton in a series of experiments between 1919 and 1923.[17] He showed that when x-ray photons collide with free electrons, the photons suffer a loss of energy, manifested as an increase in wavelength, of precisely the amount that one would calculate for an elastic collision between two particles, the photon being treated as a particle of momentum $p = E/c = h\nu/c$. It is worth noting, in this connection, how the classical relation between wavelength and frequency for electromagnetic radiation can be converted into a statement of the wavelength to be associated with photons of a given momentum:

$$\lambda = \frac{c}{\nu} = \frac{hc}{h\nu} = \frac{hc}{E} = \frac{h}{E/c}$$

That is,

$$\lambda = \frac{h}{p} \tag{1-16}$$

This is an extremely important relation; we shall see that its relevance goes far beyond photons.

Most of the above evidence concerning energy quanta followed, rather than preceded, the Bohr theory of atomic structure. Bohr's theory was built primarily on the photon concept of radiation, which was understood as an hypothesis but was not yet unambiguously verified when Bohr did his early work. Since we are not giving primarily an historical account

---

[17] A. H. Compton, Phys. Rev. **22**, 409 (1923). For details and a discussion of the relativistic dynamics of the process see, for example, the volume *Special Relativity* in this series.

here, we have, for clarity, presented the evidence for photons before discussing the Bohr theory itself.

1-7 THE RUTHERFORD-BOHR ATOM

Rutherford's discovery of the nucleus, on the basis of the famous alpha-particle scattering experiments done in his laboratory by Geiger and Marsden in 1909,[18] demanded a picture of atomic structure completely different from the Thomson model described earlier. Rutherford's paper in 1911[19] explained the scattering data in detail by assuming all the positive charge and most of the atomic mass to be concentrated in a space that was minute compared to the total volume occupied by the atom. Subsequent measurements showed that the nuclear radius was less than 10^{-4} of the atomic radius (that is, $< 10^{-12}$ cm).

Accepting this nuclear model of the atom, Niels Bohr in 1913 devised a theory that was spectacularly successful in accounting for the spectrum of atomic hydrogen.[20] He forced a connection between the Einstein relation $E = h\nu$ for photons and the Newtonian picture of electrons in orbits around the atomic nucleus. Bohr's two main postulates were the following:

1. An atom has a number of possible "stationary states." In any one of these states the electrons perform orbital motions according to the laws of Newtonian mechanics, but (contrary to the predictions of classical electromagnetism) do not radiate so long as they remain in fixed orbits.

2. When the atom passes from one stationary state to another, corresponding to a change of orbit (a "quantum jump") by one of the electrons in the atom, radiation is emitted in the form of a photon. The photon energy is just the energy difference between the initial and final states of the atom. The classical frequency ν is related to this energy through the Planck-Einstein relation:

$$E_{\text{photon}} = E_i - E_f = h\nu \qquad (1\text{-}17)$$

[18]H. Geiger and E. Marsden, Proc. Roy. Soc. A **82**, 495 (1909).

[19]E. Rutherford, Phil. Mag. **21**, 669 (1911). For a discussion of "Rutherford scattering" see, for example, the volume *Newtonian Mechanics* in this series.

[20]N. Bohr, Phil. Mag. **26**, 1 (1913).

Simple models of the atom

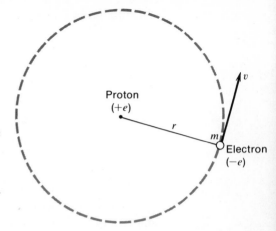

Fig. 1-10 Electron
traveling around a
proton in a circular
orbit corresponding
to one of Bohr's sta-
tionary states of the
hydrogen atom.

We shall not reproduce here the line of argument that Bohr followed in seeking the consequences of these postulates. Rather, we shall make use of the (experimentally obtained) Balmer formula (Eq. 1-7) in deriving the stationary orbits of hydrogen.

To begin, we apply classical mechanics to an electron in an orbit (assumed circular) corresponding to one of Bohr's stationary states (see Figure 1-10). The electron (charge $-e$) moves at speed v in a circle of radius r around an effectively fixed central proton (charge $+e$). Then by Newton's second law we have[21]

$$F = ma = \frac{mv^2}{r} = \frac{e^2}{r^2} \text{ (cgs)} \qquad \frac{e^2}{4\pi\varepsilon_0 v^2} \tag{1-18}$$

The total energy of the electron is negative (the zero of energy being taken as that of a stationary electron at $r = \infty$) and is given by

$$E = \tfrac{1}{2}mv^2 - \frac{e^2}{r} \qquad \tfrac{1}{2}mv^2 - \frac{e^2}{4\pi\varepsilon_0 v^2}$$

Using Eq. 1-18 we can rewrite this in the following way:

$$E = -\tfrac{1}{2}mv^2 = -\frac{e^2}{2r} \qquad -\frac{e^2}{8\pi\varepsilon_0 v^2} \tag{1-19}$$

[21]See *Newtonian Mechanics*, p. 198. In the analysis that follows we use cgs units only.

The negative value of the total energy corresponds, of course, to the fact that energy must be supplied to remove the electron from the atom.

Any quantization of the energy of the system (restriction to certain allowed values) implies, through Eq. 1-19, a corresponding restriction of the orbit radius to discrete values. We therefore label the energies E_n and radii r_n by integers n in order of increasing energy, starting with $n = 1$ at the lowest (most negative) energy state which, by Eq. 1-19, has the smallest orbit radius.[22]

Now we consider a quantum jump from an initial orbit of energy E_i and radius r_i to a final lower-energy orbit of radius r_f and energy E_f. This transition is, according to the Bohr model, accompanied by the emission of a photon of energy equal to the energy difference between the two levels:

$$E_{\text{photon}} = \frac{hc}{\lambda} = E_i - E_f = \frac{e^2}{2}\left(\frac{1}{r_f} - \frac{1}{r_i}\right) \tag{1-20}$$

This result is consistent with Eq. 1-7 for the Balmer series if radius r_i of the *initial* orbit is proportional to n^2. Such a correspondence demands that the final orbit in the Balmer transitions is the $n = 2$ orbit ($2^2 = 4$). Then we obtain a direct interpretation of that series as arising from transitions to the state $n = 2$ from the various higher-energy states $n = 3$, 4, 5 .... If this interpretation is correct, the radii of the stationary orbits in the hydrogen are given by

$$r_n = n^2 a_o \qquad n = 1, 2, 3 \ldots \tag{1-21}$$

where a_o represents the radius of the smallest orbit and is called the *Bohr radius*. Moreover, the quantized values of energy in the atom, according to Eq. 1-19 are

$$E_n = -\frac{e^2}{2a_o n^2} \qquad n = 1, 2, 3 \ldots \tag{1-22}$$

Indeed, by comparing this to the second term in the Balmer expression 1-7, we can obtain a value for the Bohr radius:

[22]For the following argument we are indebted to Paul A. Tipler, *Foundations of Modern Physics*, Worth, New York, 1969, pp. 163 ff.

Simple models of the atom

$$a_o = \frac{e^2}{2hcR_H} \tag{1-23a}$$

Using the experimentally determined value for R_H, the atomic radius of hydrogen in its lowest energy state can be calculated. We shall calculate a_o below in terms of fundamental physical constants.

Thus, beginning with Bohr's postulates, a classical analysis buttressed by Balmer's empirically derived formula leads to values for the energies of stationary states in the hydrogen atom and for the radius of hydrogen in its lowest or ground state. Bohr himself is reported to have said, "As soon as the Balmer formula was shown to me, the whole thing became clear in my mind."

Bohr's first solution of the hydrogen-atom problem in his historic 1913 paper depended on a rather arbitrary combination of classical and quantum ideas. Subsequently, he made extensive use of a very powerful general principle (known as the *correspondence principle*) that requires classical and quantum predictions to agree in the limit of large quantum numbers (in our case the number n). To see how this works, we derive a theoretical value for a_o by applying the correspondence principle to the frequency of the emitted radiation. For large n, the energy of the photon emitted as the atom drops from state $n + 1$ to state n can be written as a simple approximation from the Bohr theory. The exact equation is

$$E_{\text{photon}} = \frac{e^2}{2}\left[\frac{1}{n^2 a_o} - \frac{1}{(n+1)^2 a_o}\right] = \frac{e^2}{2a_o}\left[\frac{2n+1}{n^2(n+1)^2}\right]$$

For large n, the last expression gives, approximately,

$$E_{\text{photon}} = h\nu \approx \frac{e^2}{2a_o} \cdot \frac{2}{n^3} \qquad \text{(Bohr theory, large } n\text{)}$$

that is,

$$\nu^2 \approx \frac{e^4}{a_o^2 h^2 n^6} \qquad \text{(Bohr theory, large } n\text{)}$$

In the *corresponding classical analysis* the frequency of the radiation is equal to the orbital frequency of the electron. Using

1-7   The Rutherford-Bohr atom

Eqs. 1-18 and 1-21, we find for the square of this frequency

$$\nu^2 = \left(\frac{v}{2\pi r}\right)^2 = \frac{1}{(2\pi)^2}\frac{e^2}{mr^3} = \frac{e^2}{(2\pi)^2 ma_o^3 n^6} \qquad \text{(classical)}$$

The correspondence principle demands agreement between these two predictions for the square of the radiation frequency provided n is very large:

$$\frac{e^4}{a_o^2 h^2 n^6} = \frac{e^2}{(2\pi)^2 ma_o^3 n^6} \qquad \text{(large } n)$$

This gives a value for a_o evaluated directly from fundamental physical constants:

$$a_o = \frac{h^2}{4\pi^2 me^2} = 0.529 \text{ Å} \qquad \text{(1-23b)}$$

This value corresponds roughly to other independent measures of the radius of hydrogen. For example, the separation of hydrogen atoms in the hydrogen molecule H_2 is 0.74 Å, consistent with a radius of 0.37 Å for each constituent atom.

Using this theoretical value of a_o, we can calculate a value for the Rydberg constant R_H from Eq. 1-23a:

$$R_H = \frac{e^2}{2hca_o} = \frac{2\pi^2 me^4}{h^3 c} = 109,500 \text{ cm}^{-1} \qquad \text{(theoretical)}$$

This is very close to the experimental value 109,700 cm^{-1}. The agreement is very impressive because the calculation of a_o using the correspondence principle involves only constants whose values are already known.[23]

Bohr's scheme of energy states predicted the existence of many spectral lines other than those of the Balmer series. In fact, there would be lines at all wavelengths λ defined by

$$\frac{1}{\lambda} = R_H\left(\frac{1}{n_f^2} - \frac{1}{n_i^2}\right) \qquad \left(\begin{array}{c} n_f = 1, 2, 3\ldots \\ n_i > n_f \end{array}\right) \qquad \text{(1-24)}$$

[23]We are ignoring, among other things, a small correction due to the fact that the proton is not infinitely heavy, so that both the proton and the electron orbit about their common center of mass. This makes R_H a little smaller in value than the Rydberg constant R_∞ for an infinitely heavy nucleus. See the exercises.

Simple models of the atom

The Balmer series consists of those lines with $n_f = 2$. The series of lines corresponding to $n_f = 3$ had already been observed in the infrared region by Paschen.[24] Soon after Bohr published his theory, Lyman[25] looked for and found the series for $n_f = 1$, which had not been observed before because it lies in the ultraviolet region. The series for $n_f = 4$ (Brackett series) and for $n_f = 5$ (Pfund series) were also found in accordance with the theoretical prediction.[26] These observations lent great credibility to Bohr's model.

Furthermore, the Bohr theory predicted a value for the radius of the hydrogen atom in the ground state ($a_o = 0.53$ Å) that was of the proper order of magnitude, and a value for the energy required to remove the electron entirely from the lowest-energy state of hydrogen (the so-called ionization energy $-E_1 = e^2/2a_o = 13.6$ eV) that agreed well with observation.

1-8 FURTHER PREDICTIONS OF THE BOHR MODEL

Bohr noted, in his first paper on the theory, that Balmer spectra in the light from stars showed as many as 30 lines or more, whereas only a dozen lines appeared when laboratory sources were used. He could account for this by assuming that lower pressures, and hence greater interatomic distances, exist in the luminous atmosphere of a star than in a gas discharge tube. (Note that, by Eq. 1-21, the orbit radii grow in proportion to n^2. Thus for large n the effective size of the atom becomes relatively enormous, and very low pressures are needed to avoid interatomic collisions that would interfere with the spontaneous radiation process.) More recently, transitions between high states of hydrogen atoms, with both n_i and n_f greater than 100, have been recorded from tenuous sources in space (see the exercises). This radiation is in the radio-frequency range ($\approx 10^{10}$ Hz).

An important further extension of the Bohr model appears if the theory is rewritten for a single electron in the field of an arbitrary central charge Q. Equation 1-24 is then replaced by the more general relation

[24]F. Paschen, Ann. Phys. **27**, 565 (1908).

[25]T. Lyman, Phys. Rev. **3**, 504 (1914).

[26]F. Brackett, Nature **109**, 209 (1922); H. A. Pfund, J. Opt Soc. Am. **9**, 193 (1924).

$$\frac{1}{\lambda} = \frac{2\pi^2 m e^2 Q^2}{ch^3}\left(\frac{1}{n_f^2} - \frac{1}{n_i^2}\right) \quad \text{(cgs)} \qquad (1\text{-}25)$$

Bohr realized that the spectrum of singly ionized helium (He$^+$) could be calculated from this, putting $Q = 2e$, and that the He$^+$ series for $n_f = 4$, $n_i = 5$, 6, 7 ... would have alternate lines (for $n_i = 6$, 8, 10 ...) coincident with the Balmer series. Such a spectrum had been observed from stars but not previously associated with a substance known on earth.

1-9   DIRECT EVIDENCE OF DISCRETE ENERGY LEVELS

Bohr's theory hypothesized the quantized energy levels, but the evidence for them through spectra was only indirect (although of course very compelling). But then, beginning with an experiment performed by J. Franck and G. Hertz in 1914, more direct evidence for these energy levels began to be gathered.

The Franck-Hertz Experiment[27]

In this experiment (see Figure 1-11a), electrons from a hot cathode C were accelerated through mercury vapor at low pressure toward a grid G. Those electrons that passed through the grid could reach the anode A if they had enough energy to overcome a small retarding potential. The accelerating voltage V between cathode and grid was continuously adjustable. The results (Figure 1-11b) showed that electrons accelerated through mercury vapor could be given as much as 4.9 eV of kinetic energy without difficulty, but then had a high probability of losing all their energy to mercury atoms, as shown by a sudden drop of collector current for the successive increases of V by 4.9 V. This increase of electron energy agreed with the quantum energy for one of the most prominent lines in the mercury spectrum—an ultraviolet line of wavelength 2537 Å:

$$E_{\text{photon}} = \frac{hc}{\lambda} = \frac{1.24 \times 10^4}{2537} \text{ eV} \approx 4.9 \text{eV}$$

[27]James Franck and Gustav Hertz, Verh. Dtsch. Phys. Ges. **16**, 512 (1914). A modern version of this experiment, with a delightful epilogue spoken by James Franck, is available on film (Educational Development Center, Newton, Mass., 1961). Franck points out that when he and Hertz did the original experiment they had not even heard of the Bohr theory!

Simple models of the atom

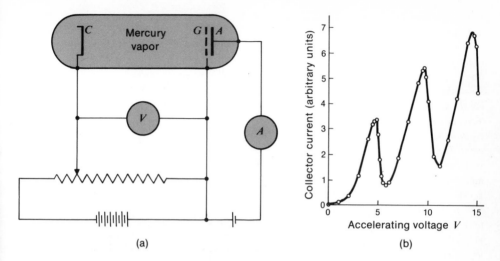

Fig. 1-11   The Franck-Hertz experiment.
(a) Schematic diagram of experimental arrangement.
(b) Collector current versus accelerating voltage.
The sharp decreases in current near V = 10 V and again
near V = 15 V result when electrons undergo multiple
(two and three, respectively) inelastic collisions.

Thus one could picture the mercury atoms being driven from
their state of lowest energy (*ground state*) up to an excited
state 4.9 eV higher in energy, absorbing all the kinetic energy
of the incident electron—with the implication that the atoms
subsequently fell back to their ground state by emitting photons of 2537 Å.

In a later experiment, Gustav Hertz (not to be confused
with the Heinrich Hertz who first observed the electromagnetic waves predicted by Maxwell) looked for the sharp spectral
emission lines that should be expected to follow the excitation
of mercury atoms by electron impact.[28] He showed that up to
4.9 V accelerating voltage there was no subsequent radiation
at all. Just above 4.9 V the 2537 Å line appeared. For still higher voltages further lines were added to the spectrum in a manner completely consistent with the energy level interpretation
of the term diagram (see Figure 1-12).

Observation of Inelastically Scattered Electrons

Using incident electrons of far greater kinetic energy
than is needed to excite atoms of a given kind, one can

[28]G. Hertz, Z. Phys. **22**, 18 (1924).

1-9   Direct evidence of discrete energy levels

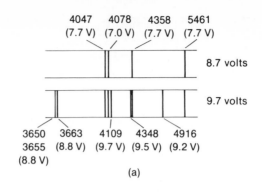

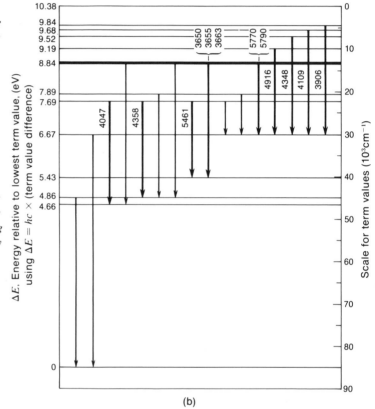

Fig. 1-12 Stimulation of spectral emission in mercury vapor by electrons. (a) Sharp spectral lines following the excitation of mercury vapor by electron bombardment. The upper spectrum was obtained with electrons accelerated through 8.7V in the vapor; the lower spectrum is for 9.7V. Wavelengths are given in Å. (After G. Hertz, 1924.) (b) Partial term diagram for mercury. The term values (found by applying the Rydberg-Ritz principle to the mercury spectrum) can be read from the scale at the right. By tentatively identifying the terms with energy levels, the energy needed to excite spectral lines involving a given term can be calculated from the difference between the bottom term value and the particular higher-term value. These calculated energies appear at the left of Figure 1-12(b) and parenthetically, as voltages, in Figure 1-12(a). [Note: the term at 8.84 eV is actually several closely spaced terms.]

analyze the energy distribution of the scattered electrons coming off in some chosen direction. Figure 1-13a indicates the kind of experimental arrangement by which this can be studied, using a magnetic field to curve the paths of the electrons and a photographic plate to record their arrival—the radius of curvature of the paths being proportional to the electron momentum. Figure 1-13b shows the results of such

Simple models of the atom

measurements, using helium as the target gas. In addition to the elastically scattered electrons, there are other groups of electrons of lower energy, corresponding to scattering processes in which the bombarding electrons lose kinetic energy and the electrons of the target atoms are excited to higher energy levels. Since kinetic energy is not conserved in such interactions, we call them *inelastic collisions*.

1-10   X-RAY SPECTRA

Some features of x-ray spectra provide striking confirmations and extensions of the Bohr theory. We mentioned earlier the continuous spectrum of x rays that is produced when ener-

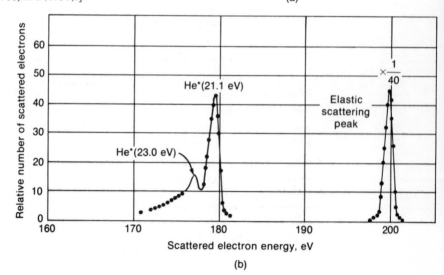

Fig. 1-13   (a) Experimental arrangement for electron scattering experiments.   (b) Results on elastic and inelastic scattering processes, showing the production of characteristic excited states of sharply defined energy in atoms of the target helium gas. The scattering peaks associated with the production of excited helium states are marked He, with the excitation energy in parentheses. [After L. C. Van Atta, Phys. Rev. 38, 876 (1931).]*

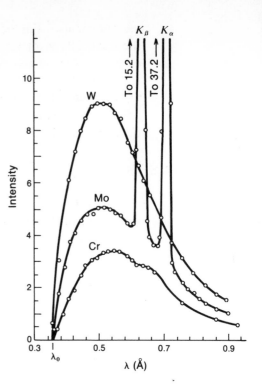

Fig. 1-14 X-ray spectra produced by 35-keV electrons striking tungsten, molybdenum, and chromium targets. The three spectra exhibit the same short-wavelength limit, as demanded by the Duane-Hunt law. Contrast this feature with the two very sharp and intense lines which appear only in the spectrum obtained with the molybdenum target. [After C. T. Ulrey, Phys. Rev. 11, 405 (1918).]

getic electrons (10–100 keV) strike a target (see Figure 1-9a). However, in addition to this continuous (bremsstrahlung) radiation, the x-ray spectrum may also contain a few sharp lines, as illustrated in Figure 1-14. (For a brief description of how x-ray spectra are analyzed, see the next section.) The wavelengths of these x-ray lines are characteristic of the bombarded target atoms, but (in contrast to the great diversity of optical spectra) there is a remarkable family likeness between the x-ray line spectra of all the different elements. Also, the x-ray spectra are far simpler than most optical spectra. The systematics of these x-ray spectra were explored by the British physicist H. G. J. Moseley in a brilliant and intensive period of research in 1913–1914 (just before he went off to World War I and his own death).[29] Figure 1-15 shows a sample of Moseley's results for a number of different atomic species. And what Moseley also did was to show how these results were to be understood in terms of an extension of the Bohr theory.

[29]H. G. J. Moseley, Phil. Mag. 26, 1024 (1913); 27, 703 (1914). The story of Moseley and his work is well told in Bernard Jaffé, *Moseley and the Numbering of the Elements*, Doubleday Anchor, New York, 1971, and in J. L. Heilbron, *H. G. J. Moseley*, University of California Press, Berkeley, 1974.

Simple models of the atom

Fig. 1-15   Spectra showing the charac-
teristic x-ray lines for several elements.
In this composite photograph, the ab-
scissa is wavelength, which increases
toward the right. The two strongest lines
in each spectrum were designated K_α
and K_β, and the spectra were arranged
so that these line wavelengths increase
from bottom to top. [Figure reproduced
from H. G. J. Moseley, Phil. Mag. 26,
1024 (1913). Reprinted with the permis-
sion of the Philosophical Magazine.]

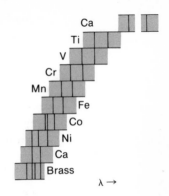

The basis of the theoretical analysis is a simple model of a many-electron atom. We begin with a central nuclear charge, and consider the kind of Bohr orbits that one would have for a single electron. But now we introduce the idea that each orbit can accommodate only a certain number of electrons. This leads to a picture of an atom as built up of successive "shells" of electrons, as indicated in Figure 1-16. By long-established tradition these shells are labeled by the successive letters $K, L,$ $M, \dots.$ In fact the periodic similarity of chemical properties, summarized in the periodic table of the elements, is associated with the recurrence of similar electron configurations in the outermost shells as one goes from lighter to heavier elements. For example, beryllium, which has its two outermost electrons in the L shell, labeled by the Bohr quantum number $n = 2$, is chemically similar to magnesium, which has its two outermost electrons in the M shell, belonging to $n = 3$.

Suppose now that a many-electron atom is hit so hard (by

Fig. 1-16
(a) Shell model of
many-electron
atom, showing the
four innermost
shells. Each shell is
taken to have a lim-
ited capacity for
electrons.   (b)
Standard nota-
tion for x-ray
spectral lines.
[Note: This figure is
purely schema-
tic—the radii used
in the figure were
arbitrarily chosen.]

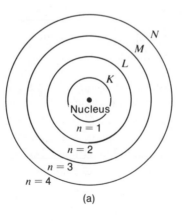

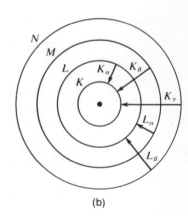

(a)   (b)

an energetic electron from outside, for example) that one of the electrons in the filled innermost orbit ($n = 1$) is knocked out of the atom completely. The ejected electron leaves a vacancy into which an electron from one of the larger orbits (for example, $n = 2$ or $n = 3$) can fall, emitting a photon in the process. The energy of the emitted photon will be the energy difference between the initial and final states of this many-electron atom. In principle, this energy difference must take into account the interactions among *all* of the electrons in the atom. However, an easily visualized approximate model which makes use of the Bohr theory ignores these multiple interactions and analyzes the downward transition as that of an electron moving under the influence of a single effective central charge. By Gauss's law, the effective central charge is that of the nucleus diminished by the charge of any electrons remaining in orbits which lie inside that of the radiating electron. (This reduction of effective charge is called *screening*.) Call the effective central charge Q. Then from the Bohr formula, Eq. 1-25, with $n_f = 1$, who have

$$\nu = \frac{c}{\lambda} = \frac{2\pi^2 m e^2 Q^2}{h^3} \left(1 - \frac{1}{n_i^2}\right) \tag{1-26}$$

If we take $n_i = 2$ in Eq. 1-26, we define what Moseley recognized as the line known as K_α in the characteristic x-ray spectra. It is the exact counterpart of the first (longest-wavelength) line in the Lyman series of hydrogen. (The x-ray lines are coded according to the shell on which the quantum jump terminates. A companion line, called K_β, corresponds to $n_i = 3$ and has a shorter wavelength.) Thus we have

$$\nu(K_\alpha) = \frac{3}{4} \cdot \frac{2\pi^2 m e^2 Q^2}{h^3} \tag{1-27}$$

Then if one measures the frequency of the K_α transition for all the different elements, the value of $\sqrt{\nu}$ should be a measure of the effective central charge Q in each case. And what Moseley found was that the graph of $\sqrt{\nu}$ against the chemical atomic number Z was a straight line (Figure 1-17a). In Moseley's own words: "We have here a proof that there is in the atom a fundamental quantity, which increases by regular steps as we pass

Simple models of the atom

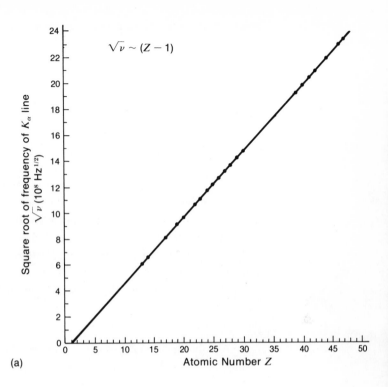

(a)

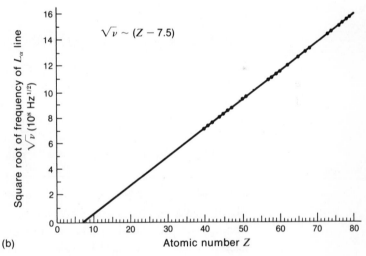

(b)

Fig. 1-17 The physical basis for chemical atomic number. (a) Square root of K_α transition frequency as a function of chemical atomic number Z. (The line is adjusted to intercept the horizontal axis at Z = 1.) (b) Square root of L_α transition frequency as a function of Z. [Graphs based on data from H.G.J. Moseley, Phil. Mag. 27, 703 (1914).]

from one atom to the next. This quantity can only be the charge on the atomic nucleus."[30]

The graph of $\sqrt{\nu}$ against Z actually has an intercept at about 1 on the Z axis. This implies that the effective central

[30] H. G. J. Moseley, Phil. Mag. **26**, 1031 (1913).

charge for the K_α transition is equal to $(Z - 1)e$, and hence that one electron was left in the $n = 1$ orbit when its companion was originally removed. The implication is that the inner orbit normally contains two electrons.

In a similar way it was possible to analyze the so-called L transitions—those that end on the $n = 2$ orbit. (These correspond to lines in the Balmer series of hydrogen.) For these L transitions the effective central charge was found to be equal to about $(Z - 7.5)e$, indicating that the shielding of the nucleus represented by the electrons remaining in the $n = 1$ and $n = 2$ shells together was the equivalent of between 7 and 8 electrons (Figure 1-17b). This suggests that 8 or 9 electrons populate the inner two orbits when filled. We shall see later that a more realistic value is 10 electrons. Evidently we need a better theory to account accurately for all the details.

Let us illustrate the above discussion with one typical example.

Molybdenum $(Z = 42)$ has a characteristic K_α line of about 0.71 Å. The corresponding photon energy is given by

$$E_{\text{photon}} = \frac{1.24 \times 10^4}{\lambda(\text{Å})} \text{ eV} \approx 17.5 \text{ keV}$$

Theoretically, by Eq. 1-27, the photon energy should be given by

$$h\nu = \frac{3}{4} (41)^2 \left(\frac{2\pi^2 me^4}{h^2}\right) \approx 1260 \left(\frac{2\pi^2 me^4}{h^2}\right)$$

But the quantity $2\pi^2 me^4/h^2$ is the ionizaton energy of the hydrogen atom, that is, 13.6 eV, so we would expect the energy of the molybdenum K_α radiation to be about 1260×13.6 eV, or 17.1 keV, which checks pretty well.

We can recognize that the way in which these x-ray spectra are generated is very similar to the stimulation of optical emission lines following electron bombardment in the Franck-Hertz experiment. The only basic difference is that in the Franck-Hertz experiment the radiation comes from electrons

Simple models of the atom

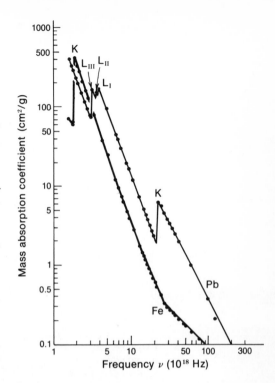

Fig. 1-18 Log-log plot showing the frequency dependence of x-ray absorption in lead and in iron. For increasing frequency, the absorption coefficient shows a general decline but exhibits an abrupt increase when the x-ray photon energy becomes large enough to remove an electron from an inner shell. [From data of S. J. M. Allen, Phys. Rev. 27, 266 and 28, 907 (1926).]

falling back into their original state after being temporarily dislodged, whereas the x-ray emission comes from an outer electron falling into the vacancy created by the complete ejection of an inner electron from the atom. In connection with this last point, it is clear that the initial process of removing an electron from the K or L shell requires a definite minimum energy equivalent to the ionization energy of that shell. This is manifested in the "absorption edges" that appear when a continuous x-ray spectrum is passed through a material—the absorption as a function of photon frequency ν shows a sharp increase at each value of ν corresponding to removal of an electron from a particular shell. (See Figure 1-18.)

1-11 A NOTE ON X-RAY SPECTROSCOPY

The early study of x rays was one of the most exciting and important episodes in the history of physics. It was in 1895 that W. K. Roentgen first observed the production of penetrating radiation when cathode rays (themselves not yet fully understood) were stopped in the walls of a discharge tube.[31] Sev-

[31]W. K. Roentgen, Sitzungsber, Wurzburger Phys.-Med. Ges., December 28, 1895; translated, Nature 53, 274 (1896).

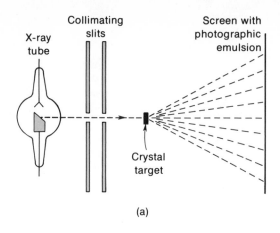

Collimating slits

X-ray tube

Screen with photographic emulsion

Crystal target

(a)

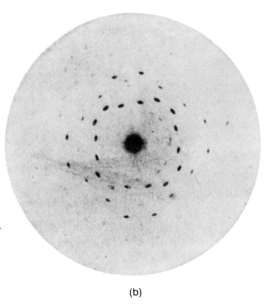

Fig. 1-19 (a) Experimental arrangement used by Friedrich and Knipping. (b) X-ray diffraction pattern produced using a thin crystal of zinc blende (ZnS). This is a reproduction of a photograph published by Friedrich, Knipping, and von Laue in 1912.

(b)

eral years went by before x rays were identified as a form of electromagnetic radiation. The first evidence was of very slight diffraction effects when they passed through a very narrow slit; a wavelength of the order of 1 Å was indicated, but the results were not entirely convincing. In 1912, however, Max von Laue realized that if x rays were indeed electromagnetic waves of this wavelength, it might be possible to observe their diffraction by the regular array of atoms in solids—which, as we have seen, involve interatomic distances of a few angstroms. He promptly arranged for two experimentalists, W. Friedrich and P. Knipping, to try it out. The result (Figure 1-19) was a striking confirmation of the ideas, and the opening of a research field that illuminated, at one stroke, both the

Simple models of the atom

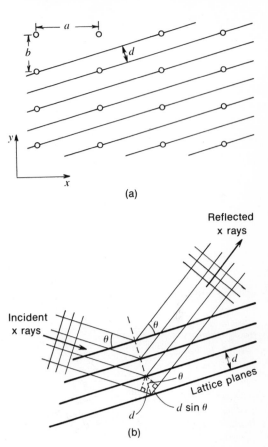

Fig. 1-20 (a) Side view of a rectangular lattice showing a set of parallel planes (other than x and y) which are rich in atoms. (b) Bragg reflection. Constructive interference among atoms in a single plane requires the angle of reflection to equal the angle of incidence. Constructive interference among planes requires $n\lambda = 2d \sin \theta$. Thus x rays of given wavelength will be strongly reflected by the array only for particular values of θ.

precise character of x rays and the atomic lattice structure of crystals.[32]

The original analysis by von Laue treated the problem, as basically one must, in terms of the scattering of an incident x-ray beam by all the individual atoms in the three-dimensional lattice that makes up a crystal. This is somewhat formidable, but very soon afterwards (in 1913), W. L. Bragg showed that the results of the analysis are the same as if one regarded the x rays as being reflected at sets of parallel planes that include many atoms.[33] Consider, for example, an atomic array that has atoms spaced by equal distances a along the x direction and equal distances b along the y direction (Figure 1-20a). Then in addition to the principal directions x and y themselves, there will be other planes, such as those shown in the figure, that are rich in atoms. Bragg then showed that strong reflections of x rays will occur if, as indicated in Figure 1-20b, the rays reflected from successive planes have path differences of a

[32]W. Friedrich, P. Knipping, and M. Laue, Sitzungsber, Math.- Phys. Kl. Bayer. Akad. Wiss., 303 (1912).

[33]W. L. Bragg, Proc. Roy. Soc. A **89**, 248 (1913).

1-11   A note on x-ray spectroscopy

whole number of wavelengths. If the spacing of the planes is d, we have the so-called *Bragg condition*:

$$n\lambda = 2d\sin\theta \qquad (n = 1, 2, 3 \ldots) \tag{1-28}$$

where θ is the angle between the incident or reflected x rays and the direction of the planes in question. For x rays of a given wavelength, this defines particular angles θ_n for which strong reflections from the chosen planes will occur.

On the basis of the above analysis, W. L. Bragg and his father (W. H. Bragg) constructed an x-ray crystal spectrometer (see Figure 1-21a) in which the crystal and the detector were placed on separate turntables, geared to make the crystal turn always half as much as the detector so that the reflection condition was always maintained. The detector was an ionization chamber that recorded an ionization current proportional to the intensity of the x rays entering it. Figure 1-21b shows an example of the results obtained when the x-ray source was emitting characteristic x rays of the type studied by Moseley. It was a fortunate circumstance that this important research tool became available just when Moseley needed it. The Braggs themselves received the Nobel Prize in 1915 for their work.

Fig. 1-21 (a) Schematic drawing of the Bragg spectrometer. (b) Variation of intensity of reflected x rays with angle. The crystal used was sodium chloride. Graphs I and II were obtained by using two different faces of the crystal, but with the same x-ray source, which produced characteristic lines denoted by A, B, and C. Notice that maxima corresponding to $n = 2$ and $n = 3$ were obtained, in addition to the $n = 1$ maxima. [The graphs are reproduced from W. H. Bragg and W. L. Bragg, Proc. Roy. Soc. A88, 428 (1913).]

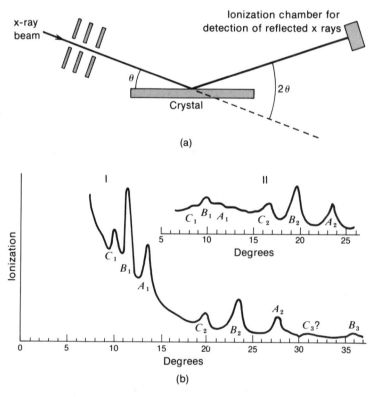

Simple models of the atom

The impression given by this chapter may be that atomic theory was relatively ineffectual until the advent of the Rutherford-Bohr model, but that the latter achieved such dazzling successes that the problem of atomic structure was essentially solved. The successes were undeniable but they were far from complete, and some very worrying questions remained. There was, in the first place, the sheer arbitrariness of Bohr's postulates concerning stationary states and quantum jumps. Furthermore, although the theory worked exceedingly well for the hydrogen atom, and even for "hydrogen-like" systems in general (as in the analysis of x-ray spectra), it was notably unsuccessful in dealing with problems explicitly involving two or more electrons. Bohr had almost no success in accounting for the energy-level structure of even the next simplest atoms after hydrogen. Moreover, even his theory of the hydrogen atom embodied a feature that proved to be at variance with the facts. This was the angular momentum associated with every Bohr orbit. Let us discuss this a little more fully.

After Bohr had first obtained his solutions to the hydrogen atom problem, he looked for a more fundamental condition to define the permitted orbits,[34] and found it in the consideration of the orbital angular momentum L ($= mvr$). From the basic statement of Newton's law for a circular orbit,

$$\frac{e^2}{r^2} = \frac{mv^2}{r}$$

we have

$$L^2 = (mvr)^2 = me^2r$$

However, for the orbit of quantum number n, the Bohr theory (Eqs. 1-21 and 1-23b) gives us

$$r_n = \frac{n^2h^2}{4\pi^2me^2}$$

[34]N. Bohr, Phil. Mag. **26**, 1, 576 (1913).

Hence

$$L_n{}^2 = \frac{n^2 h^2}{4\pi^2}$$

or

$$L_n = \frac{nh}{2\pi} = n\hbar \qquad\qquad (1\text{-}29)$$

where $\hbar\ (= h/2\pi)$ appears as a natural atomic unit of angular momentum. (When speaking of $\hbar$ one says "h bar.") In many presentations of the Bohr theory, Eq. 1-29—the quantization of orbital angular momentum—is used as the starting point of the analysis. In any event, according to the Bohr theory, the stationary states all possess orbital angular momentum, with the ground state ($n = 1$) having one unit of $h/2\pi$. Unfortunately, however, the ground state of the hydrogen atom has *zero* orbital angular momentum, as we shall see in Chapter 12. Thus Bohr's theory conflicts with the facts in a very crucial point, and cannot be accepted at its face value. It is an interesting dilemma. When Bohr's theory, on its first appearance in 1913, was being assailed by the classical physicists who could not stomach it, Albert Einstein remarked, "There must be something behind it. I do not believe that the derivation of the absolute value of the Rydberg constant is purely fortuitous."[35] Yet, as we shall see, the same results that Bohr obtained emerge from a totally different approach to the problem, embodying a feature that carries quantum theory even further away from classical physics. This new feature is the existence of a wave property for electrons and all other particles, and out of it comes the new and far more powerful form of atomic dynamics known as wave mechanics. In the next chapter we shall discuss the experimental basis on which this whole development rests.

[35]For this and many other documentations of the history of quantum theory, see Max Jammer, *The Conceptual Development of Quantum Mechanics*, McGraw-Hill, New York, 1966.

Simple models of the atom

EXERCISES

1-1 *Benjamin Franklin and the size of molecules.* In 1773 Benjamin Franklin, investigating the calming effect of oil on the water of a pond near London, observed that a teaspoonful of oil (a few cm³) would cover about half an acre (2000 m²) of the surface. In principle (if atomic theory had been developed in his time) he could have inferred the linear dimension of an oil molecule, assuming the oil layer to be monomolecular. (Similar methods were used for this purpose by Kelvin and others a century later.)

Use Franklin's observations to deduce the approximate length of an oil molecule. (We know now that oil molecules are linear molecules that align themselves side by side with their long dimension vertical on a water surface.)

1-2 *Atoms abounding.* Given the knowledge that atomic diameters are of the order of 1 or 2 Å, estimate the order of magnitude of the number of atoms in (a) a pin's head; (b) a human being; (c) the whole earth.

1-3 *An early method for estimating molecular diameters.* In 1858, J. J. Waterston devised a very clever method for deducing molecular sizes from macroscopic properties of a liquid, roughly as follows:

Picture a liquid as a cubic array of molecules with n molecules per cm length in each of the three principal directions. Each molecule has bonds to its 6 nearest neighbors. Suppose that it takes an amount of energy ϵ to break one bond. Then the complete vaporization of 1 cm³ of liquid involves the breaking of $6\,n^3$ bonds; the required energy $6\,n^3\epsilon$ is the latent heat of vaporization L in erg/cm³.

Now consider a molecule in the surface of the liquid. It has only 5 bonds—four to its nearest neighbors in the surface layer and one to its nearest neighbor in the next layer below the surface. Thus to create 1 cm² of fresh surface requires the breaking of one bond for each of n^2 molecules—a total energy of $n^2\epsilon$, which is measured by the surface tension S in erg/cm².

For water, $L \approx 2 \times 10^{10}$ erg/cm³ and $S \approx 75$ erg/cm². Use this theory to deduce the diameter d of a water molecule assuming that the molecules are closely packed so that $d \approx 1/n$.

1-4 *Thomson's determination of e/m.* The text outlines the procedure by which J. J. Thomson obtained a value for the charge-to-mass ratio of the electron. However, the e/m value quoted in the text (Eq. 1-2) is the modern value. Use the following typical data (from Thomson, Phil. Mag. **44**, 293, 1897) to compute e/m:

Length of deflecting plates: $l = 5$ cm

Electric field strength: $E = 0.5$ statvolt/cm
$(1.5 \times 10^4$ V/m$)$

Magnetic field strength: $B = 5.5$ gauss
$(5.5 \times 10^{-4}$ Tesla$)$

Angle of deflection with
electrostatic field only: $\theta = 8/110$ rad

The rather large discrepancy between the value you will obtain and the modern value for e/m is due at least in part to the magnetic field in the region outside the deflecting plates. This field was directed opposite to that in the deflecting region, lowering the effective value of B and causing Thomson to underestimate e/m.

1-5 *The spacing of atoms in a sodium chloride crystal.* The crystal lattice of sodium chloride has ions of sodium (Na^+) and ions of chlorine (Cl^-) arranged alternately in a simple cubic array. Thus the crystal volume per ion is d^3, where d is the distance between the centers of adjacent ions. The molecular weight of NaCl is 58.45 and its density is 2.164 g/cm³.

(a) Calculate the volume of 1 mole of crystalline NaCl.

(b) Using the fact that 1 mole of NaCl contains Avogadro's number N of sodium ions and N chloride ions, calculate the spacing d of the ions in the lattice. Compare your result with the handbook value of 2.82 Å.

1-6 *Elliptic orbits in the Thomson atom.* In Section 1-4 on the Thomson atom it is stated that any orbit of the electron within the positive-charge sphere has the same frequency as that of the linear oscillation discussed in the text.

(a) Generalize the analysis of Section 1-4 to show that the vector force on the electron inside the positive-charge sphere is given by the expression $\mathbf{F} = -(qQ/R^3)\,\mathbf{r}$ with symbols defined in that section.

(b) From the result of (a), show that for motion in the xy plane the x and y displacements are independent harmonic functions of time, each with the same frequency as that for linear displacement discussed in the text, and hence that the path is in general an *elliptic* orbit centered on the center of the sphere (as long as the electron stays inside the positive-charge sphere!).

1-7 *Quantization of circular orbits in the Thomson atom.* In the Thomson model of the hydrogen atom, the electron moves about *inside* a uniformly charged sphere of positive charge, which has a resulting restoring force $\mathbf{F} = -K\mathbf{r}$ on the electron and a potential energy function (inside the charge distribution) of $V = \frac{1}{2}Kr^2$. The text analyzes the resulting harmonic oscillation through the center of the atom. Instead of this, consider *circular* orbits of the electron (while it remains inside the positive charge distribution) and impose a quantum condition that the angular momentum mvr be an integral multiple of $h/2\pi$, say $nh/2\pi$. Show that the quantized energies for such circular orbits would be

$$E_n = n \left(\frac{h}{2\pi}\right) \omega \qquad n = 1, 2, 3, \ldots$$

where $\omega = (K/m)^{1/2}$ is the classical angular frequency of the harmonic oscillator. If such a system existed, what would its spectrum look like?

1-8 *Photons and radio waves.* A radio station broadcasts at a frequency of 1 MHz with a total radiated power of 5000 watts.

(a) What is the wavelength of this radiation?

(b) What is the energy (in electron-volts) of the individual quanta that compose the radiation? How many photons are emitted per second? per cycle of oscillation?

(c) A certain radio receiver must have 2 microwatts of radiation power incident on its antenna in order to provide intelligible reception. How many 1-MHz photons does this require per second? per cycle of oscillation?

(d) Do your answers to parts (b) and (c) indicate that the granularity of electromagnetic radiation can be neglected in these circumstances?

1-9 *Time delay in the photoelectric effect.* A beam of ultraviolet light of intensity 1.6×10^{-12} watts ($= 10^7$ eV/sec) is suddenly turned on and falls on a metal surface, ejecting electrons through the photoelectric effect. The beam has a cross-sectional area of 1 cm², and the wavelength corresponds to a photon energy of 10 eV. The work function of the metal is 5 eV. How soon might one expect photoelectric emission to occur?

(a) The text (Section 1-6) suggests a classical estimate based on the time needed for the work-function energy (5 eV) to be accumulated over the area of one atom (radius $\approx$ 1 Å). Calculate how long this would be, assuming the energy of the light beam to be uniformly distributed over its cross section.

(b) Actually, as Lord Rayleigh showed [Phil. Mag. **32**, 188 (1916)] the estimate from (a) is too pessimistic. An atom can present an effective area of about λ^2 to light of wavelength λ corresponding to its resonant frequency. Calculate a classical delay time on this basis.

(c) On the quantum picture of the process, it is possible for photoelectron emission to begin immediately—as soon as the first photon strikes the emitting surface. But to obtain a time that may be compared to the classical estimates, calculate the *average* time interval between arrival of successive 10 eV photons. This would also be the average time delay between switching on the beam and getting the first photoelectron.

1-10 *The photoelectric effect in lanthanum.* The threshold wavelength of light required to eject electrons from the surface of the metal lanthanum ($Z = 57$) is 3760 Å.
    (a) What is the work function of the metal in electron-volts?
    (b) What is the maximum kinetic energy of photoelectrons emitted by this metal when it is illuminated with ultraviolet light of wavelength 2000 Å?

1-11 *Determination of Planck's constant.* The clean surface of sodium metal (in a vacuum) is illuminated with monochromatic light of various wavelengths and the retarding potentials required to stop the most energetic photoelectrons are observed as follows:

Wavelength (Å)	2536	2830	3039	3302	3663	4358
Retarding potential (V)	2.60	2.11	1.81	1.47	1.10	0.57

Plot these data in such a way as to show that they lie (approximately) along a straight line as predicted by the photoelectric equation, and obtain a numerical value for Planck's constant h.

1-12 *X-ray tube voltage and short-wavelength cutoff.* A continuous spectrum of x rays is often produced using a tube in which electrons accelerated through a large potential difference V_o strike an anode made of a heavy metal. As discussed in the text (and shown in Figure 1-9) the resulting continuous x-ray spectrum has a sharp cutoff: below a certain wavelength λ_o, whose value depends on V_o, no radiation is produced. For $V_o = 40$ kV, calculate the value of λ_o. How much is the result modified if the metal's 4-eV work function is taken into account?

1-13 *Rutherford scattering and the size of the nucleus.* The point-nucleus model Rutherford used to describe the alpha-particle scattering results of Geiger and Marsden fails to describe correctly the results for collisions in which the minimum separation is smaller than the nuclear radius. These close encounters occur for high-energy alpha particles and for collisions which are nearly head-on and result in scattering through large angles. Significant departures from the Rutherford scattering formula occur in such collisions for alpha particles of 32-MeV energy scattered by gold.

48   Simple models of the atom

(a) As a crude indication of the size of a nucleus, calculate the distance of closest approach of a 32-MeV alpha particle in a head-on collision with a gold nucleus ($Z = 79$). Neglect the recoil of the gold nucleus. Give a qualitative argument why we may ignore in this calculation the electrons surrounding the gold nucleus—or should we?

(b) What is the ratio of the radius of the gold atom to the radius of the gold nucleus deduced in (a)? Compare this to the ratio of the average distance of the earth from the sun (1.5×10^{13} cm) to the radius of the sun (7.0×10^{10} cm).

(c) In a sketch of the atom in which the nucleus is represented by a circle 1 mm in radius, how large a circle is needed to represent a typical Bohr electron orbit in correct relative scale?

1-14 *Transition energies in ionized lithium.* A photon with energy in the visible region (between about 4000 and 7000 Å) causes the transition $n \rightarrow n + 1$ in doubly ionized lithium, Li^{++}—a hydrogen-like system with a single electron and a central charge equal to $3e$. What is the lowest value of n for which this could occur?

1-15 *Identifying a hydrogren-like system.* In the spectrum from a certain star is observed a regular series of spectral lines in the ultraviolet, one in every four of which coincides with a line in the Lyman series of hydrogen.

(a) Assuming that this spectrum is produced by an atom stripped of all but one of its electrons, what is the element in question?

(b) What is the longest-wavelength line in this particular spectral series?

1-16 *A one-electron uranium atom.* A neutral uranium atom has 92 electrons surrounding a nucleus containing 92 protons.

(a) Given that the Bohr radius for the lowest energy in hydrogen is $a_o = 0.5$ Å, derive an approximate numerical value on the basis of Newtonian mechanics for the radius of the smallest Bohr orbit about the uranium nucleus.

(b) In a violent nuclear event a uranium nucleus is stripped of all 92 electrons. The resulting bare nucleus captures a single free electron from the surroundings. Given that the ionization energy for hydrogen is 13.6 eV, derive an approximate numerical value for the maximum energy of the photon that can be given off as the uranium nucleus captures this first electron.

(c) Calculate the value of v/c for the first Bohr orbit in U^{92} according to Newtonian mechanics. This will show you that relativistic dynamics should really be used in this problem.

1-17   *The sizes of highly excited atoms.*

(a) Calculate the radius r_n of a hydrogen atom whose electron is in Bohr orbit n. Evaluate your result for $n = 50$, 100, and 500.

(b) As mentioned in the text, transitions between such highly excited states of hydrogen have been observed. For instance, an emission line has been observed with radio telescopes at a frequency of 5009 MHz.[36] Show that this is the frequency of the transition from $n = 110$ to $n = 109$ in the Bohr model of hydrogen.

1-18   *Elastic collisions and the Franck-Hertz experiment.* In the Franck-Hertz experiment one observes the energy lost by electrons that suffer inelastic collisions with atoms—collisions in which the atoms change internal energy. Far more common are *elastic* collisions, in which there is no change in the internal energy of the atoms. In analyzing the results of the Franck-Hertz experiment, no account is taken of the electron energy loss in elastic collisions. Why not? Use classical mechanics to show that the maximum kinetic energy loss by an electron in an elastic collision with a mercury atom (atomic weight: 201) is very much smaller than the energy losses relevant to the Franck-Hertz experiment. Assume that the mercury atoms are at rest before the collisions.

1-19   *Discrete energy levels of the nucleus.* The first excited state of a lithium-7 nucleus is about 478 keV above the ground state. What is the least kinetic energy of incident protons that would be able to excite this state through inelastic scattering? Recoil of such a light nucleus is important in this problem. Consider conservation of both momentum and total energy in a head-on collision.

1-20   *Characteristic x-ray wavelengths.* The wavelength of the K_α x-ray line for a certain element is 3.36 Å. Using Eq. 1-27 deduce what is the element.

1-21   *Absorption edges and x-ray emission wavelengths.* Figure 1-18 shows how the absorption of x rays by iron and lead exhibits sharp increases when the x-ray energy becomes sufficient to eject from the atom an electron in a particular "shell."

For the element tungsten ($Z = 74$) the wavelength corresponding to the K-absorption edge is 0.178 Å and the wavelength corresponding to the L-absorption edges is about 1.14 Å.

(a) What value would you predict for the wavelength of the K_α x-ray line emitted by tungsten? (The observed figure is about 0.21 Å.)

(b) What value would you predict for the K_β wavelength for this element? (See Figure 1-16.)

---

[36]B. Höglund and P. G. Mezger, Science **150**, 339 (1965).

Simple models of the atom

1-22 *X-ray determinations of N and e.* In Section 1-3 we describe how Avogadro's number N was found from the ratio of the Faraday (F) to the elementary charge e (obtained from Millikan's experiments). Knowing N, the interatomic distances in crystals could be deduced.

Subsequently J. A. Bearden (1931) and others completely reversed this procedure. They made direct and very precise measurements of x-ray wavelengths, using ruled gratings. The Bragg equation (Eq. 1-28) could then be used to find absolute values of interatomic spacings d and hence N. Finally, e was obtained as the ratio F/N, with a precision an order of magnitude better than Millikan's.

The most remarkable feature is the possibility of making very precise measurements (to about 1 part in 10^5) of x-ray wavelengths (1–10 Å) using ruled gratings with spacings of the order of 0.001–0.01 mm (10^4–10^5 Å). The trick is to use glancing incidence (see the figure). If x rays of a certain wavelength approach the grating at an angle α to the surface, and leave it at an angle $\alpha + \theta$, the path difference δ is given by

$$\delta = d[\cos\alpha - \cos(\alpha + \theta)]$$

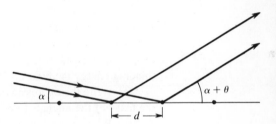

(a) Show that, if α and θ are very small angles, this gives $\delta \approx d(\alpha\theta + \theta^2/2)$, and hence, for grazing incidence ($\alpha = 0$) we have

$$\delta \approx \frac{d}{2}\theta^2$$

(b) Putting $\delta = \lambda$ we have

$$\theta = \sqrt{\frac{2\lambda}{d}}$$

Find the magnitude of θ, in minutes of arc, for x rays of wavelength 2 Å incident on a ruled grating with 100 lines per mm.

(c) If the position of a photographic image of diffracted x rays can be measured to 0.001 mm, how far away from the grating must a photographic plate be placed to allow the x-ray wavelength to be determined with an accuracy of 1 part in 10^4?

1-23 *The spectrum of hydrogen with a nucleus of finite mass*. The analysis in Section 1-7 of the text assumes that the nucleus remains fixed as the electron orbits about it. This corresponds to assuming that the proton mass is effectively infinite. More correctly, we should picture the proton and electron as orbiting about their common center of mass. The effects of this can be incorporated in the analysis of Section 1-7, but a simpler approach is to use the quantization of angular momentum (Section 1-12) and assume that the permitted circular orbits are those for which the *total* angular momentum of proton and electron together about the center of mass is $nh/2\pi$ (Eq. 1-29).

(a) Show that the corrected value of the Rydberg constant R_H is given by

$$R_H = \frac{M_p}{M_p + M_e} R_\infty$$

where R_∞ is the value for an infinite-mass nucleus.

(b) What is the *difference* in wavelength for the $n = 2 \rightarrow n = 1$ "Lyman alpha" transition ($\lambda \approx 1216$ Å) calculated using R_∞ and R_H, respectively?

1-24 *The spectrum of deuterium*. Deuterium is the isotope of hydrogen that has an atomic weight 2. The nucleus of deuterium, called the *deuteron*, consists of a proton and a neutron tightly bound together (binding energy is about 2 MeV).

(a) Write the expression for the Rydberg R_D of deuterium. Take the mass of the nucleus to be $2m_p$ (see Exercise 1-23).

(b) What is the fractional difference $(R_D - R_H)/R_H$ between the Rydberg for hydrogen (Exercise 1-23) and the deuterium Rydberg found above? What is the fractional difference in wavelength between any ordinary hydrogen line and the corresponding line in the deuterium spectrum?

1-25 *Energy of rotating molecules*. As a model for a diatomic molecule (e.g., H_2 or N_2 or O_2) consider two point particles, each of mass m, connected by a rigid massless rod of length r_0. Suppose that this molecule rotates about an axis perpendicular to the rod through its midpoint. Show that, if the Bohr quantization condition on angular momentum (Eq. 1-29) is applied, the calculated quantized values of the energy of rotation are given by

$$E_n = \frac{n^2 h^2}{4\pi^2 m r_0^2} \qquad n = 1, 2, 3 \ldots$$

[*Note*: The correct quantum-mechanical treatment of this system is discussed in Sections 11-2 and 11-3.]

Simple models of the atom

1-26 *Muonic atoms and the size of the nucleus.* The negative muon (symbol μ^-) is a particle with the same charge as the electron but with a larger mass ($m = 207\ m_e$). High-speed muons are produced in violent nuclear collisions. These muons can be slowed down in matter and captured into orbits around the nuclei of atoms in the material. The resulting system, with a negative muon in the inner orbit, is called a *muonic atom*. After about 2 microseconds (on the average) the muon decays into an electron and two neutrinos, destroying the muonic atom. Despite this brief existence, the muonic atom can be considered stable since its lifetime corresponds to very many periods of revolution of the muon in a Bohr orbit about the nucleus.

Because of its large mass relative to the electron, the inner orbit of the muon lies very much closer to the nucleus than that of any electron. This makes it possible to use muonic atoms to probe the size and structure of nuclei.

(a) For a central charge Ze, obtain an expression for the radius r_n of the nth muonic orbit. Express this as a multiple of the radius a_o of the first Bohr (electron) orbit in hydrogen.

(b) For the muonic orbit of (a), what is the energy E_n expressed as a multiple of the energy of the lowest state of an electron in the hydrogen atom?

(c) For $Z = 13$ (aluminum) calculate the energy in keV and the wavelength in Angstroms of the $n = 2 \rightarrow n = 1$ muonic x-ray transition.

(d) For copper ($Z = 29$) the energy of the $n = 2 \rightarrow n = 1$ muonic x-ray transition is significantly lower than one would calculate for a point nucleus; this implies that the nuclear charge distribution reaches out to the $n = 1$ orbit. What does this tell us about the radius of the copper nucleus?

After the end of World War I, I gave a great deal of thought to the theory of quanta and to the wave-particle dualism . . . It was then that I had a sudden inspiration. Einstein's wave-particle dualism was an absolutely general phenomenon extending to all physical nature . . . The new concept also gave the first wave interpretation of the conditions of quantizing the momenta of atomic electrons.

L. DE BROGLIE, *New Perspectives in Physics* (1962)

We think we understand the regular reflection of light and x-rays—and we should understand the reflection of electrons as well if electrons were only waves instead of particles . . . It is rather as if one were to see a rabbit climbing a tree, and were to say, "Well, that is rather a strange thing for a rabbit to be doing, but after all there is really nothing to get excited about. Cats climb trees—so that if the rabbit were only a cat, we would understand its behavior perfectly."

CLINTON J. DAVISSON, *Franklin Institute Journal* (1928)

2

The wave properties
of particles

2-1 DE BROGLIE'S HYPOTHESIS

The fact that light has some of the properties of ordinary particles led Louis de Broglie, in 1924, to construct a theory on the proposition that photons *are* particles. In a paper entitled "A Tentative Theory of Light Quanta" he developed the consequences of combining the Planck-Einstein relation $E = h\nu$ for photons with the generalized statement of the mass-energy equivalence, $E = mc^2$, of special relativity.[1] Building on the results of this analysis, he then speculated that his theory might be applicable to *all* particles, in which case wave behavior would be a universal property of moving objects. It was a hypothesis made without benefit of supporting evidence, and indeed it seemed to fly in the face of the long-accepted descriptions provided by particle dynamics. But, as was quickly realized, the wave behavior that de Broglie postulated would not become apparent except in circumstances not previously explored. Then, when the appropriate conditions were set up, the wave behavior was duly observed, as we shall describe shortly.

The central result of de Broglie's analysis was that *any*

[1]L. de Broglie, Phil. Mag. **47**, 446 (1924).

particle with momentum p has an associated wavelength λ given by

$$\lambda = \frac{h}{p} \qquad (2\text{-}1)$$

For photons, as we saw in Chapter 1, Eq. 1-16, this result comes directly from the Planck-Einstein relation $E = h\nu$, coupled with the relation $E = cp$ for classical electromagnetic waves. We have

$$E = cp = h\nu = \frac{hc}{\lambda}$$

from which $\lambda = h/p$ follows at once. De Broglie's hypothesis was that Eq. 2-1 applies to *all* particles.

De Broglie's own derivation of Eq. 2-1 was less direct than the simple one given above, but it has some features that will be of interest to us later. He began by assuming that every particle of light, whatever its quantum energy, has a certain rest mass m_0. This proposed rest mass would necessarily be exceedingly small, since the speeds of photons of all energies (or momenta) are equal within the limits of observational accuracy. [From relativistic mechanics the relation between speed and momentum for a particle of rest mass m_0 is $p = m_0 v (1 - v^2/c^2)^{-1/2}$ giving $v/c = \{1 + (m_0 c/p)^2\}^{-1/2}$.] Thus $v \approx c$ for any particle having $p \gg m_0 c$. In fact de Broglie, by assuming that the speeds of visible light and long radio waves do not differ by more than 1 percent, deduced that m_0 must be less than 10^{-44} g (see the exercises). He supposed, nonetheless, that this rest mass existed and hence that sufficiently precise measurements would reveal that the speed v of photons of a given energy (or frequency) was always less than c, even if only by a small amount. (This particular aspect of his theory has now been abandoned. There is no evidence, either experimental or theoretical, that the rest mass of photons is anything other than zero.[2] One cannot, of course, rule out the possibility that future measurements of sufficient refinement might require us to revise our ideas.)

---

[2]For discussion citing observations that place strict upper limits on photon rest mass, see A. S. Goldhaber and M. M. Nieto, Rev. Mod. Phys. **43**, 277 (1971); and Scientific American **234**, 86 (May 1976).

The wave properties of particles

For a photon as observed in the laboratory frame, we have $E = h\nu$. De Broglie argued that we can write a similar equation for a photon as observed in its rest frame (achievable for $m_o \neq 0$, $v < c$). In this frame we can associate with the photon a characteristic frequency ν_o such that the energy $h\nu_o$ is equal to the rest energy, that is

$$h\nu_o = m_o c^2 \tag{2-2}$$

The existence of ν_o implies that the photon has some characteristic vibration described by the equation

$$\xi_o \sim \sin 2\pi\nu_o t_o$$

where t_o is time as measured in the rest frame.

Now consider this same vibration as observed from the standpoint of the laboratory frame, relative to which the photon is moving with a speed v. Using the Lorentz transformation,[3] we have

$$t_o = \frac{1}{(1 - \beta^2)^{1/2}} \left(t - \frac{vx}{c^2} \right)$$

(where $\beta = v/c$) and hence a sinusoidal disturbance described by the equation

$$\xi(x, t) \sim \sin \frac{2\pi\nu_o}{(1 - \beta^2)^{1/2}} \left(t - \frac{vx}{c^2} \right)$$

or

$$\xi(x, t) \sim \sin 2\pi\nu \left(t - \frac{x}{w} \right) \tag{2-3}$$

This is the equation of a wave of frequency ν and wave speed (phase velocity) w where

$$\nu = \frac{\nu_o}{(1 - \beta^2)^{1/2}} \tag{2-4}$$

---

[3]See, for example, the volume *Special Relativity*, in this series.

2-1 De Broglie's hypothesis

and

$$w = \frac{c^2}{v} \tag{2-5}$$

Note that the value of $h\nu$ as given by Eq. 2-4 is equal to the total energy $E = mc^2$ of the photon as measured in the laboratory frame, for we have

$$h\nu = \frac{h\nu_0}{(1 - \beta^2)^{1/2}} = \frac{m_0 c^2}{(1 - \beta^2)^{1/2}} = mc^2$$

The wave disturbance described by Eq. 2-3 has a characteristic wavelength λ defined by the ratio of the wave speed w to the frequency ν:

$$\lambda = \frac{w}{\nu} = \frac{c^2}{v} \cdot \frac{(1 - \beta^2)^{1/2}}{\nu_0} \tag{2-6}$$

Making use of Eq. 2-2 and the relation $m = m_0/(1 - \beta^2)^{1/2}$, we finally obtain the de Broglie relation between wavelength and linear momentum for these particles. For, from Eq. 2-6, we then have

$$\lambda = \frac{hc^2}{v} \cdot \frac{(1 - \beta^2)^{1/2}}{h\nu_0} = \frac{h(1 - \beta^2)^{1/2}}{m_0 v} = \frac{h}{mv} = \frac{h}{p} \tag{2-6a}$$

Thus we have recreated Eq. 2-1, but in a way that makes it applicable to particles with rest mass. We shall return to this very shortly, but first we must deal with a vital point of detail in de Broglie's theory.

2-2 DE BROGLIE WAVES AND PARTICLE VELOCITIES

The analysis that we have presented, although it is simple and straightforward, appears on the face of it to have a flaw that would disqualify it as a physical theory. If the particle velocity is v ($< c$), the velocity of its associated wave w, being given by c^2/v, is greater than c. Thus it appears that the particle and its wave must inevitably part company. De Broglie showed, however, that this is not so. In the theory of wave motions, there is in general a difference between the speed of waves of some uniquely defined frequency, and the speed of a

The wave properties of particles

localized pulse such as de Broglie proposed to associate with a moving particle. If one watches waves on the ocean, for example, one can see regions over which the individual wave amplitudes grow to a maximum and then dwindle away again, as indicated in Figure 2-1a. Careful observation will then show that the maximum of an overall disturbance (the "group") moves with a speed that is different from the speed of the individual crests within it. In fact, with water waves the individual crests can be seen to pass rapidly forward through the group, rising and then falling in amplitude as they do so.

The analysis of this phenomenon builds on the fact that any localized disturbance or pulse can be regarded as synthesized from a number of pure sinusoidal waves of different frequencies.[4] It then develops that if the pure sinusoidal waves have speeds that vary with wavelength, the synthesized pulse travels with a speed (the *group velocity*) that may be quite different from the characteristic speeds (the *phase velocities*) of the individual waves. To illustrate the basic result in the simplest possible way, consider the result of superposing just two sine waves of slightly different wavelengths and frequencies. At any given instant the combined disturbance has a succession of maxima, as shown in Figure 2-1b. These maxima move along at the group velocity, which we can calculate directly by writing down the mathematical expression for the combined disturbance.

The simplest mathematical description of a sinusoidal wave (traveling along the positive x axis) is through the equation

$$y = A \sin (kx - \omega t)$$

where k is equal to $2\pi/\lambda$ (and is usually called the *wave number*) and ω is the *angular frequency*, equal to $2\pi\nu$.

Our two combining waves can then be written

$$y_1 = A \sin (kx - \omega t)$$
$$y_2 = A \sin [(k + \Delta k) x - (\omega + \Delta\omega)t]$$

---

[4]This is a basic result of Fourier analysis. See, for example, the book *Vibrations and Waves* in this series. Also see the corresponding analysis of quantum wave packets in Chapter 8 of the present book.

2-2   De Broglie waves and particle velocities

Fig. 2-1 (a)
Sketch showing a
group of waves. The
group is the region
of large amplitudes
outlined by the
dashed line (which
is called the enve-
lope). Individual
crests within the
group move at a
speed which differs
from that of the
group. (b) Suc-
cession of wave
groups produced by
the superposition of
two pure sinusoids.
The wave groups
travel at the group
velocity; individual
crests travel at the
phase velocity.

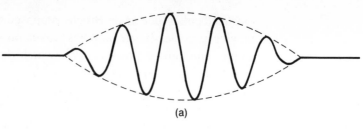

(a)

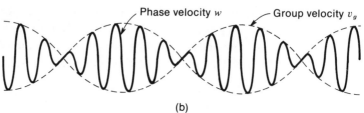

Phase velocity w — Group velocity v_g

(b)

The individual phase velocities are given by

$$w_1 = \frac{\omega}{k} \qquad w_2 = \frac{\omega + \Delta\omega}{k + \Delta k}$$

For $\Delta\omega \ll \omega$ (and $\Delta k \ll k$) these phase velocities are almost equal. The combined disturbance can then be written[5]

$$y = y_1 + y_2$$
$$= 2A \sin\left[\left(k + \tfrac{1}{2}\Delta k\right)x - \left(\omega + \tfrac{1}{2}\Delta\omega\right)t\right] \cos\left(\frac{\Delta k}{2}x - \frac{\Delta\omega}{2}t\right) \qquad (2\text{-}7)$$

The last factor, $\cos\tfrac{1}{2}(x\Delta k - t\Delta\omega)$, describes the modulation of
amplitude of the resultant waveform. The velocity with which
this modulation envelope moves is the group velocity v_g (see
Figure 2-1b). This velocity is given, as for any progressive
wave, by the coefficient of t divided by the coefficient of x.
Thus we have

$$v_g = \frac{\Delta\omega}{\Delta k}$$

In the limit of infinitesimally small differences in frequency and
wave number between the combining waves, we thus have

[5]This uses the trigonometric identity
$$\sin A + \sin B = 2 \sin \frac{A + B}{2} \cos \frac{A - B}{2}$$

60   The wave properties of particles

$$\text{(Phase velocity)} \ w = \frac{\omega}{k}$$

$$\text{(Group velocity)} \ v_g = \frac{d\omega}{dk} \tag{2-8}$$

Let us now apply these results to a superposition of de Broglie waves. From Eqs. 2-4 and 2-6 we have

$$\omega = 2\pi\nu = \frac{\omega_o}{(1-\beta^2)^{1/2}}; \qquad k = \frac{2\pi}{\lambda} = \frac{\beta\omega_o}{c(1-\beta^2)^{1/2}}$$

where we have put $2\pi\nu_o = \omega_o$. By direct division, the phase velocity ω/k is equal to c/β or c^2/v, which agrees with Eq. 2-5, as it must. To calculate the group velocity, it is simplest to find $d\omega/d\beta$ and $dk/d\beta$ separately:

$$\frac{d\omega}{d\beta} = \frac{\beta\omega_o}{(1-\beta^2)^{3/2}}; \qquad \frac{dk}{d\beta} = \frac{\omega_o}{c(1-\beta^2)^{3/2}}$$

Then, by division, we get

$$\text{Group velocity} \ (v_g) = \frac{d\omega/d\beta}{dk/d\beta} = \beta c = v$$

Thus the velocity of the associated wave group is exactly the same as that of the particle itself. Therefore the particle and its wave group move along together and we need not worry that they will part company.

We mentioned earlier that de Broglie, having developed this theory for photons, was then so bold as to suggest that its application was universal. In his 1924 paper he wrote:

> We are then inclined to suppose that *any* moving body may be accompanied by a wave and that it is impossible to disjoin motion of the body and propagation of the wave.[6]

He presented the theory and this suggestion in his Ph.D. thesis at the University of Paris in November 1924. And in a reply to a question from a member of his skeptical examining committee, he suggested that his waves should be detectable through diffraction experiments for electrons scattered by crystals.[7]

[6]L. de Broglie, *op. cit.*, p. 450.

[7]This was done later by Davisson and Germer. See Section 2-4 below.

61   2-2   De Broglie waves and particle velocities

Nothing was done about it at the time, however, and the theory remained just an intriguing speculation. An interesting detail of this story is that Einstein was consulted on the merits of de Broglie's hypothesis; he recommended award of the doctorate even though the applicability of the hypothesis to real particles was untested and a matter for skepticism. A few years later (in 1929) de Broglie was awarded the Nobel Prize for his work.

2-3 CALCULATED MAGNITUDES OF DE BROGLIE WAVELENGTHS

In this section we use the basic relation $\lambda = h/p$ to derive simplified formulas for de Broglie wavelengths of various particles. As we shall see later, the magnitude of this wavelength can often tell us whether the description of motion can be classical or requires quantum physics.

Most often, one wishes to know the de Broglie wavelength for particles of a given kinetic energy. For low energies, such that relativistic efforts can be ignored, the momentum p of particles of kinetic energy K and mass (rest mass) m_o is given by

$$p = (2m_oK)^{1/2}$$

so that the equation for λ is simply

$$\lambda = \frac{h}{(2m_oK)^{1/2}} \qquad (2\text{-}9)$$

In order to develop a formula that embraces both relativistic and nonrelativistic situations, we can use the relativistic connection between total energy E and linear momentum:

$$E^2 = E_o^2 + c^2p^2$$

where E_o is the rest energy of the particle ($= m_oc^2$). Now the total energy of the particle is, by definition, equal to the kinetic energy plus the rest energy. Thus we have

$$(E_o + K)^2 = E_o^2 + c^2p^2$$

from which

The wave properties of particles

$$p = \frac{1}{c} (2E_oK + K^2)^{1/2}$$

and so

$$\lambda = \frac{hc}{(2E_oK + K^2)^{1/2}} \tag{2-10}$$

For particles of very high kinetic energy ($K \gg E_o$) Eq. 2-10 reduces to the simple expression

$$\lambda \approx \frac{hc}{K} \tag{2-11}$$

By means of a simple piece of algebra, we can convert Eq. 2-10 into a sort of universal formula for calculating de Broglie wavelengths. A natural unit of energy for any particle is its rest energy E_o ($= m_oc^2$). Dividing both numerator and denominator of Eq. 2-10 by E_o, we obtain

$$\lambda = \frac{h/m_oc}{[2 (K/E_o) + (K/E_o)^2]^{1/2}}$$

We then recognize the quantity h/m_oc as a characteristic unit of wavelength λ_o for a particle of rest mass m_o. It is in fact known as the *Compton wavelength* because it appears as an important parameter in the theory of the Compton effect (Section 1-6). We can then put

$$\frac{\lambda}{\lambda_o} = \frac{1}{[2 (K/E_o) + (K/E_o)^2]^{1/2}} \tag{2-12}$$

Figure 2-2 shows a graph of this relationship, using logarithmic scales for both axes. This gives us two sections of constant slope for the regions $K \ll m_oc^2$ and $K \gg m_oc^2$, connected by a curve over the intermediate region defined approximately by $0.1E_o < K < 10E_o$. The straight-line sections have slopes differing by a factor of two, corresponding to $\lambda \sim K^{-1/2}$ for $K \ll E_o$ (Eq. 2-9) and $\lambda \sim K^{-1}$ for $K \gg E_o$ (Eq. 2-11).

In Table 2-1 we list a few examples of the numerical values of de Broglie wavelengths. See the end of the chapter for practice problems involving such calculations.

2-3 Magnitudes of de Broglie wavelengths

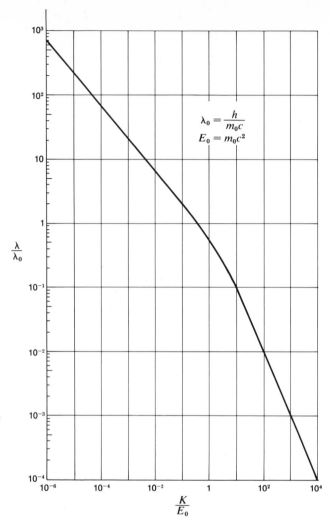

Fig. 2-2 De Broglie wavelength as a function of kinetic energy. For electrons, $\lambda_0 = 0.0243 \, \overset{\circ}{A}$ and $E_0 = 0.511$ MeV; for protons and neutrons, $\lambda_0 \approx 1.32 \times 10^{-5} \, \overset{\circ}{A}$ $= 1.32 \, F$ and $E_0 \approx 939 \, MeV$.

2-4 THE DAVISSON-GERMER EXPERIMENTS

In 1925 C. J. Davisson and L. H. Germer at Bell Telephone Laboratories were making studies of electron scattering by crystal surfaces using polycrystals of nickel when they had a famous disaster that led them to triumph. The vacuum system broke open to the air while the nickel target was hot, thereby oxidizing the target. They tried to change the nickel oxide back to nickel by heating it in hydrogen and then in vacuum. In the process they incidentally changed it from a polycrystalline

The wave properties of particles

TABLE 2-1 De Broglie Wavelengths

Particle		Value of λ^a
Electrons of kinetic energy	1 eV	12.2 Å
	100 eV	1.2 Å
	10^4 eV	0.12 Å
Protons of kinetic energy	1 keV	0.009 Å (= 900 F)
	1 MeV	28.6 F
	1 GeV	0.73 F
Thermal neutrons (at 300 K)		1.5 Å (average)
Neutrons of kinetic energy	10 MeV	9.0 F
He atoms at 300 K		0.75 Å (average)

[a] $1 \text{ Å} = 10^{-8}$ cm; $1 \text{ F} = 10^{-13}$ cm $= 10^{-5}$ Å

aggregate into a few large crystals. The character of the electron scattering was drastically changed, showing new strong reflections at particular angles. Taking the hint, they then deliberately used single crystals as targets, and proceeded to discover just the kind of behavior that de Broglie's hypothesis of particle waves predicted (although they had not known of de Broglie's theory at the time they made the initial observations).[8]

The basic features of the experimental arrangement are shown schematically in Figure 2-3a. An electron beam was directed perpendicularly onto a face of a nickel crystal. A collector accepted electrons coming off at some angle ϕ to the normal (or $\pi - \phi$ to the incident beam). The crystal could be rotated about the beam axis. The electrons were accelerated through a few tens of volts.

Davisson and Germer found a particularly strong reflection at $\phi = 50°$ for electrons accelerated through 54 V, when the crystal was cut and oriented so that the atoms in its surface were known to be in equidistant and parallel rows perpendicular to the plane containing incident and reflected beams (that is, in rows perpendicular to the paper in Figure 2-3b). Such

[8]C. Davisson and L. H. Germer, Phys. Rev. **30**, 707 (1927). See the film *Matter Waves* by Alan Holden (with Lester Germer) produced by Education Development Center, Inc., Newton, Mass. (1965).

2-4 The Davisson-Germer experiments

Fig. 2-3 (a) Sche-
matic diagram of
apparatus used by
Davisson and
Germer. (b) End
view of parallel
rows of surface
atoms. The rows are
separated by dis-
tance D and are
perpendicular to the
plane containing the
incident and scat-
tered electron
beams. The condi-
tion for constructive
interference is
λ = D sin φ.

data are often presented graphically as polar diagrams, where
the distance from the origin to the curve in a given direction is
a relative measure of the scattered intensity in that direction.
This is shown in Figure 2-4, together with a more conventional
graph of scattered intensity versus φ.

This strong reflection is a wave-like phenomenon totally
unexpected from the classical picture of electrons. Verification
of de Broglie's hypothesis and his expression for electron
wavelength followed an analysis similar to that for x rays
presented in Section 1-11. It was known from x-ray scattering
experiments that the spacing D between the parallel rows of
atoms was 2.15 Å, and the calculation of electron wavelength
then followed from an analysis similar to that for x rays (Eq.
1-28). The condition for reinforcement of waves scattered
from adjacent atoms in the crystal surface (Figure 2-3b) is

$$\lambda = D \sin \phi \qquad (2\text{-}13)$$

Fig. 2-4 Scattered
intensity as a func-
tion of detector
angle for electrons
accelerated through
54 V, based on data
of Davisson and
Germer. (a) Po-
lar diagram, in
which the distance
from the origin to
the curve in a given
direction is propor-
tional to the inten-
sity in that direc-
tion. (b) Car-
tesian graph of the
same data. The in-
tensity scale is arbi-
trary but is the same
for both graphs.

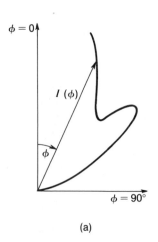

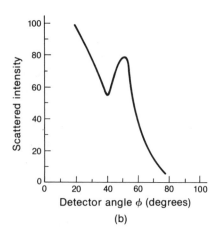

(a)   (b)

The wave properties of particles

Putting in the numerical values, this gives

$$\lambda = 2.15 \sin 50° = 2.15 \times 0.766$$
$$= 1.65 \text{ Å}$$

Now if electrons are accelerated through a potential difference V, they acquire a kinetic energy K equal to Ve, where e is the elementary charge. Thus for low-energy electrons, using Eq. 2-9, we have

$$\lambda = \frac{h}{(2m_oVe)^{1/2}}$$

Putting in the numerical values, one finds

$$\lambda \text{ (Å)} = \left(\frac{150}{V}\right)^{1/2} \qquad (V \text{ in volts})$$

For $V = 54$ V this gives

$$\lambda = \sqrt{\frac{150}{54}} = \sqrt{2.78} = 1.66 \text{ Å}$$

The agreement with the experimental result is excellent! Figure 2-5 shows the results of the systematic study that Davis-

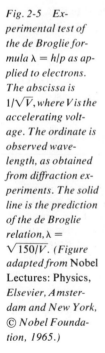

Fig. 2-5 Experimental test of the de Broglie formula $\lambda = h/p$ as applied to electrons. The abscissa is $1/\sqrt{V}$, where V is the accelerating voltage. The ordinate is observed wavelength, as obtained from diffraction experiments. The solid line is the prediction of the de Broglie relation, $\lambda = \sqrt{150/V}$. (Figure adapted from Nobel Lectures: Physics, Elsevier, Amsterdam and New York, © Nobel Foundation, 1965.)

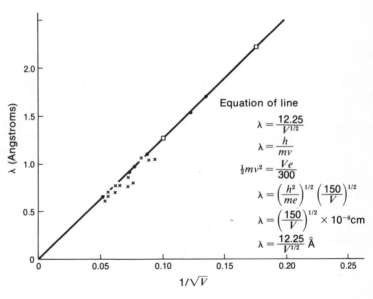

Equation of line

$$\lambda = \frac{12.25}{V^{1/2}}$$

$$\lambda = \frac{h}{mv}$$

$$\tfrac{1}{2}mv^2 = \frac{Ve}{300}$$

$$\lambda = \left(\frac{h^2}{me}\right)^{1/2} \left(\frac{150}{V}\right)^{1/2}$$

$$\lambda = \left(\frac{150}{V}\right)^{1/2} \times 10^{-8}\text{cm}$$

$$\lambda = \frac{12.25}{V^{1/2}} \text{ Å}$$

2-4   The Davisson-Germer experiments

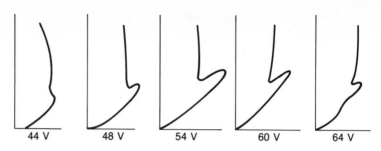

Fig. 2-6 Polar
plots of scattered in-
tensity showing de-
pendence of diffrac-
tion pattern on ac-
celerating voltage.
(Based on data of
Davisson and
Germer.)

44 V   48 V   54 V   60 V   64 V

son and Germer subsequently made to test the de Broglie relationship through the dependence of measured electron wavelength on accelarating voltage. This study verified the de
Broglie relation within experimental error.

There is more in this than meets the eye, however. If the
effect depended only on the rows of atoms in the surface of the
crystal, it would be possible to accelerate electrons through
any arbitrary voltage V, and find a strong diffraction peak at
the value of ϕ equal to $\sin^{-1}(\lambda/D)$. In fact, however, the intensity of the diffraction peak is very sensitive to accelerating
voltage and therefore to electron energy. Use of electrons of
energies very different from 54 eV leads to much reduced
diffraction intensities, as indicated in Figure 2-6. To understand the results more fully, one should consider the process as
involving the lines of atoms below the surface as well; like
x-ray diffraction, it is really a problem in scattering by a three-
dimensional lattice. In the next section (written for those
readers who may wish to delve a little deeper into the problem)
we discuss this and some other details of the process.

2-5 MORE ABOUT THE DAVISSON-GERMER EXPERIMENTS

In describing x-ray diffraction in Chapter 1, we pointed out
that the scattering of waves in certain preferential directions
from an atomic lattice involves the cooperation of successive
parallel planes that are rich in atoms, and that the observed
scattering corresponds to reflections from these planes under
the conditions of reinforcement described by the Bragg equation

$$2d \sin\theta = n\lambda \qquad (n = 1, 2, 3 \ldots)$$

On this basis we can now develop a clearer understanding of the
results of the Davisson-Germer experiment. In that experiment, the incident electron beam was perpendicular to the crys-

The wave properties of particles

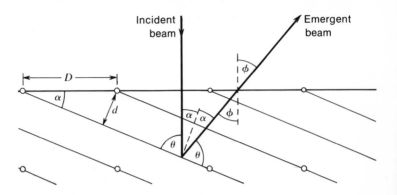

Fig. 2-7 End view of a set of parallel planes which intersect the crystal surface at angle α.

tal surface. We now picture the beam as penetrating into the crystal and encountering a set of reflecting planes (see Figure 2-7).[9] Each atom in the surface layer of the crystal lies in one of these planes. Thus, if the planes make an angle α with the surface, then the spacing d between planes is related to the interatomic distance D in the surface layer through the equation

$$d = D \sin \alpha$$

Now the angle θ between the incident beam and the reflecting planes is just $\pi/2 - \alpha$ (because of the perpendicular incidence of the electron onto the crystal surface), and the Bragg condition can therefore be written[10]

$$2d\sin\left(\frac{\pi}{2} - \alpha\right) = 2d\cos\alpha = n\lambda$$

Substituting for d from the previous equation, we have

$$2D\sin \alpha \cos \alpha = n\lambda$$

or

$$D\sin 2\alpha = n\lambda$$

But 2α is just the angle ϕ between the incident and reflected electron beams. Thus we can put

$$D\sin \phi = n\lambda \qquad (2\text{-}14)$$

which (for $n = 1$) reproduces our original equation (Eq. 2-13). The earlier equation was derived using only the surface atoms, while Eq. 2-14 takes account of the three-dimensional structure below the surface. It is interesting, though, to see how the

[9]Actually, these low-energy electrons do not penetrate very far at all, with consequences that we shall discuss later.

[10]We are for the moment ignoring any change of the wavelength of the electrons as they enter the crystal; see discussion at the end of this section.

2-5   More about Davisson-Germer experiments

dependence on the inner structure of the crystal drops out of sight in the final form of this equation, leaving only the dependence on the separation D between atoms on the surface.

The fact that the inner structure *is* crucial shows up, as we mentioned earlier, when the electron energy is varied. This causes a change in the wavelength λ. On the basis of Eq. 2-14 alone, we should expect that this would merely lead to a change in the angle ϕ that satisfies the equation $D \sin \phi = n\lambda$. In terms of our picture of reflecting planes, however, we see that the value of ϕ, for normal incidence, is set by the angle of the planes. If no crystal plane lies at the angle $\alpha = \phi/2$ for which $\phi = $ arc sin (λ/D) then no strong reflection will be observed for that wavelength and angle. This is the reason why the reflection maxima shown in Figure 2-6 change magnitude with accelerating voltage. Thus, because of the importance of atoms under the surface, Eq. 2-14 is a *necessary* but *not sufficient* condition for strong reflected beams. Instead of searching for the correct angle for a given wavelength, we can keep the angle fixed and change wavelength. Equation 2-14 can be written

$$\lambda = \frac{D\sin\phi}{n} = \frac{D\sin2\alpha}{n}$$

Fig. 2-8   Variation in intensity of reflected electron beam with wavelength. Actually, these data were obtained with source at 10° to the normal, with the detector positioned to receive regularly reflected electrons. The slight discrepancies between predicted and observed maxima are due to refraction at the crystal surface, which does have an effect on the locations of the maxima when the incident beam is not normal to the surface. [After C. Davisson and L. H. Germer, Proc. Natl. Acad. Sci. 14, 619 (1928).]

Since λ is proportional to $V^{-1/2}$, where V is the accelerating voltage, a graph of the reflected intensity (always at the angle ϕ) plotted against $V^{1/2}$ ($\sim 1/\lambda$) should show a series of equally spaced sharp peaks corresponding to the successive values of the integer n. Davisson and Germer verified this (see Figure 2-8).

It is actually rather surprising that the influence of the deeper layers inside the crystal shows up so clearly in the Davisson-Germer experiments, since electrons of the energies

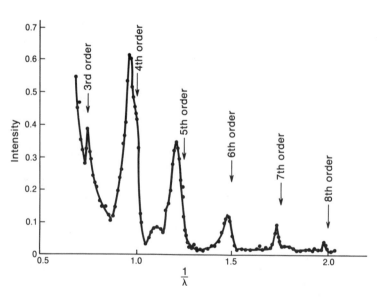

The wave properties of particles

they used ($\approx$30–600 eV) are unable to penetrate through many interatomic distances in a solid. This is in contrast to x rays. with their high penetrating power. One can guess that the extent to which the angle of Bragg reflection is sharply defined by the collection of successive reflecting planes must depend on the number of planes involved. The angular spread represented by a single spot in an x-ray diffraction pattern (for the same wavelength and crystal) is far less than that of the electron diffraction peak in the Davisson-Germer experiment, and at least part of this difference can reasonably be ascribed to the vastly different numbers of atomic planes involved. (You may wish to consider this problem in greater theoretical detail yourself.)

Our analysis of the Davisson-Germer experiment is still not quite right as given above. It ignores a feature that deserves mention, even though, again, it does not affect the particular final result expressed by Eq. 2-14. This feature is the increase in kinetic energy of the electrons when they enter the nickel crystal. It occurs because the interior of a metal is a region of negative potential energy for electrons. One manifestation of this is that, as we saw in connection with the photoelectric effect (Section 1-6), electrons inside a metal have a hard time getting out; the work function is a quantitative measure of the difficulty. Now an increase of kinetic energy of the incident electrons means a decrease of their de Broglie wavelength. But a change of wavelength in going from vacuum into a medium implies refraction, just as in the analogous case of light entering glass from air. Except at normal incidence, the direction of a beam entering or leaving the medium is changed in the manner described by Snell's law and an index of refraction. The following calculation shows how the final form of Eq. 2-14, and its use to infer the de Broglie wavelength of the incident electrons, is not affected by the refraction phenomenon, so long as the incident beam falls normally on the crystal surface.

Since the incident beam enters the crystal at normal incidence, it is undeviated (see Figure 2-9). The Bragg reflection produces a reflected beam that makes an angle ϕ' ($= 2\alpha$) with the normal to the surface of the crystal. Refraction then occurs so that the emergent diffracted beam makes the angle denoted by ϕ (as before) with the incident beam. Now the Bragg reflec-

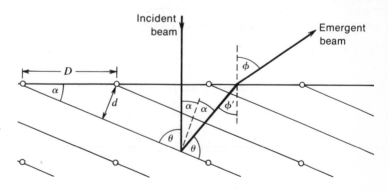

Fig. 2-9 Analysis of Bragg reflection, taking into account the change in wavelength which occurs when the beam crosses the crystal surface.

2-5   More about Davisson-Germer experiments

tion condition depends on the wavelength λ' *inside* the crystal; thus, we put

$$2d \sin\theta = n\lambda'$$

or

$$2d \cos\alpha = n\lambda'$$

Now $d = D \sin \alpha$ as before; hence we have

$$n\lambda' = 2D \sin\alpha \cos\alpha = D\sin 2\alpha$$

But

$$2\alpha = \phi'$$

Therefore

$$n\lambda' = D\sin \phi'$$

However, Snell's law of refraction (Figure 2-10) relates the change of beam direction at the surface to the wavelengths in the two media according to the formula

$$\frac{\sin \phi}{\sin \phi'} = \frac{\lambda}{\lambda'} \qquad \text{or} \qquad \frac{\sin \phi'}{\lambda'} = \frac{\sin \phi}{\lambda}$$

and so $D \sin \phi = n\lambda$, exactly as before. Hence the change in wavelength of electrons as they enter a solid does not need to be included in the analysis of the reflected diffraction pattern in the case of normal incidence. When the incidence is not normal, refraction effects do affect observed results. The slight discrepancies in Figure 2-8 between predicted and observed positions of intensity maxima are due to refraction of electrons at the surface of the nickel crystal.

In summary, the Davisson-Germer experiment verified the de Broglie relation for the wavelength of electrons at low energies, even though some care was required in interpreting their observations.

2-6 FURTHER MANIFESTATIONS OF THE WAVE PROPERTIES OF ELECTRONS

G. P. Thomson's Experiments

We described in Chapter 1 how, in 1897, the British physicist J. J. Thomson measured the value of e/m for free electrons and thereby identified them as a species of particle with well-defined properties. In 1927 and 1928 his son, G. P. Thomson, published the results of experiments that equally

The wave properties of particles

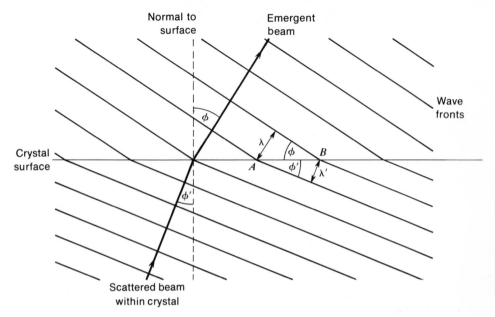

Fig. 2-10 Refraction of electron de Broglie waves at crystal surface. The wave fronts are bent as they cross the crystal surface because the phase velocity is higher outside the crystal than inside. The line segment AB is equal to $\lambda'/\sin \phi'$; it is also equal to $\lambda/\sin \phi$. Thus we have

$$\frac{\sin \phi}{\sin \phi'} = \frac{\lambda}{\lambda'}.$$

clearly showed the wave properties of these particles. His experiments differed from those of Davisson and Germer in two principal ways:

1. The electron energies were 10–40 keV instead of 30–600 eV. This meant de Broglie wavelengths of about 0.1 Å instead of 1 Å. The higher energy also meant that the electrons were quite able to pass completely through thin foils (~ 1000 Å) of solid material and could be observed by transmission rather than reflection.

2. The diffracting material was not a single crystal but a microcrystalline aggregate in which the individual crystals are randomly oriented so that more or less all possible orientations are represented. Thus, for a given electron energy or wavelength, some microcrystals were always present at the right orientation to satisfy the Bragg condition for any given reflecting plane. This resulted in diffraction patterns in the form of

2-6   The wave properties of electrons

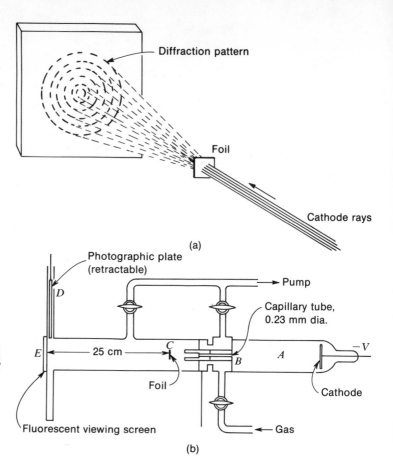

*Fig. 2-11 (a)
Because Thomson's
foil consisted of por-
tions of many indi-
vidual crystals, the
diffraction pattern
consisted of concen-
tric circles rather
than the individual
spots obtained when
a slice of a single
crystal is used.
[Recall the single-
crystal x-ray diffrac-
tion pattern of Fig-
ure 1-19(b).]
(b) G.P. Thom-
son's electron
diffraction appara-
tus. Cathode rays
generated in tube A
passed through col-
limating tube B be-
fore striking thin foil
C. The transmitted
electrons struck the
fluorescent screen
E, or a photo-
graphic plate D
which could be low-
ered into the path.
The entire appara-
tus was evacuated
during the experi-
ments. [After G. P.
Thomson, Proc.
Roy. Soc. A 117,
600 (1928).]*

concentric circles (see Figure 2-11a). The method used here is patterned on a widely used technique, known as the powder method, for the x-ray analysis of materials in powdered or microcrystalline form.

The actual arrangement (Figure 2-11) contrasted greatly with that of Davisson and Germer in that the diffracted beams were received on a photographic plate on the far side of the scattering foil. The order of magnitude of the angular deviations (approximated by λ/d) was about 0.1 rad or 5°, so that the radii of the diffraction rings formed on a plate placed, say, 10 cm beyond the foil could be expected to be about 1 cm. Thomson's experiments were done under just such conditions, and Figure 2-12a shows an example of his results. The extremely close similarity between the results of electron and x-ray diffraction is shown in Figure 2-12b, which is a composite pic-

The wave properties of particles

*Fig. 2-12 (a)
Electron diffraction
pattern obtained by
G. P. Thomson
using a gold foil
target. (b) Com-
posite photograph
showing diffraction
patterns produced
with an aluminum
foil by x rays and
electrons of similar
wavelength. (Cour-
tesy of Film Studio,
Education Develop-
ment Center, New-
ton, Mass.)
(c) Diffraction
pattern pro-
duced by 40 keV
electrons passing
through zinc oxide
powder. The distor-
tion of the pattern
was produced by a
small magnet which
was placed between
the sample and the
photographic plate.
An x-ray diffraction
pattern would not
be affected by a
magnetic field!
(Photograph made
by Dr. Darwin W.
Smith; reproduced
by permission of R.
B. Leighton.)*

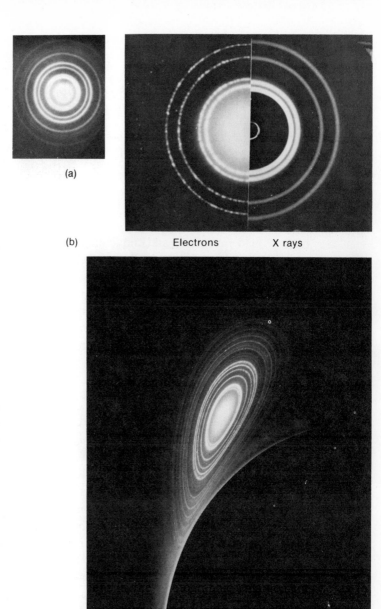

(a)

(b)

Electrons   X rays

(c)

ture constructed from the ring patterns produced by a given
foil with x rays and electrons of comparable wavelengths (with
some adjustment in the photographic enlargements so as to
bring them to the same scale). Figure 2-12c shows the distor-
tion produced in an electron diffraction pattern by a magnetic

75   2-6   The wave properties of electrons

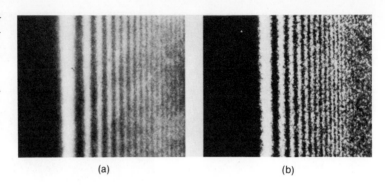

(a)   (b)

field—a striking and convincing demonstration that the diffraction phenomenon is directly related to the charged electrons and is not due to an electromagnetic or other wave that might be thought to accompany the particles.

G. P. Thomson's results verified to within 1 percent the applicability of the de Broglie relation to electrons at these higher energies. Thomson shared the 1937 Nobel Prize with Davisson for these researches.

Diffraction from Edges and Slits

The development of electron microscopy, with its use of extremely well collimated, highly monoenergetic electron beams, has led to some beautiful demonstrations of the wave properties of electrons in analogs of optical diffraction phenomena. Figure 2-13 shows a direct comparison of the straight-edge diffraction patterns of electrons and visible light. In contrast to Figure 2-12b, the two original diffraction patterns here are greatly different in size since the electron wavelength was very much smaller than the corresponding wavelength of light. The electron pattern has been magnified by a factor of the order of 1000, and the graininess of the photograph shows that the original fringes were near the limit of resolution (≈0.2 micron) of the photographic emulsion.

Some of the most impressive demonstrations of electron wave behavior are the interference patterns obtained by C. Jönsson using tiny slit systems formed in copper foil.[11] The slits were 0.5 micron wide, spaced 1–2 microns apart, and the

---

[11]Claus Jönsson, Z. Phys. **161**, 454 (1961). An edited translation by D. Brandt and S. Hirschi is available: Am. J. Phys. **42**, 5 (1974).

     The wave properties of particles

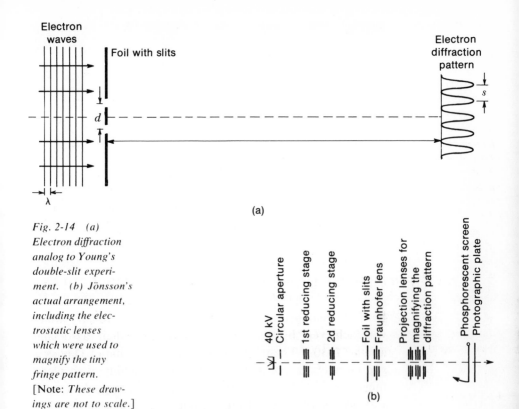

Electron waves

Foil with slits

Electron diffraction pattern

s

d

λ

(a)

Fig. 2-14 (a)
Electron diffraction
analog to Young's
double-slit experi-
ment. (b) Jönsson's
actual arrangement,
including the elec-
trostatic lenses
which were used to
magnify the tiny
fringe pattern.
[Note: *These draw-*
ings are not to scale.]

40 kV
Circular aperture
1st reducing stage
2d reducing stage
Foil with slits
Fraunhofer lens
Projection lenses for magnifying the diffraction pattern
Phosphorescent screen
Photographic plate

(b)

experiments were performed with 50-keV electrons having a wavelength of about 0.05 Å—smaller by a factor of about 10^5 than that of visible light. Interference fringes were formed at a distance of 35 cm from the slits. Except for the reduction in wavelength and slit spacing, the arrangement comes close to duplicating that with which Thomas Young, in his famous double-slit interference experiment in 1803, obtained the first indisputable evidence of the wave character of light. If we call the slit separation d and the distance from the slits to the plane of the fringe pattern D (see Figure 2-14a), then the spacing s of adjacent fringes (see the exercises) is given by

$$s = \frac{D\lambda}{d}$$

Putting $D = 35$ cm, $\lambda \approx 0.05$ Å, and $d \approx 2 \times 10^{-4}$ cm, one finds $s \approx 10^{-4}$ cm $= 1$ micron. Thus, the fringe pattern is exceedingly small. Therefore, in Jönsson's actual experiment,

2-6   The wave properties of electrons

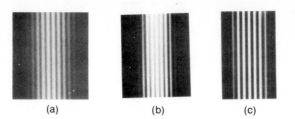

Fig. 2-15 Multislit diffraction patterns for electrons obtained by Jönsson. (a) two-slit pattern, (b) three-slit pattern, (c) four-slit pattern, (d) five-slit pattern. (From Jönsson, op. cit. Reproduced with permission of the author)

(a) (b) (c) (d)

electrostatic lenses were placed between the slits and the screen in order to magnify the fringes (see Figure 2-14b).

Jönsson obtained electron interference fringe patterns of this kind for systems of 2 to 5 slits, and a selection of his results is shown in Figure 2-15.

2-7 WAVE PROPERTIES OF NEUTRAL ATOMS AND MOLECULES

The demonstration of wave properties of ordinary atoms was achieved by Estermann, Stern, and their associates around 1930.[12] The technique was basically that of Davisson and Germer, except that there was no possibility of accelerating a beam of atoms to some single chosen energy. Instead, it was necessary to start from the equilibrium thermal distribution of molecular velocities. The energies in question were thus only of the order of 0.03 eV; we have seen (Section 2-3) that this implies a wavelength of the order of 1 Å for atoms of low atomic weight (just about the same as for 100-eV electrons or 10-keV x rays).

Experiments were done with beams of helium atoms or neutral hydrogen molecules incident on a single crystal of lithium fluoride. The detector was an extremely sensitive manometer that recorded an increase of pressure when the atomic beam (traveling through the vacuum system) entered it. The detector was set so as to receive atoms that left the crystal surface at an angle θ (actually 18.5°) equal to the angle of incidence (see Figure 2-16a), but the detector could be moved around a vertical axis that passed through the crystal, thus varying the azimuthal angle ϕ. Thus, for $\phi = 0$ the incident and reflected beams are in the same vertical plane, making equal

---

[12]I. Estermann and O. Stern, *Z. Phys.* **61**, 95 (1930); and O. Stern, *Phys. Z.* **31**, 953 (1930).

The wave properties of particles

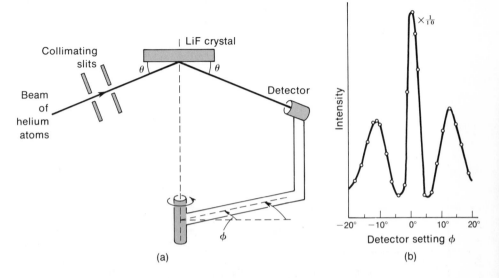

Fig. 2-16   (a) Experimental arrangement used by Stern et al. to investigate crystal diffraction of neutral helium atoms.   (b) Experimental results showing central reflection peak ($\phi = 0°$), plus first-order diffraction peaks ($\phi = 11°$). In the experiment, $\theta = 18.5°$.

but opposite angles θ to the horizontal. The graph of scattered (reflected) intensity versus ϕ, keeping θ constant, was as shown in Figure 2-16b.

The penetration into the crystal by atoms of thermal energy is quite negligible. Thus, in this case, in contrast to the Davisson-Germer experiment, the scattering really *is* due just to the two-dimensional array of atoms forming the top surface of the crystal (see Figure 2-17). It follows that diffraction peaks can be obtained for any given λ, and that a change of λ (that is, of atomic velocity) causes the angular deviation of the diffraction maximum to change. With the small value of θ used in the experiment (with $\cos \theta \approx 0.95$) the condition for a diffraction peak is given very nearly by $\lambda = D \sin \phi$.

In some later experiments, Stern *et al.* refined their apparatus by incorporating a velocity-selector for the atoms. This consisted of two identical wheels, each cut with over 400 radial slits near the edge, mounted on the same shaft a short distance apart (see Figure 2-18a). By varying the speed of rotation of the wheels, the speed of the atoms that could pass freely through the system could be prescribed and the diffraction of

2-7   Wave properties of atoms and molecules

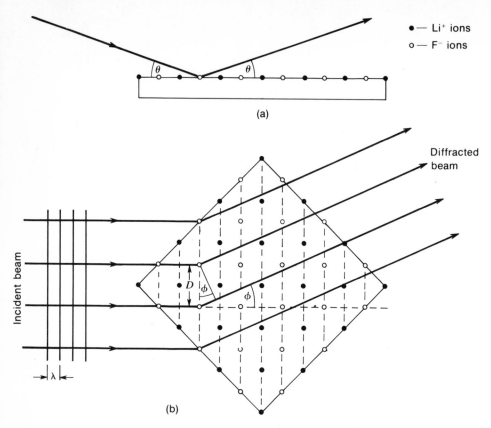

Fig. 2-17 Surface of lithium fluoride crystal used in experiments of Estermann, Stern, et al. (a) Side view,
showing the incident and reflected beams of helium
atoms. (b) Top view, showing the particular orientation
of the crystal lattice relative to the incident wavefronts. It
can be seen from the figure that, for small values of θ, a
diffraction peak occurs for λ = D sin φ.

atoms by a crystal could then be studied as a function of velocity. Figure 2-18b shows the azimuthal position of the diffraction peak as a function of the rotational frequency of the
mechanical beam chopper. As this frequency was increased,
higher velocity beams were selected, thus leading to shorter
wavelengths and hence to a shift of the peak to smaller values
of ϕ. The de Broglie relation was verified for helium atoms to
within the experimental error of 1–2 percent.

This verification of the de Broglie relation for neutral

The wave properties of particles

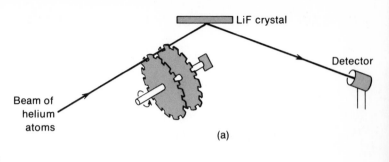

Fig. 2-18 (a) Schematic diagram of helium diffraction apparatus with velocity selector. The slotted wheels allowed only particles with a narrow range of velocities to pass through and strike the crystal. (b) Results obtained using velocity selector. When the rotational frequency (and hence the selected velocity) was increased, the diffraction peak shifted to smaller values of φ, just as expected. [After I. Estermann, R. Frisch, and O. Stern, Z. Phys. 73, 348 (1931).]

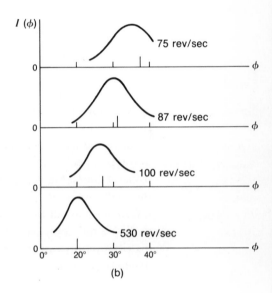

atoms, besides being a tour de force experimentally, can be regarded as of real importance philosophically. Free electrons are not part of our normal experience, and it is perhaps not too hard to stomach the discovery that they have certain attributes that we had never associated with particles before. But when it comes to bits of ordinary neutral matter, as represented by atoms of helium or molecules of hydrogen, we have nowhere to hide. In terms of all that was known prior to de Broglie's hypothesis, the collisions of neutral atoms with crystal surfaces at ordinary thermal velocities should surely be simply a problem in classical particle dynamics. Yet this was just not the case; the atomic beam experiments made the wave nature quite inescapable as a basic feature of all moving particles.

2-7 Wave properties of atoms and molecules

Slow Neutrons[13]

As you may know from the history of twentieth-century physics, neutrons (despite being building blocks of all nuclei) were discovered rather late, in 1932, by James Chadwick. It was subsequently pointed out that low-energy ("thermal") neutrons, with energies of the order of 1/40 eV, should have wavelengths similar to those of x rays (that is, of the order of 1 Å) and, because of their neutrality and consequent penetrating power, might provide a powerful supplementary tool for the study of crystal structures. This has proved to be the case. Neutron diffraction does not simply duplicate x-ray diffraction because neutrons and x rays respond to different things. X rays are scattered from the periodic electronic charge distribution of the atoms in a crystal; neutrons scatter, in most materials, from the atomic nuclei alone. Since atoms and nuclei in a crystal have the same periodicity of position, the locations of the diffraction peaks appear at the same angles for x rays and neutrons of the same wavelength, but the relative magnitudes of the peaks are quite different, leading to rather different detailed information about the crystal. Neutrons, possessing a magnetic dipole moment, can also scatter from the magnetic dipole distribution prevalent in magnetic materials and from nuclear magnets. This additional "magnetic scattering" is used to great advantage in the study of various problems of magnetism. Of course, our present concern is not primarily with such details, but simply with the confirmation that neutrons, like all other particles, have the wave properties that are predicted for them by the de Broglie relationship. Figure 2-19 presents some data that show the essential similarity of neutron diffraction to that of x rays or electrons. Slow neutrons are typically obtained from a nuclear reactor in which high-energy neutrons given off in a self-sustaining nuclear reaction are slowed down by repeated collisions with "moderator" atoms in the reactor.

[13]As general references on this topic, see G. E. Bacon, *Neutron Diffraction*, Clarendon Press, Oxford, 1955; D. J. Hughes. *Neutron Optics*, Interscience, New York, 1954; and D. J. Hughes, *The Neutron Story*, Doubleday Anchor, New York, 1959.

The wave properties of particles

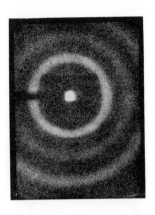

(a)

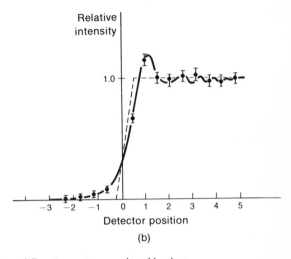

(b)

Fig. 2-19 (a) Ring diffraction pattern produced by slow neutrons with a polycrystalline iron target. (Courtesy of Dr. C. G. Shull.) (b) Straight-edge diffraction pattern for "cold" (20° K) neutrons of wavelength 4.3 Å. The straight-edge was made of gadolinium, which absorbs slow neutrons very strongly. This pattern may be compared to the straight-edge diffraction patterns for light and electrons shown in Figure 2-13. Broken line shows expected pattern in absence of diffraction. [A. G. Klein, L. J. Martin, and G. I. Opat, Am. J. Phys. 45, 295 (1977).]

Fast Neutrons

Fast neutrons, such as those from many nuclear reactions, have energies of the order of 1–10 MeV and hence wavelengths of the order of 10^{-12} cm. These wavelengths are comparable to nuclear diameters, so that the elastic scattering of fast neutrons by nuclei exhibits important diffraction effects, somewhat similar to the diffraction of light by an aperture (or better, by an obstacle), as indicated in Figure 2-20a. The angular width of the central maximum of the diffraction pattern is of the order of λ/d in each case, where d is the dimension of the obstacle. One might think that the extreme smallness of nuclei would make the effect hard to observe, but this is not necessarily so. Diffraction is basically a change of *direction*. The angle λ/d is not small, and at distances of several meters from the position of the scattering material the diffraction pattern is of macroscopic dimensions. Figure 2-20b shows the results of

2-8 Wave properties of nuclear particles

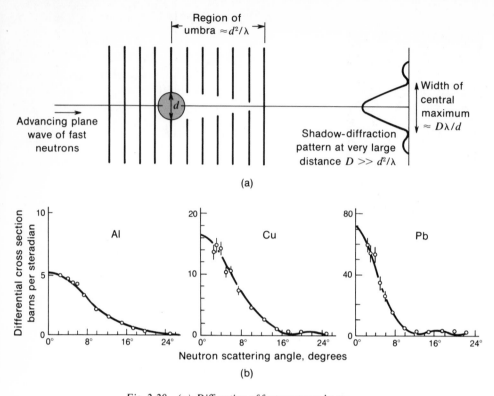

Fig. 2-20 (a) Diffraction of fast neutrons by nuclei. (b) Experimental results showing diffraction of 84-MeV neutrons from three different nuclei. [Data from A. Bratenahl et al., *Phys. Rev.* **77**, 597 (1950).]

some experiments on the diffraction of 84-MeV neutrons from different nuclei.

Alpha Particles

The scattering of energetic ($\sim$ 10-MeV) alpha particles by nuclei involves just the same kind of wavelengths as for fast neutrons. The problem is complicated by the Coulomb repulsion between alpha particles and nuclei, but the ability to select alpha particles of very precisely defined energy (experimentally much easier to achieve for charged particles than for neutrons) has resulted in scattering data that exhibit the wave diffraction behavior in an especially beautiful way, with numerous subsidiary peaks in the diffraction patterns. Figure 2-21 shows one example.

All the experiments on diffraction and interference effects with particles show that a definite wavelength is associated with particles of a given momentum. What does this mean? Does it mean that we were wrong in the first place to use the word "particle" in connection with such things as atoms and electrons? Do we have to abandon the idea that these objects are of some definite small size or that they can be localized? That would seem quite unacceptable in view of good evidence, such as we cited in Chapter 1, concerning the measured sizes of atoms, etc., and their particulate character generally. Must we now replace the picture of any such particle moving along by the picture of something smeared out over the extent of its de Broglie wave? What *is* this wave, whose reality seems undeniable? The debate on these questions, and on the apparent contradictions and paradoxes that they raised, continued fiercely for many years after the wave nature of matter had been discovered. Some would even say that the situation has still not been resolved. De Broglie himself, who opened the Pandora's box with his speculations founded on the dual (wave versus photon) character of light, wrote as follows:

> ... I shall suppose that there is reason to associate wave propagation with the motion of all the kinds of corpuscles whose existence has been revealed to us by experiment.... I shall take the laws of wave propagation as fundamental, and seek to deduce from them, as consequences which are valid in certain cases only, the laws of dynamics of a particle.[14]

In some respects this statement comes close to being an acceptable description of the views held today, although it is probably fair to say that our own present insight into the problem is clearer and is certainly based on a firmer experimental foundation. There is no doubt that the discovery of wave properties in what all physicists had hitherto regarded as particles in the classical sense led to a profound change, both conceptually and philosophically, in our description of nature. Indeed, it was probably the biggest revolution in the whole history of physical theory—far greater, for example, than was

[14]L. de Broglie, J. Phys. Rad. **7**, 1 (1926).

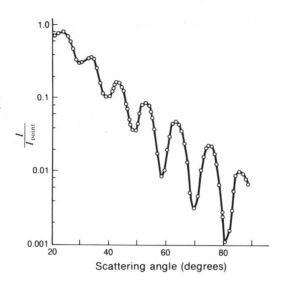

Fig. 2-21 Angular distribution of 40-MeV alpha particles elastically scattered from niobium nuclei. The ordinate is the ratio of observed intensity to that predicted on the assumption that the nucleus and the alpha are point particles. The general decrease of this ratio with increasing scattering angle can be attributed to the finite size of the nucleus without invoking the wave nature of the alphas. However, the striking pattern of strong maxima and minima is a definite indication of diffraction. [Figure after G. Igo, H. E. Wegner, and R. M. Eisberg, Phys. Rev. 101, 1508 (1956).]

demanded by the change from Newtonian mechanics to special relativity. In the discussion below we try to indicate something of the nature of this re-examination of fundamental ideas.

Two main lessons emerge when one considers the so-called wave-particle duality. The first is that the classical physics of macroscopic objects had generated a very clear—deceptively clear—understanding of the meaning of the words "particle" and "wave." The word "particle" conjured up a picture of the discrete, massive object of Newtonian dynamics, with sharply defined shape and size, as typified by billiard balls and similar objects. The word "wave," on the other hand, evoked the picture of some kind of disturbance in a continuous material medium, with water waves as the most familiar and vivid example. This picture of waves had already run into difficulties when applied to electromagnetic radiation; the medium whose vibrations were presumed to carry the wave was utterly elusive. Nonetheless, the description of the transmission of water waves, sound, light, and radio waves with the help of classical wave equations worked extremely well for the most part. Thus the exploration of the atomic world had as background this classical dichotomy of objects and phenomena. Atoms and electrons clearly belonged to the category of particles, and light emitted by excited atoms or accelerated electrons had an undeniable wave character. But the discovery

The wave properties of particles

of the granular character of photons and the wave properties of electrons undermined this dichotomy from both sides. Reluctantly, but inevitably, physicists had to acknowledge that the tidy classifications of the classical, macroscopic world could simply not be imposed at the atomic level. The whole question of what one meant by a particle or a wave had to be reopened.

The other lesson was that one must accept experimental results at their face value. One must not read more into them than the facts justify. It is necessary to ask rigorously, "What precisely do we observe, under what specific conditions?" When we do this carefully, we begin to be able to identify the circumstances under which, say, electrons display behavior like that of classical particles, and other circumstances in which they seem to behave like classical waves.

To sharpen up this discussion, we shall consider in detail some specific situations in which the two aspects of behavior can be separately observed.

2-10 THE COEXISTENCE OF WAVE AND PARTICLE PROPERTIES

The Formation of Optical Images

Optical image formation displays both the wave and particle properties of light. The production of a more or less faithful photographic image of an object is a typical example of the success of classical optical theory. The path of light through a lens system can be calculated using methods developed from the wave theory of light. The analysis is based on a picture of spherical waves of light spreading out from each point of the object, then being modified and ultimately reversed in curvature by the lens system, so that a convergent spherical wave comes to a focus on the image plane at a point that is, in the ideal case, uniquely associated with an individual point on the object. In practice things do not work quite this well because of small aberrations in the lens system and (more fundamentally) because the wavelength of light, although very small, is not completely negligible compared to the dimensions of typical optical systems. However, as everyone knows, the sharpness and accuracy of photographic images can be very impressive indeed.

But now consider more carefully the way in which the photographic image is formed. The photographic emulsion is a carefully prepared sheet of material containing individual grains of silver halide crystal. A typical grain is a micron or less in linear dimensions, containing perhaps 10^{10} silver atoms. Now the initiation of the photographic image is a photochemical process in which a single photon interacts with a single ion in the silver halide crystal. The development process is a chemical amplification of this initial atomic event—an amplification by a factor of 10^9 or more, so that all the silver atoms in a single grain of the emulsion can be developed out (that is, deposited as metallic silver) if just a few of them are activated in the original photographic exposure. And if a photograph is taken with exceedingly feeble light, one can verify that the image is built up by *individual* photons arriving independently and, it would seem at first, almost randomly distributed in position. This is beautifully illustrated in Figure 2-22, which shows how a picture begins to take shape, and to correspond more and more closely with the expectations of classical optics, as the total number of photons contributing to it is increased.

This example shows the complementary nature of wave and particle descriptions of optical image formation: at low levels and short exposures the statistical photon theory best describes the situation, while at high intensity or long exposures the wave theory is adequate to describe the results.

Double-Slit Interference Experiments[15]

Consider the experimental arrangement shown in Figure 2-23a, which (like Figure 2-14a) shows the basic elements of Young's double-slit experiment. The following analysis holds for both photons and electrons, although the equipment used in the electron experiment is much more sophisticated, requiring sub-miniature slits, a vacuum system, electron-microscope lenses, and so forth. A beam of light (or electrons) falls from the left onto a screen containing a single slit S_o. If this slit is very narrow, its diffraction pattern gives rise to cylindrical waves spreading out from it as shown. These waves are ob-

---

[15]See also *The Feynman Lectures on Physics*, Vol. I, Chap. 37, Addison-Wesley, Reading, Mass., 1963 and the B. B. C. film *Probability and Uncertainty—the Quantum Mechanical View of Nature*, part 6 of the Messenger Lectures delivered by Feynman at Cornell University in 1964. (The text of these lectures has been published as a book, *The Character of Physical Law*, B.B.C. Publications, 1965; MIT Press, Cambridge, Mass. 1967.)

The wave properties of particles

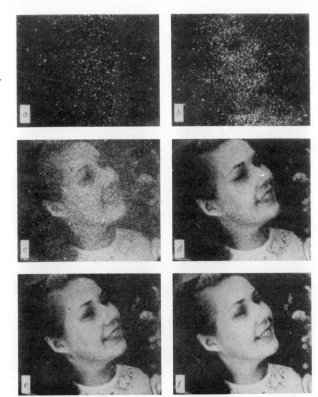

Fig. 2-22 Series of photographs showing how the quality of an image improves as the number of photons contributing to it increases. The approximate numbers of photons involved in each exposure were: (a) 3×10^3 (b) 1.2×10^4 (c) 9.3×10^4 (d) 7.6×10^5 (e) 3.6×10^6 (f) 2.8×10^7 [Figure reproduced from A. Rose, J. Opt. Sci. Am. 43, 715 (1953).]

structed by the second screen, which contains two parallel slits, S_1 and S_2. Diffraction at these two slits, in turn produces a double set of expanding cylindrical waves which proceed to overlap and superpose in the region beyond the slits. [A ripple-tank photograph (Figure 2-23b) gives a vivid picture of the wave pattern from an analogous double source using water waves.] Because both slits are illuminated from the same primary slit they are always in fixed phase relative to one another. The classical calculation of the familiar double-slit interference pattern can be made as follows. If the interfering slits are of equal width, they contribute equal amplitudes A_o at every point of the detector plane. (This is very nearly true, at least, because in the actual arrangement the fringe pattern extends over a distance that is very small compared to the distance from slits to detector, so that effects on the *individual* amplitudes due to differences of distance and direction from the separate slits are almost negligible.) However, if we consider the resultant effect at a transverse distance y from the

2-10   Coexistence of wave and particle properties

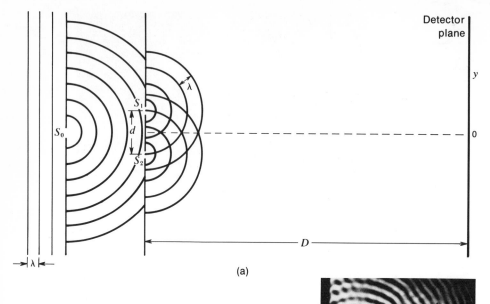

(a)

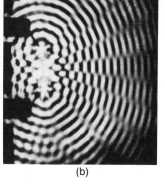

Fig. 2-23 (a) Young's double-slit experiment.
(b) Ripple-tank photograph showing pattern
of water waves produced by two sources oscillating in
phase. (Reproduced from PSSC Physics, 3rd ed., Ray-
theon Publishing Co, Boston, Mass. 1971. Reproduced
by permission of Education Development Center.)

(b)

center of the detector, there is a difference in the path lengths
from the two slits given very nearly by yd/D where d is the distance between slits and D is the distance between slits and detector (see Figure 2-24). For light of wavelength λ this results
in a phase difference δ given by

$$\delta = \frac{2\pi}{\lambda} \cdot \frac{yd}{D}$$

The combination of two waves of equal amplitude with this
constant phase difference has an amplitude A given by

$$A(y) = 2A_o \cos \frac{\delta}{2} = 2A_o \cos \left(\frac{\pi d}{D\lambda} \cdot y \right)$$

90   The wave properties of particles

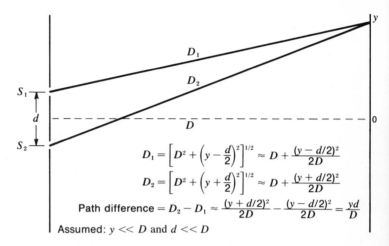

Fig. 2-24 Calcula-
tion of path dif-
ference in two-slit
interference experi-
ment.

$$D_1 = \left[D^2 + \left(y - \frac{d}{2}\right)^2\right]^{1/2} \approx D + \frac{(y - d/2)^2}{2D}$$

$$D_2 = \left[D^2 + \left(y + \frac{d}{2}\right)^2\right]^{1/2} \approx D + \frac{(y + d/2)^2}{2D}$$

Path difference $= D_2 - D_1 \approx \frac{(y + d/2)^2}{2D} - \frac{(y - d/2)^2}{2D} = \frac{yd}{D}$

Assumed: $y \ll D$ and $d \ll D$

The resultant *intensity* I is proportional to A^2, and if I_o is the intensity at any point on the detector when only one slit is open,
then we have

$$I(y) = 4I_o \cos^2\left(\frac{\pi d}{D\lambda} y\right) \tag{2-15}$$

Thus we arrive at the classical intensity variation shown in
Figure 2-25. Actually, the values of A_o and I_o are not independent of y because of the variation of intensity due to the
diffraction pattern of each individual slit. But since the slits are
close together, the y variations of A_o and I_o are nearly the same
for both slits.

Fig. 2-25 Two-slit
interference pat-
tern. The full line
shows the theoreti-
cal intensity varia-
tion, ignoring the y
variation of the
component single-
slit amplitudes.
(This is equivalent
to assuming that the
individual slits have
zero width.) The
broken line shows
the intensity due to
either slit alone on
the same assump-
tions.

But now, what is the situation if we investigate it in terms
of individual photons? This can be done experimentally if we
use a very weak light source and have as detector a photomul-

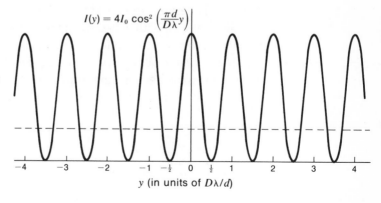

$$I(y) = 4I_o \cos^2\left(\frac{\pi d}{D\lambda}y\right)$$

y (in units of $D\lambda/d$)

2-10   Coexistence of wave and particle properties

tiplier tube that is sensitive enough to record and count the arrival of single photons.[16] The observations show that the classical pattern is gradually built up (like the photographic image discussed above) by the arrival of these photons, which are numerous at the maxima of intensity in the interference fringe pattern and very scarce at the minima. Indeed, quantitative measurements would show that the *numbers* of photons arriving at various points on the screen are proportional to the classical intensities $I(y)$ of Eq. 2-11. Notice, however, that again there is a randomness and an unpredictability about where the photons arrive when small numbers of photons are involved, which means that the correspondence between the classically calculated pattern and the observed distribution of photon counts becomes good only when the numbers of photons become very large. We shall return to this point in Chapter 6. The same unpredictability characterizes two-slit *electron* interference effects at low intensity.

Granted, however, the experimental fact that the distribution of large numbers of photons on the screen must follow the well-established intensity pattern predicted by wave optics, we are still faced with some extremely tough questions. Mustn't light—as photons—travel in straight lines? Clearly not, for the pattern is nothing like a geometrical shadow of the screen containing S_1 and S_2. Indeed, the spreading of the light by diffraction at the individual slits, so that the two sets of waves overlap, is an essential feature of the experiment. But this spreading does not account for the *interference* character of the fringe pattern. The distribution obtained in this experiment is completely different if, instead of having both slits open all the time, they are alternated so that while one is open the other is closed. If this is done, the fringe pattern disappears and is replaced by the diffraction pattern of an individual slit—which, if the slit is very narrow, is almost independent of y.

Since having both slits open simultaneously is essential to the interference phenomena, does this mean that the pattern is caused by pairs of photons, one passing through each slit? This might seem a tenable and appealing hypothesis. But it must be rejected. For experiments show that the interference pattern

---

[16]See, for example, J. G. King, *Interference of Photons*, PSSC film, Education Development Center, Inc., Newton, Mass., 1959.

The wave properties of particles

still develops even for extremely low light intensity—so low that, assuming photons arrive randomly, not more than one photon is likely to be in transit through the system at one time. This was established in a experiment by G. I. Taylor as long ago as 1909.[17]

Faced with the above results, one's first impulse is perhaps to suggest that each photon divides into two parts, with one-half going through each of the slits S_1 and S_2. But what would be meant by halving a photon? This would necessitate making two photons, each with half the original energy. But half the energy means half the frequency and, therefore, twice the wavelength, which would make the fringe separation twice as large as it is observed to be. So the photon does not split in this manner. Yet we cannot, from this, conclude that the photon must have passed through either one slit or the other. As we have seen, if this condition is defined experimentally by blocking one slit, the two-slit interference pattern vanishes. Thus, we are driven to a remarkable conclusion: to account for the spatial distribution of the photons arriving at the screen we must still use an analysis that closely parallels the classical wave description in terms of interfering amplitudes from both slits. Instead of classical wave amplitudes we have *quantum amplitudes* that are appropriate to the prediction of interference phenomena for individual particles. To end this chapter, we shall make this discussion a little more specific.

2-11 A FIRST DISCUSSION OF QUANTUM AMPLITUDES

Suppose that we have a double-slit arrangement with two slits of equal width, and beyond them, at a certain distance, a phototube with a narrow sensitive area that serves as a photon detector and can be moved transversely across the interference pattern. Suppose further, for simplicity, that the slits are so narrow, and hence their individual diffraction patterns so broad, that if only one slit is open the photon counting rate at the detector is effectively constant over a wide region. To make the discussion quantitative, we shall suppose that, with

---

[17]G. I. Taylor, Proc. Camb. Phil. Soc. **15**, 114 (1909). The PSSC film of Ref. 16 was also performed under these conditions.

2-11   A first discussion of quantum amplitudes

only one slit open, the number of photons that arrive at the detector (wherever it is located in the target area) is a fraction F of all the photons that fall upon the slit system from the left. Thus, if N photons fall upon the slit system in a certain time interval, then with one slit closed the number NF arrive at the detector and are counted. We can then say that F is the *probability* that any photon approaching the slit system under these conditions will arrive at the detector, which scans a small area only.

Now suppose that the second slit is opened. Twice as many photons as before will be emerging to the right of the slit system in a given time. Does this mean, then, that the number of photons arriving at the detector is now necessarily $2NF$? No! We know that under these conditions the interference pattern appears and the number of photons arriving at the detector now depends on its transverse position. At some places the number of counts actually falls to zero, and between these zeros will be the interference maxima. To calculate the number of counts at an interference maximum, we will introduce the *probability amplitude* associated with the passage of a photon from the left, through either slit, to the detector. This probability amplitude, or quantum amplitude, is selected by definition to have a magnitude equal to the square root of the probability F itself, but with a phase that is a function of position. And the quantum amplitudes of the two slits together add, with a relative phase depending on position, to give a resultant quantum amplitude. This resultant amplitude, when squared, tells us what fraction of all the incident photons will reach the detector at that position. The maximum value of this combined amplitude with both slits open is clearly $2\sqrt{F}$. In order to obtain the maximum probability we must square the maximum quantum amplitude. This gives $4F$. Thus, when N photons strike the slit system, $4NF$ reach the detector when it is placed at a maximum of the classical interference pattern. This number is twice as large as one would calculate by summing the counts from the two slits separately, but it corresponds precisely to the classical intensity pattern for waves (Eq. 2-15). The zeros of the interference pattern correspond to a combination of the equal amplitudes $\sqrt{F}$ with a phase difference of 180°, hence with opposite sign. The variation between $4NF$ and zero as the detector crosses the pat-

The wave properties of particles

tern in fact gives an *average* of $2NF$, which has to be so because, on the average, one must expect that two slits will pass twice as many photons as one slit.

It thus appears that the wave-particle duality consists in the fact that, although particles are detected as localized, complete entities, we must, as de Broglie hypothesized, use a wave-amplitude analysis to calculate the way in which they distribute themselves at a detector after passing through a region that contains any array of apertures or scattering centers. However, the calculation in terms of quantum amplitudes is only a recipe for telling us how particles will be distributed *statistically* across the plane of the detector. This may seem unsatisfying, and far from being an explanation, but it is the most that is permitted by a strict adherence to what we are actually able to observe, and it is the best description that we can give of the way nature works in such situations.

The concept of quantum amplitude, introduced here to describe a limited range of interference experiments, will be generalized and used in the following chapters as a fundamental tool for describing an immense variety of systems for which a quantum description is required.

EXERCISES

2-1 *Upper limit on the rest mass of a photon.* Most physicists believe that the photon has zero rest mass, but it is difficult to prove that de Broglie was wrong in assuming that the photon has an extremely small, but finite, rest mass m_0 and travels at speeds less than, although very close to, the speed c. Nevertheless, strict upper limits can be placed on m_0.

As noted in the text, de Broglie placed an upper limit of 10^{-44} g on m_0. This was obtained by assuming that radio waves of wavelength 30 km travel with a speed at least 99 percent of the speed of visible light ($\lambda = 5000$ Å). Beginning with the equation $h\nu = m_0c^2/(1 - v^2/c^2)^{1/2}$ of de Broglie's development (Section 2-1), obtain an exact expression for v/c in terms of m_0c^2 and $h\nu$. Use this to find an approximate expression for $(c - v)/c$ in the case $m_0c^2 \ll h\nu$. Check de Broglie's calculation of the 10^{-44} g limit.

2-2 *Group velocity of localized waves.* Section 2-2 discusses the superposition of *two* classical sinusoidal waves of slightly different wavenumbers k and $k + \Delta k$ to yield a modulated result whose modulations move with a group velocity. A composition almost as simple as the two-component case is the superposition of *three* waves y_0, y_1, and y_2 with the same total spread Δk in wavenumber ($k = 2\pi/\lambda$):

$$y_0 = A \sin (kx - \omega t)$$

$$y_1 = \frac{A}{2} \sin \left[\left(k - \frac{\Delta k}{2} \right) x - \left(\omega - \frac{\Delta \omega}{2} \right) t \right]$$

$$y_2 = \frac{A}{2} \sin \left[\left(k + \frac{\Delta k}{2} \right) x - \left(\omega + \frac{\Delta \omega}{2} \right) t \right]$$

(The amplitude $A/2$ of y_1 and y_2 has been chosen to give the superposition the simplest possible form.)

(a) Express the superposition $y_0 + y_1 + y_2$ as a single product of trigonometric functions. [*Hint*: Add $y_1 + y_2$ first and use trigonometric identities.] Sketch the resultant wave for $t = 0$.

(b) If $\Delta k/k = 10^{-2}$, how many zeros does the waveform have within each region of reinforcement (between adjacent zeros of the envelope)?

(c) If the phase velocity $\omega/k = 10$ cm/sec, $\Delta k/k = 10^{-2}$, and $\Delta \omega/\omega = 10^{-3}$, then what is the group velocity of the waveform?

2-3 *Nonrelativistic treatment of de Broglie waves.* De Broglie applied his relativistic treatment of wavelength for photons (Section 2-1) to particles of nonzero rest mass. The wave nature of particles of nonzero rest mass can also be described consistently with a *nonrelativistic* theory, a fact of vital importance in allowing simple solutions of many quantum-mechanical problems. Carry out the analysis using the equations $E = h\nu = h\omega/2\pi$, but this time use the nonrelativistic expression for a free particle, $E = p^2/2m$.

(a) Find ω ($= 2\pi\nu$) as a function of k ($= 2\pi/\lambda$) and compute ω/k and $d\omega/dk$.

(b) Does the group velocity $d\omega/dk$ equal the particle velocity, as required?

(c) The phase velocity ω/k differs from the value c^2/v given by the relativistic analysis. Can experiment tell us which of these phase velocities is correct?

(d) For both the relativistic and nonrelativistic theories, sketch phase and group velocities as functions of particle velocity v.

(e) What changes occur in the velocities computed in (a) if we allow for a rest energy E_0, so that the total energy E is given by the expression $E = p^2/2m + E_0$? Sketch the graph of phase velocity versus v for this case.

2-4 *de Broglie wavelengths for a wide range of energies.* Using Figure 2-2, obtain the de Broglie wavelengths for (a) electrons, (b)

The wave properties of particles

protons at the following kinetic energies: 10^{-2} eV; 1 eV; 100 eV; 10 keV; 1 MeV; 100 MeV; 10 GeV (10×10^9 eV).

Tabulate these values, along with the corresponding wavelengths for photons at the same energies; this will emphasize the result that $\lambda \rightarrow hc/E \approx hc/K$ for any "highly relativistic" particle ($K \gg m_0c^2$).

2-5 *Resolving power of an electron microscope.* A fundamental result of physical optics is that no optical instrument can resolve the structural details of an object that is smaller than the wavelength of the light by which it is being observed. A similar analysis applies to the de Broglie wavelength of electrons in an electron microscope.

It is desired to study a virus of diameter 0.02 micron ($= 200$ Å). This is impossible with an optical microscope (wavelength about 5000 Å) but can be done with an electron microscope. Calculate the voltage through which electrons must be accelerated to give them a de Broglie wavelength 1000 times smaller than the linear dimension of the virus, so as to permit formation of a very good image.

2-6 *The domain of wave mechanics.* A rough criterion for the applicability of ordinary particle mechanics (either Newtonian or relativistic) is that the de Broglie wavelength be much less than some characteristic linear dimension l of the system. For $l \lesssim \lambda$, wave mechanics is needed. Apply these criteria to the following systems.

(a) *Electron in the atom.* Use the de Broglie relation to calculate λ/r for the electron in the first Bohr orbit of hydrogen. Does your result verify that we need to use quantum physics to describe the behavior of electrons in atoms?

(b) *Proton in the nucleus.* Assume that one of the protons in a large nucleus (nuclear radius $\approx 10^{-12}$ cm) is in a circular orbit inside the nucleus with kinetic energy 10 MeV. Calculate λ/r for the orbiting proton. Do we need to use quantum physics to describe this nuclear system?

(c) *Electrons in a TV set.* The electrons in a typical television receiver have kinetic energies of the order of 10 keV. Do the designers of the image-forming systems in television sets have to use quantum physics in their calculations?

2-7 *Refraction of de Broglie waves.* A classical wave changes direction (is *refracted*) when it crosses a surface between regions of differing wave velocity. Similarly, a classical particle changes direction when it crosses a surface between regions of differing potential energy. The de Broglie relation involves both wave-like and particle-like properties. This raises a question: Can a consistent description be given for the behavior of a de Broglie "particle wave" at the boundary between two media?

(a) *Refraction of Newtonian particles.* A stream of particles crosses a flat boundary across which the potential energy drops from $V = 0$ to $V = -U$. The initial momentum of each particle is p_i and the initial direction is at an angle θ_i to the normal. Using basic classical mechanics, show that

$$\frac{\sin \theta_i}{\sin \theta_f} = \frac{p_f}{p_i}$$

[*Hint*: Only the momentum component normal to the boundary is affected as the particle passes from one region into the other. Why?]

(b) *Refraction of a classical wave.* Consider the refraction of a plane wave as it crosses the boundary between regions of phase velocity w_i and w_f. Using the facts that the wave fronts (surfaces of constant phase: e.g., crests or nodes) are planes within each region, and that the phase must be continuous across the boundary, obtain the equation expressing Snell's law:

$$\frac{\sin \theta_i}{\sin \theta_f} = \frac{w_i}{w_f}$$

(c) *Refraction of de Broglie waves.* Consider a stream of electrons passing from a vacuum (potential energy $V = 0$) into a metal (potential energy $= -U = -$work function). Show that the particle analysis of (a) and the wave analysis of (b) completely agree in the prediction of the refraction, and obtain the final direction θ_f in terms of θ_i, p_i, and U. Remember that the phase velocity of a de Broglie wave is E/p, where E is the *total* energy.

(d) Satisfy yourself that the agreement found in (c) does not depend on the choice of the zero of the energy scale, and that it holds even for beams of particles moving at relativistic speeds.

2-8 *Refraction of electrons.* Many solid substances present an attractive potential of about 10 V to incoming electrons. Consider an electron, accelerated through 50 V in vacuum, entering such a substance at an angle of incidence of 70° to the normal. Find the wavelength and the direction of travel of the electron inside the material.

2-9 *Electron diffraction by a microcrystalline foil.* Electrons accelerated through 40 kV pass through a thin metal foil made up of randomly oriented microcrystals and fall on a photographic plate 30 cm behind the foil.

(a) Calculate the de Broglie wavelength of the electrons.

(b) The innermost ring of the electron diffraction pattern has a diameter of 1.7 cm. What is the spacing of planes of atoms in the microcrystals to which this ring corresponds?

The wave properties of particles

2-10   *Double-slit interference of electrons.*

(a) Electrons of momentum p fall normally on a pair of slits separated by a distance d. What is the distance between adjacent maxima of the interference fringe pattern formed on a screen a distance D beyond the slits?

(b) In the actual experiment performed by Jönsson (see Section 2-6), the electrons were accelerated through 50 kV, the slit separation d was about 2 microns (a micron is a millionth of a meter), and D was 35 cm. Calculate λ and the fringe spacing. You will then appreciate why subsequent magnification using an electron microscope was required.

(c) What would be the corresponding values of d, D, and the fringe spacing if Jönsson's apparatus were simply scaled up for use with visible light (all dimensions simply multiplied by the ratio of wavelengths)?

2-11   *Slit width in an atomic beam apparatus.*

(a) The diffraction of plane waves of wavelength λ by a slit of width d is characterized by an angular spread of the order of λ/d. At a distance D beyond the slit, this makes the beam broader than d by a distance of the order of $D\lambda/d$. If λ and D are given, what condition on d makes this spreading relatively unimportant?

(b) In an atomic beam apparatus, potassium metal (atomic weight 39) is heated to its boiling point ($T = 760°C \approx 1000$ K) and streams out of the oven aperture as individual atoms (at average energy $\approx kT \approx 1/10$eV, where k is Boltzmann's constant $= 1.38 \times 10^{-16}$ erg/degree Kelvin and T is in degrees Kelvin) into a vacuum chamber. The experimenter wishes to make the beam narrower and masks it down with a slit 0.1 mm wide. The detector is located 50 cm downstream from the slit. By what percentage (approximately) is the beam made broader than the geometrical image of the slit as a result of diffraction?

2-12   *Diffraction of helium atoms.*

(a) In the velocity selector (Figure 2-18a) used by Stern and his collaborators for their refined experiments on helium diffraction, each wheel had N (= 408) slots and the wheels were separated by s (= 3.1 cm). The two sets of slots were accurately aligned. Obtain an expression for the de Broglie wavelength of the fastest transmitted particles in terms of the rotational frequency (rev/sec) of the wheels.

(b) Using the data of Figure 2-18b, plot the expression for λ obtained in (a) against that obtained with the formula $\lambda(Å) = 1.8 \sin \phi$, which Stern *et al.* obtained from an analysis of their apparatus.

2-13   *The de Broglie wavelengths of visible particles.* Why does the wave nature of particles come as such a surprise to most people? If, as de Broglie says, a wavelength can be associated with *every* moving particle, then why are we not forcibly made aware of this property in our everyday experience? In answering, calculate the de Broglie wavelength of each of the following "particles":

(a) an automobile of mass 2 metric tons (2000 kg) traveling at a speed of 50 mph (22 m/sec),

(b) a marble of mass 10 g moving with a speed of 10 cm/sec,

(c) a smoke particle of diameter 10^{-5} cm (and a density of, say, 2 g/cm^3) being jostled about by air molecules at room temperature (27°C = 300 K). Assume that the particle has the same translational kinetic energy as the thermal average of the air molecules:

$$\frac{p^2}{2m} = \frac{3kT}{2}$$

with k = Boltzmann's constant = 1.38×10^{-16} erg/degree K.

(d) *Speculative question:* If the constants of nature were such that the de Broglie wavelength *was* of importance in decoding everyday experience, what forms would this experience take?

(e) *Interpretive question:* Student A says that the wavelengths calculated in this exercise are utterly meaningless, since they are incapable of being verified. Student B maintains that, although they are miniscule, these wavelengths have an indisputable meaning. What criteria would you use in judging these competing claims?

2-14 *The classical treatment of Rutherford scattering.* Rutherford successfully used classical mechanics to account for the scattering pattern of alpha particles by heavy nuclei: he treated the alpha particle as a point mass that follows a hyperbolic trajectory under the repulsion of a fixed point nucleus. But the de Broglie view of the wave nature of particles raises some awkward questions about the validity of Rutherford's analysis.

(a) Calculate the de Broglie wavelength of a 6-MeV alpha particle.

(b) Compare your result for (a) with the classical distance of closest approach in a head-on collision of a 6-MeV alpha particle with a gold nucleus ($Z = 79$). Neglect the recoil of the nucleus; see Exercise 1-13.

(c) Do you *expect* that classical mechanics will be valid to describe this encounter? Is your conclusion strengthened or reversed if one takes account of the fact that the alpha particle loses momentum as it approaches the nucleus?

[*Comment*: More careful analysis shows that the classical result and the quantum result agree only because of the $1/r$ dependence of the Coulomb potential; any other potential near the nucleus leads to conflicting classical and quantum predictions. In this sense the

The wave properties of particles

agreement between Rutherford's analysis and experiment was fortuitous.]

2-15 *The Bohr atom derived from de Broglie's relation.* Here is another development of Bohr's results for hydrogen, based directly on the de Broglie relation. If a de Broglie wavelength can be associated with an electron in orbit, then it seems reasonable to suppose that the circumference of an orbit be equal to an integral number of wavelengths. Otherwise (one might argue) the electron would interfere destructively with itself. Leaving to one side the mongrel-like nature of this argument (which employs both words like "orbit" and words like "wavelength"), apply it to rederive Bohr's results for hydrogen using the following outline or some other method.

(a) Calculate the speed, and hence the magnitude of the momentum, of an electron in a circular orbit in a hydrogen.

(b) Use the de Broglie relation to convert momentum to wavelength.

(c) Demand that the circumference of the circular orbit be equal to an integral number of de Broglie wavelengths.

(d) Solve for the radii of permitted orbits and calculate the permitted energies. Check that they conform to Bohr's results, as verified by experiments.

(e) As an alternative method, omit all mention of forces and simply use the de Broglie relation and the condition of integer number of wavelengths in a circumference to show directly that the angular momentum rp (for the circular case) equals an integer times $h/2\pi$. This is just the condition that we showed earlier leads most simply to the Bohr results. (You should keep in mind the fact, noted in Section 1-12, that the Bohr theory assigns incorrect values of the orbital angular momentum to the various energy levels of hydrogen. See Section 12-3.)

2-16 *Hydrogen: a structure of minimum energy.* A simple but sophisticated argument holds that the hydrogen atom has its observed size because this size minimizes the total energy of the system. The argument rests on the assumption that the lowest-energy state corresponds to physical size comparable to a de Broglie wavelength of the electron. *Larger* size means larger de Broglie wavelength, hence smaller momentum and kinetic energy. In contrast, *smaller* size means lower potential energy, since the potential well is deepest near the proton. The observed size is a compromise between kinetic and potential energies that minimizes total energy of the system. Develop this argument explicitly, for example as follows:

(a) Write down the classical expression for the total energy of the hydrogen atom with an electron of momentum p in a circular orbit of radius r. Keep kinetic and potential energies separate.

(b) *Failure of classical energy minimization.* Use the force law to obtain the total energy as a function of radius. What radius corresponds to the lowest possible energy?

(c) For the lowest-energy state, demand that the orbit *circumference* be one de Broglie wavelength (Exercise 2-15). Obtain an expression for the total energy as a function of radius. Note how a larger radius decreases the kinetic energy and increases the potential energy, whereas a smaller radius increases the kinetic energy and decreases the potential energy.

(d) Take the derivative of the energy versus radius function and find the radius that minimizes total energy. Show that the resulting radius is the Bohr radius a_0 and that the resulting energy is that calculated by Bohr for the lowest-energy state.

[*Comment:* This *exact* result depends on setting the de Broglie wavelength equal to the *circumference* of the orbit. In fact the argument we are following specifies only an *approximate* correspondence between system size and de Broglie wavelength. We could have chosen "system size" to mean radius or diameter and then derived a size differing from that calculated above by factors of 2 or π, which is as near to the observed values as one has any right to expect.]

The wave properties of particles

Once at the end of a colloquium I heard Debye saying something like: "Schrödinger, you are not working right now on very important problems . . . why don't you tell us some time about that thesis of de Broglie, which seems to have attracted some attention?" So, in one of the next colloquia, Schrödinger gave a beautifully clear account of how de Broglie associated a wave with a particle, and how he could obtain the quantization rules . . . by demanding that an integer number of waves should be fitted along a stationary orbit. When he had finished, Debye casually remarked that he thought this way of talking was rather childish . . . To deal properly with waves, one had to have a wave equation.

FELIX BLOCH, *Address to the American Physical Society* (1976)

3

Wave-particle duality and bound states

3-1 PRELIMINARY REMARKS

In Chapter 2 we described the theoretical and experimental bases for ascribing a definite wavelength to particles of a given momentum. This development involved the recognition of some very basic similarities between the descriptions of light propagation and the motion of material particles. We shall now take up the question of what sort of wave equation can be devised for material particles that will embody these similarities and enable us to extend particle dynamics into the atomic domain.

Our starting point is the knowledge that light waves are characterized by two parameters—wavelength λ and frequency ν. The description of light as photons relates these classical parameters to the quantized values of energy E and momentum p for individual photons:

$$\lambda = \frac{h}{p}; \qquad E = h\nu$$

Now in Chapter 2 we saw that the first of these relationships applies to material particles as well as to photons. Our next step is to introduce the *postulate* that the second relationship is equally general, so that the motion of a particle of momentum p

and energy E will be described by a wave of wavenumber k and angular frequency ω, where

$$k = \frac{2\pi}{\lambda} = \frac{2\pi p}{h}; \qquad \omega = 2\pi\nu = \frac{2\pi E}{h}$$

If we can find the value of E in any given case, we can, according to this new postulate, infer the value of ω through the relationship

$$E = \frac{h}{2\pi}\,\omega = \hbar\omega \tag{3-1}$$

Recall (Eq. 1-29) that the symbol $\hbar$ is an abbreviation for $h/2\pi$. Using this same notation, the original de Broglie relationship for moving particles can be written

$$p = \hbar k \tag{3-2}$$

Equations 3-1 and 3-2 will be the basis for constructing an equation for particle waves.

Any wave equation that we develop must embody the correct dynamical relationship between E and p. In classical (nonrelativistic) mechanics, this relationship comes directly from the statement that the total energy of a particle is the sum of its kinetic energy $p^2/2m$ and its potential energy V. Thus we require

$$E = \frac{p^2}{2m} + V$$

and hence

$$\hbar\omega = \frac{\hbar^2 k^2}{2m} + V \tag{3-3}$$

This last relationship certainly does not apply to photons, which are "completely relativistic" ($m = 0$ and $v = c$ always), but is applicable to all particles possessing rest mass m under the conditions such that the kinetic energy is much less than mc^2.

Wave-particle duality and bound states

It was Erwin Schrödinger[1] who, in 1925, discovered an appropriate form of wave equation, making use of some deep formal analogies between optics and classical particle mechanics that had been evolved in the nineteenth century by W. R. Hamilton and others. We shall not attempt to retrace here the development of these fundamental analogies, but will simply try to make the form of the equation plausible. Our main concern will be with solutions of the equation and their physical interpretation. The equation we are looking for is one that will tell us how to calculate quantum amplitudes, and hence the probability distribution, as a function of position and time for particles of a given kind in a given physical environment such as the interior of an atom.

We begin by considering free particles ($V = 0$ in Eq. 3-3) having specific values of E and p. The classical wave analog is a progressive sinusoidal wave having specific values of k and ω. Such a wave can be described by an equation of the form

$$y(x,\ t) = A \sin{(kx - \omega t)} \tag{3-4}$$

for a wave traveling in the positive x direction. The variable y here might be linear displacement, pressure, electric field, etc., expressed as a function of position and time.

Underlying this equation is a basic *differential equation* of wave motion. For classical waves we arrive at such a differential equation from a consideration of the dynamics of the system. For example, the differential equation governing transverse waves on a long string, or longitudinal sound waves in a gas, comes from the statement of Newton's second law ($F = ma$) applied to a small section of the system. For such waves the basic differential equation takes the form

$$\frac{\partial^2 y}{\partial x^2} = \frac{1}{w^2}\frac{\partial^2 y}{\partial t^2} \quad \text{(classical)} \tag{3-5}$$

[1] E. Schrödinger, Ann. Phys. (4) **79**, 361 and 489 (1925). Reprinted (English translation) in *Collected Papers on Wave Mechanics*, Blackie & Son Ltd., London and Glasgow, 1928.

3-2   The approach to a particle-wave equation

where w is the speed of propagation of the waves. For mechanical waves we can recognize that the term involving the second time derivative $\partial^2 y/\partial t^2$ is directly related to the acceleration in the equation $F = ma$. An equation of the same form applies to electromagnetic waves, although the physical basis is very different.

Once the basic differential equation 3-5 has been set up on physical grounds, we can verify that Eq. 3-4 is a possible solution to it, but there may be (and are) many other solutions. Suppose, however, that we *start* from the knowledge that Eq. 3-4 describes a possible type of wave. What can we deduce about the basic differential equation that underlies it? If we differentiate y in Eq. 3-4 with respect to x and t separately, we obtain the results

$$\frac{\partial y}{\partial x} = kA \cos(kx - \omega t)$$

$$\frac{\partial y}{\partial t} = -\omega A \cos(kx - \omega t)$$

Thus y obeys the differential equation

$$\frac{\partial y}{\partial x} = -\frac{k}{\omega}\frac{\partial y}{\partial t} = -\frac{1}{w}\frac{\partial y}{\partial t}$$

where w is the wave speed, ω/k.

This first-order differential equation might seem to be a perfectly good basis for analysis of waves in one dimension. One can verify very easily that *any* progressive disturbance of the form $f(x - wt)$ is a possible solution to it. However, it suffers from a major defect: It does *not* lead to the possibility of waves traveling along the x axis in the opposite (negative) direction. Solutions corresponding to this situation, $f(x + wt)$, would require as their basis a first-order differential equation of the form

$$\frac{\partial y}{\partial x} = +\frac{1}{w}\frac{\partial y}{\partial t}$$

Since there is nothing in our experience to suggest a fundamental difference in the physics of waves traveling in these opposite directions, we could reasonably conclude that we

Wave-particle duality and bound states

need to dig a little deeper. We do this by considering the second derivatives, instead of the first derivatives, of y with respect to x and t. We then find that waves traveling in either direction emerge as possible solutions of the second-order equation 3-5, which, as we know on other grounds, expresses the basic physics of the problem.

3-3 THE SCHRÖDINGER EQUATION

Our approach to the Schrödinger equation will have something in common with the line of thought we have just presented, although it involves a new ingredient. For the purpose of describing a progressive particle wave of wavenumber $k = p/\hbar$ and angular frequency $\omega = E/\hbar$ we might reasonably begin with a wave function that satisfies Eq. 3-4:

$$\Psi(x, t) \stackrel{?}{=} A \sin (kx - \omega t) \qquad (3-6)$$

Since k is proportional to momentum, whose sign determines the direction of propagation along the x axis, the argument we used about the essential symmetry of opposite directions for classical waves can also be applied here. Therefore we go directly to the *second* derivative of Ψ with respect to x:

$$\frac{\partial^2 \Psi}{\partial x^2} \stackrel{?}{=} - k^2 A \sin (kx - \omega t) \qquad (3-7)$$

However, in the differentiation with respect to t we shall stop at the first derivative:

$$\frac{\partial \Psi}{\partial t} \stackrel{?}{=} - \omega A \cos (kx - \omega t)$$

We stop with the first derivative because, according to Eq. 3-3, the relation between ω and k for a free particle, based on the relation between E and p, is given by

$$\omega = \frac{\hbar k^2}{2m}$$

Using this expression we can rewrite the preceding equation as

3-3   The Schrödinger equation

follows:

$$\frac{\partial\Psi}{\partial t} \overset{?}{=} -\frac{\hbar k^2}{2m} A \cos{(kx - \omega t)} \tag{3-8}$$

If only we did not have a sine function in Eq. 3-7 and a cosine function in Eq. 3-8, we could combine them beautifully into a partial differential equation of the form

$$\frac{\partial^2\Psi}{\partial x^2} \overset{?}{=} \frac{2m}{\hbar} \frac{\partial\Psi}{\partial t}$$

Clearly we cannot do this. However, there is another possible form for Ψ, often used in the purely mathematical analysis of classical wave problems, that gets around this difficulty. We try

$$\Psi(x, t) = A e^{i(kx - \omega t)} \tag{3-9}$$

In classical wave analysis the actual physical solution emerges as the real part of this complex quantity, but in quantum mechanics we do not impose this condition. This is the "new ingredient" that we mentioned earlier. And if we now adopt Eq. 3-9 as the essential form of the wave function for a free particle of unique momentum and energy traveling in a particular direction, we have

$$\frac{\partial^2\Psi}{\partial x^2} = -k^2 A e^{i(kx - \omega t)} = -\frac{p^2}{\hbar^2}\Psi$$

$$\frac{\partial\Psi}{\partial t} = -i\omega A e^{i(kx - \omega t)} = -\frac{iE}{\hbar}\Psi$$

Combining these, we arrive at the differential equation

$$\frac{\partial^2\Psi}{\partial x^2} = -\frac{2mE}{\hbar^2}\Psi = -i\frac{2m}{\hbar}\frac{\partial\Psi}{\partial t}$$

More generally, when the total energy E also includes a potential energy term, we put

$$p^2 = 2m (E - V) = \hbar^2 k^2$$

This embodies the fact (familiar from classical mechanics) that

Wave-particle duality and bound states

a particle of some definite total energy E has a momentum p that changes from one place to another as a result of the spatial variation of the potential energy V. For the quantum case, we have

$$\frac{\partial^2 \Psi}{\partial x^2} = -\frac{2m}{\hbar^2}(E - V)\Psi = -i\frac{2m}{\hbar}\frac{\partial \Psi}{\partial t} + \frac{2m}{\hbar^2}V\Psi$$

which can be written in two different forms:

$$-\frac{\hbar^2}{2m}\frac{\partial^2 \Psi}{\partial x^2} + V\Psi = E\Psi \tag{3-10}$$

$$-\frac{\hbar^2}{2m}\frac{\partial^2 \Psi}{\partial x^2} + V\Psi = i\hbar\frac{\partial \Psi}{\partial t} \tag{3-11}$$

Equations 3-10 and 3-11 represent alternative forms of Schrödinger's equation in one dimension. Equation 3-10, which does not include the time explicitly, is known as the *time-independent Schrödinger equation;* it provides the basis for analyzing the stationary states of atomic systems. The *time-dependent Schrödinger equation*, Eq. 3-11, must be used when we are dealing with such problems as the actual motion of particles from one point to another.

Clearly we have not been inexorably driven to Eqs. 3-10 and 3-11 any more than Schrödinger was in his argument from analogy. One can construct many other differential equations that embody the dynamical relations expressed in Eqs. 3-1, 3-2, and 3-3 (see the exercises). But the Schrödinger equations, besides being the mathematically simplest equations that satisfy the requirements, have other properties that cause them to be preferred over all other possibilities:

1. They have the property of *linearity*, so that if Ψ_1 and Ψ_2 are specific solutions to one of these equations, then any linear combination of them is a solution of the same equation. This property of *superposition* is one of the most basic properties of waves.

2. Their solutions are (as we shall see in Chapter 9), perfectly suited to the interpretation of Ψ as a probability amplitude.

3-3   The Schrödinger equation

Finally, there is all the accumulated evidence that the Schrödinger equations *work*; they provide the basis for a correct analysis of all kinds of molecular, atomic, and nuclear systems. Whatever questionable features there may be in the manner of their formulation are swept away in the evidence of their manifest success—a success of which we shall see many examples in this book.

3-4 STATIONARY STATES

In the present chapter our chief goal will be to learn how to calculate the energies of particles bound within certain regions of space, as, for example, the electron in a hydrogen atom. We have seen that the allowed energies of such states —exemplified by the *stationary states* of Bohr's theory —are quantized. In contrast to a free particle, a bound particle is limited to energies belonging to a set of discrete values, even though, as in the hydrogen atom, these energy states may be infinite in number. The theoretical basis for the existence of states of discrete energies emerges more or less directly from the wave-particle duality; one of the most familiar features of classical wave behavior is that of standing waves. The free vibrations of any physical system of finite dimensions—a violin string, an organ pipe, a microwave cavity—are limited to a set of characteristic frequencies. Each such frequency belongs to what is called a *normal mode* of the system, in which a standing wave is set up in the system as a whole. This standing wave may have a complicated shape (it may also be very simple, as in a vibrating string) but the central feature is that each part of the system vibrates, with some fixed amplitude, in a simple harmonic motion. For a one-dimensional classical system, therefore, the standing-wave pattern of a single normal mode can be described by an equation of the form

$$F(x, t) = f(x) \sin \omega t$$

The counterpart of this in Schrödinger's "wave mechanics" is a solution of the form[2]

$$\Psi(x, t) = \psi(x) e^{-i\omega t} \tag{3-12}$$

[2]In general, we shall use the capital Greek psi (Ψ) to stand for the complete wave function that includes time, while the lower case psi (ψ) stands for the spatial part alone.

Wave-particle duality and bound states

where the complex time factor will lead to an important physical interpretation of the wave function (see Section 3-6). Substituting $\Psi(x, t)$ into Eq. 3-10 and canceling the common exponential leads to an ordinary differential equation governing the spatial factor $\psi(x)$ alone.

$$-\frac{\hbar^2}{2m}\frac{d^2\psi}{dx^2} + V(x)\psi = E\psi \tag{3-13}$$

The potential energy $V(x)$ must be given before the equation can be solved for ψ. We then find that, in general, physically and mathematically acceptable solutions to Eq. 3-13 exist only for particular values of the total energy E. These are the quantized energy levels of the system. In the next section we consider a simple specific example of such quantization.

3-5 PARTICLE IN A ONE-DIMENSIONAL BOX

Imagine a particle that is free to move in a one-dimensional region but is confined between two rigid walls a distance L apart. (For example, if x is the direction of motion, the coordinates of the walls can be chosen as $x = 0$ and $x = L$.) One could in principle set up such a situation for a charged particle such as an electron by the arrangement shown in Figure 3-1. The inside of the one-dimensional "box" lies between two grids at ground potential. As long as the negatively charged particle remains in this region, it experiences no electrical forces. However, if it passes through grid A at either end of the box, it enters a region in which a uniform electric field exists that exerts a backward force on the particle, tending to return it toward the inside of the box. An energy diagram is shown in Figure 3-2a, where the energy of the particle, E, is represented by a horizontal line. As long as E is less than the height of the bounding potential energy V_o, the particle will remain bound and will not escape from the "box."

Fig. 3-1 Schematic diagram of an electron trapped in a one-dimensional "box" made of electrodes and grids in an evacuated tube.

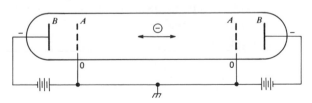

3-5 Particle in a one-dimensional box

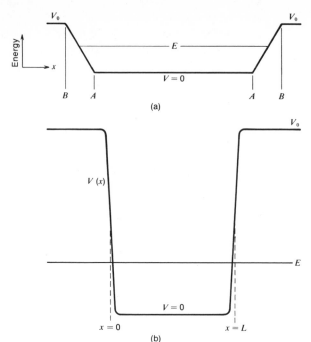

Fig. 3-2 (a) Potential-energy diagram for one-dimensional well with low "soft" walls. (b) Well of finite depth with walls that are not infinitely rigid.

The two regions AB can be thought of as "soft walls" from which the particle bounces. How can we make a *rigid* wall? One way to approach rigidity as a limiting case is by putting a higher and higher potential difference between grids A and B (Figure 3-2b). Then the particle energy E is very much less than the potential energy V_o so that the particle does not penetrate far beyond grid A before turning back. For higher and higher potential difference between grids A and B the walls become more and more rigid as the particle with small energy E penetrates less and less deeply beyond the first grid A before being reflected. As a limiting case one speaks as if the potential walls are vertical (Figure 3-3) and talks about an "infinitely deep square well," having $V_o \rightarrow \infty$.

Inside the box (between $x = 0$ and $x = L$) the potential energy is constant and can be set equal to zero. The total energy E is then entirely kinetic, $E = p^2/2m = (\hbar k)^2/2m$, where k represents a wavenumber independent of position. Within this region the differential equation 3-13 for $\psi(x)$ becomes

$$\frac{d^2\psi}{dx^2} = -\frac{2mE}{\hbar^2}\psi = -k^2\psi \qquad (k = \pm\sqrt{2mE}/\hbar) \qquad (3\text{-}14)$$

Wave-particle duality and bound states

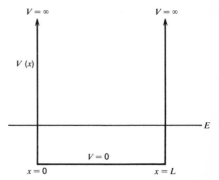

Fig. 3-3   *Potential-energy diagram for an infinite square well, characterized by walls that are impenetrable and perfectly rigid.*

As in the case of a free particle, this equation can be satisfied by the functions $\psi(x) \sim e^{\pm ikx}$. However, we now have an important new consideration: the existence of *boundary conditions* at the rigid walls.

The Schrödinger amplitude ψ is a *probability* amplitude. Now a particle confined between completely rigid walls has zero probability of being found outside those walls. This condition is satisfied if ψ falls to zero at $x = 0$ and $x = L$ and remains zero throughout the regions external to the box. As in the case of the violin string fixed at both ends, this requires that the wave function ψ be equal to zero at $x = 0$ and $x = L$.[3] But the general solution of Eq. 3-14 is

$$\psi(x) = C_1 e^{ikx} + C_2 e^{-ikx} \qquad (|k| = \sqrt{2mE}/\hbar)$$

where C_1 and C_2 are constants. Putting $\psi(0) = 0$ gives the condition $C_2 = -C_1$. Hence

$$\psi(x) = C_1(e^{ikx} - e^{-ikx})$$

which means that $\psi(x)$ has the form

$$\psi(x) = A \sin kx$$

Applying the further condition $\psi(L) = 0$ requires

$$\sin kL = 0$$

[3]The requirement that $\psi = 0$ at the walls is discussed further in Section 3-7 below.

3-5   Particle in a one-dimensional box

and hence

$$kL = n\pi$$

The values of k are thus restricted to a set of discrete values:

$$k_n = \frac{n\pi}{L} \qquad (n = 1, 2, 3, \ldots)$$

But from Eq. 3-14, we see that this condition defines a set of energies E_n that are thus the permitted quantized energies of a particle in a one-dimensional box:

$$E_n = \frac{\hbar^2 k_n^2}{2m} = \frac{n^2 h^2}{8mL^2} \tag{3-15}$$

The complete wave function $\Psi(x, t)$ is then given by the equations

$$\Psi(x, t) = A_n \sin \frac{n\pi x}{L} \, e^{-(E_n/\hbar)t} \qquad (0 \leq x \leq L) \tag{3-16}$$
$$= 0 \qquad (x < 0 \text{ and } x > L)$$

Mathematically, the problem we have just solved is closely parallel to the problem of finding the natural frequencies of a uniform string under tension between fixed supports; for example, a violin string. Each characteristic frequency belongs to a normal mode, defined by the condition that an integral number of half-wavelengths of a sine curve fit between the fixed ends (Figure 3-4). The equation of any one mode is of the form

$$F(x, t) = A_n \sin k_n x \sin \omega_n t$$

The boundary conditions give $k_n = n\pi/L$, just as in the wave-mechanical problem, and the permitted frequencies are then defined through the relationship $\omega_n = w k_n$, where w is the speed of transverse waves along the string. (In the theory of the ideal string, the wave speed w is independent of frequency or wavelength.)

It is remarkable that the *existence* of quantized energies for a bound particle should follow so simply and directly from

Wave-particle duality and bound states

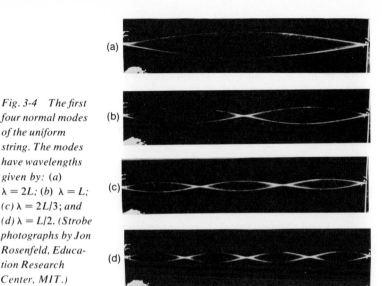

Fig. 3-4 The first four normal modes of the uniform string. The modes have wavelengths given by: (a) $\lambda = 2L$; (b) $\lambda = L$; (c) $\lambda = 2L/3$; and (d) $\lambda = L/2$. (Strobe photographs by Jon Rosenfeld, Education Research Center, MIT.)

applying boundary conditions to a wave function that satisfies the Schrödinger equation. Of course, the *values* of the quantized energies depend on details of the binding potential, and the box with rigid walls is only the crudest approximation to most real binding potentials (see the exercises.) Nevertheless, the *fact* of quantization from which quantum mechanics gets its name is neither perplexing nor obscure once one accepts the consequences of placing boundaries around particles with wave-like properties.

3-6 UNIQUE ENERGY WITHOUT UNIQUE MOMENTUM

The above analysis fails to make clear an important feature of both classical and quantum waves that may at first seem surprising: A classical vibrating string in a normal mode has a unique frequency while not possessing a unique wavelength. Similarly, a particle bound in a box in a stationary state has a unique energy while not possessing a unique momentum. We discuss this briefly here and in more detail in Chapter 8.

Think of a violin string vibrating in a normal mode. The transverse standing wave is sinusoidal along the string but is zero outside the fixed endpoints. One would be justified in saying that, considered by itself, the sinusoidal portion has a definite wavelength. However, a classical wave theory can also analyze the situation in such a way that a single prescription describes both the sinusoidal portion of the string between the fixed endpoints and the zero-displacement portion outside the endpoints: it is possible to think of the finite train of standing waves as a superposition of waves of different wavelengths, with amplitudes and phases chosen so that the disturbance cancels everywhere but in the region between the endpoints. Typically, the shorter the wavetrain the greater the range of wavelengths that must be superposed to yield the resulting profile. Only if the sinusoidal wave is infinitely long (an unbounded wave) will the superposition of wavelengths reduce to a single value.

Even though the classical description of the violin string vibrating in a normal mode makes use of more than one wavelength, the normal mode does have a unique frequency. Every part of the string vibrates harmonically, and if this vibration goes on indefinitely (so there is no bound in *time*) then the frequency is unique and single-valued.

A similar analysis can be made for a particle bound in a one-dimensional box. Suppose that this particle is in one of the unique energy states. Then Eq. 3-16 tells us that the wave function is sinusoidal within the box but is zero outside the box. Then the classical violin-string analogy leads us to believe that the confined wave function is describable in terms of a superposition of different wavelengths. Through the de Broglie relation $p = h/\lambda$ we then conclude that this particle is properly described as having a range of momenta.[4] In contrast, the unique particle energy E of this quantum state relates to a unique frequency $\omega = E/\hbar$, assuming that the particle occupies the state for an unbounded time.

In what follows, we will sometimes refer loosely to "the wavelength" of a confined particle. We shall mean by this the local wavelength which characterizes the wave function over a small region of space. However, one must acknowledge that the confined wavetrain as a whole is described more analytically in terms of a superposition of wavelengths.

There is a further analogy between the classical and quantum analyses of confined waves: the shorter the box that confines the particle, the greater will be the range of momenta that must be superposed in order to describe the wave function of

[4]See F. Landis Markley, "Probability Distribution of Momenta in an Infinite Square-Well Potential," Am. J. Phys. **40**, 1545 (1972).

Wave-particle duality and bound states

the particle. This relation between confinement in space and spread in momenta, when given quantitative expression, is a fundamental result of quantum physics known as the *position-momentum uncertainty relation*. It will be discussed in detail in Chapter 8. We shall also discuss in that chapter an *energy-time uncertainty relation* which describes the fact that a quantum state with a short lifetime has a range of energies. This contrasts with the unique energy that results in the cases treated here under the assumption of a long-lived quantum state.

3-7 INTERPRETATION OF THE QUANTUM AMPLITUDES FOR BOUND STATES

We saw in Chapter 2 that the waves associated with material particles must be interpreted as waves of probability amplitude. The evidence from interference and diffraction experiments indicated that the relative probability of finding a particle at any point is given by the square of the resultant quantum amplitude at that point. We have now developed the idea that the quantum amplitude is the solution $\Psi(x,t)$ of the Schrödinger equation. For free particles we arrived at the form given by Eq. 3-9:

$$\Psi(x, t) = A e^{i(kx - \omega t)}$$

and our analysis of bound states is based on solutions of the form given by Eq. 3-12:

$$\Psi(x, t) = \psi(x)e^{-i\omega t}$$

where the capital Greek psi Ψ stands for the complete wave function including time while the lower case psi ψ stands for the spatial part. For both free and bound particles the quantum amplitude Ψ is complex; we cannot obtain a real probability by simply squaring it. Let us assume, however, that the required probability is given by the *square of the magnitude* of Ψ. (We shall see later, in Chapter 9, that this assumption rests on good foundations.) If we write Ψ as a real amplitude A multiplied by a complex phase factor $e^{i\alpha}$, then we have

$$|\Psi|^2 = |A e^{i\alpha}|^2 = A^2$$

since the phase factor $e^{i\alpha}$ is of magnitude unity.[5]

We can see at once that this procedure for obtaining probabilities from quantum amplitudes makes good sense for free particles as described by Eq. 3-9, for we have

$$|\Psi|^2 = |Ae^{i(kx-\omega t)}|^2 = A^2 = \text{constant}$$

This corresponds to the statement that in a uniform particle beam the probability of finding a particle is the same at all points along the beam.

Correspondingly, for a bound-state wave function, we shall have

$$|\Psi|^2 = |\psi(x)e^{-i\omega t}|^2 = |\psi(x)|^2$$

This means then that in a bound state having a definite energy E associated with a single characteristic frequency $\omega = E/\hbar$, the probability distribution is independent of time—it is, in truth, a *stationary* state. To calculate this probability distribution, we have only to evaluate the square of the space-dependent factor ψ in Ψ, starting from the time-*independent* Schrödinger equation 3-10.

Let us apply this to our solutions for a particle in a perfectly rigid box. For an individual quantized state, we have

$$\psi_n(x) = A_n \sin\frac{n\pi x}{L}$$

Hence

$$|\psi_n(x)|^2 \sim \sin^2\frac{n\pi x}{L} \tag{3-17}$$

Quite apart from the restriction of the energy to discrete (quantized) values, the wave-mechanical solution to the problem thus contrasts strongly with the classical description of a bound particle. In Newtonian mechanics a particle of

---

[5]Recall that $e^{i\alpha}$ can be expressed in terms of real and imaginary components through the use of Euler's formula:
$$e^{i\alpha} = \cos\alpha + i\sin\alpha$$
We then have
$$|e^{i\alpha}|^2 = \cos^2\alpha + \sin^2\alpha = 1$$

Wave-particle duality and bound states

given energy, confined between rigid walls but otherwise free, would be equally likely to be found at any point between $x = 0$ and $x = L$. This could be verified, for example, if one took flash photographs at random intervals. The number of photographs showing the particle within some given small range of distance Δx between the boundaries would be independent of x. The wave-mechanical solution, on the other hand, leads to a probability, proportional to the square of $\psi(x)$, that varies with x and (except for the lowest state, $n = 1$) is actually zero at certain points between $x = 0$ and $x = L$. For example, for $n = 2$, $|\psi|^2 = 0$ at $x = L/2$, as shown in Figure 3-5a. This result is completely foreign to the classical description of a particle. There is, however, a clue to the way in which the quantum and Newtonian solutions join. If we consider a very large value of n, the wavelength within the well becomes exceedingly small, and in this case the result of averaging the quantum-mechanical probability over some Δx (which may now be much larger than λ) will give a result almost independent of x (Figure 3-5b). One can regard this as another example of the correspondence principle (Section 1-7.)

It is possible to study a classical system statistically in a similar way. Consider, for example, a simple pendulum of a

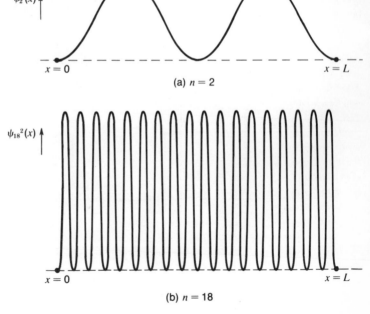

(a) $n = 2$

(b) $n = 18$

3-7   Quantum amplitudes for bound states

given length. If one such pendulum were set swinging and then photographed at random instants, the numbers of photographs showing the pendulum within a small range of distance Δx at various values of x would be distributed in a way such as that shown in Figure 3-6a. This shows clearly that the pendulum spends relatively more time near the extremes of its swing and less time near the center. (This is also shown in Figure 3-6b.) This is a statistical distribution which indicates approximately the relative probabilities of finding the pendulum at different positions. Several distributions measured independently would show differences due to the fact that each is based on a finite number of observations. The calculated distribution, which the experimental results would approach if the number of observations were made enormously large, is shown in Figure 3-6c. There is an alternative way to obtain the same statistical distribution. Instead of taking many photographs of a single pendulum, one could take a very large number of identical

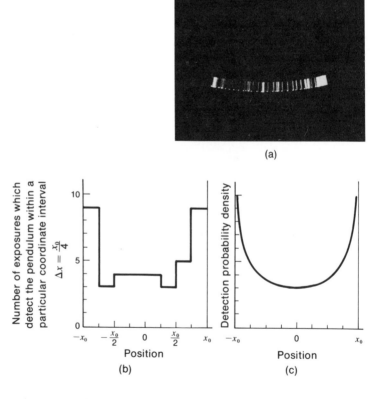

Fig. 3-6 Statistical study of a classical system. (a) Multiple-exposure photograph of an oscillating pendulum, showing the location of the pendulum at various random instants. A total of 41 exposures is included. (Photograph by Jon Rosenfeld, Education Research Center, MIT.) (b) Histogram which displays the distribution of pendulum positions in (a). (c) Calculated continuous distribution which could be verified by an experiment involving an enormous number of exposures.

Wave-particle duality and bound states

pendulums and release them at random intervals from the same initial displacement. A single photograph of the positions of all the pendulums at some instant after they had all been set in motion would represent an equivalent statistical sampling of "the system."

Actually, in quantum systems, that is, systems so small that the use of quantum principles is essential to analyze them, only the second type of observation is possible. Repeated measurements on the same particle are ruled out because the act of making an observation disturbs the system in a way that can neither be neglected nor precisely predicted. No such difficulties arise when, for example, one suddenly illuminates a pendulum with a stroboscopic flash; but an electron in an atom, struck by a single photon, may perhaps be knocked right out the atom and thenceforth be unavailable for investigation. Thus, our primary view of the wave function—the totality of quantum amplitudes for the various values of x as symbolized by $\psi(x)$ for a given state—is that it allows prediction of the results of measurements made on many identical quantum systems. It is in this sense that the value of $|\psi(x)|^2 \Delta x$ gives a relative measure of the probability that a particle described by the wave function $\psi(x)$ will be found within the range of position Δx near x. But remember, this statment, like any statement of probability, holds no meaning for a single observation. If one knows that a particle is in a certain state, and then does an experiment to detect it within Δx at x, the particle will either be detected or it will not. After the measurement is made, the answer to the question, "Was it in that small region?" will be a definite yes or no, although one cannot tell in advance which answer will be obtained. It is only when one considers independent measurements on, say, 1000 systems, each originally in a given state, that one can talk with any assurance about the numbers of cases in which the particle will be observed at various positions.

3-8 PARTICLES IN NONRIGID BOXES

The perfectly rigid box, represented by a rectangular potential well with infinitely high walls, is an ideally simple vehicle for introducing the mathematics of quantum systems. But of course no real physical system has a potential function with infinite discontinuities. Just as the end supports of a violin string vibrate a *litttle bit* when the string is bowed (transmitting the vibration to the body of the violin so that we can hear it), in the same way particles bound in real potential wells have *some* probability of being found at and just outside the boundaries, according to quantum mechanics. We shall now consider the behavior of wave functions at the boundaries and "outside" of potential wells. In the process we will dispose of the apparent arbitrariness of the boundary conditions for the rigid box, for which the wave function $\psi(x)$ goes to zero at the boundaries.

Figures 3-7a and 3-7b show two potential functions more realistic than the infinite square well, namely the parabolic potential well of a simple harmonic oscillator and the *finite* square well with a potential discontinuity that is finite. In both potentials a classical particle of total energy E (horizontal lines) moves back and forth between the boundaries $\pm x_0$. In the parabolic potential as the particle approaches one boundary its kinetic energy becomes less and less until, at the boundary, the entire energy is potential: $E = V(x_0)$, and the particle stops and reverses its direction of motion at this turning point. The walls of the "box" represented by the harmonic oscillator potential are not vertical, which has the consequence that the range of

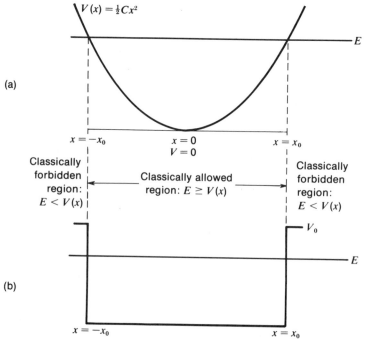

(a)

(b)

Fig. 3-7 (a) Simple harmonic oscillator potential-energy function $V(x) = \frac{1}{2}Cx^2$, showing the allowed and forbidden regions for a classical particle of total energy E. (b) A finite square well, with infinitely steep walls of height V_0.

$V(x) = \frac{1}{2}Cx^2$

E

$x = -x_0$ $x = 0$ $x = x_0$
           $V = 0$

Classically forbidden region: $E < V(x)$

Classically allowed region: $E \geq V(x)$

Classically forbidden region: $E < V(x)$

V_0

E

$x = -x_0$ $x = x_0$

Wave-particle duality and bound states

space over which a particle can oscillate increases as its energy increases. In contrast, the walls of the finite square well potential are vertical (although finite), and the classical particle will be reflected suddenly with no gradual decrease in kinetic energy as it approaches the boundary. But the essential feature of both potentials is that, for any given value of E for a bound particle, the points where $E = V(x)$ define sharp boundaries beyond which (classically) the particle cannot go. If the particle were to continue beyond such a boundary, then it would enter a region in which the potential energy is greater than the total energy $[E < V(x)]$, that is, where the kinetic energy $(K = E - V)$ is negative! But there is no way that the kinetic energy $K = \frac{1}{2}mv^2$ can be negative. Therefore, the region outside the boundary is called the *classically forbidden region* on the basis of Newtonian particle mechanics.

However, the classical analogies that are relevant in this chapter are those not of particle mechanics but of wave behavior, and even classical waves *can exist* in regions where they *cannot propagate*. We obtain a hint of the corresponding quantum result by asking what becomes of the Schrödinger equation (Eq. 3-13) for the particle's wave function if we impose the classically inadmissible condition $E < V(x)$. The Schrödinger equation then takes the form

$$\frac{d^2\psi}{dx^2} = \alpha^2\psi$$

where α^2 is a positive quantity:

$$\alpha^2 = \frac{2m(V - E)}{\hbar^2}$$

To simplify the analysis, consider the case of the finite square well (Figure 3-7b) for which the quantity α is a constant independent of position in the classically forbidden regions, since $V = V_o$ there. In this case, the solution of the equation above yields an exponential increase or decrease of ψ with x:

$$\psi(x) \sim e^{\pm\alpha x} \tag{3-18}$$

This exponential form contrasts strongly with the sinusoidal form of the wave function which, via the de Broglie wave relationship, provided our starting point. The physically meaning-

3-8   Particles in nonrigid boxes

ful solutions of the exponential type must be those that *decrease* with increasing distance from the potential well, since an exponential *increase* with distance would imply an overwhelming probability of finding the particle at great distances from the well, a result that is incompatible with the basic physical assumption that we are describing a particle bound *within* the well (see Figure 3-8). On the other hand, the

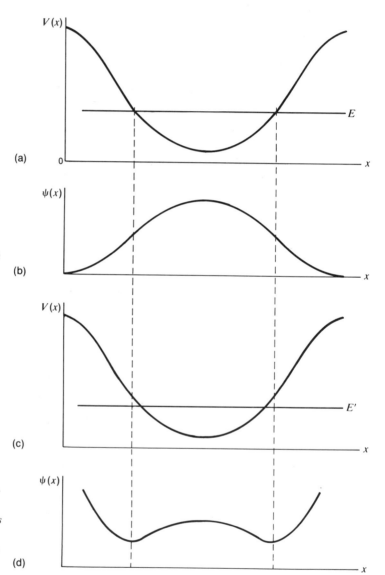

Fig. 3-8 Physical restriction on wave functions which decrease outside the well. (a) Potential-energy well with an allowed energy value E indicated. (b) Wave function for bound particle with total energy E. (c) Same potential-energy well as in (a), but with a proposed energy value E' < E. (d) Function which satisfies the Schrödinger equation for energy value E'. This function is not acceptable because it does not describe a particle confined to the well.

Wave-particle duality and bound states

decreasing exponential solutions *do* naturally describe the circumstance that, in a bound state, the particle is effectively confined within a limited region, such that the probability of finding it outside that region decreases rapidly toward zero with increasing distance from the confining well.

The form of the wave function outside the parabolic potential well of Figure 3-7a is not a simple exponential, but it does decrease with distance from the well. The exact form of the wave function will be presented in Chapter 4.

The possibility of finding a particle in a classically forbidden region is a feature of the quantum analysis wholly foreign to our direct experience. Nevertheless, many common devices (such as solid state radios and hand-held calculators) depend for their practical operation on the "tunneling" of electrons through potential barriers. Some simplified models of this behavior will be treated in later chapters.

3-9 SQUARE WELL OF FINITE DEPTH

In examining further the behavior of particles in nonrigid boxes, we will continue the analysis of the mathematically simplest case, that of the rectangular potential of Figure 3-9a, similar to that of Figure 3-7b. Mathematically this potential is characterized as follows:

$$V = V_0 \quad \text{for} \quad x < 0$$
$$V = 0 \quad \text{for} \quad 0 \leq x \leq L$$
$$V = V_0 \quad \text{for} \quad x > L$$

From the standpoint of classical particle mechanics, this well, like the infinitely deep square well, is rigid in the sense that dV/dx becomes infinitely great at $x = 0$ and $x = L$. If the particle energy E is less than V_0, the particle experiences an infinite force ($F = -dV/dx$) at these positions, keeping it strictly within these limits. In any real situation, the change of V would take place in a small but finite distance (as in Figure 3-2b), but the assumption of a finite change of V within a vanishingly small distance is a reasonable approximation to physically conceivable situations, for example, the potential experienced by a conduction electron inside a thin metal foil. At the surfaces of the foil the potential rises by something of the order of 10 eV in about one atomic diameter, a distance which is extremely

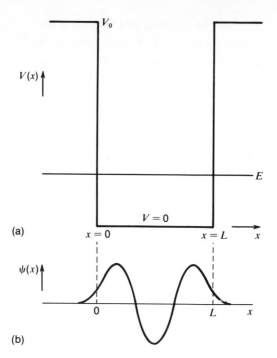

Fig. 3-9 (a) Finite
square well, with an
allowed energy
value indicated.
(b) Wave func-
tion ψ (x) corre-
sponding to the al-
lowed energy value
E shown in (a).
Since $E \ll V_o$, $\psi(x)$
decays rapidly out-
side the classically
permitted region.

small compared to the thickness of the foil. (Of course, this is really a three-dimensional system, but for electrons traveling perpendicular to the faces of the foil the potential would be well represented by Figure 3-9a.)

Let us explore qualitatively the possible bound states of a particle in such a finite square well. Within the well the form of $\psi(x)$ is sinusoidal. Outside, both right and left, $\psi(x)$ decreases exponentially, according to Eq. 3-18, with increasing distance from the walls. And here we come to the central feature of the analysis: *The characteristic quantized values of E are those that allow a smooth join of the curves that represent $\psi(x)$ in the different regions.*

The smooth join is defined by requiring $\psi(x)$ and its first derivative, $d\psi/dx$, to be continuous at $x = 0$ and $x = L$. In the case of the square well that is all the continuity we can get, since the sudden jump in $V(x)$ at the sides of the well must be accompanied by a discontinuous change in $d^2\psi/dx^2$ as required by the Schrödinger equation (see Eq. 3-13).

In more realistic situations, in which $V(x)$ itself has no discontinuities, then $\psi(x)$ and all its derivatives are continuous at the points where $E - V(x)$ changes sign. But for the purpose of

Wave-particle duality and bound states

solving the problem and finding the allowed values of E, the continuity of ψ and $d\psi/dx$ is in any case sufficient because the Schrödinger equation is only a second-order differential equation.

In Figure 3-9b we show a typical example of an acceptable wave function belonging to one of the allowed energies for a particle in a square well of finite depth (it is the third state in order of increasing energy). We see that, for square wells of finite depth, the wave function extends slightly beyond the boundaries of the well. For permitted values of E that are much smaller than the potential step ($E \ll V_o$) the exponential decrease of $\psi(x)$ with distance will be very rapid outside the classically permitted region, as indicated in Figure 3-9b. Analytically this results from the fact that the constant α in the decaying exponential $\psi(x) \sim e^{-\alpha x}$ outside the well is large ($\alpha = \{2m(V - E)\}^{1/2}/\hbar$). For such cases the sinusoidal form of $\psi(x)$ within the permitted region would, if projected into the forbidden region, intersect with the x axis slightly to the left of $x = 0$ and to the right of $x = L$. The result for the finite square well is a characteristic wavelength for each "normal mode" slightly *longer* and an energy slightly *lower* than for the corresponding infinite-well state. As V_o is imagined to become arbitrarily large the wave function $\psi(x)$ goes to zero arbitrarily close to the boundaries—a condition that we simply postulated in Section 3-5. But now we see the result as a natural consequence of applying continuity conditions that hold in general. The square-well solutions for finite V_o will be presented in more detail in Chapter 4.

3-10 NORMALIZATION OF THE WAVE FUNCTION

Up until now we have been concerned only with the *relative* probabilities of the various possible outcomes of position measurements on a quantum system, as described by values of $|\psi(x)|^2$. In order to make the statistical interpretation of the wave function more precise, we need to use the quantitative ideas of probability distributions. In any situation where our predictive ability is limited to statistical statements, we introduce the probabilities p_1, p_2, and so on, of different possible results—for example, which of the six faces of a die will appear on top when the die is thrown. (For a perfect die,

$p_1 = p_2 = \cdots = p_6 = \frac{1}{6}$.) Since it is certain that *one* of the six faces will be on top each time the die is thrown, we have $p_1 + p_2 + \cdots + p_6 = 1$, because unit probability, by definition, corresponds to certainty. Then in a large number of trials, N, our best guess of the numbers of cases in which these various outcomes will be obtained is given by Np_1, Np_2, and so on.

When the possible results form a continuous distribution, discrete probabilities can no longer be used; instead we employ what is called a *probability density*. To go directly to our quantum-mechanical problem, the value of $|\psi(x)|^2$ is just such a probability density if $\psi(x)$ is scaled in such a way that the value of $|\psi(x)|^2 \, \Delta x$ is *equal* to the probability of finding the particle between x and $x + \Delta x$. This process of scaling is called *normalization* of the wave function. Since the particle must be found *somewhere* along the x axis in each case, and since the probability of this is the sum of probabilities for all intervals $\Delta x \to dx$ along the axis, we have a normalization condition for the probability density:

$$\int_{\text{all } x} |\psi(x)|^2 dx = 1 \qquad (3\text{-}19)$$

In this way the quantum amplitudes for a particle in a bound state can be normalized. (We are, for the present, limiting our discussion to one-dimensional situations.)

As an example of normalization, consider the "violin-string" states in the infinite square-well potential of width L. We have

$$\psi_n(x) = A_n \sin k_n x = A_n \sin\left(\frac{n\pi x}{L}\right) \qquad (0 \le x \le L)$$

where A_n is a *normalization constant*. Since $\psi_n(x) = 0$ for all x outside the well, the normalization condition becomes

$$A_n^2 \int_0^L \sin^2\left(\frac{n\pi x}{L}\right) dx = 1$$

Note that the function $\sin^2\theta$ oscillates symmetrically between zero and unity, so that its average value is one-half, taken over an integer number of half-cycles. Multiplying this average value by L (the range of x), we see that the integral is equal to

Wave-particle duality and bound states

$L/2$. We thus have[6]

$$A_n^2 = \frac{2}{L}$$

The normalized probability amplitudes, and the normalized wave function composed of them, are thus described by the set of equations

$$\psi_n(x) = \sqrt{\frac{2}{L}} \sin\left(\frac{n\pi x}{L}\right) \tag{3-20}$$

For each integral value of n there is a bound state and a corresponding wave function given by this equation.

We shall discuss the properties and the significance of the quantum amplitudes more thoroughly and rigorously in Chapter 6 in the context of photon polarization states. For the moment, however, the above discussion is sufficient, since our main concern in this chapter and the next is to explore the character and the energy of bound states with the help of basic wave-mechanical principles. In the next section we return to the qualitative analysis of such problems, for which the preceding considerations of normalization of wave functions will be temporarily shelved.

3-11 QUALITATIVE PLOTS OF BOUND-STATE WAVE FUNCTIONS[7]

On the basis of the approach developed in Section 3-8, one can seek values of E that represent allowed energies of bound states in a variety of selected potentials. Such problems need not involve a lot of mathematical analysis if one has once understood the general character of the situation and can then turn to the use of analytical or numerical methods to explore trial values of E. Indeed, by knowing just a little more about the properties of wave functions, we can learn to draw qualitatively correct pencil-and-paper plots of them for bound states

---

[6]In the special case of the infinite square well, the normalization constant is the same for all energy eigenstates. For most potentials, each eigenstate has a different normalization constant.

[7]A. P. French and E. F. Taylor, Am. J. Phys. **39**, 961 (1971).

in *any* one-dimensional potential. Such a developed skill has several uses: First, it gives an appreciation of the common characteristics of all bound-state wave functions. Second, it provides a first approximate solution to a problem that can later be solved more exactly by analytic, graphical or computer methods. Third, the rough plot of a wave function for a state in a particular potential can give qualitative insight (and sometimes allow rough quantitative estimates also) concerning the behavior of a particle in that state. Fourth, one has an independent basis for judging whether or not an analytic formula or a computer-generated solution is correct and qualifies as a wave function. Finally, the analysis of many symmetrical three-dimensional systems can be reduced to an equivalent one-dimensional problem, so that the apparently artificial one-dimensional analysis is really of wide applicability.

In the paragraphs below we present the additional features of bound-state wave functions that can be utilized in making qualitative plots. We begin with the qualitative analysis of a potential well only slightly more complicated than the infinite square well, namely the *finite* square well, discussed in a preliminary way in Section 3-9. This potential well is shown again in Figure 3-10a.

Curvature of Wave Functions

An isolated system tends to drop into its lowest energy state, so we are particularly interested in the "ground state" of any system under study. What is the wave function for the lowest-energy state of a particle in a *finite* square well? From the results of Section 3-9 we conclude that this wave function is a sinusoid whose half-wavelength fits approximately into the well, matching a real exponential at either boundary of the well as shown in Figure 3-10b. A closer look at this result will provide the basis for a general conclusion. Why is this the wave function for the *lowest* energy state? Because its *curvature* is less than that of any other possible sinusoid of the same maximum amplitude that matches to exponentials at the boundary.

To outline the general argument, consider first the requirement that, inside *any* well, the lowest energy state must correspond to the lowest possible momentum at every point in the the potential. This follows since p^2 is proportional to $E - V(x)$ at any point. Now the sharpness of curvature of the wave

Wave-particle duality and bound states

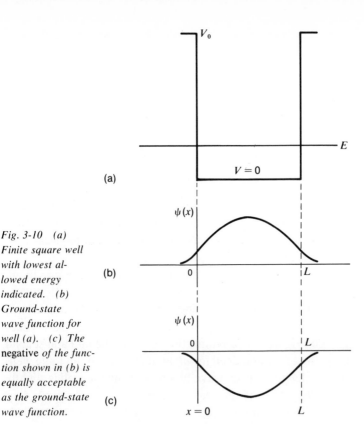

Fig. 3-10 (a)
Finite square well
with lowest al-
lowed energy
indicated. (b)
Ground-state
wave function for
well (a). (c) The
negative of the func-
tion shown in (b) is
equally acceptable
as the ground-state
wave function.

(a)

$\psi(x)$

(b)   0   L

$\psi(x)$

0   L

(c)
$x = 0$ L

function at any point is proportional to $d^2\psi/dx^2$ and hence, through Schrödinger's equation, proportional to the magnitude of the product of $E - V(x)$ with ψ itself. Thus, for a given value of $\psi(x)$, the wave function for the lowest state [with the smallest value of $E - V(x)$, at each point] has the smallest possible curvature at that point. Expressing it in slightly different terms, the lowest state has the smallest value of the ratio $(d^2\psi/dx^2)/\psi$ everywhere. (Of course, at points where ψ itself is zero, $d^2\psi/dx^2$ is also zero.)

If we consider, for example, the bound states in a finite square well (Figure 3-11), the state with the lowest energy has a certain constant value of p^2 across the whole width of the well, and this has associated with it the smallest value of wavenumber k (and the longest possible wavelength λ). However, a lower limit is placed on curvature (and an upper limit on wavelength) by the need to match the sine curve to decreasing exponentials that start with a downward slope on each side of the well (see Figure 3-11).[8] Hence the ground-state wave function for this potential is the sinusoid whose half-wavelength is slightly longer than the width of the well.

Is the *negative* function of Figure 3-10c also an acceptable wave function for the lowest energy state? Yes. Neither the de Broglie hypothesis nor the Schrödinger equation for wave functions "derived" from it determines uniquely the *sign* of the quantum amplitude. The probability density $|\psi|^2$ at any point is equal to the squared *magnitude* of the quantum amplitude, and so is the same whether the function $\psi(x)$ is positive or negative at a particular point. In general, if the wave function $\psi(x)$ represents a possible set of quantum amplitudes, then $-\psi(x)$ is equally acceptable as an equivalent and physically indistinguishable wave function.[9]

The wave function for the second-lowest energy in the finite square well is the sinusoid of second-lowest curvature that can match decreasing exponentials outside the well, namely, the one having approximately one wavelength within the well (Figure 3-12). Notice that for both the first and second energy levels (and indeed for *all* bound states in the finite well) the wavelength of each wave function inside the well is slightly *longer* than that of the wave function of the corresponding state in the infinitely deep square well (Figure 3-13). This results from the need to match the sinusoid inside the well to a

[8]Why not have for the lowest energy state a wave function that has the least possible of *all* curvatures within the well, namely a straight line (infinite wavelength)? The corresponding energy would be zero, equal to the potential energy at the bottom of the well. But the wave function for such a state cannot be constructed because there is no way to match a straight line to decreasing exponentials on *both* sides of the well without a discontinuity of slope on at least one side. Actually, there *is* a formal exception to this: a wave function for which the probability amplitude is zero everywhere. But that means a total absence of the particle from the well or its vicinity, and so for our purposes it is a nonexistent state. The lowest energy states in *all* binding potentials share this property: the energy of the lowest state is higher than the minimum of the potential energy. As a result a bound particle always has some kinetic energy, even in its lowest energy state. The difference between the lowest energy value and the minimum value of the potential energy is called the *zero-point energy*.

[9]In one-dimensional problems, the spatial factor $\psi(x)$ of the quantum amplitude for a *bound* state can always be chosen to be *real*. For *unbound* states, quantum amplitudes are in general *complex*. Nevertheless, the probability density at any point is still given by the squared magnitude $|\psi|^2$ of ψ at that point. (When ψ is complex, $|\psi|^2$ denotes $\psi^*\psi$, where ψ^* is the complex conjugate of ψ: if $\psi = a + ib$ at some point, then $\psi^* = a - ib$.) In any situation, if ψ is a possible wave function, then $\psi e^{i\delta}$ is equally acceptable as an equivalent and physically indistinguishable wave function for any real value of the constant δ.

Wave-particle duality and bound states

decreasing exponential outside the well. Accordingly, the energies for the finite well are, respectively, slightly *lower* than for the infinite well. This qualitative conclusion will have quantitative verification in a later section.

The exponential decrease of the wave function outside the square well for the second energy state is less rapid than is the corresponding decrease for the lowest energy state as indicated in Figure 3-13. This is because the quantity $(V - E)$ is smaller in the exponential decay constant $\alpha = [2m(V - E)]^{1/2}/\hbar$.

Fig. 3-11   Trial fittings to determine the lowest energy state of a particle bound in a square well. The potential-energy diagram is shown at the top with four trial values for the ground-state energy. Wave functions corresponding to the trial values E_a, E_b, E_c, and E_d. (a) The trial function for $E = E_a$ intersects the axis within the well. A lower trial energy value is needed. (b) The trial function for $E = E_b$ intersects the axis just outside the well and then diverges. A still lower trial value is needed. (c) For $E ... E_c$, the trial function turns up and diverges outside the well, never crossing the axis. A higher trial value is needed. (d) The trial function converges smoothly toward the axis. The lowest allowed energy for the given well is E_d.

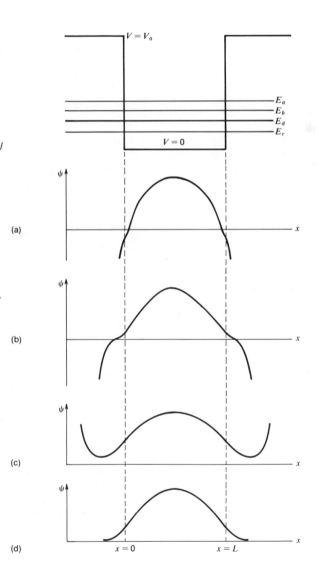

3-11   Qualitative plots

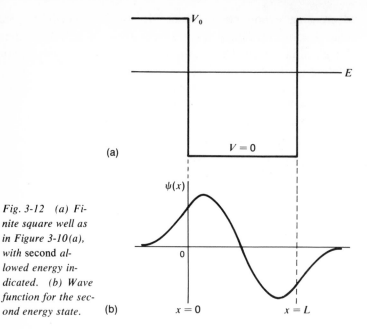

Fig. 3-12 (a) Finite square well as in Figure 3-10(a), with second allowed energy indicated. (b) Wave function for the second energy state.

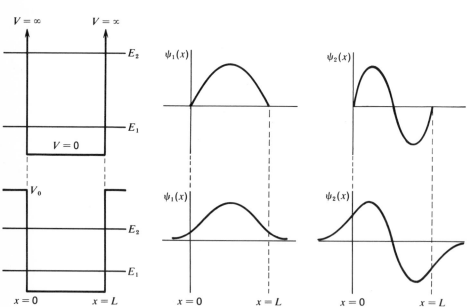

Fig. 3-13 *Infinite square well and finite square well compared. The infinite square well and the wave functions for the two lowest states are sketched across the top. Below them are the corresponding graphs for the finite square well. The de Broglie wavelength within the well is longer for any given bound state of the finite square well than for the corresponding state of the infinite square well.*

Wave-particle duality and bound states

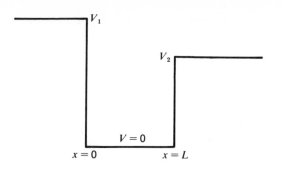

Fig. 3-14 Lop-
sided square well,
for which you are
asked to sketch the
wave functions and
probability densities
for the first three
energy states.

V_1

V_2

$V = 0$

$x = 0$ $x = L$

As an exercise, draw the *probability density* function for
the second energy level in the finite square well. What is the
physical significance of the total area under this probability
curve? Draw the wave function and the probability density
function for the *third* energy level in the finite square well, pay-
ing attention to the exponential decay constant outside the well
relative to those for the first two levels. For the "lopsided"
square well potential of Figure 3-14, sketch the wave functions
for the first three energy levels and the probability density
functions for all three states.

Potentials that Vary Continuously with Position

Experimenters have never verified discontinuities on the
subatomic scale; apparently potentials always vary continuously with position. Already in this chapter we have touched
on such cases. Now, let us see how we can analyze them in detail, both inside and outside the potential well.

Consider a well with a potential that varies continuously
with position (see Figure 3-15). If we make an arbitrary choice

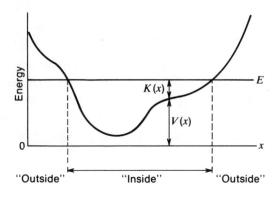

Fig. 3-15 A well
with a potential en-
ergy function that
varies continuously
with position.

E

$K(x)$

$V(x)$

0 x

"Outside"   "Inside"   "Outside"

3-11   Qualitative plots

of energy (above the minimum of potential), we automatically divide the x axis into regions "inside" and "outside" the well, the points of demarcation being at those values of x at which $E = V$. These points mark the limits of displacement of a classical particle of the chosen energy E. In constrast to the square-well situation, there is now a continuous variation with x of the difference between E and V; thus, in the central region where $E > V$, it is not possible to define a wavelength. Physically, the situation is comparable to that of a normal mode of a rope whose thickness varies in some arbitrary way along its length. The rope has a unique frequency of vibration but (in contrast to a normal mode of violin string) its deformation does not follow a sinusoidal shape with some well-defined wavelength between the supports.

The two following statements derived from the Schrödinger equation (Eq. 3-13) now provide our basic guide to the appearance of $\psi(x)$:

1. Inside the well, where $E > V(x)$, we have

$$\frac{d^2\psi}{dx^2} = -k^2\psi \qquad \left\{ k = k(x) \text{ and } k^2 = \frac{2m}{\hbar^2}[E - V(x)] \right\}$$

From this relation between ψ and its second derivative, we see that the curvature of $\psi(x)$ is always *toward* the x axis (that is, concave side faces the x axis). The greater the value of $E - V(x)$, the sharper the curvature for a given value of $\psi(x)$.

2. Outside the well, where $E < V(x)$, we have

$$\frac{d^2\psi}{dx^2} = +\alpha^2\psi \qquad \left\{ \alpha = \alpha(x) \text{ and } \alpha^2 = \frac{2m}{\hbar^2}[V(x) - E] > 0 \right\}$$

The curvature of $\psi(x)$ is always *away from* the x axis (that is, convex side faces the x axis). The greater the value of $V(x) - E$, the sharper the curvature for a given value of $\psi(x)$.

In many cases the above qualitative statements go a long way toward making possible useful sketches of acceptable forms of $\psi(x)$. To be acceptable the "outside" solutions must decrease smoothly toward zero as $x \to \pm\infty$ and must join the "inside" solution with continuity of amplitude and slope at the points where $E = V(x)$ (the boundaries).

Wave-particle duality and bound states

Number of Nodes in the Wave Function

In discussing the wave functions for the energy states in a square well, we pointed out how the successively higher states have wave functions with shorter and shorter wavelengths. This behavior can be characterized in a simple and powerful way by considering the number of times the graph of $\psi(x)$ crosses the x axis. In the *lowest* energy state it makes no such crossing at all; we say that the wave function has no *nodes*. The wave function for the *second* level has *one* node. The wave function for the *third* energy level has *two* nodes. In general, the wave function for the nth bound energy state has $n - 1$ nodes. This conclusion is of central usefulness in sketching bound-state wave functions because it proves to be valid for any form of one-dimensional binding potential, not just for the square well.

The Effect of a Varying Well Depth on the Maximum Value of the Wave Function.

An important property of quantum amplitudes in space-varying potentials can be illustrated by considering the simplified case of a "step well." In Figure 3-16 is plotted the wave function for the fifth energy level in such a well. (You should examine the figure and use the foregoing arguments to check that it indeed represents the fifth level.) One feature of this plot is new; nothing we have said thus far has prepared you for it: The *maximum values* of the quantum amplitudes—the heights of the peaks of $\psi(x)$—are greater in region II (smaller p) than in region I (larger p). Speaking classically, one may say that the particle moves more slowly in region II, because of its smaller momentum, that it does in region I. In consequence, the probability density for finding it in region II is enhanced.[10] Since the quantum amplitudes have similar sinusoidal variations in both regions, it follows that the maximum values of these sinusoids must be greater in region II than in region I. A more acceptable analysis, using the quantum-mechanical wave description, goes as follows.

---

[10]One must be very wary, however, of talking about *motion* when one is dealing with pure quantum states (states belonging to a single value of E). We shall go into this when we discuss time-dependent problems (Chapter 8).

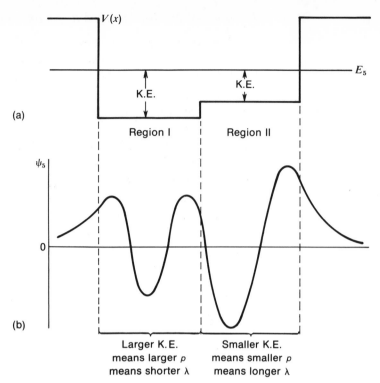

In each part of the "step well," the wave function is of the form

$$\psi(x) = A \sin (kx + \phi) \qquad (3\text{-}21)$$

The values of k are different for regions I and II, and so, in general, will be A and ϕ. We want to show that the coefficient A (the maximum value of the quantum amplitude) is larger in region II than in region I. This is done by considering the slope $d\psi/dx$ of the wave function. Calling this slope m, we have

$$\frac{d\psi}{dx} = m = kA\cos(kx + \phi) \qquad (3\text{-}22)$$

We can rewrite Eqs. (3-21) and (3-22) as follows:

$$A \sin (kx + \phi) = \psi$$
$$A \cos(kx + \phi) = \frac{m}{k}$$

Wave-particle duality and bound states

These equations can be squared and added to eliminate the phase ϕ and yield an expression for the maximum value A:

$$A = \left(\psi^2 + \frac{m^2}{k^2}\right)^{1/2} \tag{3-23}$$

Now recall that both the value of ψ and also its slope m are continuous across the step, but that the value of k changes discontinuously. The continuity of ψ and m means that infinitesimally close to the step, the values of both ψ^2 and m^2 have the same magnitude on either side. Therefore at the step, according to Eq. 3-23, the coefficient A of the wave function *increases* as k *decreases* (except for the special case $m = 0$). The momentum p is smaller in region II than in region I, so the wave number k is also smaller. Therefore, the maximum value A is greater in region II than in region I, as we set out to show.[11]

The result expressed by Eq. 3-23 is important because any potential function whatever can be approximated by a sequence of step functions. The analysis we have developed is valid for each such step, so the conclusion can be applied to a continuously varying potential: regions of smaller momentum have larger maximum values of the amplitude than adjacent regions of larger momentum.[12] As an example, Figure 3-17b shows the wave function for the fifth energy state in the ramp-bottom potential of Figure 3-17a. The dashed line in Figure 3-17a shows one possible series of steps that approximate the ramp. The word "wavelength" in the figure labels is enclosed in quotation marks because in nonuniform potentials the wave function is not sinusoidal. But, over a small part of one spatial cycle, the potential does not change much, and this is a case in which one may speak loosely of a "local wavelength." Figure 3-17c shows the wave function generated by an analog computer for the seventh energy state (six nodes) in just such a potential.

[11]In the special case that the slope m happens to be zero at the step, Eq. 3-23 says that the continuity of the quantum amplitude at the step requires the maximum value A to be the same in both regions. (Note that the possibility of this admittedly unlikely case shows that the classical argument we advanced earlier is not really sound.)

[12]The word "adjacent" is important in this generalization. You might like to think of a special case in which the amplitude A does not satisfy this condition on two sides of a "double well" with a hump in the middle, the two wells being of unequal depth.

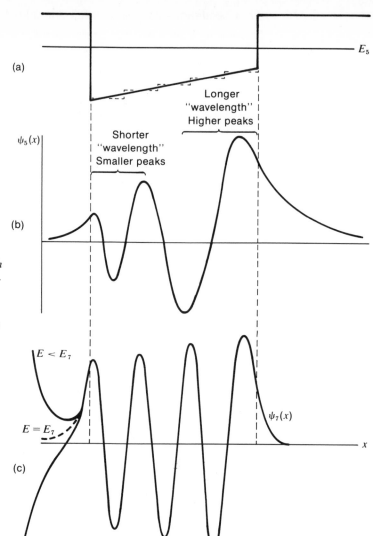

Fig. 3-17 (a) Ramp-bottom potential well, with fifth-lowest allowed energy indicated. Dashed line indicates a series of steps which could be used to approximate the ramp. (b) Wave function for the fifth energy state. (c) Trial functions generated by an analog computer. The energy values that were tried bracket the seventh allowed energy value E_7. Inside the well, the trial functions coincide to within the width of the line, so the function ψ_7 has been obtained.

(a)

E_5

(b)

$\psi_5(x)$

Longer "wavelength" Higher peaks

Shorter "wavelength" Smaller peaks

(c)

$E < E_7$

$E = E_7$

$E > E_7$

$\psi_7(x)$

x

Symmetry Considerations

Many atomic and molecular potentials have a center of symmetry. Whenever the potential energy curve $V(x)$ is symmetrical with respect to a given point, we can draw some useful conclusions about the properties of the wave-mechanical solution $\psi(x)$. Consider, for example, the potential shown in Figure 3-18a. If we choose an origin at the central point, we can argue on quite general physical grounds that a

Wave-particle duality and bound states

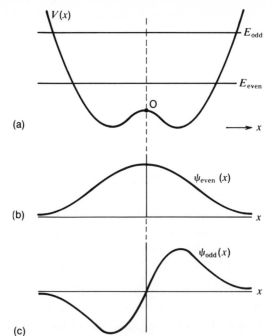

Fig. 3-18 (a) A potential function $V(x)$ which is symmetric about the point O. (b) For the potential function shown in (a), the ground-state wave function is even. (c) The second energy level in a symmetric potential has a wave function that is odd.

particle of a given energy in this potential has exactly the same probability of being found at points equidistant from the origin on either side.[13] There is no reason for the particle to prefer one side of a symmetric potential to the other. Since this probability is proportional to the square of the wave function $\psi(x)$ at any given x, we must have

$$[\psi(-x)]^2 = [\psi(+x)]^2 \qquad (3\text{-}24)$$

Hence

$$\psi(-x) = \pm\psi(+x) \qquad (3\text{-}25)$$

if

$$V(-x) = V(+x)$$

[13]The argument presented here is valid only for the "stationary" states that are the subject of this chapter. Later, in dealing with time-dependent situations, we shall find that a probability function can "slosh back and forth" from one side to the other of a symmetric well. Under these conditions Eq. 3-24 is not correct at every instant, but the *time average* of the position probability distribution will be symmetrical even in this case.

3-11   Qualitative plots

The condition represented by Eq. 3-25 allows us, if we know that $V(x)$ is symmetrical about the origin, to classify all possible wave functions $\psi(x)$ into two types:

1. *Even* functions, for which $\psi(-x) = \psi(x)$. Such functions necessarily have a first derivative equal to zero at the symmetry point ($d\psi/dx = 0$ at $x = 0$). We shall then have $\psi(0) \neq 0$ in general, as indicated in the sketch labeled "Even"—see Figure 3-18b. Having $\psi(0) = 0$ would imply, according to the Schrödinger equation, that $d^2\psi/dx^2$ as well as $d\psi/dx$ would be zero at $x = 0$. Although conceivable, this is a highly unlikely situation.

2. *Odd* functions, for which $\psi(-x) = -\psi(x)$. To be continuous through the origin, such functions must have $\psi(0) = 0$. But we shall then have a first derivative not equal to zero $[\psi'(0) \neq 0]$, as indicated in Figure 3-18c.

This symmetry property of stationary-state wave functions in a symmetric potential is called *parity*, and we say that a particular wave function has either odd or even parity.

The recognition that the wave function of every stationary state in a *symmetrical* potential must belong to one or the other of these categories brings a very great simplification into the search for the permitted energy values in such cases. For example, in programming the solution by computer, one needs only to look for those energies that cause $\psi(x)$ to decrease smoothly toward zero at large x if one starts out at $x = 0$ with *either* an arbitrary value of $\psi(0)$ and zero slope (even solutions) *or* an arbitrary value of $\psi'(0)$ and $\psi(0) = 0$ (odd solutions). The construction of rough sketches of possible wave functions is also greatly aided by the knowledge of these results.

Summary of Rules for Qualitative Plots

Using the results of this section, you can draw for yourself the wave function for, say, the seventh energy level in a given one-dimensional potential, provided only that the energy of the state is indicated on the potential plot. Notice that we have not needed to concern ourselves with the normalization of the quantum amplitudes. Our goal has been simply to find the *form* of the wave function for a possible state. Notice also that the sketch can be made without any need to consider the particle mass, or any other of the numerical magnitudes that are clearly

Wave-particle duality and bound states

relevant to a quantitative solution of the problem. The reason is that the indication of the energy [relative to $V(x)$], together with information about the state number, provide between them an implicit *dimensional scale* that allows us to ignore, for the purposes of a rough plot, the constants m and $\hbar$ in the Schrödinger equation (Eq. 3-13), along with numerical magnitudes of x, E, and $V(x)$. In constructing such a wave function, you can make use of the following check list of properties of bound-state wave functions. Check that your wave function for quantized energy states has:

1. odd or even symmetry in the case of a symmetric potential;
2. the correct number of nodes for the specified number of the energy level;
3. the correct relative wavelengths (longer or shorter) for different values of the potential at different places inside the well;
4. the correct relative maximum values of amplitudes at adjacent points of different potential inside the well;
5. the correct relative rate of decrease (more or less gradual) of the wave function with distance outside the well for different values of potential at different places and for quantum states of different energy.

In the problems at the end of this chapter you will find a number of potential wells for which you may sketch wave functions. The game of sketching wave functions for various potential wells can be fun. But beyond this, nature is so varied and prolific that almost any shape we choose for an exercise, no matter how bizarre, will resemble some real potential found in nature. So the exercises are more than a game, even though the game itself develops intuition about the formal results of quantum physics.

EXERCISES

3-1 *Alternative wave equations.* Devise at least one and preferably more than one differential "wave equation" for particles that satisfies Eqs. 3-1 through 3-3 but differs from the Schrödinger equation.

3-2 *Product solutions of the time-dependent Schrödinger equation.* Show that whenever a solution $\Psi(x, t)$ of the time-dependent

Schrödinger equation separates into a product $\Psi(x,t) = F(x) \cdot G(t)$, then $F(x)$ *must* satisfy the corresponding time-independent Schrödinger equation and $G(t)$ *must* be proportional to $e^{-iEt/\hbar}$.

3-3 *Effects of a uniform potential.* A particle is in a stationary state in the potential $V(x)$. The potential function is now increased over all x by a constant value V_0. What is the effect on the quantized energy? Show that the spatial wave function of the particle remains unchanged.

3-4 *Normal modes of a composite string.* Consider the normal modes of a vibration under uniform tension of a stretched string consisting of two sections of unequal mass density, so that the wave speed differs from one section to the next.

(a) What conditions must be satisfied by the string displacement at the junction of the two sections?

(b) Suppose that the mass density is higher (and the wave speed lower) in region A than in region B. The string is set vibrating in a high normal mode, so that there are several nodes in each region. Is the distance between nodes in region A larger or smaller than that in region B? In which region is the amplitude of vibration larger? Explain your choices.

3-5 *Another estimate of the size of atoms.* The Lyman alpha radiation for hydrogen (resulting from the transition $n = 2$ to $n = 1$) has a wavelength in the ultraviolet, 1216.0 Å. Using the crude model of the one-dimensional infinite square well, derive an estimate of the diameter of a hydrogen atom. How does this value compare with twice the Bohr radius: $2a_0 = 1.06$ Å?

3-6 *Energy of particles in the nucleus.*

(a) Suppose that the potential seen by a neutron in a nucleus can be represented (in one dimension) by a square well of width 10^{-12} cm with very high walls. What is the minimum kinetic energy of the neutron in this potential, in MeV?

(b) Can an electron be confined in a nucleus? Answer this question using the following outline or some other method.

(i) Using the same assumption as in (a)—that the nucleus can be represented as a one-dimensional infinite square well of width 10^{-12} cm—calculate the minimum kinetic energy, in MeV, of an electron bound within the nucleus.

(ii) Calculate the approximate coulomb potential energy, in MeV, of an electron at the surface of the nucleus, compared with its potential energy at infinity. Take the nuclear charge to be $+50e$.

(iii) Is the potential energy calculated in (ii) sufficient to bind the electron of the kinetic energy calculated in (i)?

Wave-particle duality and bound states

[*Note:* Although electrons cannot be *bound* in the nucleus, they can be *created* in the nucleus and then escape in the process called *beta decay.*]

3-7 *Position probability distribution for a pendulum.* Consider the classical motion of a pendulum bob which, for small amplitudes of oscillation, moves effectively as a harmonic oscillator along a horizontal axis according to the equation

$$x(t) = A \sin \omega t$$

The probability that the bob will be found within a small distance Δx at x in random observations is proportional to the time it spends in this region during each swing. Obtain a mathematical expression for this probability as a function of x, assuming $\Delta x \ll A$. Compare the result with Figure 3-6(c).

3-8 *Electron density outside the surface of a metal.* The boundary between the interior of a metal and the air or vacuum may be modeled by a rectangular potential step, as shown in the figure. The height of the potential step (V_o) exceeds the energy of the most energetic conduction electrons (E). The difference $V_o - E$ is called the work function, W.

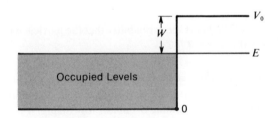

(a) What is the form of the wave function outside the metal for the electrons that are most energetic? What is the form of the function that describes the probability density of finding these electrons at a given distance outside the surface?

(b) Assume that the work function is 5 eV. Estimate the distance outside the metal at which the probability density function drops to the fraction $(1/e)^7 \approx 1/1000$ of that just inside the metal.

(c) Compare the distance estimated in part (b) with an estimate of the minimum roughness of a polished metal surface over comparable distances. How useful do you expect the "step-potential" model of a surface of a metal to be?

3-9 *Why can bound-state wave functions be chosen to be real?* The text states that, in one-dimensional problems, the spatial wave function for any allowed state can be chosen to be real-valued. Verify this using the following outline or some other method.

(a) Write the wave function $\psi_n(x)$ in terms of its real and imaginary parts: $\psi_n = \mathrm{Re}\,\psi_n + i \cdot \mathrm{Im}\,\psi_n$, and substitute this into the Schrödinger equation.

(b) Show that $\mathrm{Re}\,\psi_n$ and $\mathrm{Im}\,\psi_n$ separately satisfy the Schrödinger equation.

(c) In one dimension there is only one (linearly independent) wave function for each energy eigenvalue E_n. What does this imply about $\mathrm{Re}\,\psi_n$ and $\mathrm{Im}\,\psi_n$? Describe how you can construct a real, normalized wave function from ψ_n.

3-10 *Normalization of a wave function.* A particle bound in a certain one-dimensional potential has a wave function described by the following equations:

$$\psi(x) = 0, \qquad\qquad x < -\frac{L}{2}$$

$$\psi(x) = A e^{ikx} \cos\frac{3\pi x}{L}, \qquad +\frac{L}{2} \le x \le \frac{L}{2}$$

$$\psi(x) = 0, \qquad\qquad x > \frac{L}{2}$$

(a) Using Eq. 3-19, find the value of the normalization constant A.

(b) What is the probability that the particle will be found between $x = 0$ and $x = L/4$?

3-11 *Qualitative plots of simple harmonic oscillator wave functions.* The figure shows the parabolic potential function of the one-dimensional simple harmonic oscillator. In Chapter 4 we shall show that the permitted energy levels are equally spaced, as indicated in the figure. (For historical reasons, the lowest energy level is numbered zero.)

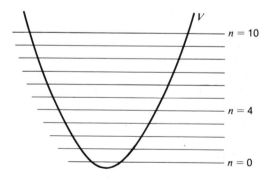

Wave-particle duality and bound states

(a) Sketch qualitative plots of the wave functions for the energy levels labeled 0, 4, and 10. In the last case, pay special attention to the maximum values of the wave function near the edges of the well (inside the well but near the classically forbidden regions on either side) compared to the maximum values near the center.

(b) Using the results of (a), sketch qualitative graphs of $|\psi|^2$ as functions of x, and compare them with the classical probability distribution for a harmonic oscillator (Figure 3-6c). You should see signs of the correspondence principle at work [cf. Exercise 3-7].

3-12 *Wave functions in a double well.*

(a) Is the function shown in diagram (a) a possible wave function for a particle bound in the one-dimensional potential shown in diagram (b)?

(b) Sketch the wave function for the lowest energy level in the potential shown in (b), assuming that the energy is less than the height of the rectangular hump.

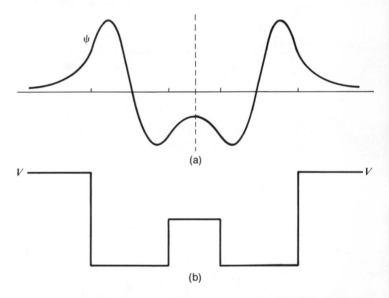

(a)

(b)

3-13 *Qualitative plot of a wave function.* A rectangular potential well in one dimension is bounded by a wall of height $5V_o$ on one side and a wall of height $2V_o$ on the other, as shown in the figure. Suppose that the well is of such a width L that the *second* energy state for a particle of mass m has energy E_2 exactly equal to V_o. (The state of *lowest* energy is defined to be the first state.)

(a) Make a qualitative plot of the wave function $\psi_2(x)$. Indicate the relative rates of decrease of $|\psi_2|$ with x in regions I and III. Does the node of this wave function occur to the right or the left of the center of the well?

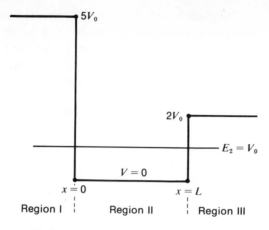

(b) By what factor is a particle in this second state less likely to be found in a small region near $x = -2L$ than in a region of equal width near $x = -L$?

(c) Comparing your qualitative plot of $\psi_2(x)$ with the wave function for the second energy level in the infinite square well, we can infer that the mass m of the particle is less than a certain quantity involving L, V_0 and h. What is this quantity?

(d) In order to satisfy the boundary conditions, the wave function must be "headed" toward the axis at $x = 0$ and at $x = L$. This means that $\lambda_2/2$ must be less than L. (Here λ_2 is the wavelength of the wave function ψ_2 in region II.) What *lower* limit does this imply for m?

(e) By extending the argument used in (d) can you rule out the existence of a fourth energy level in this potential? Can you rule out a third level?

3-14 *Criteria for the existence of bound states.* The figure presents four one-dimensional potential-energy curves.

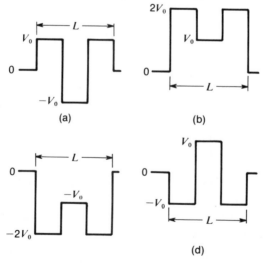

(a)　　　　　(b)

(d)

Wave-particle duality and bound states

(a) For which of these potentials can you definitely exclude the existence of a bound quantum state, regardless of the mass of the particle?

(b) For which potentials are you confident that there is at least one bound state?

(c) How would you approach the problem of determining which of these potentials *probably* have a second bound state?

3-15 *Practice in constructing qualitative plots: I.* Make qualitative plots for wave functions for several energies in each of the potentials shown in the figure.

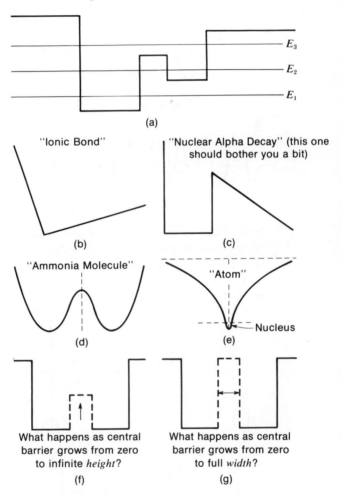

(a)

"Ionic Bond"

(b)

"Nuclear Alpha Decay" (this one should bother you a bit)

(c)

"Ammonia Molecule"

(d)

"Atom"

Nucleus

(e)

What happens as central barrier grows from zero to infinite *height*?

(f)

What happens as central barrier grows from zero to full *width*?

(g)

3-16 *Qualitative plots II: Potentials from wave functions.* Sketch the one-dimensional *potentials* that would give rise to each of the wave functions shown in the figure. Include quantitative features where possible. Mark on each of your diagrams a horizontal line to indicate an appropriate energy for the state in question. What is the *number* of this energy level, counting the lowest-energy level as one?

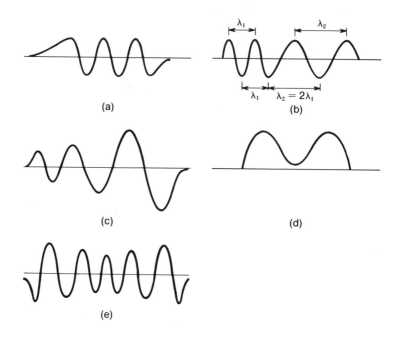

(a)

(b)

(c)

(d)

(e)

3-17 *Exposing an unsuccessful plot.* The curve in the figure is alleged to be the plot of a computer-calculated wave function for the fifth energy level of a particle in the diagrammed one-dimensional potential well. By means of arrows and labels, indicate the way or ways in which the plot *fails* to be even qualitatively correct.

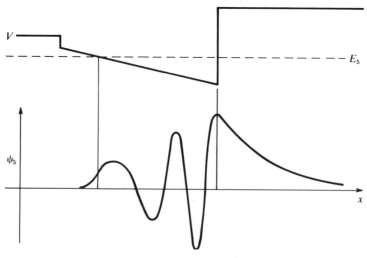

Wave-particle duality and bound states

3-18 *Characteristics of wave functions for stationary states.* Below are some general statements about wave functions for stationary states of unique energy for a particle bound in a one-dimensional potential well $V(x)$. Decide whether each statement is true or false. Name one or more counterexamples for false statements. Be careful: except where noted, these are meant to be *general* statements, true, for example, even if there is a classically forbidden region *inside* the well [a "hump" in $V(x)$]. The phrase "outside the well" for any given energy E means a continuous classically forbidden region $[E < V(x)]$ extending to infinity.

Assertions to be considered for correctness or falsehood:

(a) There are no nodes in the wave function outside the well.

(b) There are no nodes in classically forbidden regions.

(c) If the potential has only one relative minimum, the ground-state probability function $|\psi|^2$ has only one maximum.

(d) The ground-state probability function has no nodes.

(e) The ground-state probability function has only one maximum.

(f) The probability function for any state is greater at positions of higher potential than at positions of lower potential.

(g) The probability function in a classically forbidden region is greater at positions of higher potential than at positions of lower potential.

(h) For a given region outside the well, the probability function is smaller as one goes farther from the well.

The miracle of the appropriateness of the language of mathematics for the formulation of the laws of physics is a wonderful gift which we neither understand nor deserve.

EUGENE P. WIGNER, *Richard Courant Lecture, New York University* (1959)

4

Solutions of Schrödinger's equation in one dimension

4-1 INTRODUCTION

As we showed in the preceding chapter, it is possible to gain considerable insight into the bound states of quantum systems through a qualitative use of the Schrödinger equation along with considerations of continuity, curvature, and symmetry. Our next steps are to treat a wider range of examples, to refine and extend the quantum description applied to these examples, and to begin assembling mathematical and computational tools that can lead to the precise predictions required for the crucial comparisons of theory with experiment.

In Chapter 3 we arrived at the famous time-independent Schrödinger equation (Eq. 3-13) for a bound-state wave function by analogy to the classical analysis of vibrating systems. We rewrite it here:

$$-\frac{\hbar^2}{2m}\frac{d^2\psi}{dx^2} + V(x)\psi = E\psi \tag{4-1}$$

This equation has its limitations, as we shall see in later chapters, but (as remarked in Chapter 3) it is also impressively successful in analyzing and predicting a very wide range of phenomena in atomic and subatomic physics. In the present chapter we shall examine some explicitly one-dimensional systems, including the finite square well, the harmonic oscillator, and the application of harmonic oscillator theory to molecular vibrations. The chapter will conclude with an introduction to computer solutions of the Schrödinger equation.

4-2 THE SQUARE WELL

We begin by examining, more carefully than we did in Chapter 3, the bound states of a particle in a rectangular potential well of finite depth V_o, which can be used to model in one dimension the three-dimensional potential experienced by a neutron in a nucleus. The rectangular potential well also corresponds approximately to the potential experienced by an electron in linear molecules such as acetylene (H—C≡C—H). It will be convenient here to take the bottom of the well to define the zero of energy. For a given value of the energy E, we then have three distinct regions as shown in Figure 4-1, symbolized by the Roman numerals I, II, and III. A bound state will have $E < V_o$, so that in regions I and III the wave function $\psi(x)$ falls off toward zero at large distance from the origin. In region II the solution is sinusoidal.

Following the arguments of Chapter 3, we assume particular forms for the wave functions in the three regions and look for values of the constants in these functions that lead to continuity at the edges of the well. If we place the origin at the center of the well, as shown in Figure 4-1, the considerations of symmetry presented in Chapter 3 tells us that the wave functions for the energy states will be even or odd functions of x.

Even Solutions	Odd Solutions	Region
$\psi_I = A e^{\alpha x}$	$\psi_I = C e^{\alpha x}$	I: $x < -L/2$
$\psi_{II} = B \cos kx$	$\psi_{II} = D \sin kx$	II: $-L/2 \leq x \leq +L/2$
$\psi_{III} = A e^{-\alpha x}$	$\psi_{III} = -C e^{-\alpha x}$	III: $x > +L/2$

These functions satisfy the Schrödinger equation if we put

$$k = \frac{(2mE)^{1/2}}{\hbar} \quad \text{and} \quad \alpha = \frac{[2m(V_o - E)]^{1/2}}{\hbar} \tag{4-2}$$

For both even and odd solutions we must satisfy the condi-

Schrödinger's equation in one dimension

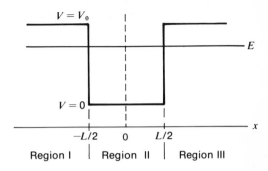

Fig. 4-1 The finite square well.

Region I | Region II | Region III

tions of continuity of ψ and $d\psi/dx$ at $x = \pm L/2$. The calculation proceeds as follows:

Even Solutions

$$\frac{d\psi_{\text{I}}}{dx} = \alpha A e^{\alpha x}; \qquad \frac{d\psi_{\text{II}}}{dx} = -kB \sin kx; \qquad \frac{d\psi_{\text{III}}}{dx} = -\alpha A e^{-\alpha x}$$

Continuity of ψ at either $x = -L/2$ or $x = L/2$ gives

$$A e^{-\alpha L/2} = B \cos\left(\frac{kL}{2}\right)$$

Continuity of $d\psi/dx$ at $x = \pm L/2$ gives

$$\alpha A e^{-\alpha L/2} = kB \sin\left(\frac{kL}{2}\right)$$

Combining the preceding two equations, we obtain

$$\tan \frac{kL}{2} = \frac{\alpha}{k} \qquad\qquad (4\text{-}3a)$$

where both α and k are functions of energy through Eq. 4-2. Therefore Eq. 4-3a defines the values of energy for which even solutions exist.

Odd Solutions

$$\frac{d\psi_{\text{I}}}{dx} = \alpha C e^{\alpha x}; \qquad \frac{d\psi_{\text{II}}}{dx} = kD \cos kx; \qquad \frac{d\psi_{\text{III}}}{dx} = \alpha C e^{-\alpha x}$$

4-2   The square well

Continuity of ψ at $x = \pm L/2$ gives

$$Ce^{-\alpha L/2} = -D \sin\left(\frac{kL}{2}\right)$$

Continuity of $d\psi/dx$ at $x = \pm L/2$ gives

$$\alpha Ce^{-\alpha L/2} = kD \cos\left(\frac{kL}{2}\right)$$

Combining these two, we have the expression corresponding to Eq. 4-3a:

$$\cot\frac{kL}{2} = -\frac{\alpha}{k} \qquad (4\text{-}4a)$$

Equations 4-3a and 4-4a provide the basis for finding the permitted values of E. When rewritten with the value of E explicitly introduced they become the following:

For even solutions:

$$\tan\left(\frac{\sqrt{2mE}}{\hbar} \cdot \frac{L}{2}\right) = \left(\frac{V_0 - E}{E}\right)^{1/2} = \left(\frac{V_0}{E} - 1\right)^{1/2} \qquad (4\text{-}3b)$$

For odd solutions:

$$\cot\left(\frac{\sqrt{2mE}}{\hbar} \cdot \frac{L}{2}\right) = -\left(\frac{V_0 - E}{E}\right)^{1/2} = -\left(\frac{V_0}{E} - 1\right)^{1/2} \qquad (4\text{-}4b)$$

Unfortunately these are transcendental equations; no algebraic solutions for E can be found. To solve them, we could plot the left and right sides of Eqs. 4-3b and 4-4b respectively as functions of E and look for the intersections, thus giving values of energy for which the equations are satisfied. The resulting energy values would, of course, depend on the particular values of L, m, and V_0. We can find a single parameter that replaces all three of these quantities—at the price of introducing a new variable θ. Set θ equal to the argument of the trigonometric functions on the left of Eqs. 4-3b and 4-4b and set θ_0 equal to the value of θ for $E = V_0$ (the top of the well):

$$\theta = \frac{\sqrt{2mE}}{\hbar} \cdot \frac{L}{2} = \frac{kL}{2} \qquad (4\text{-}5)$$

$$\theta_0 = \frac{\sqrt{2mV_0}}{\hbar} \cdot \frac{L}{2} \qquad (4\text{-}6)$$

Schrödinger's equation in one dimension

Then V_0/E on the right sides of Eqs. 4-3b and 4-4b becomes

$$\frac{V_o}{E} = \frac{\theta_o{}^2}{\theta^2}$$

and these equations take on the simple form

$$\tan \theta = \left(\frac{\theta_o{}^2}{\theta^2} - 1\right)^{1/2} \quad \text{(even solutions)} \tag{4-7}$$

and

$$\cot \theta = -\left(\frac{\theta_o{}^2}{\theta^2} - 1\right)^{1/2} \quad \text{(odd solutions)} \tag{4-8}$$

In these equations, the single parameter θ_o (which incorporates the values of L, m, and V_o) determines the permitted values of θ and, through Eq. 4-5, the permitted values of energy E. We still treat the even and odd solutions separately, as shown in Figures 4-2a and 4-2b. The circled intersections in each graph locate the points where the left side of Eq. 4-7 or Eq. 4-8 equals the right side. These points give permitted values of $kL/2$ for bound states in this potential well. The *even* solutions occur for values of $kL/2$ somewhat less than an *odd* multiple of $\pi/2$; the *odd* solutions occur for values of $kL/2$ somewhat less than an *even* multiple of $\pi/2$. Thus the lowest state is even, and the highest states are successively odd and even, giving one additional node in $\psi(x)$ for each successive energy state. This is consistent with the general rule given in Section 3-11 of Chapter 3.[1]

To see how the possible energy states depend on the well depth V_o, it is helpful to consider the dependence on both V_o and E of the function that appears on the right-hand side of both Eq. 4-3b and Eq. 4-4b:

$$\left(\frac{V_o - E}{E}\right)^{1/2} = \left(\frac{V_o}{E} - 1\right)^{1/2} = \left(\frac{\theta_o{}^2}{\theta^2} - 1\right)^{1/2} \tag{4-9}$$

---

[1]There exists an elegant and simple analysis in which *all* solutions, both even and odd, can be obtained from the points of intersection of straight lines with a single quadrant of a cosine curve. See Paul H. Pitkanen, "Rectangular Potential Well Problem in Quantum Mechanics," Am. J. Phys. **23**, 111–113 (1955).

4-2   The square well

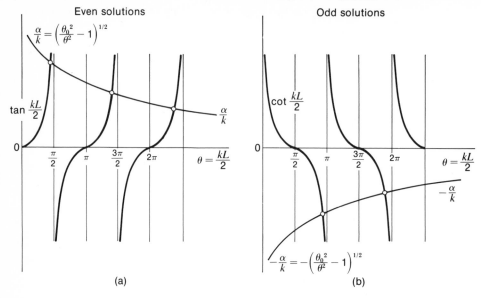

Fig. 4-2 *Graphical solution for permitted energy values
in the finite square well. The circled intersections define
the permitted values of kL/2, from which the permitted
energy values can easily be obtained. (a) The even
solutions are those values of kL/2 for which* tan $(kL/2) = \alpha/k$.
*(b) The off solutions are those values of kL/2
for which* cot $(kL/2) = -\alpha/k$. *(Recall that* $\alpha = \sqrt{2m(V_o - E)}/\hbar$.)

This is shown in Figure 4-3, together with a graph
through a number of cycles of the function tan θ
$(\theta = kL/2 = L\sqrt{2mE}/2\hbar)$ for various values of the parameter $\theta_o (= L\sqrt{2mV_o}/2\hbar)$. The function of Eq. 4-9 falls to zero with a vertical tangent at $\theta = \theta_o$—that is, at the value of θ
corresponding to $E = V_o$. No solutions exist for $\theta > \theta_o$, corresponding to the physical result that no bound state exists for
$E > V_o$. Thus, for example, the middle value of V_o in Figure
4-3 permits three energy states corresponding to even solutions of the Schrödinger equation. (How many *odd* states are
possible in addition?) For a deeper well (larger V_o and θ_o)
more bound states can be obtained. As V_o is made larger and
larger, the permitted states (taking both odd and even together)
occur at values of $kL/2$ that correspond more and more nearly
to the succession of integral multiples of $\pi/2$. Thus, if V_o is
large, we can roughly describe the whole series of solutions

Schrödinger's equation in one dimension

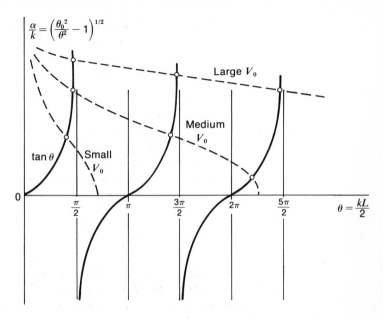

Fig. 4-3 Depen-
dence of permitted
energy values on
well depth V_0. This
graph gives permit-
ted values of
$kL/2 = \theta$ for the
even solutions for
each of three dif-
ferent values of V_0.
(Recall that $\theta_0 = L\sqrt{2mV_0}/2\hbar$.) For
each value of V_0
shown, you should
try to visualize the
graph for the odd
solutions in order to
obtain the number
of those as well.

In the figure: $\left|\dfrac{\alpha}{k} = \left(\dfrac{\theta_0^2}{\theta^2} - 1\right)^{1/2}\right|$, Large V_0, Medium V_0, tan θ, Small V_0, $\theta = \dfrac{kL}{2}$

through the inequality

$$\frac{kL}{2} \lesssim \frac{n\pi}{2}$$

Substituting $k = \sqrt{2mE}/\hbar$, one obtains

$$E_n \lesssim \frac{n^2 h^2}{8mL^2} \tag{4-10}$$

In other words, under these conditions the allowed energy values approach, as we should expect, the quantized energies of
the familiar violin-string model for an infinitely deep square
well. Short of this limiting condition the bound-state energy
values are *less* than those predicted for an infinite well, consistent with the results previously obtained in discussing the qualitative plots (Section 3-11).

One other important result emerges from a study of Figures 4-2 and 4-3. No matter how small the value of V_0 is made,
there is always at least *one* intersection in the graph of
tan $(kL/2)$ against α/k. This means that for any square well,
no matter how weak its attractive power, there is always
one bound state at least, and this lowest state is even. (In

4-2   The square well

three dimensions a shallow potential may or may not have at least one bound state, depending on the form of the potential function.)

Although there are no real one-dimensional square wells in nature, many real systems have potentials that resemble the square well. For this reason we have chosen the simple model and worked it out thoroughly, both to illustrate the method and to give you a feeling for such problems. We have by no means exhausted the treatment of the square well in the above discussion, and other details and methods of approach are suggested in the problems at the end of the chapter.

4-3 THE HARMONIC OSCILLATOR

The simple harmonic oscillator potential, $V(x) = \frac{1}{2}Cx^2$, is one of the most important potential-energy functions in physics, because it describes with high accuracy the basic behavior of many real systems as well as the consequences of small departures x from equilibrium in a huge variety of circumstances.[2] For example, the vibrational states of a diatomic molecule are accurately described using this potential, even though no simple classical picture ("mass on a spring") of the system is valid. We shall often use the abbreviation SHO for "simple harmonic oscillator."

We begin with a preliminary inspection of the SHO potential and the qualitative results that follow from its parabolic shape. For a given choice of energy above the minimum of the potential we automatically divide the x axis into three regions, the points of demarcation being at those values of x ($x = \pm x_o$) at which $E = V$ (see Figure 4-4a). These two points mark the limits of displacement of a classical harmonic oscillator of the energy E. In contrast to the square well, the width of this parabolic well, as measured by the value of $2x_o$, is greater for higher energies.

Inside any well, as we saw in Chapter 3, the wave function curves back toward the x axis as x varies, but in this case not sinusoidally, since the potential varies with position. Outside any well the wave function decreases in magnitude for greater distance from the well, but in this case not exponentially, again because the potential varies with position.

[2]See, for example, the volume *Newtonian Mechanics* in this series, p. 395.

Schrödinger's equation in one dimension

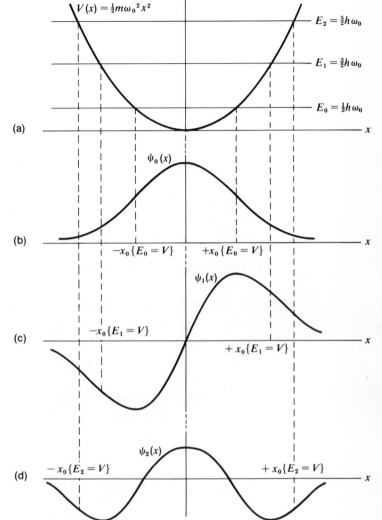

Fig. 4-4 (a) Harmonic-oscillator potential with first three permitted energy values indicated. (b) through (d) Wave functions for $E = E_0$, E_1, and E_2, respectively. (The wave functions are not normalized.) [Note: For the harmonic-oscillator potential, it is conventional to use $n = 0$ as the quantum number for the lowest state.]

You will recall from Chapter 3 other qualitative features of correct wave functions for bound states. Figures 4-4b through 4-4d show harmonic oscillator wave functions (derived below) for the first, second, and third energy levels.

We have in the SHO our first example of a potential described for all values of x by a single algebraic function of x. There is, therefore, only a single differential equation to be solved, and we should not need to look separately at the conditions at the points $x = \pm x_o$ where the E line crosses the $V(x)$

curve. The differential equation, obtained by substituting $V(x) = \frac{1}{2}Cx^2$ in the general one-dimensional Schrödinger equation, is

$$-\frac{\hbar^2}{2m}\frac{d^2\psi}{dx^2} + \frac{1}{2}Cx^2\psi = E\psi \qquad (4\text{-}11)$$

Given this single equation, our task is reduced to finding each value of E for which the associated wave function $\psi(x)$ falls toward zero at large distances from the origin. We can hope to find wave functions ψ, each of which has a single functional form over the entire range of x.

To begin looking for acceptable functions ψ, we use an often-helpful tactic: examine the form of the wave function far from the center of force. What happens to the single analytic function ψ for very large x (very deep into the region that is forbidden for classical particles)? We can obtain a hint by recalling that outside a *square well* the wave function decreases with distance according to the expression $e^{-\alpha x}$, where the constant α is

$$\alpha = \frac{[2m(V_o - E)]^{1/2}}{\hbar} \qquad \text{(square well)}$$

and V_o is the depth of the well. For the harmonic oscillator, V increases as the square of x, so the decay of the wave function with distance is not a pure exponential. A crude way of describing this is to say that for the harmonic oscillator potential, the coefficient α is not constant but increases with x, so that the wave function tends toward zero even more rapidly with distance than in the case of the square well. If we ask for the variation of α with x for very large x (at a given E), the answer is that α becomes linear in x:

$$\alpha = \frac{[2m(\frac{1}{2}Cx^2 - E)]^{1/2}}{\hbar} \to \frac{[mCx^2]^{1/2}}{\hbar}$$

$$\to (\text{constant}) \cdot x \qquad (\text{large } x) \qquad (4\text{-}12)$$

Therefore, in the limit of large x, we expect the harmonic oscillator wave function to approach zero with the exponential square of the distance from the center of attraction:

$$\psi(x) \sim e^{-\alpha x} \sim e^{-(\text{constant}) \cdot x^2} \qquad (\text{large } x) \qquad (4\text{-}13a)$$

Schrödinger's equation in one dimension

This x dependence is encountered repeatedly in physics and is known as the Gaussian function. The standard Gaussian form results if we set the constant in the exponent equal to $1/2a^2$. Then the wave function at large x takes the form

$$\psi(x) \sim e^{-x^2/2a^2} \quad \text{(large } x) \tag{4-13b}$$

where the constant a has the dimension of distance. The argument leading to Eqs. 4-13 is a crude one, and the result must be verified by substitution into the Schrödinger equation. For regions far from the center of attraction, Eq. 4-11 becomes

$$-\frac{\hbar^2}{2m}\frac{d^2\psi}{dx^2} \approx \frac{1}{2}Cx^2\psi \quad \text{(large } x) \tag{4-14}$$

Taking the first and second derivatives of ψ from Eq. 4-13b we have

$$\frac{d\psi}{dx} = -\frac{x}{a^2}e^{-x^2/2a^2}$$

$$\frac{d^2\psi}{dx^2} = -\frac{1}{a^2}e^{-x^2/2a^2} + \frac{x^2}{a^4}e^{-x^2/2a^2} \tag{4-15}$$

$$\rightarrow \frac{x^2}{a^4}e^{-x^2/2a^2} \quad \text{(for large } x)$$

Substituting the last expression into Eq. 4-11 and canceling the common exponential factor gives

$$\frac{\hbar^2}{2m}\frac{x^2}{a^4} = \frac{1}{2}Cx^2$$

This yields an expression for the constant a:

$$a^4 = \frac{\hbar^2}{mC} \tag{4-16}$$

Equations 4-16 and 4-14 tell us that *every* wave function of the simple harmonic oscillator (one for each bound-state energy) is dominated by the same exponential form in the limit of large distance from the center of force. *Near* the center of force the different wave functions must differ from one another or else they would not satisfy the Schrödinger equation for different values of energy. One way to indicate this difference is to multiply the Gaussian ($e^{-x^2/2a^2}$) by a function of x that differs for different energies. We can say more about these multiplicative functions: from our experience with qualitative plots we know that, since the SHO potential is symmetric, the energy eigenfunctions must be alternately odd and even functions of x, with an even function for the lowest energy state. Since the Gaussian is an even function of x, each function that multiplies it must be odd or even itself.

One kind of function that can satisfy all our criteria is a finite polynomial in x. A polynomial that is an *odd* function of x is easily constructed using only odd powers of x, while a polynomial that is an *even* function contains only even powers of x (possibly including the zeroth power of x, which yields a constant term). Every polynomial (with a finite number of terms) is dominated at large x by the exponential Gaussian factor, so that at large x the wave function automatically has the form $e^{-x^2/2a^2}$ if the multiplicative polynomial has a finite number of terms.[3] Since, by definition, a polynomial contains only positive powers of x, the wave function will remain finite near the center of force, as it must to be physically acceptable.

For these reasons we write the SHO wave functions in the form

$$\psi(x) = f_n(x) \cdot e^{-x^2/2a^2} \tag{4-17}$$

where $f_n(x)$ is a polynomial whose highest-order term is proportional to x^n. The single analytic function $\psi(x)$ holds over all values of x, both "inside" and "outside" the well; the polynomial consists of either odd powers of x, or else even powers of x with maybe an additive constant.

---

[3]Consider the case in which the multiplicative polynomial is just x and, for simplicity, set $x/(\sqrt{2}\,a) = z$, so that the wave function is proportional to ze^{-z^2}. How can anyone possibly say that the function ze^{-z^2} "has the form" e^{-z^2} for large z when the factor z is still present no matter how large z gets? In answer, think of the *change* in this product function as z goes from 100 to 101. The factor z increases from 100 to 101, or by just 1 percent. In contrast the factor e^{-z^2} changes from e^{-10^4} to $e^{-(10^4+201)} = e^{-10^4} \cdot e^{-201}$, that is, by the factor e^{-201} which is a very great fractional change. Now think of the change as z goes from 1000 to 1001. This time the factor z increases by only 1/10 of 1 percent, while the factor e^{-z^2} changes from e^{-10^6} to $e^{-10^6} \cdot e^{-2001}$ or by the even greater factor e^{-2001}. For larger and larger z, the change in the product function is dominated more and more by the exponential factor and influenced less and less by the polynomial factor z. Hence we say that, in the limit of large z, the function "has the form" e^{-z^2}. A similar argument can be made for *any* polynomial factor that multiplies e^{-z^2}.

Schrödinger's equation in one dimension

Now one can search systematically for polynomials such that $\psi(x)$ in Eq. 4-17 satisfies the Schrödinger equation for the SHO. This is done using polynomials with undetermined coefficients and letting the substitutions into the Schrödinger equation 4-11 determine the coefficients and the quantized values of energy for each valid wave function. The result is a set of polynomials known as the *Hermite polynomials*, symbolized H_n and named after Charles Hermite, a nineteenth century French mathematician. The complete wave function consists of a Hermite polynomial multiplied by the Gaussian exponential function and by an appropriate normalization constant. Table 4-1 lists several SHO energy eigenfunctions.

Here we shall carry out this procedure of determining coefficients for the wave functions of the two lowest energy states. We shall find only the form of each wave function; the normalization constants are easily calculated (see the exercises) and are included in Table 4-1. The procedure is simplified by substituting the constant $a^4 = \hbar^2/mC$ from Eq. 4-16 into the Schrödinger equation 4-11 to obtain

$$-a^4 \frac{d^2\psi}{dx^2} + x^2\psi = \frac{2E}{C}\,\psi \tag{4-18}$$

TABLE 4-1   Energy Eigenfunctions of the Simple Harmonic Oscillator

Quantum Number n	Energy Eigenvalue E_n	Energy Eigenfunction $\psi_n(x) = \left(\dfrac{1}{n!\,2^n a\,\sqrt{\pi}}\right)^{1/2} H_n\!\left(\dfrac{x}{a}\right) e^{-x^2/2a^2}$
0	$\frac{1}{2}\hbar\omega_0$	$\left(\dfrac{1}{a\,\sqrt{\pi}}\right)^{1/2} e^{-x^2/2a^2}$
1	$\frac{3}{2}\hbar\omega_0$	$\left(\dfrac{1}{2a\,\sqrt{\pi}}\right)^{1/2} 2\!\left(\dfrac{x}{a}\right) e^{-x^2/2a^2}$
2	$\frac{5}{2}\hbar\omega_0$	$\left(\dfrac{1}{8a\,\sqrt{\pi}}\right)^{1/2} \left[2 - 4\!\left(\dfrac{x}{a}\right)^2\right] e^{-x^2/2a^2}$
3	$\frac{7}{2}\hbar\omega_0$	$\left(\dfrac{1}{48a\,\sqrt{\pi}}\right)^{1/2} \left[12\!\left(\dfrac{x}{a}\right) - 8\!\left(\dfrac{x}{a}\right)^3\right] e^{-x^2/2a^2}$
4	$\frac{9}{2}\hbar\omega_0$	$\left(\dfrac{1}{384a\,\sqrt{\pi}}\right)^{1/2} \left[12 - 48\!\left(\dfrac{x}{a}\right)^2 + 16\!\left(\dfrac{x}{a}\right)^4\right] e^{-x^2/2a^2}$

Note: $\omega_0 = (C/m)^{1/2}$; $a = (\hbar/\sqrt{mC})^{1/2} = (\hbar/m\omega_0)^{1/2}$.

The wave function for the lowest energy in a symmetric well must be *even*. The simplest even polynomial is a constant. Therefore try the Gaussian function alone for the lowest energy wave function. For historical reasons, the lowest energy state of the harmonic oscillator is given the quantum number n equal to zero rather than one.[4]

$$\psi_o = e^{-x^2/2a^2} \qquad \text{(unnormalized)} \qquad (4\text{-}19)$$

Although this form of the wave function arose as an approximation at large x for *all* simple harmonic wave functions, we shall now show that it satisfies the Schrödinger equation *exactly* (for all values of x) in one case: that of the ground state. The second derivative of this function is given in Eq. 4-15. Substituting for ψ and $d^2\psi/dx^2$ in the Schrödinger equation 4-18 and canceling the common exponential factor we have

$$a^2 - \cancel{x}^2 + \cancel{x}^2 = \frac{2E_o}{C}$$

and this gives a value for the permitted energy E_o. Making use of the expression for a from Eq. 4-16, we have

$$E_o = \frac{1}{2}\, Ca^2 = \frac{1}{2}\, \hbar\, \sqrt{\frac{C}{m}}$$

Now $\sqrt{C/m}$ is the angular frequency ω_o of the corresponding *classical* harmonic oscillator. (Classically this angular frequency is independent of the amplitude of oscillation and therefore independent of energy.) In quantum mechanics ω_o does not have such an obvious physical interpretation. However, we can use it as a parameter for E_o:

$$E_o = \tfrac{1}{2}\hbar\omega_o \qquad (4\text{-}20)$$

You can easily show (see the exercises) that the distance a which defines the characteristic width of this lowest-energy wave function (and which appears in the exponential factor of *all* SHO eigenfunctions) is equal to the maximum displacement x_o of the classical oscillator of this same energy E_o.

---

[4]With this convention n is the highest power of x in the polynomial H_n and is also the number of nodes in the wave function.

Schrödinger's equation in one dimension

In summary, for the lowest energy state ($n = 0$) of the simple harmonic oscillator, we have

$$\psi_0(x) \sim e^{-x^2/2a^2}$$

and

$$E_0 = \tfrac{1}{2}\hbar \omega_0$$

where

$$\omega_0 = \left(\frac{C}{m}\right)^{1/2}$$

and

$$\frac{1}{a^2} = \frac{\sqrt{mC}}{\hbar} = \frac{m\omega_0}{\hbar}$$

The second energy level ($n = 1$) must have an *odd* wave function with one node. But *every* odd function has a node at $x = 0$. The simplest odd polynomial with only this one node is x itself. With this polynomial factor Eq. 4-18 becomes

$$\psi_1(x) = xe^{-x^2/2a^2} \qquad \text{(unnormalized)} \tag{4-21}$$

Taking derivatives, substituting into the Schrödinger equation 4-18, and canceling the common factor $xe^{-x^2/2a^2}$ yields the equation (as you should verify for yourself)

$$3a^2 - x^2 + x^2 = \frac{2E_1}{C}$$

so that the energy has the value

$$E_1 = \tfrac{3}{2}Ca^2 = \tfrac{3}{2}\hbar\sqrt{C/m} = \tfrac{3}{2}\hbar\omega_0$$

In the exercises you will try the following form for the wave function for the third energy level (denoted by the index $n = 2$)

$$\psi_2(x) = (1 + bx^2)e^{-x^2/2a^2} \qquad \text{(unnormalized)}$$

In doing so you will obtain the energy $E_2 = \frac{5}{2}\hbar\omega_o$, as well as the necessary value of b.

The first five harmonic-oscillator wave functions, properly normalized, are given in Table 4-1. It is striking that although the wave functions $\psi_n(x)$ are complicated, the permitted energies are given by the remarkably simple formula

$$E_n = (n + \tfrac{1}{2})\hbar\,\omega_o \qquad (n = 0, 1, 2, \ldots) \qquad (4\text{-}22)$$

Note that the energy difference between any two successive levels of the harmonic oscillator is just $\hbar\omega_o = h\nu_o$, the classical frequency ν_o of the oscillator multiplied by Planck's constant. This equal spacing of the levels is a unique property of the SHO.

The lowest energy $E_o = \frac{1}{2}\hbar\omega_o$ is the *zero-point energy* of a given harmonic oscillator. No particle in a given parabolic potential can have less energy than this, just as no particle in an infinite square well can have less than the energy $h^2/8mL^2$ of the lowest quantum state permitted by the requirements of fitting the wave function to the boundary conditions. The existence of such zero-point energies is clearly a completely nonclassical phenomenon.

In the foregoing we have not *solved* the Schrödinger equation for the harmonic-oscillator potential. Instead we have followed a common procedure used by scientists who have already solved a lot of problems. In a new situation they make an informed guess at a solution based on this previous experience. Verification then follows by substitution into the Schrödinger equation.

4-4 VIBRATIONAL ENERGIES OF DIATOMIC MOLECULES

We can make a very direct application of the theory of the preceding section to the vibrational states of diatomic molecules.[5] In such molecules there is an equilibrium distance of the order of 2 Å between the nuclei of the constituent atoms, and the restoring forces for small displacements from equilibrium are well described by a harmonic-oscillator potential. In order to confine attention to *internal* motions, we measure the motions of each atom with respect to the center of mass of the

[5]See the volume *Newtonian Mechanics* in this series, pp. 405–411.

Schrödinger's equation in one dimension

molecule. Moreover, in the present treatment we limit attention to molecules that vibrate along the line joining the nuclei (no molecular rotation).

In developing the Schrödinger equation for this system, we follow a procedure similar to that used in developing the single-particle equation. With the restrictions outlined above, the classical expression for the total energy of the diatomic system is

$$E = \tfrac{1}{2}m_1v_1{}^2 + \tfrac{1}{2}m_2v_2{}^2 + V(x)$$

where m_1 and m_2 are the masses of the respective nuclei and $V(x) = \tfrac{1}{2}Cx^2$ is the increase of potential energy for a change x in the separation of the nuclei ($x = 0$ representing the equilibrium separation). With respect to the center of mass of the molecule, the momenta p of the nuclei will always be equal and opposite. Thus the energy equation can be written

$$E = \frac{p^2}{2m_1} + \frac{p^2}{2m_2} + \frac{1}{2}\,Cx^2 \qquad (4\text{-}23a)$$

Here we can introduce the quantity μ, called the *reduced mass* of the system,[6] defined by the equation

$$\frac{1}{\mu} = \frac{1}{m_1} + \frac{1}{m_2}$$

or

$$\mu = \frac{m_1 m_2}{m_1 + m_2} \qquad (4\text{-}24)$$

Using this, Eq. 4-23a becomes

$$E = \frac{p^2}{2\mu} + \frac{1}{2}\,Cx^2 \qquad (4\text{-}23b)$$

This then is formally equivalent to the classical energy equation for a single particle of mass μ in a one-dimensional potential.[7] The wave-mechanical behavior of the system is described

[6]See the volume *Newtonian Mechanics* in this series, p. 339.

[7]For further discussion of such two-body problems in classical terms, see Chapter 10 of *Newtonian Mechanics*.

4-4   Vibrational energies of diatomic molecules

by the corresponding single-particle Schrödinger equation, identical in form to Eq. 4-11:

$$-\frac{\hbar^2}{2\mu}\frac{d^2\psi}{dx^2} + \frac{1}{2}Cx^2\psi = E\psi \qquad (4\text{-}25)$$

Thus we can expect that a diatomic molecule has a set of vibrational energy levels given by Eq. 4-22 with $\omega_o = \sqrt{C/\mu}$.

In practice the analysis is worked the other way; that is, from the observed spectrum of such a molecule, the energy differences between levels are inferred. These differences are used to determine the value of the force constant C, which in turn provides information on the molecular binding. The situation is complicated by the fact that a diatomic molecule can also rotate, and in consequence has quantized rotational states as well as vibrational states. (We shall discuss these rotational states in Chapter 11.) As a result, the system of energy levels is rather elaborate because it involves all the possible combinations of vibrational and rotational states. However, using the theory, it is possible to extract from the spectroscopic data what the system of pure vibrational levels would be like. Examples of this are shown in Figure 4-5 for the molecules CO and Cl_2. Table 4-2 lists, for a number of diatomic molecules, the frequency ν_o of radiation corresponding to a transition between adjacent vibrational levels. It may be seen that the radiation is in the infrared region of the electromagnetic spectrum. From this fact one can obtain a good estimate of the "force constant" C of a molecule, as shown in the table.

The study of molecular spectra also led to an experimental

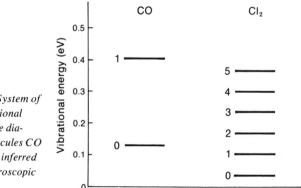

Fig. 4-5 System of pure vibrational levels of the diatomic molecules CO and Cl_2, as inferred from spectroscopic data.

Schrödinger's equation in one dimension

TABLE 4-2 Frequency ν_0 of Radiation Corresponding to a Transition between Adjacent Vibrational Levels of Some Diatomic Molecules, Plus Derived Values of the Force Constant C.

Molecule	Reduced Mass μ of molecule (in amu)[a]	Observed Frequency[b] ν_0 (in units of 10^{13} Hz)	Force Constant C Derived Using $C = \mu\omega_0{}^2 = 4\pi^2\mu\nu_0{}^2$ (in units of 10^5 dynes/cm)
H_2	0.504	12.48	5.2
D_2	1.01	8.971	5.3
HC	0.981	8.657	4.8
CO	6.85	6.429	18.6
NO	7.46	5.628	15.5
Cl_2	17.5	1.671	3.2
O_2	8.00	4.668	11.4
N_2	7.00	6.992	22.5

[a] 1 amu = 1 atomic mass unit = 1.66×10^{-24} g $= \frac{1}{12}$ the mass of C^{12}, the standard of atomic mass.

[b] Based on data presented in Gordon M. Barrow, *Introduction to Molecular Spectroscopy*, McGraw-Hill, New York, 1962.

verification of the existence of the zero-point energy. In fact, the need to describe the permitted vibrational energy levels by *odd* multiples of $\frac{1}{2}\hbar\omega_0$ (which is what Eq. 4-22 requires) rather than simply by integral multiples of $\hbar\omega_0$ (which are the *even* multiples of $\frac{1}{2}\hbar\omega_0$) was actually inferred from molecular spectra before wave mechanics had been invented. This was done by R. S. Mulliken in 1924.[8] Although the analysis was rather complicated in its details, it depended in essence on comparing the vibrational energy levels for two molecules of the same chemical type but involving different isotopes. Specifically, Mulliken made his discovery using two forms of the molecule boron monoxide, containing the abundant isotope ^{16}O in both cases, but differing in the boron isotopes (^{10}B and ^{11}B). (Both these isotopes occur in natural boron, the lighter isotope constituting nearly 20 percent of the total.) Mulliken was able to show that inclusion of the zero-point vibrational energy $\frac{1}{2}\hbar\omega_0$ at the base of the energy structure leads to a good fit with the spectral data, whereas without it there is a small but significant discrepancy. This is a particularly interesting and impressive

[8] R. S. Mulliken, Nature **114**, 350 (1924). Mulliken was awarded a Nobel Prize in 1966 for his many contributions to spectroscopy.

4-4 Vibrational energies of diatomic molecules

piece of analysis: one might be tempted to reason (incorrectly) that, since a particle in the ground state can never radiate away its zero-point energy, therefore the existence of the zero-point energy cannot be verified by experiment. Mulliken's work showed that the zero-point energy has observable consequences.

4-5 COMPUTER SOLUTIONS OF THE SCHRÖDINGER EQUATION

In the preceding sections we have seen how analytical mathematical methods can lead to solutions of the Schrödinger equation for some important forms of potential. However, in contemporary research almost all the manipulation of the Schrödinger equation is done not analytically but rather by computer using numerical methods. There are cogent reasons for this. The solution of differential equations constitutes an entire subdiscipline of mathematics. Unfortunately each different potential substituted into the Schrödinger equation typically yields a different problem, requiring a different method of solution. No single method suffices for all potentials. Moreover, as we have pointed out, for most physically realistic potentials the Schrödinger equation cannot be solved in analytic form at all. This is particularly true of real three-dimensional systems, such as a many-electron atom, for which the potential experienced by each electron is determined by the configuration of all the other electrons in the atom. For these cases—and even for the majority of one-dimensional potentials—it has become customary to resort to numerical approximation methods, employing a computer to do the repetitive calculations involved. In contrast to analytic methods, the computer solution procedures for one-dimensional potentials can be standardized. Therefore some professionals will use computer methods for *all* problems where accurate wave functions or numerical values of energy are required, even for those special cases for which analytic solutions exist.

In this section we discuss one of the simpler methods for finding computer solutions to the Schrödinger equation. (You will obtain a very much more concrete appreciation of the power of this method if you have access to a computer on which you can try it out.) The resulting computer programs are straightforward but not very efficient in the use of computer time. Many techniques exist for increasing the efficiency of

Schrödinger's equation in one dimension

computation in solving differential equations, but we feel that a discussion of these techniques is not appropriate to an introductory treatment that stresses the physical principles and results.[9]

Dimensionless Form of the Schrödinger Equation.

Suppose that we want to use a digital computer to solve Schrödinger's equation for bound states in the special case of one-dimensional potentials. The analytic form of the equation is

$$\frac{d^2\psi}{dx^2} = -\frac{2m}{\hbar^2}[E - V(x)]\psi \tag{4-26}$$

Now, the coefficient $2m/\hbar^2$ has the approximate numerical value 10^{27} in cgs units when m is taken to be the mass of the electron. Numbers of large magnitude are awkward in repetitive calculations. Computation is simpler if the equation to be solved can be reduced to a so-called dimensionless form. Such change of variables can also put into evidence natural scales of distance and energy for the particular system being analyzed. A dimensionless form of the Schrödinger equation is

$$\frac{d^2\psi}{dz^2} = -(\epsilon - W)\psi \tag{4-27}$$

Here the distance z is measured in some "natural" unit of length for the system under investigation. For example, the width L of a finite square well would be a natural unit of length, and z, measured in these units, would range from zero to unity within the well. The wave function ψ is also expressed as a function of the variable z. Similarly the total energy ϵ and potential energy W are measured in some "natural" unit of energy for the system. In the case of the finite square well, a natural unit of energy could be chosen to be the depth V_o of the well.

---

[9]For a fuller introduction, see the following publications: John R. Merrill and G. P. Hughes "Computer Solutions to Some Simple One-Dimensional Schrödinger Equations," Am. J. Phys. **39**, 1391—1395 (1971); John R. Merrill "Introductory Quantum Mechanics with the Computer," Am. J. Phys. **40**, 138—143 (1972); John R. Merrill *Using Computers in Physics*, Houghton-Mifflin, Boston, 1976.

Typically, more than one set of natural units of distance and energy is possible for a given system: for the finite square well one could just as well choose $L/2$ for the unit of distance or choose, as a unit of energy, the quantity $h^2/8mL^2$, which is the energy of the lowest state in the corresponding *infinite* square well. Different choices of natural units may put numerical constants before one or more terms in Eq. 4-27. For all reasonable choices the resulting calculations involve small numbers of a kind easily manipulated by computers. After the dimensionless equation is solved for a particular system, the permitted energies and, if desired, the wave functions can be converted back to conventional units.

In the particular case of the hydrogen atom, the Bohr analysis (and the experimental observations to which it conforms) show a natural unit of length for that system to be the Bohr radius $a_o = 0.53$ Å, and a natural unit of energy to be the "Rydberg energy" $E_R = 13.6$ eV, equal to the ionization energy of the hydrogen atom in its ground state. (The spherically symmetric solutions of the hydrogen atom are considered in Chapter 5, and the corresponding computer solutions in the last section of that chapter.)

A major advantage of the dimensionless form of the Schrödinger equation is that a single solution for one kind of system is trivially easy to extend to all similar systems. For example, once the finite square well is solved in terms of the dimensionless distance parameter z in units of L, the same solution can be extended to wells of *every* width. In this way one set of solutions can be used to model a whole range of systems of the same type, such as a neutron in nuclei of different sizes. Similarly, the spherically symmetric solutions for hydrogen can be extended to all hydrogen-like systems (such as singly ionized helium, doubly ionized lithium, the inner electron shell of a heavy element) by using a_o/Z as the natural distance unit and $Z^2 E_R$ as the natural energy unit, where capital Z is the atomic number of the central charge. The hydrogen atom then takes its place as the specific case $Z = 1$.

Figure 4-6 and its caption present a similar reduction to dimensionless form for the simple harmonic oscillator. In the exercises you will carry out such an analysis for the infinite square well.

Schrödinger's equation in one dimension

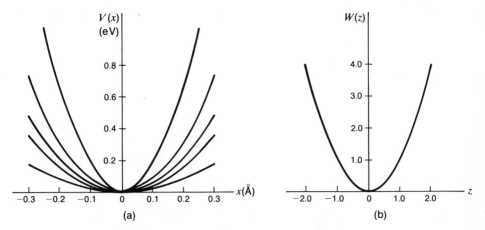

$$(a) \qquad\qquad\qquad\qquad (b)$$

Fig. 4-6   The potential function $W(z)$ for the vibrations
of a harmonic oscillator. The Schrödinger equation
(4-26) can be converted to the dimensionless form
$d^2\psi/dz^2 = -(\epsilon - z^2)\psi$ where $x = (\hbar/\mu\omega_0)^{1/2}z$ and
$E = \frac{1}{2}\hbar\omega_0\epsilon$, with $\omega_0 = (C/\mu)^{1/2}$. (You should check this conversion.) Thus the potential $V(x) = \frac{1}{2}Cx^2$ is replaced by
$W(z) = z^2$. Harmonic-oscillator potentials [such as those
shown in (a)] for various diatomic molecules are summarized in the single dimensionless form shown in (b).

Converting the Differential Equation to a Difference Equation.

The dimensionless form (Eq. 4-27) of the Schrödinger
equation can be solved numerically if it is approximated by a
so-called *difference equation*. One simple approximation
procedure is as follows: divide the distance coordinate z into a
"mesh" of small segments of equal width Δz. We use the integer j as the index number of the end of the jth segment. This
endpoint is called the jth "mesh point." It follows that the jth
mesh point is a distance $z_j = j\,\Delta z$ from the origin. The continuous coordinate z is then approximated by a discrete measure
and the elements of the Schrödinger equation are written:

$$z \rightarrow z_j = j\,\Delta z$$
$$\psi(z) \rightarrow \psi(z_j) \equiv \psi_j \qquad\qquad (4\text{-}28)$$
$$W(z) \rightarrow W(z_j) \equiv W_j$$

Then the first derivative of ψ at the point $(j + \frac{1}{2})\Delta z$ halfway between the jth and $(j + 1)$st mesh points is approximately

$$\frac{(\psi_{j+1} - \psi_j)}{\Delta z}$$

The second derivative of ψ at the point $z_j = j\Delta z$ can be approximated by the differential expression whose limit is the mathematical definition of the second derivative:

$$\frac{d^2\psi}{dz^2}\bigg|_{at\,z_j} = \frac{\dfrac{(\psi_{j+1} - \psi_j)}{\Delta z} - \dfrac{(\psi_j - \psi_{j-1})}{\Delta z}}{2\Delta z}$$

$$= \frac{\psi_{j+1} - 2\psi_j + \psi_{j-1}}{(\Delta z)^2} \qquad (4\text{-}29)$$

With these substitutions, Eq. 4-27 is converted from a differential equation to a difference equation:

$$\psi_{j+1} = + [2 - (\Delta z)^2 (\epsilon - W_j)]\,\psi_j - \psi_{j-1} \qquad (4\text{-}30)$$

Solving the Difference Equation.

Equation 4-30 can be used to generate approximate trial solutions to the Schrödinger equation. Given (1) a trial value of energy ϵ and (2) the potential $W(z_j) = W_j$ at every mesh point, one calculates the value of ψ_{j+1} from values of ψ_j and ψ_{j-1}. (Where do these two "initial" values of the wave function come from? Keep reading.) Next the value of ψ_{j+2} is calculated from ψ_{j+1} and ψ_j, and so forth, until ψ has been calculated for all desired points. This iterative process involves simple operations of addition, subtraction, and multiplication. A digital computer is well-suited to such a process. For any given potential, the smaller the mesh unit Δz, the better is the approximation to the exact solution (and the more computer time is required to generate the solution).

In carrying out the numerical method outlined above, one is faced with several important decisions. The first is what value to choose for the energy constant ϵ. Apparently the choice of the energy constant ϵ in Eq. 4-30 is entirely arbitrary. But we know that bound states exist only for certain discrete energy values. In practice, an "improper" value for ϵ will lead to a function ψ that cannot correspond to a bound state.

Schrödinger's equation in one dimension

Typically the function will diverge indefinitely from the z axis at large distances from the binding potential well (see Figure 4-7). This is physically unacceptable for the reasons we have often discussed. In order to represent physically acceptable solutions, the amplitude ψ must approach the axis for large distances from the well. By varying the value of the energy parameter ϵ one can usually find values for which divergence far from the well does not take place. Approximate trial values for ϵ can usually be estimated from a knowledge of previously solved problems. (Does the potential look most like a square well, a parabola, or a Coulomb potential?) This saves time by avoiding blind trial-and-error values for ϵ. Typically there will be a finite number of bound-state energies ϵ for which ψ does not diverge. These define the discrete energies of the bound system. It is easy to check that no wave functions have been

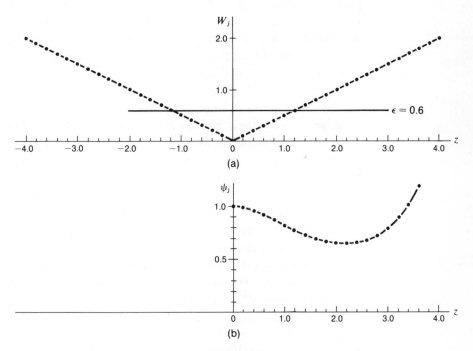

Fig. 4-7   Divergence of the trial function ψ for a trial value of ϵ which is not a permitted energy value.   (a) A set of values W_j based on the dimensionless potential $W(z) = \frac{1}{2}|z|$, with the mesh unit $\Delta z = 0.2$. Also shown is a trial energy value $\epsilon = 0.6$.   (b) Trial function values ψ_j calculated from Eq. 4-30 using W_j, Δz, and ϵ as shown in (a). The starting values used were $\psi_0 = \psi_1 = 1$.

4-5   Computer solutions

missed in a given energy range: our experience with qualitative plots tells us that the wave function for the nth energy state has $n - 1$ nodes or crossing points, so we can make sure that every number of nodes is represented among the solutions generated by this method.

A second important decision required in machine solution is the choice of starting values of ψ. Equation 4-30 is used to calculate ψ_{j+1} from the values of ψ_j and ψ_{j-1}. This calculation must start somewhere; there have to be some beginning values, say ψ_0 and ψ_1, from which the calculations of ψ_2 and values at other mesh points can commence. As with the choice of trial values of energy ϵ, prior experience can lead to a time-saving choice of starting point and starting values. From a knowledge of the potential function one can often determine the value of ψ at some mesh point. For example, if the potential includes one *rigid wall*, integration can begin at the location of the wall, setting $\psi_0 = 0$ there. The value of ψ_1 then determines the initial slope of the function and thus the vertical scale of the ψ versus z diagram printed or plotted by the computer as a solution. Therefore one chooses ψ_1 by trial and error to yield a convenient vertical scale for plotting $\psi(z)$ over the desired range of z. (Formally, ψ_1 may be chosen to normalize the wave function, but normalization is often ignored in computer solutions.) For *symmetric potentials*, as discussed in Chapter 3, the wave function for a permitted energy state is always either an odd or an even function about the point of symmetry. The *even* wave functions have zero slope at the center of symmetry, which is specified by choosing $\psi_1 = \psi_0$ if the coordinate origin is located at the center of symmetry. The value of ψ_0 is chosen to give a convenient vertical scale for the resulting plot (see Figures 4-8a and 4-8b). In the case of an *odd* solution, the amplitude is zero at the center of symmetry but the slope is not zero, which is expressed by initial values $\psi_0 = 0$ and $\psi_1 \neq 0$. Then ψ_1 is chosen to give a convenient vertical scale (Figure 4-8c). For both the odd and even cases, only the portion of ψ on one side or the other of the point of symmetry need be calculated—the other half can be constructed easily from the first half.

In most cases of interest, then, we can deduce from the symmetry or other property of the system appropriate starting

Schrödinger's equation in one dimension

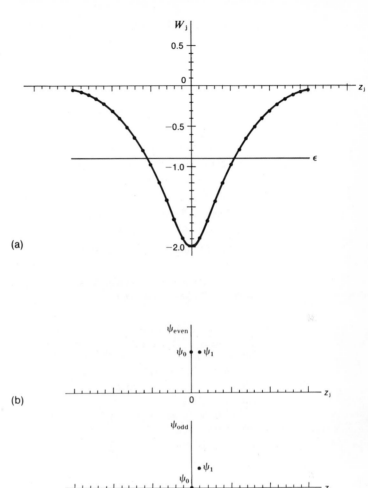

Fig. 4-8   Starting point and starting values for a symmetric potential. (a) Mesh point values W_j of a symmetric potential $W(z)$, with a trial energy value ϵ indicated. All wave functions of unique energy in this symmetric potential are either odd or even about the center of symmetry, which is thus an obvious choice for the starting point. (b) If an even solution is sought, we use $\psi_0 = \psi_1 \neq 0$. The value chosen for ψ_0 sets the vertical scale of the trial function. (c) If an odd solution is sought, we use $\psi_0 = 0$ and $\psi_1 \neq 0$. Here the value chosen for ψ_1 sets the vertical scale.

values for the machine calculation of ψ_j. For one-dimensional potential functions that are not symmetric or lack one rigid wall, the difference equation (Eq. 4-30) can still be used. One can start anywhere, choose adjacent values ψ_0 and ψ_1 for the wave function, and use them to generate a trial function. In this case one must vary at least one of these initial values along with the energy, which complicates the computer programming a bit. Fortunately, most physically interesting potentials fall into one of the simplifying classes discussed above.

A computer program that follows these procedures to integrate Eq. 4-30 for a one-dimensional simple harmonic oscillator potential is shown in Table 4-3. Computer solutions can be presented in tabular or graphical form, whether or not these solutions are expressible as analytic functions. One "run" with the program of Table 4-3 is tabulated in Table 4-4 and graphed as one of the curves in Figure 4-9 (which displays the results of a sequence of such runs).

The potential function in the program of Table 4-3 can be modified to describe the finite square well (see the exercises). The resulting computer solutions avoid the need for graphical determination of the acceptable energy values as described in Section 4-2.

Some excellent short films and film loops are available that show the trial-and-error process by which computers search out physically acceptable wave functions.[10]

TABLE 4-3 Computer Program in "Basic" for Calculating Simple Harmonic Oscillator Wave Functions

```
10 print "input energy, mesh size, zmax, psi (0), psi (1)"
20 input e, s, m, p1, p2
30 let i = 1
40 print "z", "psi (z)"
50 for z = 2*s to m step s
60 let v = z*z
70 let p3 = (2—(e—v)*s*s)*p2—p1
80 let i = i + 1
90 if i < 20 then 120
100 print z, p3
110 let i = 0
120 let pl = p2
130 let p2 = p3
140 next z
150 stop
160 end
```

[*Note*: The fundamental difference equation is found on line 70, where p3 replaces ψ_3, etc. The index i in line 90 is used to print every twentieth calculation.]

[10]See, for example, Alfred M. Bork, *Quantum-Mechanical Harmonic Oscillator* (1966); Craig R. Davis and Harry M. Schey, *Energy Eigenvalues in Quantum Mechanics* (1972). Both films are distributed by Education Development Center, Inc., Newton, Mass.

Schrödinger's equation in one dimension

TABLE 4-4   Sample Run with Program of Table 4-3

operator types: goto 10
computer responds: input energy, mesh size,
                zmax, psi (0), psi (1)
operator types: 5.04, 0.01, 6, 1, 1
computer types the following table:

z	psi (z)
0.2	0.905906
0.4	0.633892
0.6	0.240129
0.8	−0.198402
1.0	−0.603615
1.2	−0.914185
1.4	−1.09657
1.6	−1.14716
1.8	−1.0866
2.0	−0.949611
2.2	−0.774166
2.4	−0.593038
2.6	−0.429217
2.8	−0.295074
3.0	−0.194163
3.2	−0.124262
3.4	−0.080538
3.6	−0.050231
3.8	−0.054933
4.0	−0.073207
4.2	−0.125395
4.4	−0.244889
4.6	−0.514661
4.8	−1.14194
5.0	−2.66009
5.2	−6.49328
5.4	−16.5946
5.6	−44.3751
5.8	−124.101
6.0	−362.822

Stop. (at 150)

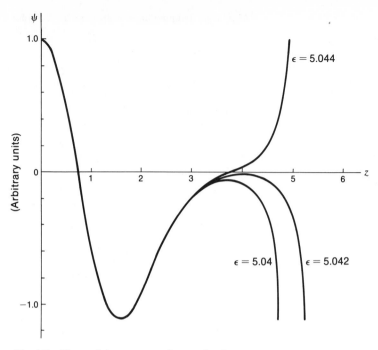

Fig. 4-9 Three trial computer solutions for the wave function for the third energy level of the SHO. The solutions were generated using the computer program of Table 4-3. The plots show only the right half; the left half, corresponding to $z < 0$, is simply the mirror image of this. For the natural energy unit used (see caption of Figure 4-6), the permitted energies are $\epsilon = 1, 3, 5, 7, \ldots$. The functions plotted here bracket the wave function for the energy value near 5. The lower curve ($\epsilon = 5.04$) is plotted directly from Table 4-4; the upper two curves ($\epsilon = 5.042$ and $\epsilon = 5.044$) are plotted from similar computer calculations. Because of the rather large mesh size ($\Delta z = 0.01$, the permitted energy value indicated by these plots ($5.042 < \epsilon < 5.044$) is too high by about 1 percent. In contrast, notice that the sensitivity of the method (the percentage change in energy that produces an obvious divergence of the wave function from one side to the other) is better than 0.1 percent (0.002 in 5.042).

EXERCISES

4-1 *Families of finite square wells.* A particle of mass m has N quantized energy levels in a one-dimensional square well of depth V_o and width L. N is more than 10.

(a) Approximately how many bound states does the particle have in a well of the same depth but width $2L$?

(b) Approximately how many states does the particle have in a well of width L but depth $2V_o$?

(c) Approximately how many states does the particle have in a well of depth $2V_o$ and width $2L$?

Schrödinger's equation in one dimension

(d) How many bound states does the particle have in a well of depth $V_o/4$ and width $2L$? In a well of depth $4V_o$ and width $L/2$?

4-2 *The "half-infinite" well.* A particle is bound in a one-dimensional well with one rigid wall whose potential is shown in the figure.

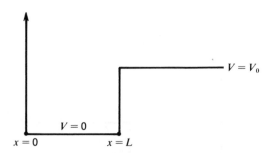

(a) For $E < V_o$, write down and solve the Schrödinger equation for the region inside the well and the region outside the well.

(b) Apply the boundary conditions at $x = 0$ and $x = L$ to obtain an equation that defines the allowed values of the energy E.

(c) Show that for V_o very large, the permitted energies approach those for the infinitely deep square well of width L.

(d) Introduce dimensionless quantities θ and θ_o according to Eqs. 4-5 and 4-6 of the text. By comparing the result with Eq. 4-8, show that the permitted energies of the "half-infinite" well of width L are exactly the energies for the odd solutions of the finite well of the same depth V_o but twice the width, $2L$. Explain how this arises out of the boundary condition at $x = 0$ for the half-infinite case.

(e) For the case $\theta_o = \pi$, or $V_o = h^2/(2mL^2)$, graphically solve the equation for the lowest permitted energy. Show that this is about three-quarters of the energy $h^2/(8mL^2)$ that one would have for the infinitely deep well of width L.

(f) How many bound states are there altogether for the case $\theta_o = \pi$?

(g) Show graphically that there are *no* bound states if $\theta_o < \pi/4$ or $V_o < h^2/(32mL^2)$.

4-3 *An asymmetric well.* [*Caution:* This problem requires either some fancy graphing or some computer programming.]

In Exercise 4-2, you found that the "half-infinite" well has no bound states if $V_o < h^2/(32mL^2)$. But we know that the symmetric finite well in one dimension always has at least one bound state (for *any* $V_o > 0$).

(a) By graphical method or computer, obtain the ground-state energy of the finite symmetrical square well for $V_o = h^2/(64mL^2)$.

(b) If we make the well of part (a) asymmetric by raising the potential on one side (see the figure), we intuitively expect that the ground-state energy will be increased. How asymmetric can we make the well before "pushing out" the bound state? For the potential shown in the figure, what is the maximum value of A for which the well possesses a bound state? [*Hint:* Use the fact that the state just ceases to be bound when $E = V_0$. This means that the wave function inside the well reaches the lower side of the well with (in the limiting case) a horizontal tangent.]

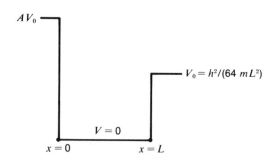

(c) If you have solved part (b) by writing a computer program, can you generalize your program to find the critical value of A as a function of the ratio $V_0/(h^2/32mL^2)$? If so, check that the ratio $\frac{1}{2}$ yields the value of A found in part (b).

4-4   *An alternative graphical solution for the finite square well.* The permitted energies for the finite square well are defined by Eqs. 4-7 and 4-8 in the text. Recall that $\theta = kL/2 = L\sqrt{2mE}/2\hbar$ and $\theta_0 = k_0L/2 = L\sqrt{2mV_0}/2\hbar$. Figures 4-2(a) and (b) indicate the graphical solution of these equations. In this exercise you will examine a slightly different approach which allows you to display all the solutions on a single graph.

(a) Use Eq. 4-7 to show that the even solutions are defined by $\theta \tan \theta = \sqrt{\theta_0{}^2 - \theta^2}$. The right-hand side of this equation is always positive. For what values of θ is the left-hand side positive?

(b) Similarly modify Eq. 4-8 so that the right-hand side is $\sqrt{\theta_0{}^2 - \theta^2}$. What is the left-hand side? For what values of θ is it positive?

(c) The equations obtained in (a) and (b) have the same right-hand side. When $\sqrt{\theta_0{}^2 - \theta^2}$ is graphed as a function of θ, what is the shape of the resulting curve?

(d) If you have answered (a) and (b) correctly, you will notice that you can plot the relevant portions of the curves for both the even and odd solutions without confusion. Make free-hand sketches of these curves on a single graph.

(e) You can now find the permitted energies for any given value

Schrödinger's equation in one dimension

of θ_o by drawing the simple curve of (c) and noting the intersections with the curves sketched in (d). Draw curves for the following values of θ_o: 0.5, 1.5, $3\pi/4$, 4, $7\pi/4$. How many odd and even bound states exist for each of these values of θ_o?

(f) Generalize the results of (e): if $(N_o - 1)\pi/2 < \theta_o < N_o\pi/2$, how many bound states exist?

(g) Use your graph to show that the energy of any given bound state increases with V_o (for given m, L). Show also that $E_n \to n^2 h^2/8mL^2$ as $V_o \to \infty$.

4-5 *The consequences of symmetry.* In Section 3-11 it was argued that the pure energy states of a symmetric one-dimensional potential $[V(x) = V(-x)]$ have either even or odd wave functions:

$$\psi_n(-x) = \pm \, \psi_n(x)$$

(This fact was used in Section 4-2 to simplify the analysis of the finite square well.) Prove this feature of the eigenfunctions of a symmetric potential, using the following outline or some other method. Use the fact (see Exercise 3-9) that $\psi_n(x)$ can be chosen to be real.

(a) Given a symmetric potential energy function $V(x) = V(-x)$, show that if $\psi_n(x)$ is a solution of the Schrödinger equation corresponding to a permitted energy E_n, then $\phi_n(x) = \psi_n(-x)$ is also a solution corresponding to energy E_n.

(b) Present an argument (which may be intuitive rather than rigorous, as long as it is convincing to yourself) to show that within a multiplicative constant only one eigenfunction can be associated with each permitted energy value. [*Hint:* The discussion in Section 4-5 of numerical solutions of the Schrödinger equation may suggest an appropriate argument.]

(c) What restrictions must be satisfied by the function $\phi_n(x)$ found in (a) in order to conform to the condition stated in (b)?

4-6 *The simple harmonic oscillator.*

(a) Show that the width parameter a [Eq. 4-16] used in solving the simple harmonic oscillator is equal to the maximum displacement of a "classical" harmonic oscillator of energy $\frac{1}{2}\hbar\omega_o$.

(b) Verify that the expression $(1 + bx^2)e^{-x^2/2a^2}$ satisfies the Schrödinger equation for the SHO with an energy $E = \frac{3}{2}\hbar\omega_o$, provided a is as given in Eq. 4-16. What is the required value of the parameter b?

4-7 *A "half-harmonic" oscillator.* In Exercise 4-2 you considered the wave functions for a rectangular potential bounded by an infinitely high wall on one side only. Consider now the corresponding problem for a particle confined to the right-hand half of a harmonic-oscillator potential:

$$V(x) = \infty, \qquad x < 0$$

$$V(x) = \tfrac{1}{2}Cx^2, \qquad x \geq 0$$

(a) Compare the allowed wave functions for stationary states of this system with those for a normal harmonic oscillator having the same values of m and C.

(b) What are the allowed quantized energies of the half-oscillator?

(c) Devise a classical mechanical system that would be a macroscopic analog of this quantum-mechanical system.

4-8 *Normalization of harmonic-oscillator wave functions.* Verify that the wave functions for the $n = 0$ and $n = 1$ states of the SHO are correctly normalized as given in Table 4-1. Use the following outline or some other method.

(a) To evaluate $\int_{-\infty}^{\infty} e^{-x^2}\, dx$, note that $\int_{-\infty}^{\infty} dx \int_{-\infty}^{\infty} dy\, e^{-(x^2 + y^2)}$ is equal to $\int_{0}^{2\pi} d\theta \int_{0}^{\infty} dr\, re^{-r^2}$. Why?

(b) Adapt your result from (a) to verify the normalization of the ground state of the SHO.

(c) Examine the normalization condition as applied to the second-state wave function. Can you see a way to apply the results of (a) and (b) to evaluate the integral? [*Hint:* What happens if you regard the SHO parameter a as a variable and differentiate the ground-state normalization integral with respect to it?]

4-9 *Classically forbidden regions for the simple harmonic oscillator.* Using the normalized wave function for the ground state of the one-dimensional simple harmonic oscillator, calculate the probability that an observation of position will detect the particle in the classically forbidden region. Your result will be in the form of a well-known integral that cannot be solved in closed form. Look up the numerical result in a table of the *error function* or *probability integral*.

4-10 *Visual observation of a quantum oscillator?* An experimenter asks for funds from a foundation to observe visually through a microscope the quantum behavior of a small harmonic oscillator. According to his proposal, the oscillator consists of an object 10^{-4} cm in diameter and estimated mass of 10^{-12}g. It vibrates on the end of a thin fiber with a maximum amplitude of 10^{-3} cm and frequency 1000 Hz. You are referee for the proposal.

Schrödinger's equation in one dimension

(a) What is the approximate quantum number for the system in the state described?

(b) What would be its energy in electron-volts if it were in its lowest energy state? Compare with the average thermal energy (1/40 eV) of air molecules at room temperature.

(c) What would be its classical amplitude of vibration if it were in its lowest energy state? Compare this amplitude with the wavelength of visible light (about $5000 \text{ Å} = 5 \times 10^{-5}$cm) by which it is presumably to be observed.

(d) Would you, as referee of this proposal, recommend award of a grant to carry out this research?

4-11 *The asymptotic behavior of harmonic-oscillator wave functions.* In Section 4-3, the behavior of the SHO wave functions for large x is found by simply inserting an x-dependent decay constant $\alpha(x)$ into the expression $e^{-\alpha x}$ which describes the decrease of a bound-state wave function in a region of *constant* $V > E$. [See Eqs. 4-12 and 4-13.] In this exercise, you will use a more refined procedure which gives not only the correct power of x, but also the correct coefficient, in the exponent.

(a) Replace the actual potential energy curve $V(x) = \frac{1}{2} Cx^2$ by a series of "steps": $V(x) = \frac{1}{2} Cx_j^2 \equiv V_j$ for $x_j \leqslant x \leqslant x_{j+1}$, where $x_j = x_o + j\Delta x$ and $\frac{1}{2}Cx_o^2 \gg E$. Denote $\sqrt{2m\,(V_j - E)}/\hbar$ by α_j. Assume that in each interval only the *decaying* exponential contributes to the wave function. What is $\psi(x_{j+1})$ in terms of $\psi(x_j)$? In terms of $\psi(x_0)$? Why is it sufficient to consider large *positive* values of x, as we have implicitly done here?

(b) Since we are interested in $\psi(x)$ for large x, neglect E in the expression for α_j and rewrite α_j in terms of x_j.

(c) Now let $\Delta x \to 0$; what does your expression for $\psi(x_j)$ become?

(d) Use the results of (b) in your equation for $\psi(x)$ to obtain the asymptotic behavior of the wave function.

4-12 *Two-particle systems in one dimension; reduced mass.* In real systems the potential energy is always a measure of the interaction between particles and typically depends on their relative positions. For two particles of masses m_1 and m_2 moving in one dimension, the motion of both particles is taken into account by using the relative coordinate $x = x_2 - x_1$ and the reduced mass $\mu = m_1 m_2/(m_1 + m_2)$ (see Eq. 4-24).

(a) Give an expression for the allowed energies of two particles whose interaction is described by an infinite square well:

$$V(x) = 0, \qquad 0 \leqslant x \leqslant L$$
$$V(x) = \infty, \qquad x < 0 \text{ and } x > L$$

(Classical analog: two gliders on a linear air track connected by a flexible but inextensible string.) By what factor do the permitted energies increase as m_2 is decreased from $m_2 = \infty$ to $m_2 = m_1$ while the value of m_1 remains constant?

(b) Perform the same analysis for a harmonic interaction:

$$V(x) = \tfrac{1}{2} C(x - x_0)^2$$

(Classical analog: two gliders on a linear air track connected by a spring of relaxed length x_0.) [*Hint:* Write the Schrödinger equation using the variable $x' = x - x_0$.]

(c) Why is the variation of energy with reduced mass different for the two potentials used above? Discuss the physical basis for the difference, not only the mathematical results.

4-13 *A simple model of ionic bonding in the sodium chloride molecule.* The molecular bonding in the compound sodium chloride (NaCl) is predominantly ionic, so to a good approximation we can model a sodium chloride molecule as consisting of two units—an Na^+ ion and a Cl^- ion—bound together. Assume that the attractive potential between the ions is electrostatic in nature: $-e^2/r$, where r is the internuclear distance. Assume that the repulsive term in the potential energy function $V(r)$, which rises steeply for contact between the electron structures of the two ions, has the form $+A/r^n$. A and n are constants to be determined.

(a) Using $V(r) = -e^2/r + A/r^n$, find the equilibrium separation r_0 of the nuclei in the NaCl molecule. (That is, find the value of r for which V is a minimum.)

(b) Find the value V_{min} of the potential energy at this equilibrium separation r_0.

(c) Evaluate the second derivative of V at $r = r_0$. This is the "spring constant" in the SHO approximation to the potential energy near the equilibrium separation.

(d) Using the reduced mass (for the isotopes Na^{23} and Cl^{35}), calculate the energy spacing of the vibrational levels in the harmonic approximation. Find also the zero-point energy of the vibrational motion.

(e) The equilibrium separation r_0 of the nuclei in $Na^{23}Cl^{35}$ is 2.51 Å, and the frequency of the radiation for transitions between adjacent vibrational levels is 1.14×10^{13} Hz. Use these experimental values to determine the parameters n and A. [*Hint:* Determine n and then A.]

(f) Using the results from (b), (d), and (e), calculate the energy (in eV) needed to dissociate an NaCl molecule from its ground state into an Na^+ ion and a Cl^- ion.

(g) The ionization energy of the Na atom is 5.1 eV, while the ionization energy of the Cl^- ion is only 3.7 eV. This means that at large separations of the two nuclei of the NaCl molecule, the neutral atoms Na and Cl are a more stable pair than the ions Na^+ and Cl^-. Use this information, together with your result from (f), to find the energy required to dissociate an NaCl molecule into an Na atom and a Cl atom. Compare your result with the experimental value of 4.3 eV. The discrepancy is not very large considering the simple model we have used.

4-14 *Computer normalization of computer-generated wave functions.* Suppose you have just used a computer program to calculate the wave function for a particle in one of the energy states of a given one-dimensional potential. The wave "function" is in the form of a set of values ψ_j, one for each point z_j on the spatial axis. Given this set of values, write a computer program to normalize the wave function. Is this normalization exact or approximate?

4-15 *Machine solution of the finite square-well potential.* Modify the computer program of Table 4-3 to calculate wave functions for the one-dimensional finite square well of width L and height V_o.

Discussion: The first step is to convert the Schrödinger equation to dimensionless units for this potential. The well width L is an obvious natural unit of length. One might think of V_o as a natural unit of energy, but this choice does not eliminate the factors $\hbar$ and particle mass m from the Schrödinger equation. Unfortunately, the lowest energy of the system depends on V_o in a complicated way, or that might be an easy unit of energy. Instead, think of using the lowest energy of the corresponding *infinite* well, namely $h^2/(8mL^2)$. Substituting this natural unit, together with $x = Lz$ for length, still leaves some factors of π in the result. The equation can be still further simplified by modifying the unit of energy to be $\hbar^2/(2mL^2)$. Derive the difference equation (4-30) for your choice of natural units and modify the program of Table 4-3 accordingly. If computer facilities are available, rewrite the program in the local computer language and run it.

. . . I wish to consider . . . the hydrogen atom, and show that the customary quantum conditions can be replaced by another postulate, in which the notion of "whole numbers," merely as such, is not introduced. Rather when integralness does appear, it arises in the same natural way as it does in the case of the node-numbers *of a vibrating string. The new conception is capable of generalization, and strikes, I believe, very deeply at the true nature of the quantum rules.*

ERWIN SCHRÖDINGER, *Annalen der Physik*
(1926)

5

Further applications of Schrödinger's equation

5-1 INTRODUCTION

The physical world is spatially three-dimensional. Therefore we cannot be content with a discussion of quantum physics that deals only with one-dimensional problems. There are certain prime features of physical systems—notably rotation and its associated angular momentum—that simply do not appear in a one-dimensional world. On the other hand, much of what we have learned in the context of one-dimensional problems can be applied directly to three-dimensional situations.

In the present chapter we shall introduce the three-dimensional Schrödinger equation and apply it to the analysis of a particle in a three-dimensional box and to the spherically symmetric wave functions of hydrogen. Although the spherically symmetric functions form only a portion of the total set of wave functions for hydrogen, they do include all the primary energy levels of hydrogen. The full set of wave functions for hydrogen will be developed in Chapters 11 and 12.

5-2 THE THREE-DIMENSIONAL SCHRÖDINGER EQUATION

The development of the Schrödinger equation in one dimension in Chapter 3 was closely related to the concept of the conservation of energy. For one-dimensional motion in the x direction the classical equation

$$\frac{p_x^2}{2m} + V(x) = E$$

corresponds to the one-dimensional Schrödinger equations (Eqs. 3-10 and 3-11)

$$-\frac{\hbar^2}{2m}\frac{\partial^2\Psi}{\partial x^2} + V(x)\,\Psi = E\Psi$$

and

$$-\frac{\hbar^2}{2m}\frac{\partial^2\Psi}{\partial x^2} + V(x)\,\Psi = i\hbar\,\frac{\partial\Psi}{\partial t}$$

Extension of this analogy to three dimensions involves simply introducing a potential energy function $V(x, y, z)$ dependent on x, y and z and considering momentum components along all three Cartesian coordinate directions. The classical conservation of energy equation becomes

$$\frac{p_x^2}{2m} + \frac{p_y^2}{2m} + \frac{p_z^2}{2m} + V(x,\,y,\,z) = E$$

and the corresponding Schrödinger equations are

$$-\frac{\hbar^2}{2m}\left(\frac{\partial^2\Psi}{\partial x^2} + \frac{\partial^2\Psi}{\partial y^2} + \frac{\partial^2\Psi}{\partial z^2}\right) + V(x,\,y,\,z)\,\Psi = E\Psi \tag{5-1a}$$

$$-\frac{\hbar^2}{2m}\left(\frac{\partial^2\Psi}{\partial x^2} + \frac{\partial^2\Psi}{\partial y^2} + \frac{\partial^2\Psi}{\partial z^2}\right) + V(x,\,y,\,z)\,\Psi = i\hbar\,\frac{\partial\Psi}{\partial t} \tag{5-2a}$$

Introducing the Laplacian operator ∇^2, we can write these equations more compactly:

$$-\frac{\hbar^2}{2m}\nabla^2\Psi + V(\mathbf{r})\Psi = E\Psi \tag{5-1b}$$

and

$$-\frac{\hbar^2}{2m}\nabla^2\Psi + V(\mathbf{r})\,\Psi = i\hbar\,\frac{\partial\Psi}{\partial t} \tag{5-2b}$$

Further applications of Schrödinger's equation

In both of these equations the expression $-(\hbar^2/2m)\nabla^2\,\Psi$ takes the place of the classical kinetic energy $p^2/2m$ where $p^2 = p_x{}^2 + p_y{}^2 + p_z{}^2$ is the square of the total linear momentum of the particle.

Equations 5-1 and 5-2 are the three-dimensional time-independent and time-dependent Schrödinger equations for a single particle that we will use from now on in our analysis of quantum systems.

In this chapter we shall be concerned only with stationary states, those having a unique energy E. For such states the wave function Ψ can be written (see Eq. 3-12)

$$\Psi(\mathbf{r}, t) = \psi(\mathbf{r})e^{-i\omega t} \tag{5-3}$$

where $\omega = E/\hbar$. With this substitution, the time terms cancel in Eqs. 5-1a and 5-1b, so they contain the space function $\psi(\mathbf{r})$ alone. Equation 5-1b becomes

$$-\frac{\hbar^2}{2m}\nabla^2\psi + V(\mathbf{r})\,\psi = E\psi \tag{5-4}$$

5-3 EIGENFUNCTIONS AND EIGENVALUES

The time-independent Schrödinger equation can be viewed in a way that gives it a family resemblance to many other differential equations of physics, and which expresses its structure in very general terms. Procedurally the potential energy function $V(\mathbf{r})$ is assumed to be given. Then the problem is to discover those values of E that permit us to have a physically acceptable wave function for the given potential energy. The analogy between Eq. 5-4 and the classical conservation of energy has already been pointed out. The left-hand side can be regarded as a set of instructions for performing certain mathematical operations on the function $\psi(\mathbf{r})$, namely (a) forming second derivatives and multiplying each of them by $-\hbar^2/2m$, and (b) adding the result to the product of potential energy and the wave function. The right-hand side is simply the function ψ multiplied by some permitted value of the energy. We can describe this formulation in somewhat abstract mathematical terms by saying that the left-hand side can be written as a certain *total energy operator* $(E)_{\mathrm{op}}$ operating on the function ψ, where

$$(E)_{\mathrm{op}} = -\frac{\hbar^2}{2m} \nabla^2 + V(\mathbf{r}) \tag{5-5}$$

The time-independent Schrödinger equation then has the following basic structure:

$$(E)_{\mathrm{op}} \psi = E\psi \tag{5-6}$$

The solution of the equation then involves finding particular functions ψ_n each of which, when operated on by the energy operator $(E)_{\mathrm{op}}$, yields some allowed energy value E_n multiplied by ψ_n: thus: $(E)_{\mathrm{op}} \psi_n = E_n \psi_n$. When the equation is satisfied in this way, then (in the jargon of such equations) the function ψ is called an *eigenfunction* of the operator $(E)_{\mathrm{op}}$, and E is called an *eigenvalue* of the energy.[1] Equation 5-6 itself is an example of what is called an *eigenvalue equation:*

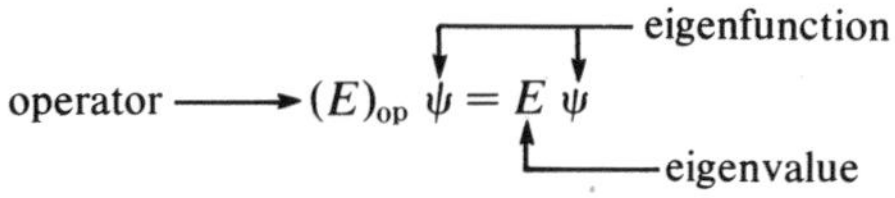

$$\text{operator} \longrightarrow (E)_{\mathrm{op}} \ \psi = E \ \psi$$

with labels: eigenfunction, eigenvalue

As we have implied, eigenvalue equations are of very wide currency in physics. All the classical problems involving calculations of the normal modes of vibrating systems require the solution of eigenvalue equations (see the exercises). Later in our study of quantum physics we shall develop eigenvalue equations for other quantities, notably angular momentum.

5-4 PARTICLE IN A THREE-DIMENSIONAL BOX

An immediate application of Eq. 5-4 can be made to the problem of a particle confined within a rectangular box with perfectly rigid walls. This is the three-dimensional equivalent of the infinite square well. Inside the box we put $V(x, y, z) = 0$, and because of the rigidity of the box we require ψ to fall to zero at every point on the walls. Our Schrödinger equation

---

[1]The words *eigenfunction* and *eigenvalue* are hybrids based on the German *eigen* meaning *own* or *special*. The names *proper function* and *proper value* are sometimes used instead.

Further applications of Schrödinger's equation

thus becomes

$$-\frac{\hbar^2}{2m}\left(\frac{\partial^2\psi}{\partial x^2}+\frac{\partial^2\psi}{\partial y^2}+\frac{\partial^2\psi}{\partial z^2}\right)=E\psi \tag{5-7}$$

If the box (Figure 5-1) has one corner at the origin and has edges of lengths a, b, and c along the x, y, and z directions, respectively, then the boundary conditions are

$$\psi(0, y, z) = \psi(x, 0, z) = \psi(x, y, 0) = 0$$

and

$$\psi(a, y, z) = \psi(x, b, z) = \psi(x, y, c) = 0$$

A professional looking at Eq. 5-7 will immediately recognize a clue to its solution: each of the operators on the left side is a function of only one coordinate, x, y, or z. When this happens the solution of the equation can often be expressed as a *product* of functions, each of which is a function of one coordinate alone. This process is called *separation of variables*, and we will use it over and over again in this text. In the present case we write

$$\psi(x, y, z) = X(x) \cdot Y(y) \cdot Z(z) \tag{5-8}$$

where X, Y, and Z are functions to be determined. Substituting

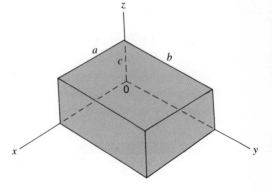

Fig. 5-1 Rectangular box with perfectly rigid walls. The potential energy is zero inside the box and infinite outside.

this into Eq. 5-7 and dividing through by the function ψ itself yields

$$-\frac{1}{X(x)}\frac{\hbar^2}{2m}\frac{d^2X(x)}{dx^2} - \frac{1}{Y(y)}\frac{\hbar^2}{2m}\frac{d^2Y(y)}{dy^2}$$
$$-\frac{1}{Z(z)}\frac{\hbar^2}{2m}\frac{d^2Z(z)}{dz^2} = E \qquad (5\text{-}9)$$

From this point the separation-of-variables argument proceeds as follows. Each of the terms on the left side of Eq. 5-9 is a function of one variable alone: x, y, or z. Each of these coordinates can be varied independently of the others. Yet the left side must add up to the constant E. The only way in which Eq. 5-9 can be valid for *every* choice of x, y, and z is for *each* term on the left to equal a constant. Call these three constants E_x, E_y, and E_z, respectively. Then Eq. 5-8 breaks down into four equations:

$$E_x + E_y + E_z = E \qquad (5\text{-}10)$$

$$-\frac{\hbar^2}{2m}\frac{d^2X(x)}{dx^2} = E_x X(x) \qquad (5\text{-}11a)$$

$$-\frac{\hbar^2}{2m}\frac{d^2Y(y)}{dy^2} = E_y Y(y) \qquad (5\text{-}11b)$$

$$-\frac{\hbar^2}{2m}\frac{d^2Z(z)}{dz^2} = E_z Z(z) \qquad (5\text{-}11c)$$

Each of the last three of these equations is simply the one-dimensional Schrödinger equation for a particle inside an infinite square well. In the three coordinate directions the well widths are a, b, and c, respectively. From Chapter 3, we know the solutions to these equations. Except for normalizing constants, they are

$$X(x) \sim \sin\frac{n_x\pi x}{a} \qquad (n_x = 1, 2, 3, \ldots)$$

$$Y(y) \sim \sin\frac{n_y\pi y}{b} \qquad (n_y = 1, 2, 3, \ldots)$$

$$Z(z) \sim \sin\frac{n_z\pi z}{c} \qquad (n_z = 1, 2, 3, \ldots)$$

and the total wave function is just the product of these solu-

Further applications of Schrödinger's equation

tions, as given in Eq. 5-8. The resulting energies are

$$E_x = \frac{h^2}{8m} \frac{n_x^2}{a^2}$$

$$E_y = \frac{h^2}{8m} \frac{n_y^2}{b^2}$$

$$E_z = \frac{h^2}{8m} \frac{n_z^2}{c^2}$$

so that the total energy given by Eq. 5-10 is

$$E(n_x, n_y, n_z) = \frac{h^2}{8m} \left(\frac{n_x^2}{a^2} + \frac{n_y^2}{b^2} + \frac{n_z^2}{c^2} \right) \tag{5-12}$$

The quantized total energies in the three-dimensional box are simply the sums of those for three independent one-dimensional wells, which we might have guessed to begin with.

Actually, the conduction electrons inside a metal can be quite well considered as particles in a three-dimensional box. Inside the metal a conduction electron is almost free, moving within a region of approximately constant potential. The physical surface of the metal represents a sharp boundary step in potential energy and as a first approximation we may assume that $\psi = 0$ at these boundaries.

With the foregoing introduction to a simple three-dimensional system, we move on to a preliminary analysis of the hydrogen atom, the most accurately and completely described system in all of physics.

5-5 SPHERICALLY SYMMETRIC SOLUTIONS

Some of the most important three-dimensional systems are those involving *central* forces, for which the potential energy of a particle can be expressed by a function $V(r)$ that depends only on the radial *distance* r of the particle from the force center, irrespective of direction. The Coulomb force of a point charge, which dominates atomic physics, is described by such a spherically symmetric potential. The Schrödinger equation for such systems can be written

$$-\frac{\hbar^2}{2m}\nabla^2\psi + V(r)\,\psi = E\psi \qquad (5\text{-}13)$$

The assumption that V is spherically symmetric does not imply that ψ has to be spherically symmetric also. (As a classical analogy recall that orbits of comets can be very elongated ellipses in the spherically symmetric gravitational field of the sun; and the orbit of each planet lies in a single plane rather than as a sphere about the sun.) Thus in general it is necessary to assume that ψ is a function of all three of the spherical polar coordinates (r, θ, ϕ) that are the most convenient coordinates to use if we have $V = V(r)$. The analysis then requires expressing the Laplacian operator ∇^2 in terms of these polar coordinates (see the Appendix to Chapter 11). Solving the resulting Schrödinger equation is rather lengthy and we shall explore this solution in Chapter 11. But there are *some* solutions in central potentials for which ψ is a function of r only; it is to these that we will restrict our attention for the present. For such cases the expression for $\nabla^2\psi$ simplifies greatly. If we limit attention to functions which have explicit dependence on r only, then we can neglect partial derivatives with respect to θ and ϕ in the spherical polar expression for the Laplacian. The result is

$$\nabla^2\psi = \frac{1}{r^2}\frac{d}{dr}\left(r^2\frac{d\psi}{dr}\right) = \frac{1}{r}\frac{d^2}{dr^2}(r\psi) \qquad (5\text{-}14)$$

Equation 5-14 can also be obtained directly from the definition of the Laplacian. (See the exercises.) If we substitute Eq. 5-14 into Eq. 5-13 and then multiply both sides by r we arrive at a very remarkable result:

$$-\frac{\hbar^2}{2m}\frac{d^2}{dr^2}(r\psi) + V(\mathrm{r})\cdot(r\psi) = E(r\psi) \qquad (5\text{-}15)$$

If for the product $r\psi$ we write the single function $u(r)$, we have an equation in the new function $u(r)$ that is *identical* in form with our previous one-dimensional Schrödinger equation:

$$-\frac{\hbar^2}{2m}\frac{d^2u}{dr^2} + V(r)u = Eu \qquad (5\text{-}16)$$

Further applications of Schrödinger's equation

where the desired wave function is

$$\psi = \frac{u}{r} \qquad (5\text{-}17)$$

The coordinate r is by definition positive and is limited to the range $0 \le r \le \infty$. Moreover, we assume as a postulate that ψ is finite everywhere, so that $u(r)$ must go to zero at $r = 0$. Now, using the same techniques we developed for the solution of truly one-dimensional problems, we can obtain certain of the wave functions for any given spherically symmetric potential.

An immediate and almost trivial application of Eq. 5-16 is to a particle confined within a rigid hollow sphere of radius R. (An atomic nucleus looks something like this to a single neutron or proton trapped inside it.) We then have $u = 0$ at $r = 0$ and $r = R$, and $V = 0$ between these limits. Thus the problem is mathematically identical with that of a rigid one-dimensional box, and the energy eigenvalues are $n^2 h^2 / 8mR^2$.

A more complicated problem is that of the lowest states of a single electron in the field of a nuclear charge of magnitude Ze. The potential energy of the electron at any point outside the nucleus is given by

We shall illustrate this by examining the lowest states of a single electron in the field of a nuclear charge of magnitude Ze. The potential energy of the electron at any point outside the nucleus is given by

$$V(r) = -\frac{Ze^2}{r} \text{ (cgs)} \qquad \left[= -\frac{1}{4\pi\epsilon_o} \frac{Ze^2}{r} \text{(SI)} \right] \qquad (5\text{-}18)$$

where the zero of potential is conventionally chosen to be at infinite separation (see Figure 5-2a). This is the potential for hydrogen ($Z = 1$) or for singly ionized helium ($Z = 2$) or for doubly ionized lithium ($Z = 3$) or for any of the other systems (Chapter 1) for which the Bohr theory gives partial insight.

If the nucleus were truly a point charge, Eq. 5-18 would describe the potential all the way down to $r = 0$, and would give a negative infinite value at the origin. This physically unreal condition does not hold in practice because the electric charge of the nucleus is spread in some fashion over the nuclear volume. Let R be the radius of the nucleus. The associated variation of $V(r)$ with r for $r < R$ depends, of course,

5-5   Spherically symmetric solutions

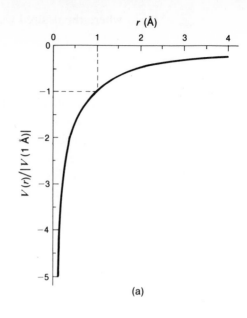

(a)

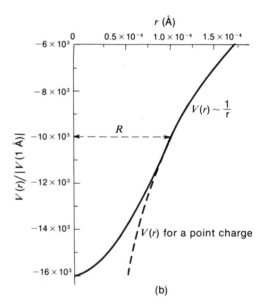

Fig. 5-2 (a) The potential energy for an electron in the Coulomb field of an atomic nucleus. With the distance scale used in this graph, the nucleus would be represented as a dot 0.0001 inch in radius! (b) The potential energy function for values of r comparable to the nuclear radius R. Note the expanded distance scale and the compressed energy scale. The value $R = 10^{-4} Å$ was used as a typical and convenient choice. Uniform charge density inside the nucleus is assumed.

(b)

on the detailed distribution of charge within the nucleus, but the essential conclusion is that the negative infinity is removed as indicated schematically in Figure 5-2b. On the scale of r shown in Figure 5-2a, it would be impossible to show this effect, because the nuclear radius ($\sim 10^{-12}$cm) is approximately

Further applications of Schrödinger's equation

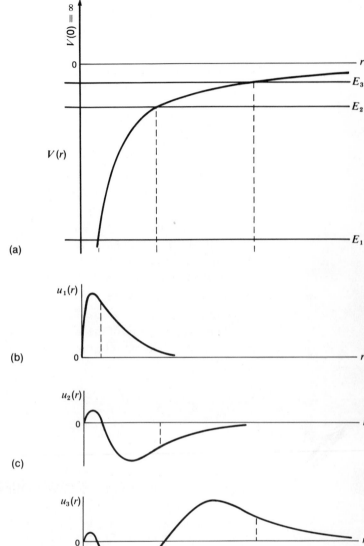

(a)

(b)

(c)

(d)

Fig. 5-3 (a) Po-
tential well for
which $V(r) = -Ze^2/r$
for $r > 0$ and
$V(0) = +\infty$. The latter feature guaran-
tees that $u(0) = 0$.
The first three per-
mitted energy val-
ues for this potential
are indicated.
(b) through (d)
Sketches of the func-
tion $u(r)$ for the first
three states. These
sketches were made
using the skills de-
veloped in Chapter 3.

one ten-thousandth of the typical atomic dimensions
($\sim 10^{-8}$cm). Thus, although the above considerations are necessary for an understanding of the true form of $V(r)$, the fact
that the nucleus is not a point charge makes very little difference in the electron wave function for $r > R$. So we proceed
to analyze the problem assuming that Eq. 5-18 describes the

potential for all r, recognizing simply that there may be problems in which this is not quite good enough.[2]

Having set the stage by the preceding discussion, we can say that the problem to be solved is equivalent to the problem of a particle in a one-dimensional potential well which, for $r > 0$, is described by Eq. 5-18, and which shoots up to a positive infinite value at $r = 0$, as shown in Figure 5-3a. This latter feature guarantees the condition $u(0) = 0$, just as it does in the case of the one-dimensional wave function at the walls of an infinitely deep square well. Now, using the skills developed in Chapter 3, it is possible to hand-draw the functions $u(r)$ for the lowest few energies of an electron in this potential. In Figures 5-3b through 5-3d are shown the forms of $u(r)$ that one can confidently predict on this basis for the first three energy levels. In addition to their approach to zero at small r, notice that the function $u(r)$ for the nth energy level has $n - 1$ nodes, the functions with several nodes have maximum values that are larger at the shallower parts of the well than at the deep parts, and so forth.

The analytic forms of these functions $u(r)$ can be obtained by solving Eq. 5-16 with the potential of Eq. 5-18 explicitly inserted:

$$-\frac{\hbar^2}{2m}\frac{d^2u}{dr^2} - \frac{Ze^2}{r}u = Eu \tag{5-19}$$

where m is the mass of the electron.[3]

As in our discussion of the harmonic oscillator in Chapter 4, we can begin the solution of this equation by ex-

[2]A prime example would be if we wanted to explore theoretically the radioactive process by which, in certain cases, a nucleus can capture one of its innermost orbital electrons ("K-capture") thereby reducing its nuclear charge by one unit. The analysis of this process requires an exact knowledge of the probability that the electron will be within the nuclear volume, and this in turn is sensitive to the precise value of $V(r)$ for $r < R$. Another example is the magnetic interaction ("hyperfine coupling") between the magnetic field of the spinning and orbiting electron and those nuclei which have magnetic moments. This interaction depends sensitively on the electron-nucleus separation.

[3]In Eq. 5-19 we are neglecting a small "correction" due to the finite mass of the nucleus. This correction involves replacing the electron mass m by the reduced mass μ of the electron-nucleus system in Eq. 5-18 and the equations which follow it. The effect of finite nuclear mass was discussed in Chapter 4 and is discussed further in Chapter 12.

Further applications of Schrödinger's equation

amining the form it takes for very large radius. At large r the potential energy term is vanishingly small and the Schrödinger equation becomes

$$-\frac{\hbar^2}{2m}\frac{d^2u}{dr^2} \approx Eu \qquad \text{(large } r\text{)} \tag{5-20}$$

Now in the Coulomb potential, bound states all have negative energy: $E < 0$. Therefore this limiting equation must be satisfied by a wave function identical in form to that for a particle in the classically forbidden region outside a *finite* square well. Therefore the solution must be an exponential that decreases with increasing distance from the center of attraction:

$$u(r) \sim e^{-br} \qquad \text{(large } r\text{)} \tag{5-21}$$

where b is a positive real constant. Substituting this form for $u(r)$ into Eq. 5-20 gives the value of the constant b:

$$b^2 = -\frac{2mE}{\hbar^2} \tag{5-22}$$

(The quantity on the right is positive, since E is negative.)

We can expect the correct solution to the full Eq. 5-19 to have a single analytic form over the entire positive range of r. The simple exponential of Eq. 5-21 cannot be the correct form for all r, since it does not go to zero for small r. As in the case of the harmonic oscillator, we try solutions in which the exponential factor is multiplied by a polynomial, this time in the variable r. Recall that the eigenfunction of lowest energy has no nodes but (in the present case) must vanish at the origin in order to satisfy the boundary condition. Now, the simple polynomial r itself vanishes at the origin and has no nodes for r greater than zero. Therefore we try a solution of the form

$$u_1(r) = re^{-br} = r \exp\left(-\frac{\sqrt{-2m\,E_1}}{\hbar}r\right) \tag{5-23}$$

Substituting this into Eq. 5-19 and dividing out the common factor e^{-br} yields

$$-\frac{\hbar^2}{2m}\left(-2\frac{\sqrt{-2mE_1}}{\hbar} - \frac{2mE_1}{\hbar^2}r\right) - Ze^2 = E_1r$$

The coefficients of r cancel, leaving an expression that fixes the lowest energy eigenvalue E_1:

$$E_1 = \frac{mZ^2e^4}{2\hbar^2} = -\frac{2\pi^2 mZ^2 e^4}{h^2} \tag{5-24}$$

This is precisely the same result as that given by the original Bohr theory for the energy of the lowest state of a "hydrogen-like" atom of central charge Ze with a single orbiting electron. From Eqs. 5-22 and 5-24 you can show also that the value of b for the lowest state is just the reciprocal of a_0/Z, where a_0 is the radius $(\hbar^2/me^2)$ of the first orbit in the old Bohr theory (Chapter 1).

In Figure 5-4 we show the form of the actual wave function $\psi_1 = u_1/r$ in comparison with the function u_1 itself. The value of the wave function is greatest at the position of the nucleus and falls off exponentially with r. In fact, for *all spherically symmetric* states in the Coulomb potential the *wave function* ψ has its greatest value at the origin. (The solutions that are not spherically symmetric—to be considered in Chapter 12—have the value zero at the origin.)

You can easily verify that the function $u_1(r)$, as given by

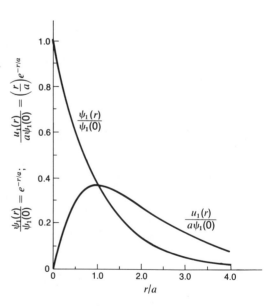

Fig. 5-4 Graphical comparison of $\psi_1(r)$ with $u_1(r)$ for the Coulomb potential. The abscissa is $br = r/a = Zr/a_0$. We use a dimensionless vertical scale for both functions by plotting $\psi_1(r)/\psi_1(0) = e^{-r/a}$ and $u_1(r)/a\psi_1(0) = (r/a)e^{-r/a}$. (The scale factor $\psi_1(0)$ is determined by the normalization condition, which is presented in the following section.)

Further applications of Schrödinger's equation

Eq. 5-23, has its point of inflection at the value of r for which $V(r) = E_1$ as it must mathematically because of the reversal of sign of $V - E$ at this radius. This value of r is closely related (but not equal) to the radius $\hbar^2/mZe^2$ as calculated from the original Bohr theory.

Table 5-1 gives the forms for $u(r)$ for the first few energy states. Note that each is a finite polynomial in r multiplied by an exponential factor. Taking the form $u(r)$ given in Table 5-1 for the second energy state, one finds, by the process of again substituting in Eq. 5-19 and matching coefficients for each power of r, that the second energy value (negative, of course) is one-quarter that of the first state. The third function, $u_3(r)$, is found in the same way to have the still higher energy $E_1/9$, and in general

$$E_n = \frac{E_1}{n^2} \tag{5-25}$$

in complete harmony with the Bohr theory. Remember that because these energies are negative, a smaller value (larger n) means a higher energy.

There are, theoretically, infinitely many bound-state solutions $u_n(r)$ that satisfy Eq. 5-19. Each of them has the asymptotic form e^{-br} at large r, with its value of the "attenuation constant" b determined, according to Eq. 5-22, by the particular

TABLE 5-1   Functions $u = r\psi$ for the First Three Spherically Symmetric States of an Electron in the Coulomb Potential $V(r) = -Ze^2/r$

Quantum number n	Energy Eigenvalue E_n	Normalized function $u_n = r\psi_n$
1	$-\dfrac{mZ^2e^4}{2\hbar^2}$	$\left(\dfrac{1}{\pi}\right)^{1/2}\left(\dfrac{Z}{a_o}\right)^{1/2}\left(\dfrac{Zr}{a_o}\right)e^{-Zr/a_o}$
2	$\dfrac{1}{4}\left(-\dfrac{mZ^2e^4}{2\hbar^2}\right)$	$\left(\dfrac{1}{4}\right)\left(\dfrac{1}{2\pi}\right)^{1/2}\left(\dfrac{Z}{a_o}\right)^{1/2}\left[2\left(\dfrac{Zr}{a_o}\right)-\left(\dfrac{Zr}{a_o}\right)^2\right]e^{-Zr/2a_o}$
3	$\dfrac{1}{9}\left(-\dfrac{mZ^2e^4}{2\hbar^2}\right)$	$\left(\dfrac{1}{81}\right)\left(\dfrac{1}{3\pi}\right)^{1/2}\left(\dfrac{Z}{a_o}\right)^{1/2}\left[27\left(\dfrac{Zr}{a_o}\right)-18\left(\dfrac{Zr}{a_o}\right)^2+2\left(\dfrac{Zr}{a_o}\right)^3\right]e^{-Zr/3a_o}$

[Note: $a_o = \hbar^2/me^2$]

5-5   Spherically symmetric solutions

value of E. The wave functions for the higher energy states fall off more gradually than those of the lower states. All solutions $u(r)$ can be expressed, over the entire range of r, by a polynomial multiplied by an exponential factor.[4] Our knowledge of qualitative plots allows us to specify the form of these polynomials somewhat: (1) There must be no constant terms (or negative powers of r), since otherwise $u(r)$ would not go to zero at small r. (2) The exponential factor e^{-br} has no nodes; therefore, to have $n - 1$ nodes for the nth energy state we need a polynomial of at least n terms, with alternating signs for increasing powers of r. For example, the wave function for the third state has two nodes, requiring a polynomial factor of the form $C_1 r - C_2 r^2 + C_3 r^3$. Substitution of this form for $u_3(r)$ into Eq. 5-19 will then yield values for two of the three constants in this polynomial. The third constant depends on the normalization of the wave function, discussed in the following section.

The above Schrödinger analysis yields quantized values of energy no different from those of the more primitive Bohr theory. However, the Schrödinger theory does not have the limitations of the Bohr theory. It incorporates the well-established wave nature of matter and can be generalized to analyze systems for which the Bohr theory has proven inadequate: multi-electron atoms, chemical properties of atoms, molecular and solid-state physics and chemistry, as well as more subtle effects in hydrogen itself.

5-6 NORMALIZATION AND PROBABILITY DENSITIES

In Chapter 3 we introduced the interpretation of the wave function of a bound state as providing a complete picture of the probability amplitude as a function of position. According to this picture, the probability of finding the particle in a range Δx around x (in a one-dimensional system) is given by $|\psi(x)|^2 \, \Delta x$. This then led to the introduction of normalization constants so that the probability density integrated over all x is unity. If

---

[4]Notice that the radial functions presented here are not alternately odd and even functions of r as we came to expect for symmetric potentials in one dimension. In the present case the position variable r can take on positive values only, so that a symmetry condition for positive and negative r cannot be applied.

Further applications of Schrödinger's equation

$\psi(x)$ is written as the product of a dimensionless function $f(x)$ [for example, $\sin(n\pi x/L)$] and a normalization constant A, the normalization condition becomes

$$\int_{\text{all } x} |\psi(x)|^2 \, dx = A^2 \int_{\text{all } x} |f(x)|^2 \, dx = 1$$

It follows from this that A must be of the dimension (length)$^{-1/2}$.

When we come to three-dimensional systems, the probability density must be integrated over all space, and so we have

$$\int_{\text{all space}} |\psi|^2 dV = A^2 \int_{\text{all space}} |f(x, y, z)|^2 dV = 1 \qquad (5\text{-}26)$$

In this case the normalization constant A must be of the dimension (volume)$^{-1/2}$ or (length)$^{-3/2}$.

When normalizing the spherically symmetric solutions for the hydrogen atom, we can take advantage of the symmetry to write a simplified integral for Eq. 5-26 in the radial coordinate alone, since the wave function $\psi(r) = u(r)/r$ is a function of r alone. As a volume element, choose a spherical shell of radius r and thickness dr centered on the origin. The area of the shell is $4\pi r^2$ and its thickness is dr; so the element of volume is $dV = 4\pi r^2 dr$. The normalization condition is then

$$\int_{\text{all space}} |\psi|^2 dV = \int_0^\infty |\psi(r)|^2 4\pi r^2 dr$$
$$= 4\pi \int_0^\infty |r\psi(r)|^2 dr = 4\pi \int_0^\infty |u(r)|^2 dr = 1 \qquad (5\text{-}27)$$

(Remember that only positive values of the radius r are meaningful.) Note that, in Eq. 5-27, the factor r^2 in the volume element allows us to use the one-dimensional function $u(r)$ when integrating, instead of the actual wave function $\psi(r) = u(r)/r$. Except for the factor 4π, and the restriction of the range of integration to $r \geq 0$, this has the same form as the one-dimensional normalization condition. As an example, we normalize the wave function for the lowest state of the hydrogen-like system. Beginning with

$$\psi_1(r) = Ae^{-br} \qquad (b = Z/a_o)$$
$$u_1(r) = Are^{-br}$$

　　　5-6　Normalization and probability densities

and substituting in Eq. 5-27 we have

$$4\pi A^2 \int_0^\infty r^2 \, e^{-2br} dr = 1$$

Integration by parts leads straight to the result

$$\int_0^\infty r^2 e^{-2br} \, dr = \frac{1}{4b^3}$$

From this we find

$$A = \frac{1}{\sqrt{\pi}} \, b^{3/2} = \frac{1}{\sqrt{\pi}} \left(\frac{Z}{a_o}\right)^{3/2}$$

The complete normalized wave function for this first state of the hydrogen-like system of central charge Ze is thus

$$\psi_1(r) = \frac{1}{\sqrt{\pi}} \left(\frac{Z}{a_o}\right)^{3/2} e^{-Zr/a_o} \tag{5-28}$$

where $a_0 = \hbar^2/me^2 = 0.53\,\text{Å}$ is the familiar Bohr radius, the radius of the smallest orbit in the Bohr model of hydrogen.

When the wave functions of spherically symmetric states in hydrogen-like atoms have been normalized in this way, the value of $4\pi|u(r)|^2\Delta r = 4\pi r^2|\psi|^2\Delta r$ is the probability that the electron will be found between r and $r + \Delta r$. The total area under the graph of $4\pi\,[u(r)]^2$ against r is equal to unity. Figure 5-5 is such a graph for the lowest state of the hydrogen atom. The peak of the curve occurs at a value of r exactly equal to

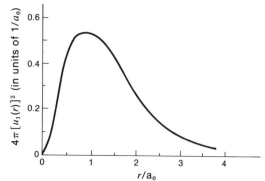

Fig. 5-5 *Normalized radial probability density for the ground state of the hydrogen atom.*

Further applications of Schrödinger's equation

the Bohr radius a_0. Thus, in the wave-mechanical description, the distance a_0, instead of being a clearly defined orbital radius, is merely the distance from the nucleus at which the electron would most often be found, for example by means of x-ray diffraction measurements carried out on a large number of hydrogen atoms all initially in the ground state. With the higher spherically symmetric states, however, this simple connection between Bohr orbit radii and the peaks of the probability distributions does not hold. We shall see later that there are states of large n that are not spherically symmetric for which the radial probability distribution function peaks at the value $n^2 a_0$ —another example of Bohr's correspondence principle that relates the results of classical physics to quantum results involving large quantum numbers.

After this discussion of normalization, let us again emphasize, as we did in Chapter 3, that for many purposes, such as determination of quantized energy values or the examination of *relative* probability densities at different points in space, normalization of the wave function for a single given state can be ignored. On the other hand, for comparison of relative probability densities for *different* states in a given potential, normalization is required. In Chapter 8 we shall deal with superpositions of wave functions in which normalization of the individual functions is essential for a correct analysis of the situation.

5-7 EXPECTATION VALUES

The goal of a quantum-mechanical analysis is often a set of probabilities for alternative outcomes of an experiment. Sometimes one wishes to make predictions in the form of specific numbers based on probability *distributions*, such as the position probability density given by the squared magnitude of the wave function itself. For example, Figure 5-5 shows that the *most probable* value of r for the electron in the ground state of hydrogen (the radius at which we will find it experimentally more often than any other) is $r = a_0$, the Bohr radius.

Another often-used number derivable from the position probability distribution is an *average* value of position that might be measured as follows: Prepare a large number of identical systems, each in a state described by the same wave function ψ. Then locate the particle in each system and average the resulting locations. The predicted value of this average is called the *expectation value of the position.*

In order to see how to calculate an expectation value of position from the simplest case of a one-dimensional probability density $|\psi(x)|^2$, think about taking the average of the following set of numbers: 7, 8, 8, 9, 9, 9, 9, 10, 10, 10, 10. The sum of the numbers is 99 and, since there are 11 items, the average value is $99/11 = 9$. Instead of summing the numbers individually, we could group them according to the number of items with each value:

$$10 \times (4 \text{ items}) = 40$$
$$9 \times (4 \text{ items}) = 36$$
$$8 \times (2 \text{ items}) = 16$$
$$7 \times (1 \text{ item}) \;\; = \;\; 7$$

Total $\overline{99}$ divided by 11 items is *9 average.*

We can generalize this algebraically by letting x be the value of each item, $n(x)$ be the number of items with the value x, and N be the total number of items. Then the average value of x, call it $\bar{x}$, is given by

$$\bar{x} = \frac{1}{N} \sum_x x \; n(x)$$

Still another way to take the average is to divide each separate contribution to the sum by 11 instead of dividing the final total:

$$10 \times \frac{(4 \text{ items})}{(11 \text{ items})} = 3.6$$
$$9 \times \frac{(4 \text{ items})}{(11 \text{ items})} = 3.3$$
$$8 \times \frac{(2 \text{ items})}{(11 \text{ items})} = 1.5$$
$$7 \times \frac{(1 \text{ item})}{(11 \text{ items})} = \underline{0.6}$$

Total   9.0 average

In other words, four-elevenths of the items have the value 10, so their contribution to the average is $10 \times \frac{4}{11} = 3.6$, and

Further applications of Schrödinger's equation

so forth. This way of calculating the average makes use of the *fraction* $n(x)/N$ of all cases that have a particular value x:

$$\bar{x} = \sum_x x\, \frac{n(x)}{N}$$

Now suppose that these were not given results but relative *probabilities*: four-elevenths of the time the number 10 is predicted to appear, and so forth. The fraction $n(x)/N$ becomes the relative probability $p(x)$ for a value to be found. Then the *predicted* average value, called the *expectation value*, is obtained by the same method as that of the last calculation, but we distinguish it from a normal average, $\bar{x}$, based on data *previously* recorded by giving it a different symbol, $\langle x \rangle$:

$$\langle x \rangle = \sum_x x\, p(x) \tag{5-29}$$

We can apply this procedure to calculate the expectation value of position for a particle in a state described by the normalized one-dimensional wave function $\psi(x)$. The probability of finding the particle in the region dx near x is $|\psi(x)|^2 dx$. The average or expectation value of position is then the sum of x times this probability. For a continuous distribution the summation becomes an integral, and the result is

$$\langle x \rangle = \int_{-\infty}^{+\infty} x\, |\psi(x)|^2\, dx \tag{5-30}$$

It is important that the wave function used in this calculation be *normalized* in order that $|\psi(x)|^2 dx$ be the appropriate absolute probability.

As a specific example, consider any one of the wave functions for an infinite square well

$$\psi_n(x) = \sqrt{\frac{2}{L}} \sin\left(\frac{n\pi x}{L}\right), \qquad 0 < x < L$$
$$= 0, \qquad\qquad\qquad \text{elsewhere} \tag{5-31}$$

We can guess that the expectation value of x for these states will be $L/2$, since the probability density $|\psi_n(x)|^2$ is symmetric about $L/2$ for all these states. To check this, substitute Eq. 5-31 into Eq. 5-30 and use the trigonometric identity $2\sin^2 z = 1 - \cos 2z$:

$$\langle x \rangle = \frac{2}{L} \int_0^L x \sin^2 \left(\frac{n\pi x}{L}\right) dx \qquad (5\text{-}32)$$

$$= \frac{1}{L} \int_0^L x \, dx - \frac{1}{L} \int_0^L x \cos \left(\frac{2n\pi x}{L}\right) dx$$

The first integral is just $x^2/2$, which gives the first term the value $L/2$. The second integral is zero, as can be seen by setting $z = 2n\pi x/L$ and using integration by parts:

$$\frac{1}{L} \int_0^L x \cos \left(\frac{2n\pi x}{L}\right) dx = \frac{L}{(2n\pi)^2} \int_0^{2n\pi} z \cos z \, dz$$

$$= \frac{L}{(2n\pi)^2} \Big[z \sin z + \cos z \Big]_0^{2n\pi}$$

$$= 0, \qquad \text{for integral values of } n$$

Therefore the expectation value of x is just $L/2$ from the first term in (5-32), as expected.

$$\langle x \rangle = \frac{L}{2} \qquad (5\text{-}33)$$

As an example with a less obvious answer, we calculate the expectation value of the radial position of the electron in the ground state of hydrogen. From the long tail in the distribution shown in Figure 5-5, we anticipate that the expectation value will be greater than the most probable value a_0. The probability that the electron is in the spherical shell between r and $r + dr$ is, from the analysis of the preceding section, $|\psi(r)|^2 \, 4\pi r^2 \, dr$. Multiplying this probability by r, inserting the expression for $\psi(r)$ from Eq. 5-28 with $Z = 1$, we can integrate from $r = 0$ to $r = \infty$ using a table of definite integrals or repeated integrations by parts:

$$\langle r \rangle = \frac{4\pi}{\pi a_0{}^3} \int_0^\infty r^3 \, e^{-2r/a_0} \, dr = \frac{4}{a_0{}^3} \frac{3!}{(2/a_0)^4} = \frac{3}{2} a_0$$

This value is greater than the most probable value a_0, as anticipated.

Thus far we have limited ourselves to calculating expecta-

Further applications of Schrödinger's equation

tion values of position itself. Actually, a similar procedure can be used to calculate the expectation value of *any function* of position. For example, we can calculate the expectation value of the potential energy of the ground state of the one-dimensional simple harmonic oscillator, because the potential energy is a function of position only: $V = \frac{1}{2}Cx^2$. By examining an argument similar to the one that introduced this section, you can convince yourself that the expectation value of potential energy will have the general form:

$$\langle V \rangle = \sum_{\text{all } x} V(x)\, p(x)$$

The normalized wave function for the ground state of the SHO is (Table 4-1)

$$\psi_o(x) = \left(\frac{1}{a\sqrt{\pi}}\right)^{1/2} e^{-x^2/2a^2}$$

where the characteristic length a is given by Eq. (4-16):

$$\frac{1}{a^4} = \frac{mC}{\hbar^2} \tag{5-35}$$

Express $V(x)$ in terms of this characteristic length:

$$V = \frac{1}{2}Cx^2 = \frac{1}{2}\frac{\hbar^2}{ma^4}x^2$$

Now we are ready to calculate the expectation value of the potential energy for the SHO ground state:

$$\langle V \rangle_o = \int_{-\infty}^{\infty} V(x)|\psi_o(x)|^2 dx$$

$$= \frac{1}{2}\left(\frac{\hbar^2}{ma^4}\right)\left(\frac{1}{a\sqrt{\pi}}\right)\int_{-\infty}^{+\infty} x^2 e^{-x^2/a^2} dx$$

Since the integrand is symmetric in x, we can integrate from zero to infinity and double the result:

$$\langle V \rangle_o = \frac{\hbar^2}{\sqrt{\pi}ma^5}\int_0^{\infty} x^2\, e^{-x^2/a^2}\, dx$$

5-7   Expectation values

From a table of definite intergrals, we find that the integral has the value $a^3 \sqrt{\pi} /4$, so that the expectation value of the potential energy becomes

$$\langle V \rangle_o = \frac{\hbar^2}{\sqrt{\pi}ma^5} \cdot \frac{a^3 \sqrt{\pi}}{4} = \frac{\hbar^2}{4ma^2}$$

This can be written in more recognizable form by again using Eq. 5-35 for a:

$$\langle V \rangle_o = \frac{\hbar^2}{4m} \left(\frac{mC}{\hbar^2}\right)^{1/2} = \frac{\hbar}{4} \left(\frac{C}{m}\right)^{1/2} = \frac{1}{4} \hbar\omega_o \qquad (5\text{-}36)$$

Thus the average or expectation value for the potential energy of the simple harmonic oscillator in its ground state is just half of the total energy of that state. We would be correct in drawing the conclusion that the average *kinetic energy* is also half of the total energy of this state. However, the procedure used above does not allow us to calculate directly the expectation value of kinetic energy, since it is not an explicit function of position. The calculation of expectation values of energy, momentum, and other quantities that are not explicit functions of position is discussed in many texts.[5]

5-8 COMPUTER SOLUTIONS FOR SPHERICALLY SYMMETRIC HYDROGEN WAVE FUNCTIONS

Even though we have developed a method for finding the spherically symmetric solutions to the Schrödinger equation for the hydrogen atom in analytic form (Section 5-5), it is instructive to show how the computer can be programmed to yield the same solutions approximately. The following analysis depends on that of Section 4-5.

A computer solution uses dimensionless variables based on natural units. In the case of the hydrogen atom, both the Schrödinger and the Bohr analysis show a natural unit of length for that system to be the Bohr radius a_0:

$$a_o = \frac{\hbar^2}{me^2} = 0.53 \text{ Å}$$

[5]See, for example, R. M. Eisberg, *Fundamentals of Modern Physics*, John Wiley, New York, 1961, pp. 201ff.

Further applications of Schrödinger's equation

For hydrogen there is also a natural unit of energy, the "Rydberg energy," E_R, equal to the ionization energy of the hydrogen atom in its ground state:

$$E_R = - E_1 = \frac{me^4}{2\hbar^2} = 13.6 \text{ eV}$$

We can measure distance in units of the Bohr radius a_0, and energy in units of the Rydberg energy E_R, by writing

$$r = a_0 z = \frac{\hbar^2}{me^2} z$$

$$E = E_R \epsilon = \frac{me^4}{2\hbar^2} \epsilon \tag{5-37}$$

$$V = E_R W = \frac{me^4}{2\hbar^2} W \tag{5-38}$$

Here z, ϵ, and W are dimensionless quantities which specify distance and energy in natural units. With these substitutions, the Schrödinger equation for the radial function $u(r) = r\psi(r)$ takes the dimensionless form of Eq. 4-27:

$$\frac{m^2 e^4}{\hbar^4} \cdot \frac{d^2 u}{dz^2} = - \frac{2m}{\hbar^2} \cdot \frac{me^4}{2\hbar^2} \cdot (\epsilon - W) u$$

or

$$\frac{d^2 u}{dz^2} = - (\epsilon - W) u \tag{5-39}$$

where the radial function u and the potential W are now functions of the dimensionless distance parameter z. For hydrogen the Coulomb potential $V(r) = -e^2/r$ (cgs units) becomes

$$V(r) = - \frac{e^2}{a_0 z} = - \frac{me^4}{\hbar^2 z} = - \frac{2E_R}{z}$$

so that

$$W(z) = - \frac{2}{z} \tag{5-40}$$

5-8   Computer solutions

TABLE 5-2 Methods of Solving the Time-Independent Schrödinger Equation for Some Important Potentials

Method of Analysis	Qualitative Plots	Analytic Solutions (mostly by trial-and-error, guesswork, etc.)	Numerical (Computer) Solutions
Binding Potential			
Infinitely Deep Square Well	Not needed: Energy eigenfunctions are sinusoidal.	Section 3-5 (sinusoidal wave functions)	Useful as a check on computer programs.
Finite Square Well	Section 3-9 (matching sinusoidal "inside functions" with exponential "outside functions" at the edge of the well).	Section 3-8 (exponential form "outside" well). Section 4-2 (graphical determination of permitted energies and thus of parameters in analytic solutions)	Ideal for computer solution (exercises, Chapter 4)
Harmonic-Oscillator Potential: One Dimension	Used to guide analytic solutions in Section 4-3	Section 4-3	Table 4-3
Harmonic-Oscillator Potential: Two and Three Dimensions	One way to write it: Product of one-dimensional SHO solutions; another way to write it: product of radial and angular functions (can do qualitative plots of radial functions).	Two-dimensional: Section 10-8 Three-dimensional: exercises, Ch. 10	Radial functions easily plotted: Simple extension of Table 4-3; angular functions usually used in analytic form

TABLE 5-2 (Continued)

Method of Analysis Binding Potential	Qualitative Plots	Analytic Solutions (mostly by trial-and-error, guesswork, etc.)	Numerical (Computer) Solutions
Coulomb Potential: One-Electron Spherically Symmetric Solutions	Used to guide analytic solution in Section 5-5.	Section 5-5, Table 5-1	Section 5-8
Coulomb Potential: One-Electron Three-Dimensional Solutions	Can be used on radial part of wave function (exercises, Chapter 11).	Chapter 11 (angular momentum) Chapter 12 (complete solutions)	Radial functions: Simple extension of Table 4-3; angular functions usually used in analytic form (Chapter 11)
Coulomb Potential: Multiple-Electron Three-Dimensional Solutions	Not applicable to three-dimensional systems of this complexity	No analytic solutions. Approximate analysis in Chapter 13 uses hydrogen-like solutions plus Pauli exclusion principle: "only one electron in a given quantum state" (see Sec. 13-5). Computer solutions require professional skills, numerous approximations, and lots of computer time.	

In carrying out the numerical solution of Eq. 5-39 with potential given by Eq. 5-40, one recognizes that the radial function $u(r) = r \psi(r)$ necessarily goes to zero at the origin in order that the wave function itself remain finite. Therefore one takes the initial value of the function $u_o = 0$ at the origin and chooses the value of the function at the first step u_1 to give a convenient vertical scale for the resulting plot.

By changing the potential function on line 60 of the program in Table 4-3 we can solve for the radial function $u(r) = r \psi(r)$ for spherically symmetric solutions in the Coulomb potential (Eq. 5-40). Figure 5-6 shows functions that bracket the radial function for the second energy level in hydrogen.

In the Chapters 3 through 5 we have examined several different binding potentials (square well, harmonic-oscillator potential, Coulomb potential) and at least three methods of solution for the wave function in these potentials—qualitative plots, analytic solutions, numerical (computer) solutions. Later chapters will extend some of these analyses to two and three dimensions and to multiparticle systems and examine other quantized quantities, especially angular momentum. Table 5-2 describes the usefulness of various methods of solution in the principal binding potentials examined in this text.

Fig. 5-6 Trial computer solutions for the hydrogen-atom radial function $u(z)$ which bracket the correct radial function for the second energy level. The solutions were generated using the program of Table 4-3, except that line 60 was altered to read "let $v = -2/z$." For the natural energy unit used (see Eqs. 5-37 ff), the permitted energy values are -1, $-\frac{1}{4}, -\frac{1}{9}, -\frac{1}{16}, \ldots$. The energy values used for these plots bracket a value near $\epsilon = -0.25$. The bracketed value $(-0.246 < \epsilon < -0.242)$ is in error because of the rather large mesh size $\Delta z = 0.01$.

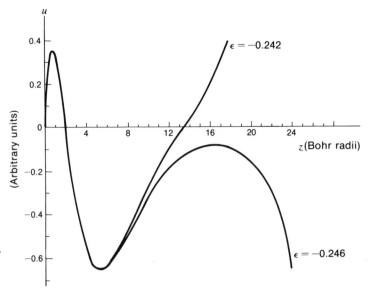

Further applications of Schrödinger's equation

EXERCISES

5-1 *Classical eigenvalue equations I: The simple harmonic oscillator.* The time-independent Schrödinger equation is an eigenvalue equation. As mentioned in Section 5-3, *classical* problems involving calculation of the normal modes of vibrating systems also require the solution of eigenvalue equations. In this and the following two exercises, we deal briefly with several classical vibrating systems and various eigenvalues that arise from their analysis.

(a) The classical one-dimensional simple harmonic oscillator is typified by a restoring force, $-Cx$, proportional to the displacement x from the equilibrium position $x = 0$. The statement of Newton's second law $F = ma$ for this system is

$$m\frac{d^2x}{dt^2} = -Cx$$

which has the form of an eigenvalue equation; x has to be some function of t such that the operator $m\ d^2/dt^2$ applied to it is equal to x multiplied by $-C$. (This is of course such a simple case that we do not usually consider it in such formal terms.) Substitute into the eigenvalue equation the trial solution (eigenfunction) $x = A\ \sin\ \omega_o t$. Derive the single eigenvalue ω_o for the angular frequency of vibration if the equation is to be satisfied. Does this procedure also determine the classical amplitude of vibration A? Substitute a second trial solution of the form $x = Ae^{\alpha t}$ and show that α must be an imaginary number. What is its value? Is the imaginary value positive or negative? What *physical* meaning can such a solution have?

(b) The *damped* simple harmonic oscillator describes a system subject to a force that resists the motion. The simplest case to analyze (which is also realistic for many slow-moving systems) is one in which the damping force is proportional (but opposite in sign) to the velocity: $-bv = -b\ dx/dt$. Then Newton's second law becomes

$$m\frac{d^2\ x}{dt^2} = -Cx - b\frac{dx}{dt}$$

which can be written in the form of an eigenvalue equation:

$$\left(m\frac{d^2}{dt^2} + b\frac{d}{dt}\right)x = -Cx$$

Substitute into this equation the trial solution

$$x = Ae^{-\gamma t/2}\ \sin\ \omega t$$

The resulting equation has both sine and cosine terms with coefficients. It must be valid for all t, and in particular when $\cos \omega t = 0$ and when $\sin \omega t = 0$. Use this requirement to derive eigenvalues for both γ and ω.

5-2 *Classical eigenvalue equations II: Flexural modes of a bar.*[6] The transverse vibration y of a rod under the action of its internal elastic forces is determined by the equation

$$\frac{\partial^4 y}{\partial x^4} = -K \frac{\partial^2 y}{\partial t^2}$$

where x is distance measured along the rod and K is a constant involving the elastic and inertial properties of the rod. Assume that the rod is of length L and is clamped at both ends. Substitute the following trial solution that satisfies these boundary conditions:

$$y = A \sin \frac{n \pi x}{L} \cos \omega t, \qquad n = 1, 2, 3 \ldots$$

Show that the resulting eigenvalues of the frequency ω are proportional to n^2.

5-3 *Classical eigenvalue equations III: Vibration of a rectangular membrane.*[7] An elastic membrane (for example, a soap film) is stretched flat across a rectangular frame. Let the x and y coordinate axes lie along two sides of the rectangle, which have lengths L_x and L_y respectively. The transverse vibration of the membrane $z(x, y, t)$ is governed by the classical wave equation

$$\frac{\partial^2 z}{\partial x^2} + \frac{\partial^2 z}{\partial y^2} = \frac{1}{v^2} \frac{\partial^2 z}{\partial t^2}$$

where v is the speed of propagation of the waves. Zero displacement at the two sides that lie along the coordinate axes is automatically satisfied by the trial solution

$$z = A \sin k_x x \sin k_y y \cos \omega t$$

(a) From the boundary condition that the displacement is zero along *all four* sides of the rectangle, find acceptable values for the wavenumbers k_x and k_y.

[6]See, for example, J. P. den Hartog, *Mechanical Vibrations*, 4th ed., McGraw-Hill, 1956, p. 148.

[7]See, for example, the volume *Vibrations and Waves* in this series, p. 181.

Further applications of Schrödinger's equation

(b) Substitute the trial solution into the wave equation to determine the eigenvalues of the angular frequency ω.

(c) If $L_x = L_y$ (a square boundary), what is the ratio of the second frequency of vibration of the membrane to the lowest frequency? The ratio of the third to the lowest? Compare these with the corresponding ratios of the first three frequencies of vibration of a violin string.

(d) You may have noticed that, in (c), there may be more than one eigenfunction for the same frequency—a condition known as *degeneracy*. Consider also the possibilities for degeneracy if $L_y = 2L_x$.

5-4 *Energy degeneracy in a cubical box.* For one-dimensional binding potentials, a unique energy corresponds to a unique quantum state of the bound particle. In contrast, a particle of unique energy bound in a three-dimensional potential may be in one of several different quantum states. For example, suppose that the three-dimensional box analyzed in Section 5-4 has edges of equal length, $a = b = c$, so that it is a cube. Then the energy states (Eq. 5-12) are given by

$$E = \frac{h^2}{8ma^2} (n_x{}^2 + n_y{}^2 + n_z{}^2), \qquad n_x, n_y, n_z = 1, 2, 3, 4 \dots$$

(a) What is the lowest possible value for the energy E in the cubical box? Show that there is only one quantum state corresponding to this energy (that is, only one choice of the set n_x, n_y, n_z).

(b) What is the second-lowest value of the energy E? Show that this value of energy is shared by three distinct quantum states (that is, three different choices of the set n_x, n_y, n_z). Given the probability function $|\psi|^2$ for each of these states, could they, in fact, be distinguished from one another? The number of distinct states corresponding to the same energy is called the *degree of degeneracy*: this energy level is threefold degenerate.

(c) Let $n^2 = n_x{}^2 + n_y{}^2 + n_z{}^2$ be an integer proportional to a given permitted energy. List the degeneracies for energies corresponding to $n^2 = 3, 6, 9, 11,$ and 12. Can you find a value of n^2 for which the energy level is sixfold degenerate? (In Chapter 10 we take up the question of degeneracy for particles in central potentials.)

5-5 *The Laplacian of a spherically symmetric function.* The Laplacian of a function $f(r)$ is defined as the divergence of the gradient of f and is written $\nabla^2 f$. In the text (Eq. 5-14) it is stated that for a spherically symmetric function $f(\mathbf{r}) = f(r)$, the Laplacian is given by

$$\nabla^2 f = \frac{1}{r} \frac{d^2}{dr^2} (rf)$$

Verify this using the following outline or some other method.

(a) Show that grad $f(r) = \hat{r}(df/dr)$, where $\hat{r}$ is a unit vector in the radial direction. [Hint: Use the fact that the direction of the gradient of any function is the direction for which the directional derivative is the greatest, and that the magnitude of the gradient is just equal to the maximum value of the directional derivative.]

(b) The divergence of any vector function w is defined in terms of a surface integral by

$$\text{div } \mathbf{w} = \lim_{\Delta V \to 0} \frac{1}{\Delta V} \iint_s \mathbf{w} \cdot d\mathbf{s}$$

where ΔV is any closed volume which contains the point at which the divergence of w is being evaluated. Apply this definition (with an appropriately chosen specific shape for ΔV) to the vector function grad f. Show that

$$\text{div (grad } f) = \frac{d^2 f}{dr^2} + \frac{2}{r}\frac{df}{dr}$$

(c) Verify that

$$\frac{d^2 f}{dr^2} + \frac{2}{r}\frac{df}{dr}$$

is identical to

$$\frac{1}{r}\frac{d^2}{dr^2}(rf)$$

5-6 *Spherically symmetric states in the hydrogen atom.* Equation 5-19 (Section 5-5) gives the Schrödinger equation for the spherically symmetric functions $u = r\psi$ for a hydrogen-like atom.

(a) In this equation, substitute an assumed solution of the form

$$u(r) = (Ar + Br^2)e^{-br}$$

and hence find the values of b and the ratio B/A for which this form of solution satisfies the equation. Verify that it corresponds to the second energy level, with $E = Z^2/4$ times the ground-state energy of hydrogen, and with $B/A = Z/2a_o$, where a_o is the Bohr radius for hydrogen. What is the value of the coefficient b in terms of a_o?

(b) If you are feeling ambitious, try the corresponding analysis for the third energy level, assuming

$$u(r) = (Ar + Br^2 + Cr^3)e^{-br}$$

Further applications of Schrödinger's equation

[The use of the symbols A, B, and b here carries no implication that their values are the same as you found in (a).]

5-7 *The spherical step-well.* In Section 4-2 we noted that the symmetrical one-dimensional finite square well has at least one bound state for any well depth $V_o > 0$, no matter how small. Moreover, in Section 5-5 we saw that the spherically symmetric wave functions for a particle in a three-dimensional central potential can be written so as to satisfy a one-dimensional Schrödinger equation. It would be natural (but mistaken) to conclude that a finite spherical step-well, defined by

$$V(r) = 0, \qquad r < R$$
$$V(r) = V_o, \qquad r \geq R$$

has at least one bound state for $V_o > 0$. In essence, the three-dimensional well has a minimum depth for a bound state because the reduced wave function $u(r) = r\,\psi(r)$ must be zero at $r = 0$.

(a) Show that the spherical step-well has no spherically symmetric bound states if $V_o < h^2/(32mR^2)$. (Actually, it has no bound states at all in this case.) You may be surprised to find that Exercise 4-4 is very germane to the present problem.

(b) Try to justify the restriction on V_o found in (a) by sketching $u(r) = r\,\psi(r)$ for various trial values of energy. With careful consideration of the conditions, the restriction *can* be obtained in this manner. Recall that the lowest energy in an *infinite* one-dimensional square well, $h^2/(8mL^2)$, corresponds to the condition that the width L is equal to half the de Broglie wavelength of the confined particle.

(c) What are the energy levels of the *infinite* spherical step-well of radius R for spherically symmetric wave functions?

5-8 *Nuclear binding in the deuteron.* Nuclear forces are exceedingly strong but of very short range. A proton and a neutron combine to form a deuteron that has only a ground state; no bound excited states. The binding energy of the ground state (the *additional* energy required to separate the particles) is 2.23 MeV. Model the interaction potential of the proton and neutron as a spherical step-well containing a single particle: $V = 0$ for $r < R$ and $V = V_o$ for $r \geq R$. Assume that the radius of the step is exactly $R = 2 \times 10^{-13}$ cm and that the wave function is spherically symmetric. Calculate the height V_o of the potential step.

5-9 *Inadequacies in the Bohr theory of hydrogen.* The observed energy levels of hydrogen are surprisingly well accounted for by both the semiclassical Bohr theory and the quantum theory. Moreover (Figure 5-5) the most probable radius for finding an electron in the ground state according to the quantum theory is just the radius a_o of the lowest Bohr orbit. Despite this similarity in some results, the Bohr theory and the quantum theory are utterly different in interpretation and lead to different predictions for the results of many experiments. Here are a few examples:

(a) Use Eq. 5-28 with $Z = 1$ to determine the probability that experiment will locate the electron *within* the nucleus. Take the nuclear radius to be $R = 10^{-13}$ cm and note that this radius is so small compared with a_o that $\psi(r)$ anywhere within the nucleus can be approximated by $\psi(0)$. The probability density of electrons in the nucleus can have observable effects, two of which are described in the second footnote in this chapter (p. 204).

(b) Calculate the probability that the electron in the ground state of hydrogen is in a classically forbidden region [where $E < V(r)$]. Such regions form an essential part of the wave-mechanical "electron cloud," and have an observable influence on the x-ray diffraction patterns of atoms.

(c) What is the expectation value of the distance of the electron from the nucleus in the second energy state of hydrogen? What is the *most probable* distance for this state? (Refer to Table 5-1 and use a table of integrals if necessary.)

(d) We shall see in a later chapter that the ground state of hydrogen has *zero* orbital angular momentum, while the Bohr theory assumes that in the smallest orbit the electron has orbital angular momentum $\hbar$. What is the shape of the *classical* orbit of zero angular momentum?

5-10 *Normalizing a hydrogenic wave function.* The result of Exercise 5-6a should have shown you that the spherically symmetric wave function for the second energy level of a hydrogen-like atom has the form

$$\psi(r) = A \left(1 - \frac{Z}{2a_o} r\right) e^{-Zr/2a_o}$$

Find by direct integration the value of the normalization factor A, and check with the value shown in Table 5-1.

5-11 *The electric field inside the hydrogen atom.* Since the electron has charge $-e$, an electron probability density $\psi^*\psi$ corresponds to a *charge* density $-e\psi^*\psi$.

(a) For the ground state of the hydrogen atom ($Z = 1$), find the electric charge (due to the electron) within a sphere of radius r centered on the proton. Now add the proton's charge $+e$ to obtain the total expected electric charge as a function of r. (Assume that the proton is a *point* charge at $r = 0$.) Sketch your result.

Further applications of Schrödinger's equation

(b) Derive the electric field as a function of r for the ground state of the hydrogen atom. Sketch your result. Does your expression take on the values you expect (i) for very large values of r and (ii) for very small values of r?

(c) In the light of the statistical, probabilistic interpretation of wave functions, what physical significance do you attach to the field as calculated in (b)? Can you suggest specific circumstances in which you would expect the results of (b) to be (i) applicable, (ii) *inapplicable*?

5-12   *Energy effect of the finite size of the nucleus of hydrogen.* The wave function for the ground state of hydrogen has been found under the assumption that the nucleus is a point charge (of zero radius). In fact, the nucleus (proton) has a radius of approximately $r_0 = 10^{-13}$ cm. Estimate the difference this makes in the ground-state energy of hydrogen using the following outline or some other method.

(a) Make the following assumptions:

The ground-state wave function is unaffected by the size of the nucleus.

The value of the wave function inside the nucleus is effectively equal to its value at $r = 0$. (This is because r_0/a_0 is very small, of the order of magnitude 10^{-5}.)

The effective charge density of the electron is given by $-e|\psi|^2$ where ψ is the *normalized* ground-state wave function.

The potential energy due to the charge contained in any volume can be found by multiplying the charge density at each point by the electrostatic potential at that point and integrating over the volume.

To make the computation easier, assume that the proton is a uniformly charged spherical *shell* of radius r_0. Then the electric potential inside the shell is given by $-e/r_0$ which is a constant.

(b) Calculate the potential energy contribution for the volume inside the nucleus for the constant potential given above and for the same volume in the case of the point charge model (electric potential $-e/r$).

(c) Subtract the expressions for the two energies calculated in part (b) to obtain an estimate of the difference in energy due to the finite size of the nucleus. By rearranging this equation, you can express the result as a fraction of the quantity $-e^2/(2a_0)$, which is the energy of the ground state: -13.6 eV. Show that, as far as the energy of the ground state is concerned, the finite size of the nucleus has negligible effect.

(d) Is the change in energy due to the finite size of the nucleus an increase or a decrease? (A simple physical argument should answer this question for you, without any need to go back over the details of your calculation.) If we had assumed the proton to have uniform charge density throughout its volume (physically simpler but computationally more awkward than the spherical shell distribution), would the change in energy have been greater or smaller than that calculated in (c)?

[*Note*: Congratulations: You have just done a problem using what is called perturbation theory, a general method covered in more advanced texts.]

5-13 *Expectation values of various quantities.*

(a) Evaluate the expectation value of position $\langle x \rangle$ for a particle in the ground state of the one-dimensional simple harmonic oscillator.

(b) Evaluate the expectation value of position $\langle x \rangle$ for a particle in the first excited state of the one-dimensional simple harmonic oscillator.

(c) Generalize about the expectation value of x for a particle in any stationary state for a symmetric potential in one dimension.

(d) Evaluate $\langle x^2 \rangle$ for the ground state of the SHO. What is the expectation value of the potential energy of the particle? Compare your result with the energy eigenvalue $E_0 = \frac{1}{2} \hbar \omega_0$.

(e) Generalize the definition of expectation value to three-dimensional systems. What is the expectation value of the potential energy of the electron for the ground state of a hydrogen-like system with $V(r) = Ze^2/r$? Compare your result with the energy eigenvalue $E_1 = -mZ^2 e^4/2\hbar^2$.

5-14 *Computer solutions for spherically symmetric potentials.*

(a) Following the argument of Section 5-8, write down a computer program to find the spherically symmetric wave functions and the corresponding energies for the spherical step potential $V = 0$ for $r < R$ and $V = V_0$ for $r \geq R$. In reducing the Schrödinger equation to dimensionless form, the discussion about the corresponding program for the finite well in Exercise 4-15 may be useful.

(b) The *Yukawa potential* is used in the analysis of nuclear forces. For small radii it has the form of the Coulomb potential, but it drops off exponentially for large radii, corresponding to the short range of nuclear forces. The Yukawa potential has the form

$$V(r) = \frac{-V_0 \, e^{-r/R}}{(r/R)}$$

where V_0 and R are constant parameters of the potential. Write a computer program to find spherically symmetric solutions and corre-

Further applications of Schrödinger's equation

sponding energies in the Yukawa potential. The natural unit of distance is clearly R. The simplest equation results if the energy is expressed in units of $\hbar^2/(2mR^2)$.

(c) If you have access to a computer facility, run your programs for the spherical step potential and the Yukawa potential.

A philosopher once said, "It is necessary for the very existence of science that the same conditions always produce the same results." Well, they don't!

RICHARD FEYNMAN, *The Character of Physical Law* (1965)

6

Photons and
quantum states[1]

6-1 INTRODUCTION

The purpose of this chapter is to refine the concept of a
quantum state. Many systems require quantum mechanics for
their description. Many different kinds of observable physical
quantities provide the basis for the designation of the quantum
states of these systems. For atoms we have already seen that
the most obvious label for states is that of energy. This is entirely appropriate, since the discovery of discrete energy
states was the starting point of quantum physics. But we can
also talk about states of linear momentum, states of angular
momentum, states of position, and so forth. In a world
designed specifically to make easy the learning of quantum
physics, the simplest example would be states of position of a
free particle or of a particle confined in an atom. But in the real
world states of position make a very complicated example,
since, as we have seen, a particle has a probability of being
found in a continuous range of positions, even if it is in a bound
state of uniquely defined energy. There is not, so far as we
know, any discreteness of allowed positions analogous to the

[1]Chapters 6 and 7 draw on previous drafts of this material composed in
collaboration with Arthur K. Kerman and Leo Sartori. These chapters are
designed to rehearse and deepen the central ideas of quantum physics. The
reader who wishes to concentrate on the practical applications of Schrödinger's equation may prefer to proceed directly to Chapter 8 and refer back to
Chapters 6 and 7 as necessary for clarification of concepts and notation.

discreteness of allowed energies in the bound states of, say, an electron in a hydrogen atom.

In order to develop in precise terms the language of quantum states we turn to an example that is very much simpler than that of position states. This example is the *polarization states of photons*. Photons have the convenient property that a complete description of any polarization state whatsoever can be made in terms of just two basic states. This allows us to develop the essential formal ideas of quantum physics with an absolute minimum of complexity. The following discussion of photon polarization states provides a kind of scaffolding for the assembly of the central logical structure of quantum physics. Once the main structure is assembled, it can be used to interpret the behavior of electrons, atoms, and other systems. Then the photon polarization scaffolding can be put in storage for use only when the general theory requires a simple example to aid interpretation.

For several reasons photon polarization states provide an attractive example. Photons in the visible part of the spectrum can be studied especially easily. They are easy to produce, easy to experiment with (with simple mirrors, lenses, polarizers, and so on), and easy to detect. The human eye is a photon detector with an enormous range of adaptation. On a sunny day, billions of photons may strike the eye each second without causing either "saturation" or discomfort, yet on a dark night a person may be able to see a star so faint that only a few hundred photons enter the eye each second. In carefully arranged experiments a human eye can detect a flash consisting of as few as 5 photons. The photomultiplier tube is a photon detector with smaller range of adaptation than the human eye but with more dependable sensitivity at low levels of illumination. A modern photomultiplier tube can detect photons of visible light with an efficiency approaching 50 percent: a count will recorded on the average for every 2 photons that fall on its photosensitive surface.

Photons do have one disadvantage as a vehicle for our preview of quantum mechanics: they are readily created and destroyed. A process may begin with one photon present and end with two, or with none. A satisfactory theory of photons requires *quantum electrodynamics*, a branch of advanced quantum theory.

Photons and quantum states

Fortunately, all that we need for our present purpose is a way of *describing* the quantum states of photons. The experimental devices described in these chapters (polarizers, diffraction gratings, calcite crystals, etc.) will be characterized by their net effect on these photon states, with no present concern for how these effects come about. By proceeding in this way we can use photons to present quantum theory in a simple way, while avoiding the complicated details of the interactions between photons and matter.

A photon quantum state can be uniquely defined by specifying three things: (1) energy, (2) direction of motion, and (3) polarization. Our main concern will be a careful study of item (3) of this statement. For simplicity we shall not mention further the energy or direction of motion of photons. Assume in what follows that mono-energetic photons $E = h\nu$ in the visible frequency range are under discussion, and that the direction of motion can be precisely enough specified to permit us to use distinct beams. Our principal preoccupation will be polarization; the phrase "state of polarization" will be understood to include the implication of a specified energy and direction of motion, so that the photon quantum state is completely specified.

6-2 STATES OF LINEAR POLARIZATION

Classical Description of Linear Polarization

The simplest photon polarization state to produce and describe is a state of "linear polarization." In terms of the classical wave description, a beam of light is linearly polarized if its electric vector lies in a single plane that includes the beam (see Figure 6-1a). The *polarization axis* is defined as the line in this plane along which the electric vector lies at any point. Because electromagnetic waves are transverse, the polarization axis is always perpendicular to the direction of propagation. A linearly polarized beam can be produced by passing an arbitrary beam of light through a sheet linear polarizer (often called "Polaroid" after one company that manufactures it). A piece of such sheet linear polarizer has a *transmission axis* (a direction in the plane of the sheet) that defines the direction of the polarization axis for the light that passes through the sheet

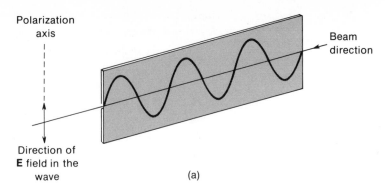

Polarization axis

Beam direction

Direction of **E** field in the wave

(a)

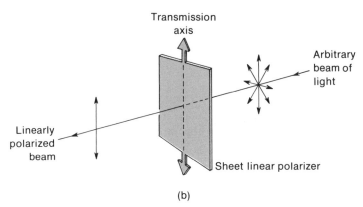

Transmission axis

Arbitrary beam of light

Linearly polarized beam

Sheet linear polarizer

(b)

*Fig. 6-1 (a) A beam of light is linearly polarized if its electric vector lies in a single plane that includes the beam.
(b) Linearly polarized light can be produced by passing any beam of light through a sheet linear polarizer ("Polaroid").*

(Figure 6-1b). A more sophisticated device for producing linearly polarized light is a piece of calcite crystal, suitably cut and polished. If a beam of light is passed through such a crystal, it will typically produce two parallel emergent beams, both linearly polarized but with their polarization axes at right angles to one another (see Figures 6-2a and 6-2b). If one of these two beams is blocked (Figure 6-2c), the calcite crystal acts as a linear polarizer, but if both are allowed to proceed, the crystal acts as what we shall call a *linear polarization analyzer*, because it divides any incident beam into separate linearly polarized parts. With a high-quality calcite crystal, well polished and coated to suppress reflection, nearly all of any incident beam will be transmitted, so that the sum of the intensities of the two separated emergent beams is effectively 100 percent of the intensity of the incident beam.

A knowledge of these few properties and definitions will be assumed in what follows. (A fuller account of polarization and polarizing devices is given in the Appendix to this

Photons and quantum states

(a)

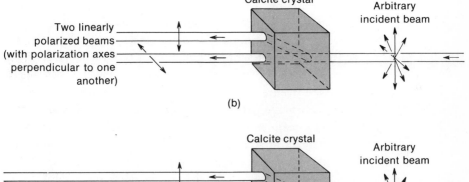

(b)

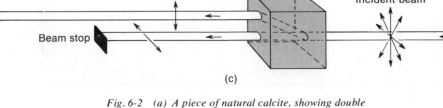

(c)

Fig. 6-2   (a) A piece of natural calcite, showing double
refraction.   (b) Schematic diagram of a calcite linear
polarization analyzer.   (c) A linear polarizer can be
constructed by blocking one of the two output beams of
the calcite analyzer.

chapter.) Our concern now is to describe and analyze the
production and properties of linearly polarized light in terms of
the states of individual photons.

Determination of a State

It is easy enough to determine whether or not the photons
in a beam are in a state of linear polarization. The simplest way
is to insert a piece of sheet linear polarizer in the beam and rotate the polarizer about the beam as axis (Figure 6-3).

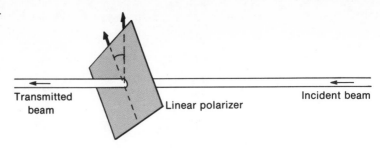

Beam photons are in a state of linear polarization if the transmitted beam is extinguished for one orientation of the sheet
polarizer and has a maximum transmission for a perpendicular orientation. For a linearly polarized beam the state can
be exactly specified in terms of the orientation of the polarizer transmission axis for maximum transmission. For example, by labeling a horizontal direction as x one can say "the
photons are y polarized" or for another beam "they are y'
polarized, where y' makes an angle of 40° with the y axis."[2]
Using a calcite analyzer we can determine a state while
reducing beam intensity by only a negligible amount. With
this analyzer in some arbitrary orientation, every photon will
emerge in one or another of the two possible exit paths,
which we shall call *channels*. (A given channel characterizes
and selects a particular polarization state.) By rotating the
analyzer one can find an orientation such that all of the incident linearly polarized light is transmitted through one
channel. Then the state is specified by the channel and the
orientation of the analyzer.

6-3 LINEARLY POLARIZED PHOTONS

How does quantum physics describe the behavior of photons known to be in a given state of linear polarization? One
simple experiment illustrates several major features of the
quantum description.

Let a linearly polarized beam produced by polarizer A
(Figure 6-4a) impinge on polarizer B whose transmission axis

[2]We are taking the z direction to correspond to the direction of the beam,
so that x and y (or x' and y') are linear polarization axes perpendicular to one
another and to z. The direction of propagation of light is taken to be from right
to left in the figures in order to simplify the relation between these figures and
the notation used to describe them, which will be introduced in Chapter 7.

Photons and quantum states

makes an angle θ with that of A. When θ is varied, by rotating either polarizer in its own plane, the intensity of the beam emerging from B varies as $\cos^2\theta$. Figure 6-4b displays the results for this fundamental experiment.

Classical Interpretation of Experiment

This experiment may be interpreted on the basis of classical wave optics if we suppose that the structure of the linear polarizer is such that it transmits only the electric field component parallel to its transmission axis (Figure 6-4c). Thus, when a wave with an electric field vector $\mathbf{E}$ along some other axis is incident on the polarizer, only the component of $\mathbf{E}$ along the transmission axis is transmitted; the component of $\mathbf{E}$ perpendicular to the transmission axis is absorbed. The electric field of the transmitted wave varies as $\cos\theta$ (Figure 6-4c). Now, according to electromagnetic theory the intensity of a wave (energy flow per unit area per unit time) is proportional to the square of the electric field amplitude of the wave. Therefore the intensity of the emergent beam is proportional to $\cos^2\theta$.

Photon Interpretation

Now we apply the photon concept to interpret this experiment. One might suppose, incorrectly, that linear polarizer B somehow splits each incident photon into two, transmitting only the member of each pair that is "polarized along the transmission axis." This explanation is defeated by the incorrectness of its predictions. For, if "part of each incident photon" is transmitted, then there are as many photons in the transmitted beam as in the incident beam. In order to explain the measured lower intensity of the emergent beam, one would be forced to say that each photon emerging from B carries (on the average) less energy than does a photon incident on B. Therefore, according to the basic relation $E = h\nu$, the frequency of the emergent beam would be correspondingly smaller than that of the incident beam. Since the light beam does not change color, this prediction is incorrect: by placing a spectrometer first in the beam incident on polarizer B and then in the transmitted beam, one verifies that no matter how the polarizers are oriented the frequencies of the two beams are exactly the same. It follows that each photon in the emergent beam has exactly the same energy as each photon in the in-

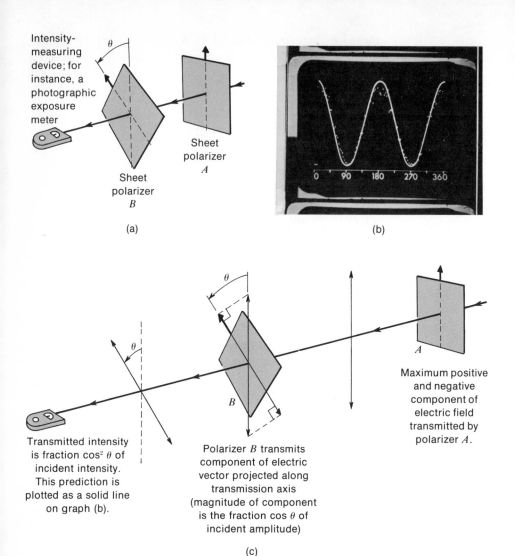

(a)

(b)

Intensity-measuring device; for instance, a photographic exposure meter

θ

Sheet polarizer A

Sheet polarizer B

0   90   180   270   360

θ

θ

B

A

Maximum positive and negative component of electric field transmitted by polarizer A.

Transmitted intensity is fraction $\cos^2 \theta$ of incident intensity. This prediction is plotted as a solid line on graph (b).

Polarizer B transmits component of electric vector projected along transmission axis (magnitude of component is the fraction $\cos \theta$ of incident amplitude)

(c)

Fig. 6-4   (a) Experimental arrangement for determining the transmission of linearly polarized light by a sheet polarizer.   (b) Results of an experiment like the one shown in (a). Bright spots are the experimental points. The curve is $I = I_0 \cos^2 \theta$, the prediction of classical wave optics, as described in (c).   (c) Interpretation of the experimental results of (b) in terms of classical wave optics. For the sake of simplicity the assumption is made that the polarizer is ideal; that is, it absorbs none of the light with vibration direction along the transmission axis. The intensity of light is proportional to the square of the magnitude of the electric vector.

cident beam. The observed decrease in intensity means, then, that the emergent beam carries *fewer* photons per second than does the incident beam. This can be verified experimentally by counting the emergent photons with a photomultiplier system. We conclude that a *fraction* $\cos^2\theta$ of the number of photons incident on polarizer B are fully transmitted by it, while a fraction $\sin^2\theta$ of the incident photons are completely absorbed. We say that, for photons in the state determined by the transmission axis of polarizer A, the *probability* of transmission by an ideal polarizer B is $\cos^2\theta$ where θ is the angle between the transmission axes of polarizers A and B. If you take a moment to think about it, you will realize that the implications of this result are quite startling. All the photons leaving A and striking B are in the *same* polarization state; yet some are stopped by B and others are not. Our classical belief that the same initial conditions always lead to the same consequences simply does not apply here. As we emphasized earlier, the observed behavior of quantum systems is fundamentally statistical.

Notice that photons transmitted by polarizer B are *not* in the same quantum state as are the photons that enter B. Photons that *enter B* are in the state defined by the transmission axis of polarizer A, as can be verified by passing them through a linear polarizer whose axis is parallel to that of A. Photons *emerging* from B are similarly verified to be in the quantum state defined by the transmission axis of B. The probability of transmission by B of photons initially in the state determined by polarizer A is given a special name: we call this the *projection probability* between the given states.

We can also describe the projection probability in a slightly different way. If photons are known to be initially in state A, then the projection probability to state B is the probability that they will be found in state B if examined by a B-polarization detector. For states of linear polarization the word "projection" can be taken as an analog of the geometrical projection of electric vectors in the classical analysis (Figure 6-4c). Later the idea of projection will be generalized to include cases in which no such geometrical analogy exists. The experiment under discussion gives the result that the projection probability between two states of linear polarization is $\cos^2\theta$, where θ is the angle between transmission axes of the respective linear polarizers.

In practice, linear polarizing sheet absorbs a substantial fraction (15–30 percent) of the photons for which it defines the state of polarization. But a calcite analyzer with one channel blocked transmits almost completely the fraction of incident photons given by the projection probability. We call such devices *x projectors* or *y projectors*, depending on which channel is blocked (Figure 6-5). Calcite devices can be made which approach the ideal performance that is assumed for simplicity of the theory.

The only projection that does not involve a change of state is a projection into the same state as that of the incident photons. Incident *x*-polarized photons are not changed in polarization when projected by an *x*-projector. Moreover, 100 percent of the incident *x*-polarized photons are transmitted by an ideal *x* projector. The formal definition of the state of *x*-polarization rests on this property of a projector:

> The photons in a given beam of intensity I_o are in the quantum state x if and only if the passage of the beam through an ideal x projector results in the same intensity I_o in the output channel.

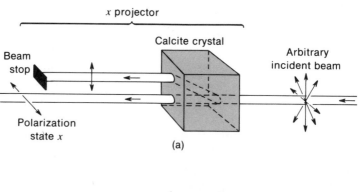

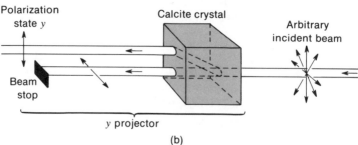

Figure 6-5 A calcite analyzer with one channel blocked transmits almost the complete fraction of incident photons given by the appropriate projection probability. Such devices are called x *projectors or* y *projectors, depending on which channel is left open. (a) Schematic drawing of an* x *projector. (b) Schematic drawing of a* y *projector.*

Photons and quantum states

The definition of a state of y polarization proceeds similarly. This way of specifying a state in terms of a device that produces particles in that state is quite general. It is a central feature of quantum physics.

6-4 PROBABILITY AND THE BEHAVIOR OF POLARIZED PHOTONS

We have already seen in earlier chapters that quantum mechanics is pervaded by one great theme: *The predictions of quantum mechanics are expressed in terms of probabilities.* The events that are the subject of these predictions are essentially statistical in nature—individually random, although conforming to a pattern in aggregate. The statistical character of the experiments described above on photon polarizations becomes quite apparent when they are carried out on beams of very low intensity according to the following "instructions": Turn the intensity of the incident beam far down. Enclose the apparatus in a light-tight box to eliminate extraneous light. Replace the exposure meter detector with a photomultiplier tube of advanced design. The measured output then becomes a random series of electrical pulses that can be heard as "clicks" over the loudspeaker and counted electronically. In a properly designed experiment most of the recorded clicks correspond to single photons incident on the face of the photomultiplier tube.

With this single-photon apparatus, repeat the experiment pictured in Figure 6-4a. Record counts for a fixed length of time for several settings of the relative orientation θ between polarizers A and B. The results of such an experiment are plotted in Figure 6-6. For a small number of counts, the $\cos^2\theta$ law of intensity is fulfilled only approximately—the sort of result which the mathematics of probability and statistics was developed to describe. Only as the counting time becomes very long—or many short determinations are averaged together—is the classical $\cos^2\theta$ result of Figure 6-4b accurately verified.[3]

Which version of the experiment—the one at high intensity or the one at low intensity—reveals more clearly the "real" nature of photons? Surely the character of individuals is best revealed by dealing with individuals. And the low-inten-

---

[3]A filmed version of exactly this experiment has been made by Stephan Berko. *Polarization of Single Photons* is available through the Education Development Center Inc., Newton, Mass.

6-4   Probability and behavior of polarized photons

sity experiments exemplify the fact that individual photons behave statistically. What, then, is the meaning of the $\cos^2\theta$ law of intensity transmitted by sequential linear polarizers? It is a law of statistical averages only. It is a well-known result of probability theory that if the average result of repeated counts on some random process (for example the number of radioactive disintegrations per minute of a small sample of uranium) is N, then the individual measurements will typically have deviations of the order of $\sqrt{N}$ from this average value. Thus, if $N = 100$, we are not too surprised if values as low as 90 or as high as 110 are recorded. The *fractional* scatter of results is $\sqrt{N}/N = 1/\sqrt{N}$, and therefore is large if N is small. If, on the other hand, we have an experiment in which N is very large, say 10^6, we arrive at a situation in which, although the absolute fluctuations are much bigger than for small N ($\sqrt{10^6} = 1000$), the *fractional* scatter, being of the order of $1/\sqrt{N}$ ($= 1/1000$), becomes exceedingly small and approaches zero in the limit of arbitrarily large N. It is in these terms that we can understand the ragged appearance of the low-intensity polarization data of Figure 6-6 as contrasted with the very smooth curve of the high-intensity measurements shown in Figure 6-4b. The only way of building up the smooth results, corresponding to a well-defined statistical average for each angle, is to accumulate very large total numbers of counts. With beams of low intensity the statistical averages can be accurately determined by running the experiment for a long

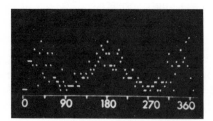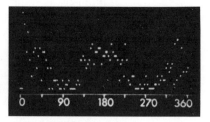

Fig. 6-6   Two "scatter diagrams" that result from separate runs of the experiment of Figure 6-4(a) at low intensity. The statistical nature of the theoretical relation $\cos^2\theta$ is obvious. (S. Berko, from the film "Single Photon Polarization," reproduced with permission of Education Development Center, Newton, Mass.)

Photons and quantum states

time, thereby recording a large number of events (a process colloquially known as "building up the statistics"). Photons have a felicitous property that allows us to build up the statistics in another way. The behavior of photons in a beam is independent of the intensity of the beam; the photons behave—most conveniently!—in a manner independent of one another. This property of photons allows a great simplification in instrumentation: We carry out experiments with a beam of visible light intense enough to be easily detected by the unaided eye. We are assured that this experiment will (for the same total number of counts) yield results equivalent to those from a low-intensity experiment—and in a much shorter time.

As we mentioned in Chapter 3, one reason why probability is so important in quantum physics is that quantum systems are typically so small that the process of observing them alters them. Since only one observation can be carried out on each system, the only way to study them is to employ a continuous production line on which systems—in this case photons—are prepared in a manner which is identical as far as we can tell. These "identical" systems are then subjected to various observations in which they are unavoidably altered or destroyed. Comparison of the results of different observations can then be used to describe the systems so produced. In each of the experiments described in this chapter the verification of a probability involves the comparison of intensities of two different beams. Separate observations are carried out on two different groups of photons. Statements as to what will happen to a particular photon are meaningless. Rather the predictions and results are always expressed in terms of relative average counts for different experimental arrangements.

6-5 STATES OF CIRCULAR POLARIZATION

A significant generalization of the idea of quantum states comes from considering other states of photon polarization besides linear states. Light can also be circularly polarized or elliptically polarized. In this section we deal with circular polarization. The Appendix to this chapter includes a classical description of circular polarization and two methods of producing circularly polarized light. According to the classical picture, a circularly polarized wave is one whose electric vector is constant in magnitude and, at any fixed point, traces out a circle at uniform angular velocity. The sense of rotation of the electric vector determines whether the polarization is called right circular or left circular.

Included in the Appendix is a description of a *right-left polarization analyzer* (Figure 6-7). This is a device with one input channel and two output channels, labeled L and R, such that

(1) a left or right circularly polarized input beam emerges with undiminished intensity in the appropriate output channel, and

(2) for an arbitrary incident beam, the sum of the intensities of the output beams equals the intensity of the input beam.

These properties are analogous to those of an *xy* analyzer. When interpreted on the basis of the photon picture, the device serves to define the quantum state of circular polarization, just as the calcite analyzer served to define states of linear polarization. The photons in a given beam of intensity I_0 are in the quantum state R if and only if the passage of the beam through an ideal RL analyzer results in intensity I_0 in the R channel and zero intensity in the L channel. Verification that a beam is in state L proceeds similarly.

If an RL analyzer is rotated about an axis along the direction of the incident light, its effect remains unchanged; the output beams are still right- and left-circularly polarized. There are no states R' or L', distinct from R and L. In this respect circular polarization differs from linear polarization.

The RL analyzer can evidently be turned into an *R-projector* or an *L-projector* by blocking the appropriate output channel. Notice that the term "projector" as used for circular polarization has a purely quantum-mechanical meaning; there is no analogous classical projection of a vector onto some axis, as is the case for a linear polarization projector. The projectors

Fig. 6-7 Schematic diagram of an RL analyzer, which analyzes an arbitrary incident beam into a right circularly polarized beam and a left circularly polarized beam. One possible design for such a device (shown here simply as a box with ports) is given in the Appendix to this chapter.

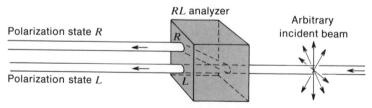

Polarization state R

R

RL analyzer

Arbitrary incident beam

Polarization state L

L

Photons and quantum states

are also polarizers—they provide a means for producing circularly polarized light.

Earlier we defined the projection probability from one quantum state to another, and considered the projection probabilities between different linear polarization states. We can now inquire as to the projection probability from a state of linear polarization to one of circular polarization, or vice versa. The experiments which measure these probabilities are straightforward: to determine the projection probability from state R to state y, for example, one merely sends a beam of R-polarized photons into a y projector (Figure 6-8), and measures the relative intensities of input and output beams. The results of all such experiments can be predicted on the basis of the classical wave picture. As remarked in the Appendix, a classical circularly polarized wave can be considered as the superposition of x and y linearly polarized waves with equal amplitudes (and a phase difference of 90°). Therefore, from the classical viewpoint, an xy analyzer separates the x and y components of an R-polarized beam and we would expect to find equal intensities in the x and y output channels of the analyzer. So the classical analysis suggests that the projection probability from state R to state y is $\frac{1}{2}$; and similar classical arguments suggest that each of the eight possible projection probabilities (x to R, x to L, R to x, L to x, and so on) has the value of $\frac{1}{2}$. These predictions are confirmed when the experiments are

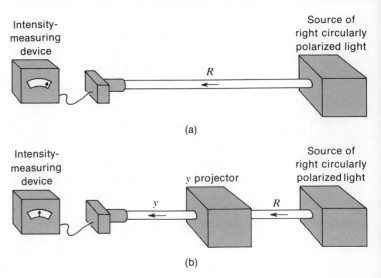

Fig. 6-8 To determine the projection probability from polarization state **R** to state **y**, one can send a beam of **R**-polarized photons into a **y** projector and compare the intensity in the output beam with the intensity in the input beam. The projection probability from **R** to **y** is just the ratio of the intensity measured in (b) to the intensity measured in (a).

TABLE 6-1   Polarization Projection Probabilities

Angle
convention:
Beam emerges
toward reader.

	From Polarization State					
To Polarization State	x	y	x'	y'	R	L
x	1	0	$\cos^2\theta$	$\sin^2\theta$	$\frac{1}{2}$	$\frac{1}{2}$
y	0	1	$\sin^2\theta$	$\cos^2\theta$	$\frac{1}{2}$	$\frac{1}{2}$
x'	$\cos^2\theta$	$\sin^2\theta$	1	0	$\frac{1}{2}$	$\frac{1}{2}$
y'	$\sin^2\theta$	$\cos^2\theta$	0	1	$\frac{1}{2}$	$\frac{1}{2}$
R	$\frac{1}{2}$	$\frac{1}{2}$	$\frac{1}{2}$	$\frac{1}{2}$	1	0
L	$\frac{1}{2}$	$\frac{1}{2}$	$\frac{1}{2}$	$\frac{1}{2}$	0	1

performed. In the analysis of photon polarization, all predictions of this kind based on the classical wave picture are experimentally verified. Table 6-1 summarizes such results for projections between different states of linear polarization and between linear and circular polarization states. Later, when we consider phenomena for which no classical picture exists, we will not have this crutch to lean on and will need to develop a purely quantum theory to predict projection probabilities.

6-6 ORTHOGONALITY AND COMPLETENESS

The polarization states of photons exhibit several formal properties that simplify and summarize our knowledge of them. Among these properties are *orthogonality* and *completeness*. The quantum states of other particles and systems also exhibit these same properties.

Orthogonality

When the projection probability for some quantum state j to some other state k is zero, we say that *state k is orthogonal to state j*. According to this definition the linear polarization state x is evidently orthogonal to state y: none of the photons in an x polarized beam are transmitted by a y projector. Also, state y is orthogonal to state x. Likewise state R is orthogonal to state L, and vice versa. These examples illustrate the general rule that orthogonality is a *reflexive* property: if (1) state k is orthogonal to state j, then (2) state j is orthogonal to state k.

Photons and quantum states

One may thus speak of two states being orthogonal to each other. Although it is not obvious in all cases that statement (1) implies statement (2), since the two statements refer to entirely separate experiments, in fact the rule is satisfied in all pertinent experiments.

In Euclidean geometry *orthogonal* is a synonym for *perpendicular*. And indeed, in the case of linear polarization, two quantum states are orthogonal if the polarization axes of the corresponding classical waves are perpendicular. This is in fact the origin of the term *orthogonal*. However, no such classical analogy exists for the statement that states R and L are orthogonal. The quantum-mechanical concept of orthogonality is defined entirely in terms of experiments with analyzers and projectors (Figure 6-9). If a particle (a photon) is definitely in one state, there is a zero probability that any measurement will show it to be in an orthogonal state. In the particular case of photon polarization states, any given state has only one orthogonal state.

Completeness

When a photon beam of arbitrary polarization enters an xy analyzer, all outgoing photons are either x polarized or y polarized, and no photons are absorbed. In this sense (to be developed in detail later) two states x and y are sufficient to describe the polarization state of an arbitrary beam. In mathematical language, the sum of the projection probabilities from any initial state to the states x and y is unity, and one can say that *the states x and y constitute a complete set of photon polarization states*. Alternatively, one says that the states x and y provide the *basis* for a *representation* of photon polarization. There are, of course, many other complete sets. In fact, any analyzer defines a complete set. For example, states L and R,

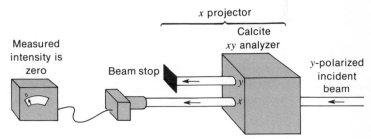

6-6   Orthogonality and completeness

TABLE 6-2 Some Concepts and Definitions of Quantum Mechanics

Concept	General Meaning	Example for Polarized Photons
Analyzer	An analyzer for a given physical system is a device with one input channel and two or more output channels such that: 1. If any beam whatever enters the input channel, the sum of the intensities of the output beams equals the intensity of the input beam. 2. If the beam in one particular output channel (say channel j) of the analyzer enters a second analyzer identical to the first in all respects, then the output of the second analyzer will be found entirely in the same channel j.	An xy analyzer made from calcite has one input channel and two output channels, x and y. Its properties closely approximate the following: 1. If any beam of photons enters the input channel, the sum of the intensities of the two output beams equals the intensity of the input beam. 2. If the output beam in channel y of the analyzer enters a second xy analyzer, then the output of the second analyzer is entirely in channel y.
Projector	A projector for a particular physical system can be formed from an analyzer by blocking all output channels but one, say the channel j. Then the device is called a j projector.	A y projector can be constructed from an xy analyzer by blocking the x output channel.
Determination of the State of a System	When the output of a j projector is equal in intensity to the input, the input systems are in state j.	When the output of a y projector is equal in intensity to the input, the input photons are said to be in polarization state y.

TABLE 6-2 Some Concepts and Definitions of Quantum Mechanics (continued)

Concept	General Meaning	Example for Polarized Photons
Projection Probability	Suppose an input beam is known to be in state k. (For example, it may have come from a k projector.) Let the beam now pass through a j projector. The relative intensity (output intensity)/(input intensity) is called the projection probability from state k to state j. *Experimental result* The projection probability from state k (input) to state j (output of a j projector) has the same value as the projection probability from state j (input) to state k (output of a k projector).	*Experimental result:* The projection probability from state y to state y' is $\cos^2\theta$. Here θ is the angle through which the y projector must be rotated to make it a y' projector. The general expression $\cos^2\theta$ implies that the "reverse" projection probability from state y' to state y also has the value $\cos^2\theta$. [$y \leftrightarrow y'$ implies $\theta \leftrightarrow -\theta$ and $\cos^2(-\theta) = \cos^2\theta$.]
Orthogonal States	Two states are orthogonal by definition if the projection probability from one state to the other is zero.	Polarization states x and y are orthogonal since the output of a y projector is zero when the input photons are known to be in state x, and vice versa.
Complete Set of States	The set of states defined by an analyzer is complete if the total number of particles emerging in the various output channels is equal to the number entering the input channel.	The set of polarization states x and y is complete because the number of photons emerging in the output channels of an xy analyzer is equal to the number entering the input channel.

as well as states x' and y', constitute complete sets. But all the complete sets have one thing in common: for photon polarization each consists of exactly two states. Each analyzer has exactly two output channels; no third or fourth channel is needed. This fact reflects a fundamental property of photons. There are many pairs of orthogonal photon polarization states, but one cannot find three photon polarization states that are mutually orthogonal. This remark has a geometric analog: in a space of n dimensions, at most n vectors can be mutually orthogonal (that is, "perpendicular"). Later on we shall study systems for which a complete set of states consists of larger numbers (even an infinite number) of states, each of which can be chosen to be orthogonal to every other state in the set.

Table 6-2 contains brief definitions of the major quantum concepts that have been introduced in this chapter, together with specific examples for photon polarization to illustrate these concepts.

6-7 QUANTUM STATES

Using photon polarization states as an example, we can now define with some precision the meaning of *quantum state*. This definition is deeply rooted in the properties of atomic, subatomic, and "elementary" particles and in our experience of investigating them. Two properties of particular importance that most such systems share are (1) *discreteness* (one class of particles can be distinguished from all other classes) and (2) *identity* (all particles in a given class are basically identical to one another). The definition of quantum state rests principally on these two properties.

Suppose that we set up a "factory" (a production line) to produce x-polarized photons. Using the factory we carry out experiments on photons of this polarization. Neither the factory not the later experiments could function if only single photons were available; we absolutely depend on having a flood of identical photons from the factory.

The concept of a "factory" producing many identical systems may seem artificial but is in fact typical of real experiments in quantum physics. Most experiments on the atomic scale involve a large number of particles. To study electron-proton scattering, for example, it would not suffice to project a single electron at a single proton (even if this were feasible) and observe how the electron is deflected. Moreover, one cannot pick up the same electron and throw it at the same proton over and over, as one could in a similar experiment with macroscopic bodies. What one does, instead, is to prepare a *beam* of electrons, aim the beam at a target that contains many protons, and observe how many electrons are deflected through various angles. How can such an experiment tell us anything about the features of a *single* electron-proton encounter? The answer is that, as far as we can tell, every electron is basically identical to every other electron; each electron has, within the accuracy of the best experiments, exactly the same mass, exactly the same charge, and so on, as every other electron.[4] Protons are also identical to one another. Therefore if the beam has been carefully prepared, an experiment in which, say, 10^{10} electrons of a certain energy are hurled at the target and 10^6 of these are detected conveys information about the interaction between one electron and one proton.

The concept of *quantum state* builds directly on the existence and availability of almost limitless numbers of basically identical particles of a given kind. However, to specify a quantum state we need additional information. Two free electrons are in different quantum states if they have different kinetic energies. It would be easy to devise an experiment that would distinguish between a sample of, say, a billion electrons, each with a kinetic energy of 1 MeV, and another sample of a billion electrons each with a kinetic energy of 10 MeV. And it is only when we have specified the conditions applying to a batch of electrons to the point where it is unique, and distinguishable from all others, that we can say that the electrons are in a definite quantum state.

The energy is almost always an important parameter in characterizing a quantum state, but it is seldom sufficient. For example, electrons of equal kinetic energy are in different, distinguishable states if they are moving in different directions. Is it sufficient, however, to use the energy together with the direction of motion? No! Electrons have a property called *spin*, an intrinsic angular momentum that can take two different orien-

[4]Many people would say that this is a tautology in that a name such as "electron" is limited, *by definition*, to a class of particles that are identical with one another. The point here is that the definition corresponds to reality: nature *exhibits* classes of identical particles.

tations with respect to a physically specified axis. And electrons having different spin orientations are physically separable. However, as far as we know, there is no way whatsoever by which we can distinguish between electrons all having the same energy, direction of motion, and spin orientation. For electrons, at least, these three parameters define a class of electrons all in the same quantum state. These features are thus closely parallel to the parameters that permit us to define a unique quantum state for photons in terms of their photon energy, direction, and polarization.

The probabilistic interpretation is very much a part of our definition of a quantum state. As an example, recall photon polarization states. Although one may be able to say that all the photons produced by a given source are in a certain polarization state, there is no way to determine the polarization state of a single photon. For if we let a photon impinge on, say, an x-polarizer and a photomultiplier on the other side registers a count, we cannot conclude that the original photon was x-polarized. It may in fact have come from a factory that produces circularly polarized photons, elliptically polarized photons, or photons that are linearly polarized along any axis except the y axis. Only if an entire beam of photons emerging from the x-polarizer is just as intense as the beam that enters can we conclude that the entering photons were x-polarized. The measurement of intensity necessarily involves the detection of many photons. Thus in general one determines the quantum state of every member of a large set of identically prepared particles. Such a determination involves making the same measurement on many members of the set. That is why the identity of atomic particles plays such an important role in quantum physics. The measurements are made on individual particles one at a time. With an intense beam of photons we do not need a photomultiplier tube, and we may not be explicitly aware that we are measuring the properties of individual photons. However, the fact that exactly the same results are obtained by measuring a very weak beam for a long time indicates that the basic phenomenon involved is the detection of individual particles.

The concepts of identity and quantum state are in general not nearly so clear or sharp when one considers macroscopic objects—even small ones such as grains of sand. A small grain of sand contains about 10^{20} atoms; the likelihood that two such

Photons and quantum states

grains, however seemingly alike, actually contain exactly equal numbers of atoms is quite negligible; we have left the realm of identity in the strict quantum-mechanical sense. However, macroscopic size in itself in not sufficient to ensure that a system may be described classically. Superconductivity and the operation of a transistor are two examples of macroscopic phenomena that can be explained only by quantum physics.

6-8 STATISTICAL AND CLASSICAL PROPERTIES OF LIGHT

The classical theory of electricity and magnetism, as summarized by Maxwell, regards light as an electromagnetic wave consisting of oscillating electric and magnetic fields propagating through space. We know that this makes good sense, in terms of the observable and measurable effects that light produces. If instead of light we consider electromagnetic waves of lower frequency, for example radio waves, the reality of the electric field is undeniable. This oscillating field can take hold of the mobile electrons in a radio antenna, for example, and force them to oscillate back and forth at its own frequency, causing a small current signal which, when amplified, brings us the broadcast program.

The discussions of this chapter, however, make it clear that the description of things becomes utterly different if we make it in terms of individual photons. To take the phenomenon of polarization for a start, we have seen that instead of describing an oscillating electric field in a certain direction, we speak of a polarization state of individual photons. It is meaningless to discuss the electric field vector of an individual photon in a given polarization state; instead, we have only an operational definition of polarization in terms of the transmission of photons, as individual particles, through analyzing devices of various kinds. The classical electric field and its direction are properties that we cannot determine unless we are dealing with enormous numbers of photons. We saw this explicitly in our account of the experiments to test the $\cos^2\theta$ law of transmission for linearly polarized light through a sheet of linear polarizer.

We chose to discuss this relation between classical and photon properties in terms of polarization because of the great simplicity of the phenomena—only two basic polarization states. But if we ask about the other classical properties that we associate with electromagnetic waves in general (with light as our chosen example) we find that they, too, assume meaning and measurability only when vast numbers of photons are involved. In particular this applies to the most basic properties that we associate with any wave—its frequency and its wavelength. An individual photon has no color. Any experiment to measure the wavelength of "monochromatic" light—that is a beam of photons all having nearly the same quantum energy—involves the passage of astronomical numbers of photons through a diffraction grating or other spectroscopic device that spreads out the spectrum over the face of a detector. As this detector we may choose a photographic plate to integrate the results of the experiment. A measurement of exposure density along the plate then gives a smooth density curve whose profile characterizes the spectrum of the light. Measuring "*the* wavelength" reduces to locating the maximum in this density profile. Instead of using a flood of photons, we could slowly build up the profile from counting individual photons, in which case the profile emerges slowly as the counting time increases. Such an experiment confirms that the well-defined spectral shape is in fact the product of photons whose individual behavior is random, in the sense that the only statements we can make about them are statistical.

When we come to particles other than photons, the wavelength again is a well-defined property, but only in terms of a large statistical sample. And for these other particles, we do not even have a seemingly concrete macroscopic property to associate with the wave, equivalent to the electric or magnetic field of a beam of light. We arrive at the conclusion that the wave property is an expression of the probabilistic or statistical behavior of large numbers of identically prepared particles—*and nothing else!* This is a profoundly important aspect of our understanding of the wave nature of matter.

6-9 CONCLUDING REMARKS

To end this chapter, we review what we have learned about quantum states in general: The quantum state refers to *individual particles*. Therefore it is not strictly correct, for example, to speak of a *beam* in the quantum state of *x* polarization. On the other hand, to be able to assert that "a photon is in a state of x polarization," one must have previously verified this polarization for many photons in the same beam from which that particular photon is taken.

An objection can be raised against the probabilistic interpretation. Return once more to the $\cos^2\theta$ law of transmission by two sequential linear polarizers (Figure 6-4a). Suppose for concreteness that the transmission axis of the first polarizer A, is parallel to the x axis, and the transmission axis of the second polarizer B is oriented at some other direction (but not perpendicular to the x axis). Now, the beam incident on B is composed of x-polarized photons, allegedly identical particles identically prepared. Yet, as we know, not all of these incident particles emerge from B; some are transmitted by polarizer B whereas others are absorbed. How can we call these particles "identical" when they do not behave identically when subjected to identical treatment?

The term identity is appropriate in the sense that there exist experiments for which the particles *do* behave identically. It is just such experiments that we use to define quantum states. Particles in a given quantum state bear witness to that fact by behaving identically when tested for that quantum state. In many other experiments, however, particles in a given quantum state do *not* behave identically. In these experiments we simply cannot make predictions with respect to individuals. But if we know the quantum state, we *can* make definite statements regarding the *probabilities* of various possible outcomes of an experiment.

Let us return now to the problem of the general definition of a quantum state. How does one know when a sufficient description has been given to define unambiguously a quantum state for a given system? How can we be sure that some crucial characteristic—as yet undetected by experiment—has not been left out of the description? We cannot be sure. New classifications are sometimes found by the combination of an educated guess and a confirming experiment. The discovery that electrons have "spin" angular momentum is an example of this process. The historical development of quantum mechanics has been in large part the search for physical quantities (sometimes called *observables*) that define quantum states and for values that these observables take for particular states. We shall see later that angular momentum is an important quantity in specifying atomic states. Angular momentum for bound states, is, like energy, quantized (that is restricted to definite discrete values): The particular values of angular momentum are specified by "angular momentum quantum numbers." Some observables required to define quantum states correspond to familiar classical quantities, such as angular momentum; others involve new, strictly quantum-mechanical concepts. Occasionally, properties initially thought to be part of the description of a quantum state turn out to be irrelevant or without meaning. Sometimes one or more properties of a quantum state are not discrete but can take on any value in a continuous range (example: the kinetic energy of a free particle).

In summary, one cannot be sure in advance what properties are involved, and what ones are excluded, in a description sufficient to determine quantum states unambiguously. For a given system these classifications are obtained by guess, by intuition, and by inspiration; from formalisms and theories and symmetry arguments. No matter how the classifications originate, they are all tested ultimately in the arena of experience.

Appendix: Polarized light and its production

6A-1   THE PRODUCTION OF LINEARLY POLARIZED LIGHT[5]

How does one *produce* photons in a given state of linear polarization? Here we treat this question using the electric and magnetic field vectors that constitute the classical description of light. In the main text we apply the methods of production and detection to compare the classical analysis (fields) with the quantum analysis (photons) of a few simple experiments with polarized light.

Classical Description of a Linearly Polarized Wave

According to the classical description, an *electromagnetic wave* consists of electric and magnetic fields. In air or vacuum

[5]An excellent brief introduction to the classical theory and practical applications of polarized light is presented in William A. Shurcliff and Stanley S. Ballard, *Polarized Light*, Momentum Book #7, D. Van Nostrand, Princeton, N.J., 1964.

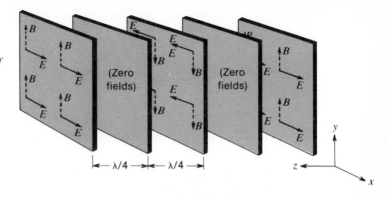

Fig. 6A-1 Portion of a linearly polarized plane wave at a single instant of time, showing electric fields (solid arrows) and magnetic fields (dashed arrows) in selected planes of constant phase. The entire pattern propagates in the positive z direction.

and far from the source, the electric and magnetic field vectors are transverse to the direction of propagation of the wave. Some special cases: A *plane wave* is one for which the "wave fronts" (surfaces of, say, maximum electric field at a given instant) are planes. A *linearly polarized plane wave* is a plane wave in which the electric vector is everywhere parallel to a line fixed in space. Figure 6A-1 represents a portion of a linearly polarized plane wave. At every point in space the magnetic vector is perpendicular to the electric vector. In what follows we describe linear polarization in terms of the electric vector alone, without mentioning the magnetic vector always associated with it. At every fixed location in space, the electric vector of a linearly polarized plane wave oscillates back and forth with time along a line whose orientation we call the *polarization axis*. So much for the classical discription of a linearly polarized wave.

Linear Polarization by Reflection

A simple practical method for *producing* light linearly polarized along a prescribed polarization axis is based on the fact that waves of different polarization have different reflection characteristics at the surface of a dielectric material such as glass. An incident ray of unpolarized light reflected from a glass surface results in a partially polarized beam (Figure 6A-2). In the reflected beam the predominant electric field is parallel to the surface of the glass. For a particular angle of incidence, such that the reflected and refracted rays are perpendicular, the reflected beam is entirely polarized with its polarization axis parallel to the surface of the glass. This characteristic angle of incidence for total polarization of the reflected

6A-1   The production of linearly polarized light

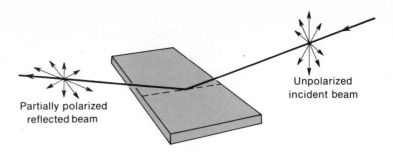

Fig. 6A-2   Partial linear polarization by reflection from a dielectric surface.

Partially polarized reflected beam

Unpolarized incident beam

beam is called *Brewster's angle* (Figure 6A-3). For glasses of different composition (index of refraction from 1.49 to 1.90) the Brewster angle is near 60°. Classical electromagnetic theory applied to the boundary conditions at the interface between air and glass is able to account for the polarization phenomenon. In this sense polarization by reflection, although somewhat inconvenient, is the most thoroughly understood of the methods listed here.

Linear Polarization with Sheet Polarizer

A commercial plastic sheet ("Polaroid") is available that transmits only light polarized along a particular polarization axis. When unpolarized light is incident on this sheet polarizer, the transmitted light is linearly polarized with its electric vector along a direction called the *transmission axis* of the polarizer. Let linearly polarized light be incident on the sheet. Then a maximum intensity is transmitted when the transmis-

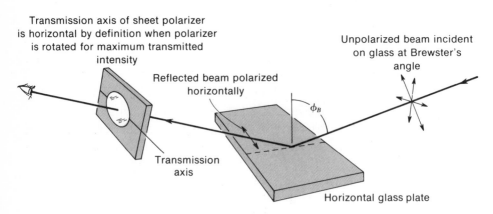

Transmission axis of sheet polarizer is horizontal by definition when polarizer is rotated for maximum transmitted intensity

Unpolarized beam incident on glass at Brewster's angle

Reflected beam polarized horizontally

ϕ_B

Transmission axis

Horizontal glass plate

Fig. 6A-3   Total linear polarization by reflection at Brewster's angle. Determination of transmission axis of sheet linear polarizer.

Photons and quantum states

sion axis is parallel to the polarization axis of the incident light (Figure 6A-3). In this case, for commercial sheet polarizer, typically 70–85 percent of the incident intensity is transmitted. A minimum intensity is transmitted when the transmission axis is perpendicular to the polarization axis of the incident light. Typically less than 1 percent is transmitted in this case. At intermediate angles between polarization axis and transmission axis the fraction of incident intensity transmitted is between these two extreme cases (see Figure 6-4b of the main text). Commercial sheet polarizer comes with the transmission axis marked. If this marking sticker is lost, one can determine the transmission axis given a source of polarized light with known polarization axis—as, for example, by reflection (Figure 6A-3).

Sheet polarizer is the simplest and most convenient material for qualitative experiments with linearly polarized light. One-inch squares cost a few pennies each. However, commercial sheet polarizer has several disadvantages for quantitative study: it absorbs a considerable fraction (15–30 percent) of the intensity of an incident beam linearly polarized parallel to the transmission axis of the polarizer. Moreover, even in ideal operation, sheet polarizer absorbs light, rather than separating out the component polarized perpendicular to the transmission axis. These disadvantages are largely overcome by using the more expensive calcite crystal.

Linear Polarization Using a Calcite Crystal

A clear crystal called *calcite* has the property that its index of refraction is different for two perpendicular components of polarization. This property is called *linear birefringence*. A calcite crystal can be cut and polished so that an incident beam of arbitrary polarization is divided ("analyzed") into two beams perpendicularly polarized that travel in different directions through the crystal (Figure 6A-4). The axes of polarization are determined by the crystallographic axes of the crystal. By using a narrow incident beam of light, one can separate the two components. Such a device, which analyzes an incident beam into two beams perpendicularly polarized, is called a *linear polarization analyzer* (Figures 6A-5 through 6A-7). An analyzer can be made into a linear polarizer by placing a stop in one of the transmitted beams (Figures 6A-8 and

6A-1   The production of linearly polarized light

Fig. 6A-4 Photograph of a piece of natural calcite showing double images due to birefringence.

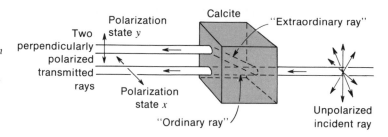

Fig. 6A-5 Schematic diagram of a calcite analyzer. In this orientation the device is called an xy analyzer.

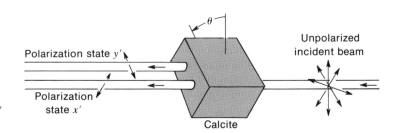

Fig. 6A-6 An x'y' analyzer.

Fig. 6A-7 Photographs of a calcite analyzer. The photograph at the right shows the two separate beams (separated horizontally in this particular analyzer).

Photons and quantum states

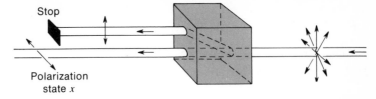

*Fig. 6A-8
Schematic diagram
of a calcite* x *po-
larizer. (Also called
an* x projector *in
the text.)*

Stop

Polarization
state *x*

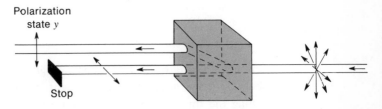

*Fig. 6A-9
Schematic diagram
of a calcite* y
*projector. (Also
called a* y projector
in the text.)

Polarization
state *y*

Stop

6A-9). If the incident ray is linearly polarized along the polarization axis of one of the transmitted beams, then the incident beam is transmitted entirely in that channel.

The fraction of incident intensity transmitted in both output beams of a calcite analyzer can be made nearly 100 percent by optical grinding, polishing, and coating the surfaces of the crystal. The cost of such an analyzer is comparable to that of a high-quality wristwatch. (In contrast, qualitative experiments can be carried out with a small rhomboid of calcite with cleaved surfaces that costs less than one dollar.) A high-quality analyzer provides all of the incident light at the two output channels: calcite itself divides (analyzes) without destroying.

6A-2 THE PRODUCTION OF CIRCULARLY POLARIZED LIGHT

A classical circularly polarized wave is one whose electric vector is constant in magnitude and, at any fixed point, traces out a circle at a uniform angular velocity. (The magnetic vector does likewise.) The two types of circular polarization, called right- and left-circular, correspond to the two possible senses of rotation. Unfortunately, there is not a universally accepted convention as to which sense of rotation should be called left and which right. The convention we adopt is the following: we call right-circular the wave in which the electric vector at a fixed point rotates clockwise with time if one looks *toward* the source (Figure 6A-10). The opposite rotation is called left-

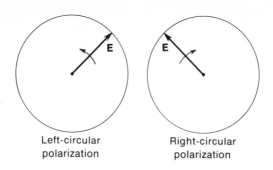

Left-circular
polarization

Right-circular
polarization

circular. With these definitions, a right-circularly polarized beam carries a negative component of angular momentum along the direction of motion, and a left-circularly polarized beam carries a positive component.[6]

A circularly polarized beam can be considered classically as a superposition of two beams linearly polarized along mutually perpendicular axes (for example x and y), with a 90° phase difference between the vibrations (Figure 6A-11). This property forms the basis for one simple method of producing a circularly polarized beam, using a linear polarizer and a device called a *quarter-wave plate*.

The Quarter-Wave Plate

In the first section of this Appendix, we examined some of the properties of a birefringent calcite crystal, and described one way of cutting such a crystal that makes possible the construction of a linear polarization analyzer. Different crystallographic cuts of the same material lead to quite different

Fig. 6A-11   Circular polarization as a superposition of two linearly polarized electric vectors 90° out of phase in time.

[6]This definition agrees with the original one of classical optics, and is used, for example, by Shurcliff and Ballard (reference above). The opposite definition is used in most current research in particle physics, and is also used in the Feynman Lectures, Vol. III. Our main concern is simply with the existence of two basic polarization states. Thus, although one should be alerted to this difference of conventions, it is not central to our present discussion.

transmission properties that are useful in other applications. In particular, a cut can be found such that for normal incidence, both rays continue unrefracted, but with *different velocities of propagation*. The polarizations of the two rays are perpendicular, as in the previous application. One ray travels faster. Its polarization axis is therefore called the *fast axis* of the crystal and the polarization axis of the other ray is called the *slow axis*. A crystal cut in this manner makes possible the construction of extremely useful devices known as quarter-wave plates and half-wave plates. We shall describe these devices entirely in classical language.

Consider a linearly polarized ray normally incident on a crystal cut in the manner referred to above. The component of the electric field of this ray along the "slow" axis emerges from the crystal retarded in time relative to the component along the "fast" axis, the magnitude of the retardation depending on the thickness of the slab. Corresponding to this relative retardation in *time* is a relative difference in *phase* between the emerging rays. Thus the net effect of the calcite is to change the relative phase of the two perpendicularly polarized components. By suitably choosing the thickness of the calcite slabs, one can obtain any desired value for this phase shift for light of a given frequency. The two most-used thicknesses are the so-called "half-wave plate" which produces a phase angle shift of 180° and the so-called quarter-wave plate which produces a phase angle shift of 90°. For light of a different frequency the relative phase shift will have some other value. Quarter-wave plates are effective over a much narrower range of frequencies than sheet polarizer and calcite analyzers. But this is not a serious drawback because, if necessary, the experiments can be carried out with nearly monochromatic beams.

Let a quarter-wave plate be oriented so that its fast and slow axes point in the y and x directions, respectively, and let an incident beam be linearly polarized at 45° (Figure 6A-12). After the beam has passed through the quarter-wave plate, the "fast" ray has acquired some extra phase ϕ because its velocity in the crystal is not equal to c. But the slow ray has acquired an additional phase. The outgoing wave is right-circularly polarized (Figure 6A-13). Thus the quarter-wave plate can be used to convert a particular linearly polarized wave to circular polarization. In conjunction with a linear polarizer, it can be used to convert an unpolarized beam into a circularly polarized

6A-2   The production of circularly polarized light

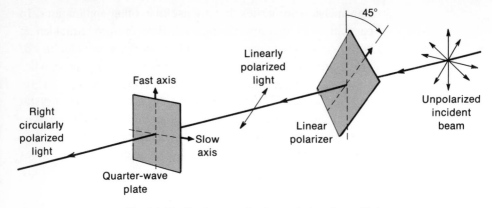

Fig. 6A-12   Production of right circularly polarized light
using a linear polarizer and a quarter-wave plate.

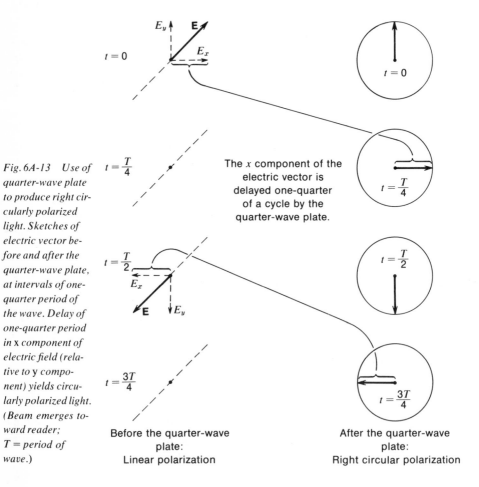

*Fig. 6A-13   Use of
quarter-wave plate
to produce right cir-
cularly polarized
light. Sketches of
electric vector be-
fore and after the
quarter-wave plate,
at intervals of one-
quarter period of
the wave. Delay of
one-quarter period
in x component of
electric field (rela-
tive to y compo-
nent) yields circu-
larly polarized light.
(Beam emerges to-
ward reader;
T = period of
wave.)*

The x component of the
electric vector is
delayed one-quarter
of a cycle by the
quarter-wave plate.

Before the quarter-wave
plate:
Linear polarization

After the quarter-wave
plate:
Right circular polarization

Photons and quantum states

beam. One kind of circular polarizer available commercially consists of a sheet of quarter-wave plate bonded to a sheet of linear polarizer with properly oriented transmission axis.

Circular Polarization Analyzer

A quarter-wave plate can also be used to convert circularly polarized light to linearly polarized light. When a circularly polarized beam falls on a quarter-wave plate the output is linear light, the axis of polarization differing by 90° in orientation depending on whether the incident light is right- or left-circularly polarized. This easy transformation of polarization between linear and circular forms allows us to design a circular polarization analyzer (Figure 6A-14). In this analyzer an initial quarter-wave plate converts R light to y light and L light to x light. A calcite xy analyzer then physically separates the two beams. Finally, a second quarter-wave plate reconverts the y light back to R light and x light back to L light. The overall effect is to separate any incident light into its R and L components. You can verify for yourself that x-polarized photons incident on the RL analyzer will, on the average, be equally divided into the two output channels.

We have used the linear birefringence of calcite in constructing analyzers. There exist also crystals that are *circularly birefringent*; such crystals can split an incident beam into two beams, one of which is left-circularly polarized and the other right-circularly polarized. Quartz is a common example of circularly birefringent material. Right- and left-circularly po-

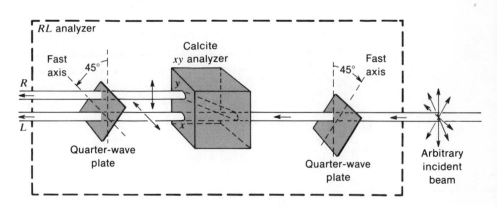

Fig. 6A-14 One possible design for a circular polarization analyzer.

6A-2 The production of circularly polarized light

larized light follow different paths through a piece of quartz properly cut and oriented with respect to the incident beam. From such a piece of quartz one can construct an "*RL* analyzer." In practice the angle between the separated right- and left-circularly polarized beams in quartz is very small. Therefore adequate separation between beams can be achieved only by using a piece of quartz which is long in the beam direction. Therefore it is less expensive to use quarter-wave plates and calcite to construct an *RL* analyzer.

The preceding discussion has been entirely classical. However, when interpreted on the basis of the photon picture, the *RL* analyzer serves to define the quantum states of circular polarization, just as the *xy* analyzer serves to define states of linear polarization. The photons in a given beam of intensity I_o are in the quantum state *R* if and only if the passage of the beam through an *RL* analyzer results in intensity I_o in the *R* channel and zero intensity in the *L* channel. Verification that a beam is in state *L* proceeds similarly.

Suggested experiments with linearly polarized light

The simplicity with which photon polarization states can be described is matched by the ease with which beams of polarized photons can be observed and manipulated. The equipment is simple, inexpensive, and fun to play with. Carrying out some of the experiments suggested below, and others which will occur to you as you go along, will improve your understanding of the material in this chapter. Most of the experiments can be carried out using three pieces of linear polarizing sheet: a piece 2 cm on an edge costs a few pennies. The last two suggested experiments require the use of one or two pieces of natural calcite: a "cleaved rhomb" 1 cm on an edge costs about a dollar. William A. Shurcliff is co-author (with E.F.T.) of these suggested experiments.

1. *Glare reduction.*

Look through a sheet of linear polarizer at light reflected

obliquely, say at 45°, from any reasonably flat surface. Rotate the polarizer slowly in its own plane.

What happens to the reflected light?

How can this effect be used in glasses to reduce road glare for those driving with the sun in their eyes?

Experiment to determine whether or not such glasses will be effective in reducing glare from the painted surfaces of automobiles, from glass surfaces, from unpainted metal surfaces.

2. *Disappearing water.*

Spill a little water on the top of a table and notice the large amount of light reflected from the small puddle of water. Look at the puddle through a linear polarizer.

Can you rotate the polarizer so that the puddle seems to disappear?

Does it disappear most completely when looked at from directly overhead, or when looked at almost horizontally along the edge of the table, or at some intermediate angle?

3. *Polarized skylight.*

Look through a linear polarizer at a bright blue sky. Rotate the polarizer slowly in its own plane.

Does the apparent brightness of the sky change as the polarizer is rotated?

Does your answer depend on what portion of the sky you are looking at? If so, how can one part of the sky possibly be different from another part of the sky in this respect?

Can a linear polarizer be used to improve cloud pictures?

4. *Transmission axis.*

Recall that the electric vector of linearly polarized light oscillates along a fixed line perpendicular to the direction of the light beam. A line drawn on a sheet of linear polarizer par-

Suggested experiments with linearly polarized light

allel to the electric vector of the transmitted light is called the transmission axis of the polarizer.

> *How can you ascertain the transmission axis of a sheet of linear polarizer?* [Hint: *Use your results on glare reduction. Recall that light reflected obliquely from a horizontal dielectric surface is partially polarized with a predominantly* horizontal *component of electric vibration.*]
> *In what direction is light from a blue sky polarized?*

5. "Haidinger's brushes."

Look through a linear polarizer at a bright sky (preferably a bright blue sky). Hold the polarizer fixed and stare through it steadily for 10 sec. Then very suddenly rotate it 90° in its own plane. During the first half-second after rotation do you see a faint yellowish "bow tie" in the direction in which you have been looking? If not, ask a friend to try: Not everyone is able to see this effect, called Haidinger's brushes.

> *When the polarizer is rotated again an additional 90° are the brushes differently oriented?*
> *Can you use Haidinger's brushes to determine the transmission axis of a linear polarizer directly by eye?*

6. A variable-transmission window.

Place two pieces of linear polarizer over one another. Rotate one of the polarizers slowly in its own plane. Explain the varying light transmission in terms of the transmission axes of the polarizers and direction of electric vibration of the light.

7. A paradox.

Place two linear polarizers over one another and orient them so that their transmission axes make an angle of 90°, that is, so that little or no light comes through. Now insert a third polarizer, with transmission axis at 45°, between the other two—like a slice of ham in a sandwich.

> *How can* adding *a light-absorbing polarizer result in* more *transmitted light?*
> *Can you explain changes in transmitted light as the*

Photons and quantum states

middle (ham) polarizer is rotated in its plane while you hold the outer two pieces (bread) fixed in orientation?

8. *Polarized light as a diagnostic tool.*

Try inserting transparent objects between two linear polarizers that are crossed (90° between transmission axes) or uncrossed. Suggested items: plastic rulers, triangles, protractors, Scotch tape, wallet windows, eyeglass frames.

What happens when you rotate these plastic pieces between crossed polarizers? (Some of these effects can be explained only in terms of circularly polarized light.)

What happens when you bend a thick piece of plastic while it is between crossed polarizers? (Glassblowers use crossed polarizers to detect unrelieved stresses in their glassware.)

9. *Calcite.*

Make a small black dot on a piece of white paper—or better, make a white dot on a sheet of black paper. Lay a calcite crystal on the paper over the dot. Analyze what you see by viewing the transmitted light through a linear polarizer.

What is the effect of calcite on unpolarized light incident upon it?

10. *Two calcite crystals.*

You are asked to look through *two* pieces of calcite (arranged one on top of the other, that is, in series) at a dot on a piece of paper. How many dots would you expect to see?

SMITH SAYS: *"The second piece of calcite will separate further the two beams separated by the first piece of calcite. Therefore you will still see two dots through the two pieces of calcite."*

BROWN SAYS: *"Smith is almost right. The second piece of calcite transmits the two beams from the first calcite, but only if the two pieces are oriented the same way. Otherwise the*

Suggested experiments with linearly polarized light

second calcite will not transmit either *beam from the first calcite. I conclude that at one orientation of the second piece of calcite you will see* two dots; *at all other orientations,* no dots."

JONES SAYS: *"Nonsense. The second calcite will split in two each of the beams transmitted by the first piece of calcite. Therefore through the two pieces of calcite you will see a total of* four dots."

GREEN SAYS: *"You are all wrong. The second piece of calcite simply recombines the two beams separated by the first piece of calcite. Through both pieces of calcite you will see only* one dot."

Can you decide how many dots (zero, one, two, or four!) will be seen through two pieces of calcite—without actually obtaining a second piece of calcite to try it out? Verify intermediate stages in your argument using one *piece of calcite and one or more linear polarizers.*

EXERCISES

6-1 *Identifying photon quantum states.* A beam of visible light is emitted from a complicated optical apparatus. When you place a piece of ideal linear sheet polarizer in the beam you find an orientation of the polarizer for which the light it transmits is *blue*. When the polarizer is rotated through an angle of 90° from this orientation about the beam as axis, the transmitted beam is *red*. A photomultiplier verifies that the intensity of the blue beam plus the intensity of the red beam is equal to the intensity of the original beam.

(a) Are the photons in this beam in a single quantum state?

(b) What is the simplest description of the photons in the beam consistent with the results given above?

(c) By what additional experiments could you verify whether or not your description for part (b) is correct?

(d) Assume that your description given for part (b) *is* correct. Sketch a possible design for the "complicated optical apparatus" that produces this beam. Using any additional equipment you care to specify, can you separate the output beam into two or more beams, each of which contains photons in a single quantum state?

Photons and quantum states

6-2 *Diagnosis using an xy analyzer.* A calcite *xy* analyzer is placed in various beams of monochromatic photons (all photons of the same energy). The analyzer is rotated about the beam as axis.

(a) For beam A, there is one orientation of the analyzer for which the output of channel *y* has intensity I_0 and the output of channel *x* is zero. Predict the intensities of the outputs of *both* channels as the analyzer is rotated about the beam as axis.

(b) For beam B both output beams of the *xy* analyzer have equal intensities for all orientations of the analyzer. What conclusion(s) can you draw about the beam incident on the analyzer?

(c) For beam C the outputs of the *x* and *y* channels each vary with orientation of the analyzer, but there is no orientation for which the output of either channel is zero. What conclusion(s) can you draw about the beam incident on the analyzer?

6-3 *The photon flux at some detectors.* The visible light from a distant point source can be crudely measured by the "photon flux": the rate at which photons in the visible region of the spectrum cross a unit area normal to the line of sight to the source. The cgs unit for photon flux is photons cm^{-2} sec^{-1}. The response of the eye is crudely logarithmic, so that equal photon flux *ratios* are perceived as equal *differences* in apparent brightness. This is the basis of the *astronomical apparent magnitude* scale, in which a *decrease* of one magnitude corresponds to an *increase* in photon flux by a *factor* of $10^{0.4}$. (Thus a difference of five magnitudes corresponds to a *factor* of 100 in photon flux.) Although the scale was invented for use in the observation and description of point sources (namely stars and planets) by the unaided eye, it has since been made very precise and has been extended to objects too dim for the eye to perceive and also to the moon and the sun, which the unaided eye can perceive as extended (that is, they are *not* unresolved "point" sources). Two objects whose visible photon fluxes are J_1 and J_2 are assigned visual apparent magnitudes m_1 and m_2 which satisfy the equation $m_1 - m_2 = -2.5 \ \log(J_1/J_2)$, or $J_1/J_2 = 10^{-0.4(m_1 - m_2)}$. The apparent magnitude scale is convenient because a tremendous range in photon fluxes is accommodated by rather modest magnitude differences, as we shall see in this exercise.

(a) The apparent magnitude *m* of the sun is -27, and that of the moon is -11. What is the ratio of the visible photon flux from the moon to that from the sun?

(b) We can calibrate the magnitude scale using the sun. In the visible wavelength region, the sun delivers about 1 calorie/min to each square centimeter of the earth's surface. (This is about 4.4×10^{17} eV cm^{-2} sec^{-1}.) Convert this to flux of visible photons. Use 5000 Å as an average wavelength in the visible region of the spectrum. Using the ratio obtained in part (a), find the photon flux from the full moon.

(c) Among the objects for which the magnitude scale was first used, the planet Venus (at its brightest) has $m = -4$; a typical bright star has $m = +1$, and the dimmest stars visible to the unaided eye have $m = +6$. Find the photon fluxes corresponding to each of the magnitudes. If the pupil of the human eye dilates to a diameter of 6 mm ($\sim \frac{1}{4}$ in.) for nighttime viewing, how many photons per second enter the eye from a barely visible star? If the eye accumulates photons over a 0.05 sec "integration time" before processing the signal, how many photons contribute to the "detection" of a barely visible star?

(d) The faintest objects that can be detected using the 200-in. telescope on Mt. Palomar have $m = +24$. For such objects, how many photons per second strike the 200-inch main mirror? If the detector used with the telescope accumulates photons over a 10^3 sec (about 15 min) integration time, how many photons contribute to the detected signal?

6-4 *The classical description of polarized light.* In this exercise we examine some features of the classical description of polarized monochromatic light. [*Note:* You may find it useful to refer to the Appendix to this chapter in doing this exercise.]

(a) The electric field of an x-polarized plane wave of angular frequency ω propagating in the $+z$ direction is given by

$$\mathbf{E}(z, t) = \hat{x} E_{xo} \cos (kz - \omega t + \delta_x)$$

where $\omega = ck$. Now according to classical electromagnetic theory, the intensity of any plane wave is proportional to $\overline{E^2}$, the time average of the square of the electric field vector. Show that for an x-polarized plane wave, the instantaneous value $E^2 (z, t)$ depends on z but that $\overline{E^2}$ is independent of z. Verify that $\overline{E^2} = E_{xo}{}^2/2$.

(b) The electric field of a wave can be written in terms of its components $E_{x'}$ and $E_{y'}$ along the x' and y' directions. Write the electric field for the x-polarized wave of (a) in terms of its x' and y' components. Compute $\overline{E_{x'}{}^2}$ and $\overline{E_{y'}{}^2}$ and verify that $\overline{E_{x'}{}^2} + \overline{E_{y'}{}^2} = E_{xo}{}^2/2$.

(c) As you may have recognized, the results obtained in (a) and (b) are simple examples of general rules. The most general monochromatic plane wave propagating in the $+z$ direction has an electric field given by

$$\mathbf{E}(z, t) = \hat{x} E_{xo} \cos (kz - \omega t + \delta_x) + \hat{y} E_{yo} \cos (kz - \omega t + \delta_y)$$

where E_{xo} and E_{yo} are positive (or zero). For this general plane wave, show that (i) $E^2(z, t) = E_x{}^2(z, t) + E_y{}^2(z, t)$ and (ii) $\overline{E^2} = (E_{xo}{}^2 + E_{yo}{}^2)/2$. Thus the intensity of any wave depends only on the *sum* of the squared amplitudes of the electric field components

Photons and quantum states

along two orthogonal directions. The intensity of the wave depends on *neither* the relative phase of the two orthogonal oscillating field components *nor* the ratio of the amplitudes E_{xo} and E_{yo}.

(d) It is the *polarization state* itself which depends on the detailed relationship between $E_x(z, t)$ and $E_y(z, t)$. Describe as fully as you can the polarization corresponding to each of the following choices in the above equation for $\mathbf{E}(z, t)$. These choices all correspond to linear or circular polarization states. (The general case would correspond to elliptical polarization.)

(i) $E_{xo} \neq 0$; $E_{yo} = 0$
(ii) $E_{xo} = 0$; $E_{yo} \neq 0$
(iii) $E_{xo} = E_{yo}$; $\delta_y = \delta_x$
(iv) $E_{xo} = E_{yo}$; $\delta_y = \delta_x + \pi$
(v) $E_{xo} \neq E_{yo}$; $\delta_y = \delta_x$
(vi) $E_{xo} = E_{yo}$; $\delta_y = \delta_x + \pi/2$
(vii) $E_{xo} = E_{yo}$; $\delta_y = \delta_x - \pi/2$

6-5 *Sequential projections.* A beam of y-polarized photons is incident on two ideal linear polarizers in sequence, as shown in the figure. The first polarizer has its transmission axis oriented at an angle θ_1 with respect to the y axis and the second at an angle θ_2.

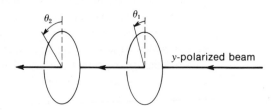

(a) What is the transmission probability through the *first* polarizer for $\theta_1 = 0$? for $\theta_1 = \pi/2$?

(b) What is the net transmission probability through the system of *two* polarizers:

for $\theta_1 = 0$ and $\theta_2 \neq 0$?
for $\theta_1 = \theta_2$?
for $\theta_1 = \theta_2/2$?

(c) Find an expression for the net transmission probability through the system as a function of θ_1 for a given fixed value of θ_2. For what value(s) of θ_1 is the transmission a maximum?

6-6 *Linear polarization rotator.* Consider a system of N *ideal* linear polarizers in sequence, as shown in the figure. The transmission axis

Exercises

of the first polarizer makes an angle of θ/N with the y axis. The transmission axis of every other polarizer makes an angle θ/N with respect to the axis of the preceding one. Thus the transmission axis of the final polarizer makes an angle θ with the y axis. A beam of y-polarized photons is incident on the first polarizer.

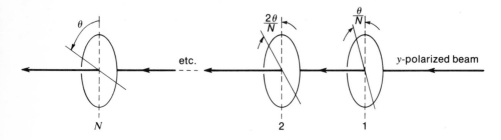

(a) What is the polarization state of the photons that emerge from the final polarizer?

(b) What is the probability that an incident photon is transmitted by the array?

(c) For a given angle θ, find the probability of transmission in the limit of large N. (Remember, these are assumed to be ideal polarizers.)

(d) Suppose the angle $\theta = 90°$. Does the result of (c) violate our understanding that the x-polarization state is orthogonal to the y-polarization state, as embodied in the statement, "If a y-polarized beam is projected into x there will be zero output"?

(e) Suppose, more realistically, that each polarizer in the array passes only the fraction f of light polarized along its transmission axis. (But still assume that all of the light perpendicular to this axis is absorbed.) How will this change the predictions of parts (b) and (c)? Typically f has the value 0.5 to 0.9 and depends on wavelength. (The fraction transmitted for the perpendicular direction is typically 10^{-2}–10^{-6} and also depends on wavelength.)

(f) (Optional) For $\theta = \pi/4$ and $f = 0.8$, find the value of N for which the transmission probability is a maximum.

6-7 *"Building up the statistics."* The percentage accuracy of the result of a counting experiment typically increases as the number of counts increases. (See Section 6-4.) An experimenter desires to verify the $\cos^2\theta$ law of transmission through two linear polarizers in sequence by counting single photons in a low-intensity experiment. For a given source intensity, how many times as long must his experiment run at $\theta = 75°$ than at $15°$ in order to yield the same percentage accuracy in the two cases?

Photons and quantum states

6-8 *Verifying a polarization state.* In a low-intensity counting experiment a weak beam of right-circularly polarized light enters an xy analyzer. In a given experimental run 54 photons are counted in the x-output channel and 46 in the y-output channel. The experimenter concludes that the incident beam cannot be R polarized since, he says, an R-polarized beam should yield equal intensities in the x and y channels.

(a) Make a brief but, if possible, quantitative argument that the experimental result is not inconsistent with the incident beam being in the R-polarized state.

(b) What alternative experiment would you propose in which detection of 100 photons would provide more convincing evidence that the incident beam is R polarized? What outcome would you expect for your proposed experiment? Would such an outcome make it *certain* that the incident beam is R polarized?

6-9 *Obtaining an L-polarized beam from an R-polarized beam.* An experimenter wishes to obtain an L-polarized beam from an R-polarized beam.

(a) If an RL analyzer is inserted in the beam, what fraction of the initial R-polarized photons will emerge from the L channel?

(b) If a linear polarizer is inserted in the R-polarized beam and then the resulting beam passed through an RL analyzer, what fraction of the original photons, on the average, will emerge from the L channel?

(c) A piece of *sheet L* polarizer (similar to the setup in Figure 6A-12 but with 45° angle changed to $-45°$) is inserted in the R beam with the linear polarizer side in the input direction. What fraction of incident photons will be transmitted by this device?

(d) If you have not done so already, describe the results of (b) and (c) in terms of the classical theory of electromagnetic waves.

(e) (Optional) If you understand the principle of operation of the sheet circular polarizer, you can devise a method using both an R-sheet polarizer and an L-sheet polarizer, each appropriately oriented, so that in principle *all* of the initial R-polarized photons emerge L polarized. Do it. This is one example of a general result that *any* polarization state can be converted to any other with no loss of intensity if ideal devices can be assumed.

6-10 *Do polarization states y and R form a complete set of states?*
(a) A quarter-wave plate is placed in a beam of x-polarized photons. By consulting Figure 6A-12 in the Appendix to this chapter, determine the angle of orientation of the quarter-wave plate such that the light emerging from it is R polarized. Draw a diagram of the rela-

tive orientation. If the quarter-wave plate is ideal, does it transmit the same number of photons as are incident upon it?

(b) Now suppose that the quarter-wave plate, oriented as determined in part (a), is placed in the *x*-output channel of an *xy* analyzer. The *y*-output beam is left undisturbed. Considering the *xy* analyzer and the quarter-wave plate as a single device, is the number of photons transmitted by the device equal to the number incident upon it?

(c) Lucy claims that the pair of states *y* and *R* that constitute the two outputs of the device described in (b) form a *complete set* of photon polarization states. Is Lucy right? Discuss your opinion with others: Are the two states *y* and *R* orthogonal to one another? Does this matter? What happens when either output beam of the device described in (b) enters the input of another identical device? Does this result matter to the argument about completeness?

6-11 *Who goes there?* On the basis of the experimental results below, try to classify each of the beams 1 through 5 under one of the following headings:

> unpolarized light
> partially polarized light
> linearly polarized light
> circularly polarized light
> elliptically polarized light

If the given experiments are not sufficient to determine classification into one of the above headings, list the alternative polarization states consistent with the given results and describe experiments that would distinguish among these alternatives.

BEAM 1: A linear polarizer placed in beam 1 transmits maximum intensity when its transmission axis is at $\theta = 45°$ to the *x* axis and essentially zero intensity when its transmission axis is at $\theta = -45°$ (note minus sign). The *x* axis is transverse to the beam direction.

BEAM 2: A linear polarizer placed in beam 2 and rotated about the beam as axis transmits the same intensity for all orientations.

BEAM 3: A linear polarizer inserted in beam 3 and rotated about the beam as axis transmits a maximum intensity in one orientation and a minimum intensity when rotated 90° from this maximum, but does not extinguish the beam for any orientation.

BEAM 4: A piece of naturally cleaved calcite with one face perpendicular to beam 4 transmits two separated beams for most

Photons and quantum states

orientations when the calcite is rotated about the beam as axis, but only a single beam for particular orientations that are 90° apart.

BEAM 5: A quarter-wave plate placed in beam 5 transmits linearly polarized light. (For a description of a quarter-wave plate, see the Appendix to this chapter.) As the quarter-wave plate is rotated about the beam as axis, the axis of linear polarization of the transmitted light rotates, but the intensity of the transmitted light does not vary.

If we want to describe what happens in an atomic event, we have to realize that the word "happens" can apply only to the observation, not to the state of affairs between two observations.

WERNER HEISENBERG, *Physics and Philosophy* (1958)

7
Quantum amplitudes and state vectors

7-1 INTRODUCTION

In the preceding chapter we defined a quantum state and examined some properties of photon quantum states. Photon states illustrate the general result that the predictions of quantum physics must be expressed in terms of probabilities. In particular, we defined the projection probability from one state to another as the fraction of incident particles in the first state that appear in the output beam of a projector for the second state.

In the present chapter photon states are used further to sharpen up a feature of quantum physics with which Chapter 2 has already made us very familiar: probabilities alone are not sufficient to describe interference experiments with particles. In order to describe the results of interference experiments we were led to define a new quantity, the *quantum amplitude*. The discussions in this chapter will show that the quantum amplitude is in general a complex number—a property strongly suggested by our initial discussion of the wave-function Ψ in Chapter 3. We shall see how, in any interference experiment, we combine these complex amplitudes to form a resultant quantum amplitude, the squared magnitude of which is a probability. The quantum *amplitudes* are necessary for analyzing an interference experiment, even though it is a *probability* that embodies the experimentally testable result.

At the end of the chapter we shall look beyond the quantum amplitudes for a given state to an even more abstract and

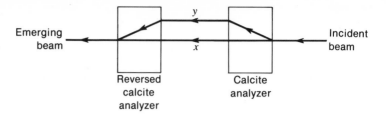

Fig. 7-1 Schematic diagram of xy analyzer loop constructed of two pieces of calcite.

Emerging beam

Reversed calcite analyzer

Calcite analyzer

Incident beam

powerful description of the state. This description, called a *state vector*, permits the same kind of freedom in thinking about quantum states that ordinary velocity and acceleration vectors, for example, provide in thinking about the classical trajectory of a particle in Newtonian mechanics.

We shall develop all these concepts in the relatively simple context of photon polarization states, but the essential results are completely general.

7-2 THE ANALYZER LOOP

The experiments discussed in this chapter involve a piece of equipment we call an *analyzer loop*. This is a two-part device of which the first part is just an analyzer as defined in Chapter 6. The second part of the analyzer loop is a "reversed" analyzer of the same type, which recombines the beams separated by the first analyzer in such a way as to reconstruct the original beam in every detail (see Figure 7-1).

An analyzer loop is not a device likely to be found in the average laboratory; nor, in fact, is it referred to by this name in the literature.[1] Figure 7-2 schematically illustrates how such a device has been incorporated in a commercial form of interference microscope. Our present concern is with the conceptual use of an analyzer loop, as illustrated in Figure 7-1, to make clear some basic properties of quantum states. An actual working device corresponding as closely as possible to Figure 7-1 has been made; it is shown in Figure 7-3.[2]

[1]See, however, *The Feynman Lectures on Physics*, Vol. 3, Addison-Wesley, Reading, Mass., 1965, Chap. 5.

[2]In constructing an analyzer loop it is not enough simply to superpose the separated beams physically. Each part of the cross-sectional area of the original beam must be restored to the same position relative to the other parts, and the optical paths traveled by the separated beams must be equal or at most differ by an integral number of wavelengths. In fact, the analyzer loop is a type of *interferometer* as we shall shortly show, and its construction requires the precision typical of interferometry.

Quantum amplitudes and state vectors

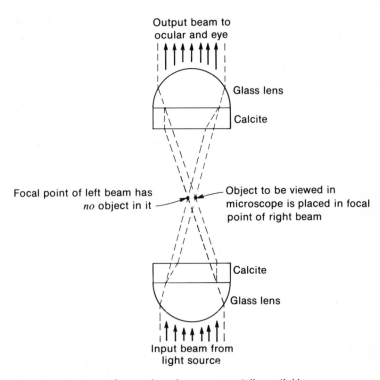

Output beam to
ocular and eye

Glass lens

Calcite

Focal point of left beam has *no* object in it

Object to be viewed in microscope is placed in focal point of right beam

Calcite

Glass lens

Input beam from
light source

Fig. 7-2   Diagram of an analyzer loop commercially available as a portion of one kind of interference microscope. When the loop is used in the microscope, the sample under observation is placed in one of the separated beams. Interference effects in the recombined beam allow determination of thickness and optical properties of the sample. The first published account of the construction of what we call an analyzer loop described a device similar to the one pictured here. [A. A. Lebedev, Revue d'Optique, **9**, 385 (1930). See also John Strong, Concepts of Classical Optics, Freeman, San Francisco, 1958, p. 388.]

Fig. 7-3   Photograph of linear analyzer loop constructed after the design of Figure 7-1. Calcite crystals are contained in metal boxes at either end. Knob controls vanes (not shown) to stop one or the other of the separated beams when desired. This device was constructed by J. L. Burkhardt.

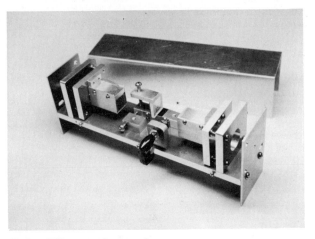

7-2   The analyzer loop

Following the nomenclature of the preceding chapter, we label an analyzer loop by the set of states associated with the analyzers that make up the device. Thus we speak of an xy analyzer loop or an $x'y'$ analyzer loop (the same device with a different orientation) or an RL analyzer loop. All three of these devices have the same overall effect as long as both channels are open: each device reconstructs any incident beam without change. Figure 7-4 shows simple experiments illustrating that

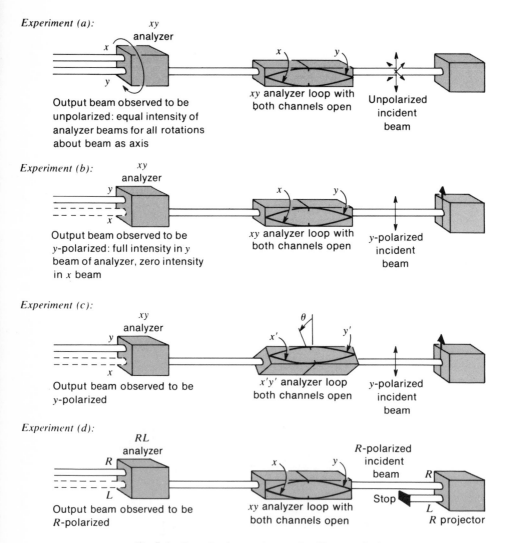

Experiment (a):

Output beam observed to be unpolarized: equal intensity of analyzer beams for all rotations about beam as axis

xy analyzer loop with both channels open

Unpolarized incident beam

Experiment (b):

Output beam observed to be *y*-polarized: full intensity in *y* beam of analyzer, zero intensity in *x* beam

xy analyzer loop with both channels open

y-polarized incident beam

Experiment (c):

Output beam observed to be *y*-polarized

x'y' analyzer loop both channels open

y-polarized incident beam

Experiment (d):

Output beam observed to be *R*-polarized

xy analyzer loop with both channels open

R-polarized incident beam

Stop

R projector

Fig. 7-4   *Four simple experiments that illustrate the fundamental property of an analyzer loop: any incident beam is transmitted unchanged.*

for a number of different inputs the analyzer loops are indeed operating as required. Notice that the xy analyzer transmits unchanged incident beams in *any* state of linear polarization or circular polarization—or indeed in any polarization state whatever! See Figure 7-4d.

7-3 PARADOX OF THE RECOMBINED BEAMS

A simple extension of the experiment of Figure 7-4c demonstrates the inadequacy of using projection probabilities alone to describe all polarization experiments with photons, and leads to the idea of quantum amplitudes. The complete experiment consists of three parts, as diagrammed in Figure 7-5. A y-polarized beam passes through the $x'y'$ analyzer loop A. For concreteness, let the angle θ between the x and x' axes be 30°. The output of A passes into the x projector B, and the intensity of the beam that emerges from B is measured.[3]

In the first part of the experiment the x' channel of the analyzer loop has been blocked (Figure 7-5a). When one interior channel of any analyzer loop is blocked, the beam that passes through the open channel is unaffected by the second stage of the loop; there is nothing left for this beam to recombine with. Hence the device as a whole acts merely as a *projector* for the state labeled by the open channel. Therefore the outcome of the experiment in Figure 7-5a can be predicted on the basis of the results of Chapter 6. The output beam of A is y' polarized, and its intensity is the intensity I_o of the input beam times the projection probability from state y to state y', which has the value $\cos^2 30° = \frac{3}{4}$. When the beam passes through projector B, its intensity is further diminished by the projection probability from state y' to state x; that is by the factor $\cos^2 60° = \frac{1}{4}$. The intensity of the final beam that emerges from B should therefore be $I_o(\frac{3}{4})(\frac{1}{4}) = \frac{3}{16}I_o$. This is indeed the observed outcome of the experiment.

If a stop is placed in the y' channel of analyzer loop A instead of the x' channel (Figure 7-5b), the experiment can be analyzed in precisely the same way as the preceding one: The projection probabilities are now $\frac{1}{4}$ for the first stage and $\frac{3}{4}$ for the second, and the final intensity is again $\frac{3}{16}I_o$.

[3]A short film that demonstrates a similar experiment is available under the title *Interference in Photon Polarization* from the Education Development Center Inc., Newton, Mass.

7-3 Paradox of the recombined beams

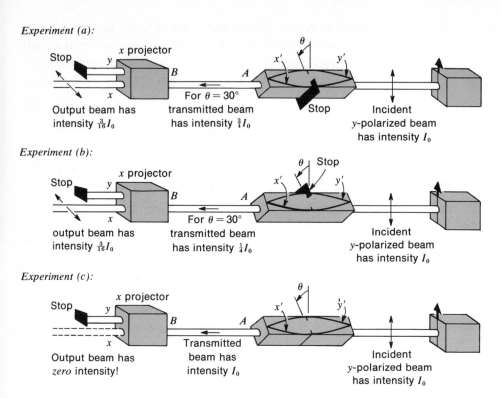

Experiment (a):

Stop

x projector

x projector *B* *A*

For $\theta = 30°$

x' θ y'

Stop

Output beam has intensity $\frac{3}{16}I_0$

transmitted beam has intensity $\frac{3}{4}I_0$

Incident *y*-polarized beam has intensity I_0

Experiment (b):

Stop

x projector

θ Stop

x' y'

B *A*

For $\theta = 30°$

output beam has intensity $\frac{3}{16}I_0$

transmitted beam has intensity $\frac{1}{4}I_0$

Incident *y*-polarized beam has intensity I_0

Experiment (c):

Stop

x projector

θ

x' y'

B *A*

Transmitted

Output beam has *zero* intensity!

beam has intensity I_0

Incident *y*-polarized beam has intensity I_0

Fig. 7-5   Opening both channels of the analyzer loop A *rather than just one channel results in a* decrease *in the intensity of the photon beam that emerges from the projector* B.

In the third part of the experiment (Figure 7-5c), both channels are open. In analyzing this experiment we might argue (erroneously) as follows. A photon that emerges from *B* must have followed one of the two possible paths through the loop, either through the *x'* channel or through the *y'* channel. The experiments of Figures 7-5a and b show that the probability for a photon to get through via either channel is $\frac{3}{16}$. Therefore the probability to get through when both channels are open should be the sum of the individual probabilities, or $\frac{3}{8}$. That is, the intensity of the emerging beam should be $\frac{3}{8}I_0$. But the actual result of the experiment is that *no photons whatever* emerge when both channels are open. When the number of paths available to a photon is increased, the probability that a photon be transmitted is decreased! How can the concept of individual photons be consistent with this experimental result?

An alternative analysis of the same experiment leads very

Quantum amplitudes and state vectors

simply to the correct result. The analyzer loop is *defined* by the property that when both channels are open the output beam is identical to the input beam. In Figure 7-5c the *input* is y polarized; therefore the *output* of the analyzer loop is also y polarized. The y-polarized beam then falls on an x projector. Since the projection probability from state y to state x is zero, no photons are transmitted by the final x projector. We thus arrive at the correct prediction, zero, instead of the incorrect prediction $\frac{3}{8} I_0$ given by the earlier line of argument. But what is wrong with the first line of argument? It is based on the assumption that the resultant probability with both paths open is the sum of the probabilities for travel by each alternative path. This apparent contradiction is of just the same kind encountered in attempting to analyze Young's double-slit interference experiment in terms of two independent paths for the rays of light and corresponding interference experiments with particles described in Chapter 2. The next section discusses such problems collectively.

7-4 INTERFERENCE EFFECTS IN GENERAL

Interference experiments with light have a straightforward explanation on the basis of classical electromagnetic wave theory. In fact, the successful interpretation of interference experiments was instrumental in bringing about the acceptance of the wave theory during the nineteenth century. According to classical wave theory, if light arrives at a given point by way of two or more paths, the total electric field is the *vector sum* of the *fields* associated with each contributing wave. The total intensity is proportional to the square of the magnitude of the *resultant* electric field vector, which can be shorter than the field vector of any of the contributing waves and can even be zero for complete "destructive interference." Crucial to the possibility of such a cancellation is the fact that one *first* adds the vectors and *then* squares the magnitude of the resultant vector. In Young's two-slit interference experiment, for example (Figure 7-6), opening slit 2 causes an additional field to arrive at point A which "interferes destructively with" (that is, points always in the direction opposite to) the field associated with the wave from slit 1. The total electric field therefore vanishes and zero intensity results. A similar argument explains the result of zero intensity in the "recom-

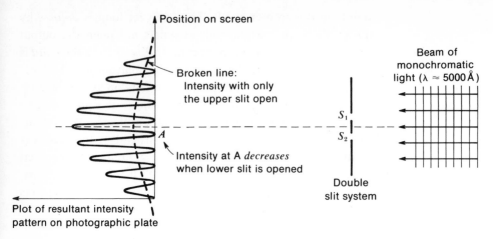

Fig. 7-6 Schematic diagram of Young's two-slit interference experiment with light.

bined beams" experiment of Figure 7-5c. Figure 7-7 illustrates the classical wave analysis of that experiment.

We saw in Chapter 2 that there is compelling evidence that interference effects strictly comparable to those for light occur when electrons, neutrons, and even ordinary atoms are scattered by crystals. In such cases, separate beams derived from a common source reach the screen or detector after being scattered from individual atoms or nuclei of the crystal. Figure 7-6 can also represent the analytically simpler case in which electrons pass through a pair of slits. "Dark" portions of the resulting interference pattern exhibit the same property that is demonstrated in Young's experiment for light (Figure 7-6) or in the experiment of Figure 7-5c; point A, for example, is illuminated *less* when both beams are superposed than when either is present alone. And, as we concluded in Chapter 2, we can obtain the cancellation needed to describe destructive interference, such as that in Young's experiment—or, now, the interference in the experiment of Figure 7-5—by making the following two assumptions:

1. The probability for incident photons to be counted in each experiment of Figure 7-5 is the squared magnitude of the *quantum amplitude* for that measurement. This amplitude is in general (see Chapter 3, Section 3-6) a complex quantity expressible in the form $Ae^{i\phi}$, where A is positive real and $e^{i\phi}$ is a phase factor of magnitude unity. The squared magnitude $|Ae^{i\phi}|^2$ is then simply equal to A^2.

Quantum amplitudes and state vectors

Experiment (a): y′ channel of analyzer loop open

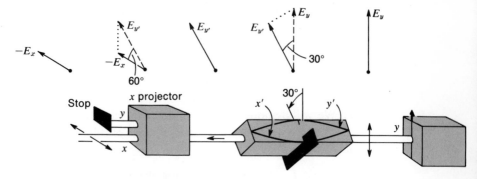

Experiment (b): x′ channel of analyzer loop open

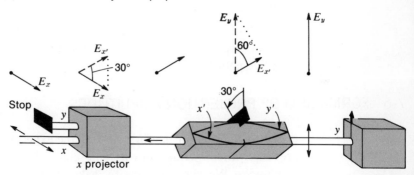

Experiment (c): Both channels of analyzer loop open

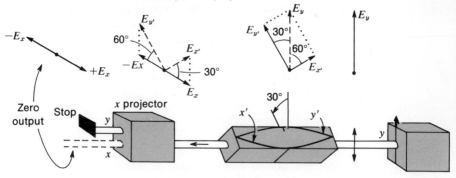

Fig. 7-7 *Wave-optical explanation of the result of experiments of Figure 7-5. In experiment (c) electric field amplitudes for alternative paths interfere to give zero resultant.*

2. The resultant quantum amplitude for the experiment of Figure 7-5c is the sum of the amplitudes for the experiments of Figures 7-5a and b.

7-4   Interference effects in general

That is, when a photon can follow two or more alternative paths, we add the *amplitudes* associated with the alternative paths rather than adding the *probabilities*. The probability is then the squared magnitude of the resultant amplitude. This hypothesis accounts for the observed results if the amplitudes for the experiments of Figure 7-5a and 7-5b have equal magnitudes but opposite sign (corresponding to a phase difference of π). We will demonstrate this in detail in the following section.

These assumptions are special cases of rules true for all of quantum physics:

Rule 1. The probability for any experimental outcome is the squared magnitude of a number called the quantum amplitude for that outcome.

Rule 2. For an interference experiment, the resultant quantum amplitude is the sum of the amplitudes for each alternative path.

7-5 FORMALISM OF PROJECTION AMPLITUDES

We cannot go further in analyzing the experiment of Figure 7-5 without writing down some equations for quantum amplitudes. We need symbols to make these equations concise. The "elementary event" in the experiments of Figure 7-5 is a projection: Each of the experiments of Figure 7-5a and 7-5b consists of two projections in sequence; the relevant quantum amplitude is a *projection amplitude*. We write projection amplitudes using a convenient "bracket notation" due to Dirac.[4] The projection amplitude from some initial state, labeled i, to some final state, labeled f, is written

$$\langle f | i \rangle$$

(Be sure not to confuse such expressions with expressions such as $\langle x \rangle$ that denote expectation values of individual quantities—see Section 5-7.) Notice that by convention the initial state is written at the right side of the bracket and the

[4]P. A. M. Dirac was one of the founders of modern quantum theory around 1926, when he was 24 years old. He shared the Nobel prize (with Schrödinger) in 1933. The first chapter of his book, *The Principles of Quantum Mechanics* (Oxford, Clarendon Press) makes a thoroughly readable and very relevant survey of this chapter's subject matter.

Quantum amplitudes and state vectors

final state at the left. Because of this convention it is convenient to draw our figures with the beams moving from right to left, as we have in fact been doing since the beginning of Chapter 6.

The two projection amplitudes in the experiment of Figure 7-5a can be written in the bracket notation as $\langle y'|y \rangle$ and $\langle x|y' \rangle$. The *projection probability* for a single projection is, according to Rule 1, the squared magnitude of the corresponding projection amplitude. For example,

$$|\langle y'|y \rangle|^2 = \cos^2 \theta \tag{7-1}$$

We cannot yet say what the algebraic sign (or, possibly, phase) of the amplitude $\langle y'|y \rangle$ is; at present we know only its magnitude, $|\cos \theta|$.

With this new notation we return to the analysis of the experiments in Figure 7-5. What are the amplitudes for each of the experiments (a) and (b)? Experiment (a) involves two projections in sequence: a projection from y to y' followed by a projection from y' to x. Figure 7-8 shows the equivalent experiment with projectors. We know the *probability* for each of the projections in the sequence. Moreover, we correctly calculated above the overall probability to be the *product* of the probabilities for each projection in the sequence. We can make this result conform to Rule 1 if we let the corresponding *amplitude* be the product of amplitudes for the individual projections. This is a third assumption:

3. The quantum amplitude for a series of projections is the product of the quantum amplitudes for each projection in the series.

This is a special case of a general result, call it Rule 3:

Rule 3.   The quantum amplitude for a given path is the product of amplitudes for each step in the path.

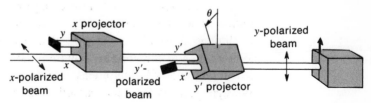

7-5   Formalism of projection amplitudes

The separate projections in the experiment of Figure 7-5a have amplitudes symbolized by $\langle y'|y \rangle$ and $\langle x|y' \rangle$. The resultant amplitude is the product of these two: $\langle x|y' \rangle \langle y'|y \rangle$. We write the amplitudes in the same (right-to-left) order as the projectors appear in the figure. The overall probability (Rule 1 again) is simply the squared magnitude of this product:

$$|\langle x|y' \rangle \langle y'|y \rangle|^2 = |\langle x|y' \rangle|^2 |\langle y'|y \rangle|^2 \tag{7-2}$$

The squared magnitude of a product is equal to the product of squared magnitudes (for complex as well as for real numbers). The right side of the equation gives us once again the product of probabilities that correctly predicted the results of the experiment of Figure 7-5a in the first place.

The experiment of Figure 7-5b can be analyzed in the same way: the amplitudes for the separate projections are $\langle x'|y \rangle$ followed by $\langle x|x' \rangle$. The overall amplitude is their product: $\langle x|x' \rangle \langle x'|y \rangle$.

7-6 PROPERTIES OF PROJECTION AMPLITUDES

How much do we know about the values of individual projection amplitudes $\langle x|y' \rangle$, $\langle y'|y \rangle$, $\langle x|x' \rangle$, $\langle x'|y \rangle$? We know their squared magnitudes, since we know the corresponding projection probabilities. But we do not know the algebraic *signs* of the amplitudes: whether they are positive or negative, or even whether they are real. For experiments with linearly polarized photons, we will find it possible to assign positive and negative real values to projection amplitudes. However, as we asserted earlier, quantum amplitudes are in general complex numbers. Indeed, we will describe below experiments with *circularly* polarized photons for which projection amplitudes are unavoidably complex numbers. In the experiments of Figures 7-5a and 7-5b, which involve real amplitudes, no sequence of projections can give information concerning the algebraic signs of projection amplitudes: when the squared magnitude is taken, all such information is lost. Information about signs comes only from experiments in which amplitudes interfere, that is, add algebraically. Only in such addition, before the absolute square is taken, can amplitudes cancel one another, thus giving information about their relative signs. The experiment that led us to introduce the idea of

Quantum amplitudes and state vectors

amplitudes, that of Figure 7-5c, is just such an experiment. By applying Rules 2 and 3 to the analysis of this and similar experiments, we shall obtain information concerning the relative signs (or, more generally, relative complex phase factors of the form $e^{i\phi}$) of various projection amplitudes. Such determinations yield only *relative* signs; we shall not be able to derive a unique value for each sign or phase. Nevertheless, this information on relative signs or phases is sufficient to predict correctly the outcome of all related experiments.

In the experiment of Figure 7-5c, there are two paths, with two steps in each path. Each step is a projection; therefore, using Rules 2 and 3, we can write the resultant quantum amplitude:

$$\text{Quantum amplitude} = \langle x|y'\rangle \langle y'|y\rangle + \langle x|x'\rangle \langle x'|y\rangle \qquad (7\text{-}3)$$

The observed experimental result (zero counting rate with both channels open) indicates that the squared magnitude of the resultant quantum amplitude of Eq. 7-3 vanishes, which implies that the amplitude itself vanishes:

$$\langle x|y'\rangle \langle y'|y\rangle + \langle x|x'\rangle \langle x'|y\rangle = 0 \qquad (7\text{-}4)$$

The square of the magnitude of each projection amplitude that appears in Eq. 7-4 is the projection *probability* between the designated states. According to the results of Chapter 6, each projection probability is the squared cosine of the angle between the axes that define the two states involved. We can express the *magnitudes* of all the amplitudes in terms of the angle θ between the y and y' axes (Figure 7-9).

$$|\langle x|x'\rangle| = |\langle y'|y\rangle| = |\cos\theta|$$
$$|\langle x'|y\rangle| = \left|\cos\left(\frac{\pi}{2} - \theta\right)\right| \qquad (7\text{-}5)$$
$$|\langle x|y'\rangle| = \left|\cos\left(\frac{\pi}{2} + \theta\right)\right|$$

Figure 7-9 Some angles used in writing amplitudes for projection from one linear polarization state onto another. (Beam emerges toward reader.)

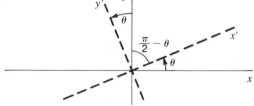

7-6   Properties of projection amplitudes

Assume, as the most convenient possibility, that the amplitudes are all real. This corresponds to the phases ϕ in the factors $e^{i\phi}$ being all either zero or π in this case.[5] Substitute the amplitudes of Eq. 7-5 into Eq. 7-4, leaving the choice of sign still to be made.

$$\left[\pm \cos\left(\frac{\pi}{2}+\theta\right)\right]\left[\pm \cos\theta\right]+\left[\pm \cos\theta\right]\left[\pm \cos\left(\frac{\pi}{2}-\theta\right)\right]=0$$

$$(7\text{-}6)$$

The experiment under study provides only one relation, Eq. 7-6, among the four amplitudes; any set of signs that satisfies Eq. 7-6 leads to a correct account of the result of this experiment. Consequently we have not determined all of the algebraic signs uniquely. However, no experiment provides any further information on these signs. The results of all experiments with linearly polarized photons are consistent with any set of signs that satisfies Eq. 7-6. This is an example of a general result in the study of quantum amplitudes: only limited information concerning signs (more generally, complex phase factors) may be obtained from experiment. Moreover, this information always concerns *relative* signs (or phases) between amplitudes rather than the absolute sign (phase) of any one amplitude. But such information is always sufficient for predicting the results of other experiments in which the same amplitudes appear.

For the present problem a particularly simple choice of signs is possible: We take all signs positive in Eq. 7-6. Since $\cos(\pi/2 - \theta) = \sin\theta$ and $\cos(\pi/2 + \theta) = -\sin\theta$, this choice leads to

$$-\sin\theta\cos\theta + \cos\theta\sin\theta = 0$$

This equation is obviously satisfied for all angles θ.

[5]If one or more of the amplitudes were necessarily nonreal complex numbers, Eq. 7-6 would have no solution and our assumption of real amplitudes would be proved wrong. This happens later (Eq. 7-10) with projection amplitudes between states of linear and circular polarization.

Quantum amplitudes and state vectors

In summary the sign convention gives

$$\langle x|x'\rangle = \langle y'|y\rangle = \cos\theta$$

$$\langle x'|y\rangle = \cos\left(\frac{\pi}{2} - \theta\right) = \sin\theta \qquad (7\text{-}7)$$

$$\langle x|y'\rangle = \cos\left(\frac{\pi}{2} + \theta\right) = -\sin\theta$$

There is a simple way to summarize the combination of experiment and convention by which we have fixed the projection amplitudes for linearly polarized photon states. Experiment determines the polarization axis of a given beam, with no unique direction either way along this axis. The sign convention amounts to adding an arrow that specifies a particular direction along this axis. We can, if we like, permanently inscribe such coordinate arrows on the analyzer that produces a state (Figure 7-10a). Then the projection amplitudes for linearly polarized states (according to the convention we have adopted) can be specified by one simple rule. *For states defined by particular channels of particularly oriented analyzers, the projection amplitude is just the cosine of the angle between corresponding arrows.* One example is shown in Figure 7-10b. A positive angle is defined by a counterclockwise rotation from the initial to the final state in the projection

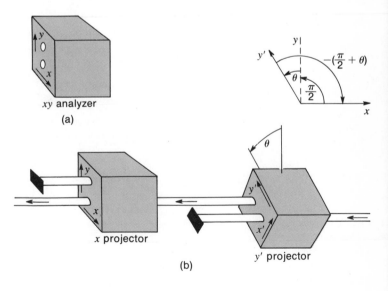

xy analyzer
(a)

x projector

y' projector

(b)

7-6   Properties of projection amplitudes

(counterclockwise looking toward the source). It may be verified directly that the amplitudes given by Eq. 7-7 are all consistent with the simple rule. The rule also implies that for our phase convention the following relation holds for any two states of linear polarization j and k:

$$\langle j|k \rangle = \langle k|j \rangle \qquad \text{(linear polarization states)} \qquad (7\text{-}8)$$

This equation results from the fact that interchanging initial and final states is equivalent to changing the sign of the angle between the arrows defining the polarization states involved, and the cosine is an even function.

We have constructed a quantum-mechanical description of an interference experiment with linearly polarized photons. The essence of this description has been the introduction of the quantum amplitude.

7-7 PROJECTION AMPLITUDES FOR STATES OF CIRCULAR POLARIZATION

In this section we investigate the projection amplitudes from states of linear polarization to states of circular polarization, and vice versa. From this will come the important result that some quantum amplitudes are necessarily complex numbers. The desired amplitudes can be determined with the help of an RL analyzer loop, a device constructed from two RL analyzers in the same way that the xy analyzer loop is constructed from two xy analyzers. An RL analyzer loop is shown schematically in Figure 7-11. If the experiments of Figure 7-5 are repeated with an RL loop replacing the $x'y'$ loop, the results are similar: with either channel of the analyzer loop

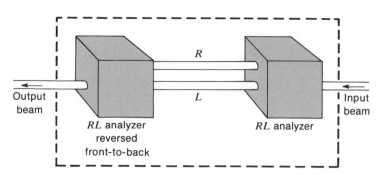

Fig. 7-11 Schematic diagram showing an RL analyzer loop constructed of two RL analyzers.

Quantum amplitudes and state vectors

blocked a beam emerges from the final projector, but with both channels open no photons emerge. It follows, just as in the original experiment, that the probability of detecting a photon with both channels open cannot be the sum of the probabilities associated with the alternative paths; once again we clarify the situation by introducing amplitudes.

Now consider the experiment of Figure 7-12. The rules for combining amplitudes, enunciated in earlier sections, lead to the following expression for the overall quantum amplitude for this experiment:

$$\text{Quantum amplitude} = \langle y'|R \rangle \langle R|y \rangle + \langle y'|L \rangle \langle L|y \rangle \qquad (7\text{-}9)$$

Since by definition the open analyzer loop has no effect on the beam, the overall amplitude for the experiment must be just $\langle y'|y \rangle$, which according to the phase convention of the preceding section has the value cos θ:

$$|\langle y'|R \rangle||\langle R|y \rangle| + |\langle y'|L \rangle||\langle L|y \rangle| = \cos \theta \qquad (7\text{-}10)$$

We already know the squared magnitudes of the four projection amplitudes on the left side of this equation: projection probabilities between states of linear and circular polarization all have the value $\frac{1}{2}$ (Table 6-1). So we know the absolute magnitudes of the projection amplitudes themselves:

$$|\langle y'|R \rangle| = |\langle R|y \rangle| = |\langle y'|L \rangle| = |\langle L|y \rangle| = \frac{1}{\sqrt{2}} \qquad (7\text{-}11)$$

Equations 7-10 and 7-11 taken together show that at least some of the amplitudes in Eq. 7-10 must be complex. For since the angle θ does not appear in Eq. 7-11, no way exists to satisfy Eq. 7-10 by choosing all amplitudes to be simply positive or negative *real* numbers. Real amplitudes will work only in the

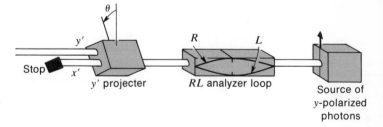

Fig. 7-12 An experiment whose interpretation gives projection amplitudes between linear and circular polarization states.

7-7   Projection amplitudes for circular polarization

special case for which cos θ is equal to zero or ± 1. For all other values of θ it is necessary to use *complex* amplitudes. From Eq. 7-11 we know the magnitudes of these complex numbers, and Eq. 7-10 contains information on the phases. To obtain this information, we give each amplitude a complex phase factor of the form $e^{i\phi}$. Let us assume a separate phase for each projection amplitude in Eqs. 7-10 and 7-11 as follows:

$$\langle y'|R\rangle = \frac{e^{i\alpha}}{\sqrt{2}}$$

$$\langle R|y\rangle = \frac{e^{i\beta}}{\sqrt{2}}$$

$$\langle y'|L\rangle = \frac{e^{i\gamma}}{\sqrt{2}} \tag{7-12}$$

$$\langle L|y\rangle = \frac{e^{i\delta}}{\sqrt{2}}$$

where α, β, γ, and δ are real. Substitution into Eq. 7-10 gives

$$\tfrac{1}{2}(e^{i\alpha}e^{i\beta} + e^{i\gamma}e^{i\delta}) = \cos \theta = \tfrac{1}{2}(e^{i\theta} + e^{-i\theta}) \tag{7-13}$$

where the expression on the right is the exponential form of the cosine function. Equation 7-13 is the only one that can be found relating the four phase factors. Just as in the analogous situation discussed in the preceding section, there are many possible solutions, each of which defines a phase convention. However, except in the special cases $\theta = 0$ or $\theta = \pi/2$, it is not possible to find a solution in which all the phase factors are real: *some projection amplitudes that relate states of linear polarization to states of circular polarization are necessarily complex!* Taking advantage of the freedom of choice in fitting Eq. 7-13, we put $\alpha = \theta$, $\beta = \delta = 0$, $\gamma = -\theta$, which gives the following simple phase convention:

$$\langle y'|R\rangle = \frac{e^{i\theta}}{\sqrt{2}} \tag{7-14a}$$

$$\langle y'|L\rangle = \frac{e^{-i\theta}}{\sqrt{2}} \tag{7-14b}$$

$$\langle R|y\rangle = \langle L|y\rangle = \frac{1}{\sqrt{2}} \tag{7-14c}$$

Once these amplitudes have been fixed, all others are uniquely

Quantum amplitudes and state vectors

determined. Notice first that since y' denotes an arbitrary direction, Eqs. 7-14a and 7-14b specify all projection amplitudes of the form $\langle\psi|R\rangle$ and $\langle\psi|L\rangle$, where ψ denotes an arbitrary state of linear polarization. For example, it follows immediately on putting $\theta = 0$ that

$$\langle y|R\rangle = \frac{1}{\sqrt{2}}$$

$$\langle y|L\rangle = \frac{1}{\sqrt{2}} \tag{7-15}$$

and on putting $\theta = 3\pi/2$ that

$$\langle x|R\rangle = \frac{-i}{\sqrt{2}}$$

$$\langle x|L\rangle = \frac{+i}{\sqrt{2}} \tag{7-16}$$

The only amplitudes that remain to be determined are those of the form $\langle R|\psi\rangle$ and $\langle L|\psi\rangle$, where ψ again denotes a state of arbitrary linear polarization. These follow from a general relation

$$\langle\psi|\phi\rangle = \langle\phi|\psi\rangle^* \tag{7-17}$$

which holds for an arbitrary pair of states.[6] Equation 7-17 can be proved using only the fact that $\langle\psi|\psi\rangle = 1$. The argument goes as follows: If a beam of photons in *any* polarization state ψ enters an RL analyzer, the number emerging from the R channel plus the number emerging from the L channel will equal the number entering the analyzer.

$$|\langle R|\psi\rangle|^2 + |\langle L|\psi\rangle|^2 = 1$$

which may be rewritten

$$\langle R|\psi\rangle^* \langle R|\psi\rangle + \langle L|\psi\rangle^* \langle L|\psi\rangle = 1$$

Now we restate the property of an RL analyzer loop: a beam in

[6]Reminder: If z is a complex number, $z = a + ib = Ae^{i\alpha}$, then its complex conjugate is $z^* = a - ib = Ae^{-i\alpha}$.

7-7   Projection amplitudes for circular polarization

any state ψ will pass through unchanged! Projection of the output beam into state ψ must yield unity.

$$\langle\psi|R\rangle\,\langle R|\psi\rangle + \langle\psi|L\rangle\,\langle L|\psi\rangle = \langle\psi|\psi\rangle = 1$$

Comparing the first term on the left of each of the preceding two equations yields Eq. 7-17, with a corresponding result from comparing the second term in each equation. The result is general not only because ψ is an arbitrary polarization state but also because *any* complete set of two polarization states could have been used instead of R and L in the argument.[7]

Notice that Eq. 7-8 for the (real) projection amplitudes between states of linear polarization is a special case of Eq. 7-17.

Table 7-1 contains a summary of all the results of this chapter concerning projection amplitudes. Compare this table with the similar but simpler table (Table 6-1) that displays the corresponding projection *probabilities*.

7-8 THE STATE VECTOR

Introduction of the State Vector

A powerful concept for discussing quantum states is provided by what is called the *state vector*. The state vector receives its name from the strong analogy between its properties and those of a geometrical vector of unit length—a *unit vector*. The *components* of an ordinary unit vector are equal to its projections onto a given set of mutually perpendicular axes. The *components* of a state vector are its projection amplitudes onto a given set of mutually orthogonal basis states (such as xy or RL). The ordinary unit vector provides an abstract description of a directed quantity independent of the particular axes along which its components are measured. The state vector provides an abstract description of a quantum state independent of the basis states onto which its projection amplitudes are taken.

To introduce the state vector we utilize a familiar thought-experiment as shown in Figure 7-13. Let ψ denote any state of photon polarization (that is, ψ can stand for y or y' or R, and so

---

[7] It is possible to construct an analyzer loop whose channels correspond to any complete set of photon polarization states.

Quantum amplitudes and state vectors

TABLE 7-1 Projection Amplitudes for Photon Polarization States

				From State			
		$\lvert x\rangle$	$\lvert y\rangle$	$\lvert x'\rangle$	$\lvert y'\rangle$	$\lvert R\rangle$	$\lvert L\rangle$
	$\langle x\rvert$	1	0	$\cos\theta$	$-\sin\theta$	$\dfrac{-i}{\sqrt{2}}$	$\dfrac{i}{\sqrt{2}}$
	$\langle y\rvert$	0	1	$\sin\theta$	$\cos\theta$	$\dfrac{1}{\sqrt{2}}$	$\dfrac{1}{\sqrt{2}}$
To State	$\langle x'\rvert$	$\cos\theta$	$\sin\theta$	1	0	$\dfrac{-i}{\sqrt{2}}e^{i\theta}$	$\dfrac{i}{\sqrt{2}}e^{-i\theta}$
	$\langle y'\rvert$	$-\sin\theta$	$\cos\theta$	0	1	$\dfrac{1}{\sqrt{2}}e^{i\theta}$	$\dfrac{1}{\sqrt{2}}e^{-i\theta}$
	$\langle R\rvert$	$\dfrac{i}{\sqrt{2}}$	$\dfrac{1}{\sqrt{2}}$	$\dfrac{i}{\sqrt{2}}e^{-i\theta}$	$\dfrac{1}{\sqrt{2}}e^{-i\theta}$	1	0
	$\langle L\rvert$	$\dfrac{-i}{\sqrt{2}}$	$\dfrac{1}{\sqrt{2}}$	$\dfrac{-i}{\sqrt{2}}e^{i\theta}$	$\dfrac{1}{\sqrt{2}}e^{i\theta}$	0	1

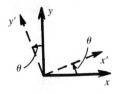

Angle convention:
Beam emerges toward reader.

Notice that with our phase conventions, all amplitudes satisfy the condition $\langle j\lvert k\rangle = \langle k\lvert j\rangle^*$

on). With the help of the open analyzer loop, we can express the projection amplitude from the state ψ to the state of linear polarization y' as

$$\langle y'\lvert\psi\rangle = \langle y'\lvert y\rangle\,\langle y\lvert\psi\rangle + \langle y'\lvert x\rangle\,\langle x\lvert\psi\rangle \tag{7-18}$$

This experiment could be altered by replacing the final y' projector with, say, an R projector or an x projector or any other projector at all. All experiments of this type can be summarized in a single equation with the final state unspecified:

$$\left\langle{\text{Final}\atop\text{state}}\middle\lvert\psi\right\rangle = \left\langle{\text{Final}\atop\text{state}}\middle\lvert y\right\rangle\langle y\lvert\psi\rangle + \left\langle{\text{Final}\atop\text{state}}\middle\lvert x\right\rangle\langle x\lvert\psi\rangle \tag{7-19}$$

Alternatively, one can simply leave the left side of the bracket

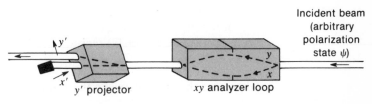

Fig. 7-13 One of a collection of experiments that suggest the introduction of a state vector.

y' projector xy analyzer loop

Incident beam (arbitrary polarization state ψ)

7-8   The state vector

blank:

$$\langle \ |\psi\rangle = \langle \ |y\rangle \langle y|\psi\rangle + \langle \ |x\rangle \langle x|\psi\rangle \qquad (7\text{-}20)$$

For any given experiment, the observer can enter the symbol for the final state in the blank left sides of the brackets—as long as he enters the *same* symbol in all three blanks.

A final step in abbreviation leads to the representation of the state vector which we seek. Remove the angular braces that enclose the blanks in Eq. 7-20:

$$|\psi\rangle = |y\rangle \langle y|\psi\rangle + |x\rangle \langle x|\psi\rangle \qquad (7\text{-}21)$$

The symbol $|\psi\rangle$ is called the *state vector* for the state ψ. Because the $|\ \rangle$ is the right half of the Dirac *bracket* $\langle\ |\ \rangle$, it is sometimes called the *ket* or *ket vector*.

One may of course regard Eq. 7-21 as merely a shorthand form of Eq. 7-19, which in turn describes a series of possible experiments such as the one pictured in Figure 7-13. However, a deeper meaning can be attached to Eq. 7-21. This equation corresponds to the right-hand half of Figure 7-13: a beam in state ψ and an analyzer loop The analyzer loop has no net effect on the beam, merely analyzing the beam into the x, y basis states and then recombining them. In some sense Eq. 7-21, therefore, is a description of the state ψ itself. We regard the state vector $|\psi\rangle$ as an abstract symbol for the state ψ. The right side of Eq. 7-21 expresses this state vector in terms of the state vectors $|x\rangle$ and $|y\rangle$ that symbolize the basis states x and y.

For example, if ψ stands for the state R, Eq. 7-21 reads (see Table 7-1)

$$|R\rangle = |x\rangle\langle x|R\rangle + |y\rangle\langle y|R\rangle$$
$$= |x\rangle \left(\frac{-i}{\sqrt{2}}\right) + |y\rangle \left(\frac{1}{\sqrt{2}}\right) \qquad (7\text{-}22)$$

Or if ψ stands for state y', the equation reads

$$|y'\rangle = |x\rangle \langle x|y'\rangle + |y\rangle \langle y|y'\rangle$$
$$= |x\rangle (-\sin\theta) + |y\rangle (\cos\theta) \qquad (7\text{-}23)$$

where θ is the angle between the y and y' reference directions.

Quantum amplitudes and state vectors

We can just as well use some other analyzer loop in Figure 7-13; for instance, an $x'y'$ analyzer loop or an RL analyzer loop. In this way we can express the state vector $|\psi\rangle$ in terms of the set of state vectors $|x'\rangle$ and $|y'\rangle$ or the set of state vectors $|R\rangle$ and $|L\rangle$. Repeating the argument that leads to Eq. 7-21 we have, in these two cases,

$$|\psi\rangle = |x'\rangle \langle x'|\psi\rangle + |y'\rangle \langle y'|\psi\rangle$$
$$|\psi\rangle = |R\rangle \langle R|\psi\rangle + |L\rangle \langle L|\psi\rangle \tag{7-24}$$

The set of orthogonal states in terms of which the state vector is expressed is called the set of *basis states* or the *representation*. Thus one speaks of the "*xy* representation" or the "basis states *RL*."

Comparison of State Vector with Ordinary Unit Vector

The expansion of $|\psi\rangle$ in various representations is analogous to the expansion of an ordinary vector in terms of unit vectors that point along mutually perpendicular axes. The analogy is most complete if we limit consideration to unit vectors in a plane. We designate an ordinary unit vector with a "hat"; thus $\hat{A}$. This unit vector can be written in terms of its components A_x and A_y with respect to the set of perpendicular axes x and y:

$$\hat{A} = \hat{x} A_x + \hat{y} A_y \tag{7-25}$$

Here $\hat{x}$ and $\hat{y}$ denote unit vectors in the x and y directions respectively. By using the dot product, the components A_x and A_y can be written in a form similar to that of projection amplitudes

$$A_x = (\hat{x} \cdot \hat{A}) = \sin \theta$$
$$A_y = (\hat{y} \cdot \hat{A}) = \cos \theta \tag{7-26}$$

where θ is the angle between the unit vector $\hat{A}$ and the y axis. Then the expansion of the vector $\hat{A}$ in the xy coordinate system is analogous to the expansion of the state vector ψ in terms of the basis states x and y.

$$\hat{A} = \hat{x} (\hat{x} \cdot \hat{A}) + \hat{y} (\hat{y} \cdot \hat{A}) \tag{7-27}$$

Alternatively one can expand the same unit vector $\hat{A}$ in terms of some other pair of perpendicular unit vectors $\hat{x}'$ and $\hat{y}'$, analogous to the expansion of the state ψ in the $x'y'$ basis:

$$\hat{A} = \hat{x}' \, (\hat{x}' \cdot \hat{A}) + \hat{y}' \, (\hat{y}' \cdot \hat{A}) \tag{7-28}$$

Table 7-2 elaborates the analogy between state vectors and unit vectors. Observe in particular the two central properties, *orthogonality* and *completeness*, that are embodied in both schemes. The main difference between ordinary vectors and state vectors is that the components of ordinary vectors are always *real numbers*, whereas the "components" (projection amplitudes) of state vectors are sometimes unavoidably *complex numbers*.

Use of the State Vector in Quantum Mechanics

The state vector plays the same powerful role in quantum mechanics that the ordinary vector does in classical mechanics. The state vector $|\psi\rangle$ denotes the state of a system independent of the basis or representation in which one chooses to express it, just as the vector **a**, for example, denotes the acceleration of a particle without specifying the orientation of the coordinate system with respect to which the acceleration is measured. In each case one is freed to think more generally about nature than would be possible in the absence of vector notation.

What we write inside a ket to represent the state of a system is often determined by the extent of our knowledge of the system or the stage we have reached in the analysis of the system. In this sense the ket provides a note pad for our own convenience. For example, we may write $|\psi\rangle$ to denote the as-yet-unspecified state of a hydrogen atom—similar to the "unknown x" of algebra. As the analysis proceeds we will come to recognize that the specification of the state of a hydrogen atom requires three integers: n, l, and m. Thereafter we can specify a particular state by writing three numbers inside the ket for that state: $|n, l, m\rangle$. Similarly, the momentum state of free electrons in a directed beam may be symbolized by the ket $|p\rangle$. These examples are considered more fully in later chapters.

Quantum amplitudes and state vectors

TABLE 7-2 Analogy between Unit Vectors and State Vectors

Unit Vector $\hat{A}$ (in two-dimensional xy plane)	*State Vector* $|\psi\rangle$ (2-state case: photon polarization)
Representation Expand $\hat{A}$ in xy coordinates	*Representation* Expand $|\psi\rangle$ in xy representation
$\hat{A} = \hat{x}\,(\hat{x}\cdot\hat{A}) + \hat{y}\,(\hat{y}\cdot\hat{A})$	$|\psi\rangle = |x\rangle\,\langle x|\psi\rangle + |y\rangle\,\langle y|\psi\rangle$
or expand $\hat{A}$ in $x'y'$ coordinates	or expand $|\psi\rangle$ in $x'y'$ representation
$\hat{A} = \hat{x}'\,(\hat{x}'\cdot\hat{A}) + \hat{y}'\,(\hat{y}'\cdot\hat{A})$	$|\psi\rangle = |x'\rangle\,\langle x'|\psi\rangle + |y'\rangle\,\langle y'|\psi\rangle$
In general, expand in any complete set of orthogonal coordinates (for example, x'', y'')	In general, expand in any complete set of orthogonal states (for example, $x''y''$, RL, etc.)
$\hat{A} = \displaystyle\sum_{\text{both } i} \hat{i}\,(\hat{i}\cdot\hat{A})$ \boxed{\text{Components } (\hat{i}\cdot\hat{A}) \text{ always real}}$	$|\psi\rangle = \displaystyle\sum_{\text{both } i} |i\rangle\,\langle i|\psi\rangle$ \boxed{\text{"Components" } \langle i|\psi\rangle \text{ may be complex}}$
Orthogonality The unit vectors in each set of coordinates are perpendicular. For example	*Orthogonality* The basis states in each representation are orthogonal. For example
$(\hat{x}\cdot\hat{y}) = 0 \qquad (\hat{x}'\cdot\hat{y}') = 0$	$\langle x|y\rangle = 0 \qquad \langle R|L\rangle = 0$
Completeness The sum of the squared components of $\hat{A}$ in any coordinate system adds up to unity. For example,	*Completeness* The sum of the squared magnitudes of the projection amplitudes of $|\psi\rangle$ into any complete set of states adds up to unity. For example
$(\hat{x}\cdot\hat{A})^2 + (\hat{y}\cdot\hat{A})^2 = \sin^2\theta + \cos^2\theta = 1$	$|\langle x|\psi\rangle|^2 + |\langle y|\psi\rangle|^2 = 1$

7-9 THE STATE VECTOR AND THE SCHRÖDINGER WAVE FUNCTION FOR BOUND STATES

In discussing the Schrödinger equation and its application to bound states in Chapter 3, we already introduced the idea that the value of a normalized wave function $\psi(x)$ at any given x is the quantum amplitude associated with finding the particle at that x. With the aid of the discussions in the present chapter we can see how this fits into the larger scheme of quantum mechanics. Establishing this connection should be helpful in lending a sense of reality to the formal considerations that we have been engaged in.

Energy as a Representation of the State of a Bound Particle

The time-independent Schrödinger equation, applied to the bound states of a particle in a given potential, yields two kinds of results—a set of eigenvalues of the energy and a corresponding set of wave functions. By fixing attention on a particular allowed energy, say E_1, we select (at least in one-dimensional systems) a unique quantum state of the system. In the language of the present chapter, we can label this state as ψ_1 and say that it has a state vector denoted by $|\psi_1\rangle$. Now we can ask about possible representations of this state vector in terms of various sets of basis states. One extremely simple representation immediately suggests itself: the set of the energy states themselves! This means forming the projection amplitudes $\langle E|\psi_1\rangle$ for all the permitted values of E (the eigenvalues). However, in the case that we have chosen—a pure energy state—there is only one nonzero projection amplitude. There is unit probability associated with finding the particle with this particular energy, say E_1. We can say, in formal terms, that the state vector $|\psi_1\rangle$ lies entirely along the "direction" of the basis vector $|E_1\rangle$ and is orthogonal to all the basis vectors $|E_n\rangle$ for other values of n. The projection amplitude $\langle E_1|\psi_1\rangle$ is of magnitude unity, and can thus be written $e^{i\phi}$; lacking any reason to the contrary we can set $\phi = 0$ so that $\langle E_1|\psi_1\rangle$ is just equal to 1. In fact $|\psi_1\rangle = |E_1\rangle\langle E_1|\psi_1\rangle = |E_1\rangle$. Of course, this particular example represents a rather cumbersome way of describing an extremely simple situation, but it does illustrate the use of the

　　Quantum amplitudes and state vectors

state vector formalism. Also, it points the way to something that we shall discuss in detail later on (Chapter 8)—the thoroughly nonclassical circumstance that a particle may be in a combination of different energy states. But we shall not pursue that intriguing question yet.

Position as a Representation of the State of a Bound Particle

To return to the pure eigenstate ψ_1, associated with the particular energy E_1, let us now consider describing it in terms of position states. By a position state, in a one-dimensional situation, we mean simply a state corresponding to having the particle at some unique value of x, say x_i. We denote this particular state by a vector $|x_i\rangle$. There is an infinity of such position states because any value within the continuous range of x represents a possible position for a particle. The position state vectors form a suitable basis for a representation of the state ψ_1, for they have the two properties of orthogonality and completeness. They are *orthogonal* because if a particle is definitely at the position x_i, there is zero probability of its being at any other position x_j; all projection amplitudes $\langle x_j|x_i\rangle$ are zero. They form a *complete set* because, in a one-dimensional world, they exhaust all the possibilities for locating the particle. Thus we can define projection amplitudes of the type $\langle x_i|\psi_1\rangle$ that represent the quantum amplitudes associated with finding the particle at every possible value of x. But these correspond exactly to the information expressed in the Schrödinger wave function $\psi_1(x)$: the value of $\psi_1(x)$ at a given value of x is again the quantum amplitude associated with finding the particle at that x. It is just the same character in a different suit of clothes, and we can write an identity:

$$\psi_1(x_i) \equiv \langle x_i|\psi_1\rangle \tag{7-29}$$

The Schrödinger spatial wave function is then the whole collection of individual quantum amplitudes obtained by projecting the state vector $|\psi\rangle$ onto position states $|x_i\rangle$. The Schrödinger description probably seems more natural than the state vector because it is a more straightforward way of describing the probability amplitude (quantum amplitude) as a continuous function of position. The description in terms of state vectors involves imagining an abstract "hyperspace"

7-9 State vector and Schrödinger wave function

based upon many (in this case infinitely many) different unit vectors, each of which is associated with a given position and is orthogonal to all the others. But although the final message is exactly the same, the description in terms of state vectors is broader and more powerful. We now recognize that the description of a given state vector $|\psi\rangle$ in terms of position (the Schrödinger representation) is only one of a variety of ways of describing the quantum state ψ.

EXERCISES

7-1 *Effects of sequential analyzer loops.* A beam of L-polarized photons is incident on three open analyzer loops in sequence, as shown in the figure. The output intensity is monitored with a detector. Assume that all devices are ideal.

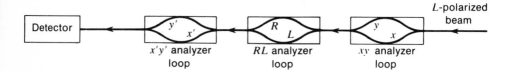

(a) What fraction of the incident beam falls on the detector?

(b) What is the polarization state of the photons entering the detector?

(c) Are there any channels within the string of analyzer loops that can be blocked without reducing the intensity at the detector?

7-2 *Combination of quantum amplitudes.* In the experiment diagrammed in the figure, what fraction of the incident light in beam A is transmitted in beam B on the average in the following cases? Answer this question first without using any formalism and then by making the proper combinations of projection amplitudes from Table 7-1.

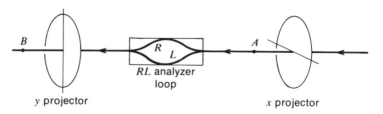

(a) The R channel is blocked.

(b) The L channel is blocked.

(c) Both channels are open.

Quantum amplitudes and state vectors

(d) For a more complicated-seeming problem, repeat steps (a)–(c) using x' and y' projectors in place of the x and y projectors, respectively.

7-3 *Paradox of the recombined beams: I.* Carry out a general analysis, analogous to that of Section 7-3, giving the transmitted intensity that one would expect (incorrectly) for an analyzer loop between crossed linear polarizers if the separated beams were independent of one another. Assume an arbitrary angle θ between the axis y of the first polarizer and the y' axis of the analyzer loop. Sketch the transmitted intensity as a function of θ predicted by this incorrect analysis.

7-4 *Paradox of the recombined beams: II.* Consider the transmission of light through a system composed of two linear polarizers, with an angle θ between their transmission axes, and between them an analyzer loop with one of its axes at an angle α to that of the first polarizer.

(a) Calculate the (incorrect) transmitted intensity, treating the separated beams in the analyzer loop as independent.

(b) Show that the correct expression for the transmitted intensity, $I(\theta) = I_o \cos^2\theta$, is obtained by the correct combination of sequential projection amplitudes for the two paths.

(c) What does the transmitted intensity become if one of the channels of the analyzer is blocked? For a given value of θ, under what conditions is this transmitted intensity greatest? How does it compare with the transmitted intensity with the analyzer completely absent?

7-5 *Laboratory identification of a polarization state.* Suppose it is known that all of the photons in a given beam are in a single state of either linear or circular polarization. Call this as-yet-undetermined state ψ. How much can one learn about ψ by making just *two* measurements of different projection probabilities? For example, suppose one measures the values of the two projection probabilities given in (a) below. If one of them is zero, then the photons in the beam must be in the *other* state. (In this case, can you predict the value of the *other* projection probability?) On the other hand, if the two projection probabilities are unequal, then (since we know the photons are either circularly or linearly polarized) the beam must be linearly polarized at some angle other than zero or 90° from the y-axis. (Can the angle be determined more precisely than this?) Finally, if the projection probabilities are equal, then the beam is either linearly polarized at 45° or circularly polarized, but, if it is circularly polarized, we cannot tell, from these two determinations, whether it is R or L polarized. Carry out a similar analysis assuming one knows only the pair of projection probabilities given in each of the cases (b) through (d) below. Then redo the analysis if, in addition to linear or circular polarization, the beam may be unpolarized. What characterizes the instances in which a *single* projection probability suffices to determine the state ψ?

(a) $|\langle x|\psi\rangle|^2$ and $|\langle y|\psi\rangle|^2$
(b) $|\langle R|\psi\rangle|^2$ and $|\langle L|\psi\rangle|^2$
(c) $|\langle y'|\psi\rangle|^2$ and $|\langle y|\psi\rangle|^2$
(d) $|\langle L|\psi\rangle|^2$ and $|\langle y|\psi\rangle|^2$

7-6 *Reversing the order of projections.*

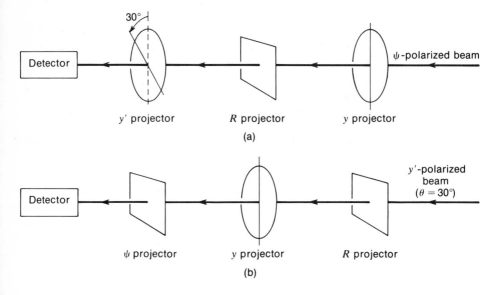

y' projector R projector y projector

(a)

ψ projector y projector R projector

(b)

(a) Write down the overall quantum amplitude for photons initially in state $|\psi\rangle$ to pass through the three projectors shown in figure (a). Given that $|\langle y|\psi\rangle| = 1/\sqrt{5}$ what is the overall transmission probability?

(b) Relate the overall amplitude (and the transmission probability) for the experiment of figure (b) to that obtained in (a).

(c) How are the results of (a) and (b) altered if the R projector is replaced by an open R-L analyzer loop?

7-7 *Identifying polarization states.* Consider the following state vector:

$$|\psi\rangle = |R\rangle(1 - i)/2 + |L\rangle(1 + i)/2$$

(a) Is this state circularly polarized? If so, is it R or L polarization?

     Quantum amplitudes and state vectors

(b) Is this state linearly polarized? If so, find the orientation of the axis of polarization.

Discussion: One way to answer parts (a) and (b) is to put the coefficients in the equation into some kind of standard form and compare them with entries in Table 7-1. Another way is to carry out the mathematical analog of experiments with polarizers; for example, project the state vector into state y' and see if there exists an angle θ such that the projection probability is equal to unity.

(c) Answer parts (a) and (b) for the following state vectors. At least one of them represents elliptical polarization; in this case, simply demonstrate that it is neither linearly nor circularly polarized.

$$|\psi\rangle = |x\rangle e^{-i\pi/2}/\sqrt{2} + |y\rangle e^{i\pi/2}/\sqrt{2}$$
$$|\psi\rangle = |x\rangle(1 - i)/2 + |y\rangle(1/\sqrt{2})$$

7-8 *Formal properties of a polarization state vector.* In this exercise, the symbol $|\psi\rangle$ stands for any one of the photon polarization states $|x\rangle, |y\rangle, |x'\rangle, |y'\rangle, |R\rangle$, or $|L\rangle$. For each of the proposed formal properties of state $|\psi\rangle$ on the following list, decide whether it is true of *all* of these states or *some* of the states or *none*. For the cases of *some* or *none*, list one or more counterexamples in each instance. Which of your answers depend on an arbitrary assignment of phase?

(a) $|\langle x|\psi\rangle|^2 + |\langle y|\psi\rangle|^2 = 1$
(b) $\langle x|\psi\rangle$ is real.
(c) $\langle y|\psi\rangle$ is real.
(d) $\langle x|\psi\rangle$ *and* $\langle x'|\psi\rangle$ are real (angle θ between x and x' axes).
(e) $\langle x|\psi\rangle$ *and* $\langle R|\psi\rangle$ are real.
(f) There is another state $|\phi\rangle$ in the set for which $\langle\phi|\psi\rangle = 0$.
(g) $|\langle x|\psi\rangle|^2 + |\langle R|\psi\rangle|^2 = 1$
(h) If $|\langle x|\psi\rangle|^2 = |\langle y|\psi\rangle|^2$, then $|\langle x'|\psi\rangle|^2 = \frac{1}{2}$ for all θ.

7-9 *Properties of a given photon polarization state.* A photon polarization state is described by the state vector

$$|\psi\rangle = |x\rangle\frac{3}{5} + |y\rangle\frac{4i}{5}$$

A beam of photons transmits N *photons per second* in this state. Answer the following questions about this beam.

(a) What fraction of the photons will pass, on the average, through a y projector?

(b) What fraction of the photons, on the average, will pass through an x' projector (note prime!)? The direction of the x' axis makes an angle θ with the direction of the x axis.

(c) An L-polarized photon has an angular momentum $\hbar$ *along* its direction of motion, and an R-polarized photon has an angular momentum of the same magnitude but pointing *opposite* to its direction of motion. [*Note*: For a discussion of this see Section 14-8. The nomenclature used here is that of classical optics. The modern convention interchanges the labels L and R.] If the beam described by the particular polarization state ψ above is totally absorbed by a surface, what angular momentum is added to the surface per second, on the average?

7-10 *Deriving a complete orthogonal set of states.* Given a photon polarization state $|\psi_1\rangle$, it is possible to construct a second state $|\psi_2\rangle$ that, together with $|\psi_1\rangle$, forms a complete orthogonal set of states. Trace through this procedure in the examples that follow.

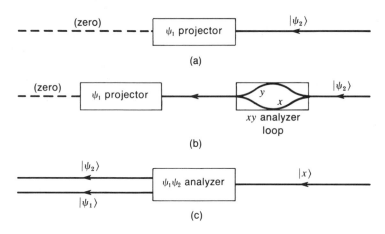

(a) Suppose we are given the following state $|\psi_1\rangle$ specified in the xy basis:

$$|\psi_1\rangle = |x\rangle(i/\sqrt{2}) + |y\rangle(1/\sqrt{2})$$

where $i/\sqrt{2} = \langle x|\psi_1\rangle$ and $1/\sqrt{2} = \langle y|\psi_1\rangle$. [*Note*: This is really polarization state $|L\rangle$, but we do not need to know it in order to derive the orthogonal state, which will turn out to be $|R\rangle$ in what follows.] We look for a state that is, first of all, *orthogonal* to state $|\psi_1\rangle$. Let

$$|\psi_2\rangle = |x\rangle a + |y\rangle b$$

and look for values of the constants $a = \langle x|\psi_2\rangle$ and $b = \langle y|\psi_2\rangle$. If $|\psi_2\rangle$ is orthogonal to $|\psi_1\rangle$, that is, if $\langle \psi_1|\psi_2\rangle = 0$, then there will be zero output in the experiment diagrammed in figure (a). In order to express the null result in terms of projection amplitudes involving x and y, introduce an open xy analyzer loop into the incident beam

Quantum amplitudes and state vectors

(which has no net effect on this beam), as indicated in figure (b). Show that the expression for $\langle \psi_1 | \psi_2 \rangle$ in terms of the alternative paths through the system is

$$\langle \psi_1 | \psi_2 \rangle = \langle \psi_1 | x \rangle \langle x | \psi_2 \rangle + \langle \psi_1 | y \rangle \langle y | \psi_2 \rangle$$

(b) Now demand orthogonality and substitute from the coefficients in the defining equations for $|\psi_1\rangle$ and $|\psi_2\rangle$. Use the relation $\langle j | i \rangle = \langle i | j \rangle^*$ where necessary. Show that the resulting relation between the constants a and b is

$$0 = (-i/\sqrt{2})a + (1/\sqrt{2})b$$

or

$$b = ia$$

(c) States $|\psi_1\rangle$ and $|\psi_2\rangle$ must be not only orthogonal but also *complete*. In terms of experiment, a $\psi_1|\psi_2$ analyzer (figure c) must pass in its two output beams every incident photon. In particular, an incident x-polarized beam has total probability unity of being projected into either the ψ_1 beam or the ψ_2 beam. Show that completeness leads to the condition

$$a^2 = \tfrac{1}{2}$$

Devise and analyze a similar experiment to show that

$$b^2 = \tfrac{1}{2}$$

(d) Now combine the conditions on the two constants derived in parts (b) and (c) to find their values. An overall phase is arbitrary; choose b to be real and positive so that $b = \langle y | \psi_2 \rangle = 1/\sqrt{2}$. From this find the value for $a = \langle x | \psi_2 \rangle$. Compare these projection amplitudes with those in Table 7-1 to confirm that, indeed $|\psi_2\rangle = |R\rangle$. Would this be true if we had chosen the constant a to be real?

(e) Now apply the above procedure to construct the state $|\psi_2\rangle$ which forms a complete orthogonal set with the elliptical polarization state $|\psi_1\rangle$ defined by the equation

$$|\psi_1\rangle = |x\rangle (i\sqrt{3}/2) + |y\rangle (1/2)$$

7-11 *Formal properties of a complete set of polarization states.* Two photon states $|\psi_1\rangle$ and $|\psi_2\rangle$ are proposed as a complete and orthogonal set of polarization states (sometimes called an *orthonormal basis*). Are they? (They could be any of the states we have considered in the past two chapters—or they might be elliptical polarization states.) Suppose this set of states has, in turn, each of the following properties. Classify each property into one of the following:

The property is *sufficient* by itself to establish them as an orthonormal basis.
The property is *necessary* for an orthonormal basis but not by itself sufficient.
The property is *irrelevant* to the question of orthonormal basis.
The property is *impossible* for an orthonormal basis.

(a) $\langle \psi_2 | \psi_1 \rangle = 0$

(b) $|\langle \psi_1 | x \rangle|^2 + |\langle \psi_2 | x \rangle|^2 = 1$

(c) $|\langle \psi_1 | \phi \rangle|^2 + |\langle \psi_2 | \phi \rangle|^2 = 1$ for *all* polarization states $|\phi\rangle$.

(d) There is a state $|\phi\rangle$ such that $\langle \psi_1 | \phi \rangle = 0$ and $\langle \psi_2 | \phi \rangle = 0$.

(e) There is a state $|\phi\rangle$ such that $|\phi\rangle = |\psi_1\rangle a + |\psi_2\rangle b$, where $|a|^2 + |b|^2 = 3$.

(f) For *every* state $|\phi\rangle$ one may find constants a and b such that $|\phi\rangle = |\psi_1\rangle a + |\psi_2\rangle b$

(g) $\langle \psi_1 | R \rangle$ and $\langle \psi_2 | R \rangle$ are both *complex* numbers.

7-12 *A photon as a three-state system?* Pandora claims that photons are really a three-state system: She can find *three* states of polarization that are orthogonal and form a complete set. In support of her claim, Pandora exhibits a device, Pandora's Box, which has three output channels, labeled A, B, and C [Figure (a)]. In reality, Pandora's Box consists of an ordinary xy analyzer with an $x'y'$ analyzer inserted in the y beam, as shown in Figure (b). Analyze Pandora's claim using the following outline or some other method.

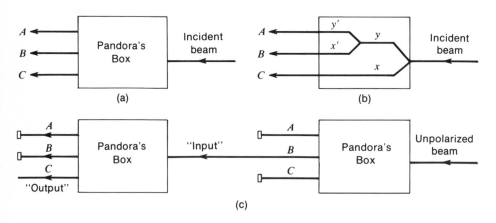

(a) In what channel or channels of the box do *nonzero* outputs appear when the incident beam is: (i) x polarized, (ii) y polarized, (iii) x' polarized?

(b) Show that Pandora's Box does *not* satisfy all properties of an analyzer, as defined in Table 6-2.

Quantum amplitudes and state vectors

(c) Suppose the squares of the amplitudes $\langle A|A \rangle$, $\langle A|B \rangle$, $\langle C|B \rangle$, etc., are measured in the conventional way by means of two sequential Pandora's Boxes. For example, the magnitude $|\langle C|B \rangle|^2$ can be measured as (output)/(input) in the experiment shown in Figure (c). Which of the following fundamental properties of complete orthogonal sets will be satisfied among states A, B, and C and which will not be satisfied? For each of the properties *not* satisfied, give a particular example which violates this property. (Symbols i and j independently take on the values A, B, and C.)

    (i)   normalization: $|\langle i|i \rangle|^2 = 1$ for all i

   (ii) orthogonality: $|\langle j|i \rangle|^2 = 0$ for $i \neq j$

 (iii) "reciprocity": $|\langle j|i \rangle|^2 = |\langle i|j \rangle|^2$

 (iv) Completeness over final states $\sum\limits_{\text{all } j} |\langle j|i \rangle|^2 = 1$ for all i

"Plus ça change, plus c'est la même chose."

ALPHONSE KARR, *Les Guêpes; Les Femmes* (1849)

8

The time dependence of quantum states

8-1 INTRODUCTION

The results of the preceding chapter pave the way for us to consider particles in motion—particles actually free to go from one place to another instead of being confined to a bound state within a potential well of some kind. The study of classical Newtonian mechanics almost always proceeds the other way round: the mathematical description of motion is applied first to particles moving through space under the action of forces of various kinds. Only later does one tackle the more sophisticated problems of the oscillatory or orbital motion of particles bound to some center of force. In contrast, quantum mechanics is most easily started from the experimentally observed discrete energy states of bound particles. The *free* particle may take on any energy whatever, which makes the quantum description of a free particle more awkward conceptually and mathematically than the quantum description of a bound particle. Our own development of the subject underscores this fact, for we shall approach the analysis of free-particle motion by means of bound states.

We saw in Chapter 3 that a general form for the total wave function for a particle of a certain energy E confined to motion along the x axis can be written

$$\Psi(x,\ t) = \psi(x)e^{-i\omega t} \tag{8-1}$$

where

$$\omega = \frac{E}{\hbar}$$

The quantum amplitude $\Psi(x, t)$ as given by Eq. 8-1 is always complex and time-dependent. However, as we discussed in Chapter 3, the time factor $\exp(-iEt/\hbar)$ for a pure stationary-state wave function (whether a bound state or a free-particle pure-momentum state) does not result in any time variation of spatial probability or any other observable; the probability expressed by the square of the modulus of Ψ is independent of t, since $|e^{-iEt/\hbar}|^2 = 1$. There is no possibility of obtaining, from such a pure energy state, a concentration of probability that moves from place to place to correspond to the motion of a particle. Motion of probability takes place when states of different energy are superposed. A simple case will be analyzed in Section 8-3. Before doing this we need to discuss briefly the meaning of such a superposition of states.

8-2 SUPERPOSITION OF STATES

In our earlier discussion we became accustomed to speaking of a bound particle in a quantum state of unique energy. The construction of *moving* quantum spatial probabilities will require a *superposition* of stationary state wave functions of different energies. What does this superposition mean? Can a particle be in more than one energy state at the same time? The answer is "yes," in the sense discussed below.

To begin with, one must recognize that the question of superposition of states is not new, but has been implicit in our earlier analysis of various systems. In discussing photons (Chapters 6 and 7), we dealt with certain superpositions of states, not states of different energy but states of different polarization. Recall the expansion of photon polarization state $|x\rangle$ in the complete set $|R\rangle$ and $|L\rangle$:

$$|x\rangle = |R\rangle\langle R|x\rangle + |L\rangle\langle L|x\rangle$$
$$= |R\rangle\left(\frac{i}{\sqrt{2}}\right) + |L\rangle\left(\frac{-i}{\sqrt{2}}\right) \tag{8-2}$$

Now the state vector $|x\rangle$ describes a single photon in the pure polarization state x. If a beam of such photons impinges on an

The time dependence of quantum states

xy analyzer, they will *all* exit from the *x* channel. The right-hand side of Eq. 8-2 tells us that this state *x* can also be described as a certain superposition of states *R* and *L*. If the beam impinges on an *RL* analyzer, half of the photons, on the average, will exit from the *R* channel and half from the *L* channel. This prediction is carried by the squared magnitudes of the coefficients in the equation, namely $|\langle R|x\rangle|^2 = |\langle L|x\rangle|^2 = \frac{1}{2}$. But the coefficients tell us more: it is their *relative phase* that determines the character of the superposition. If the coefficient of $|L\rangle$ were changed from $(-i/\sqrt{2})$ to $(+i/\sqrt{2})$, the superposition would describe not *x* polarization but *y* polarization. (Actually the resultant state vector in that case would be $|y\rangle i$, which is not experimentally distinguishable from $|y\rangle$.) Equation 8-2 describes a *coherent superposition*,* one with fixed relative phase. Conversely, *any* coherent superposition of the photon polarization states $|R\rangle$ and $|L\rangle$ defines a possible polarization state of a photon.

Similarly, any coherent superposition of energy eigenstates for a particle represents a possible quantum state of that particle. In this sense a particle *can* be in more than one energy state at the same time. An experiment to measure the energy will (like a measurement of photon polarization) yield a particular value. However, until and unless such a measurement is made, the description of the state is given by a wave function built up from different energy eigenfunctions in a definite phase relationship.

8-3 AN EXAMPLE OF MOTION IN A BOX

To begin to explore quantum time dependence, let us consider the simplest possible time-dependent superposition, that of just two energy states. We shall take the two lowest states of the infinitely deep square well (Figure 8-1a). For simplicity, set the coefficients of the two components equal to one another. Then the superposition has the form

$$\Psi(x, t) = A \sin\left(\frac{\pi x}{L}\right) e^{-i\omega_1 t} + A \sin\left(\frac{2\pi x}{L}\right) e^{-i\omega_2 t} \qquad (8\text{-}3)$$

At $t = 0$ this superposition is as shown in Figure 8-1b. Recall (Chapter 3) that the energy values for the infinite square well are proportional to the *square* of the number *n* of the level (tak-

8-3   An example of motion in a box

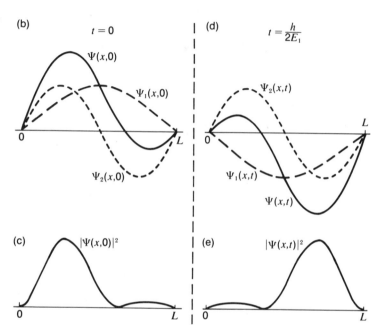

Fig. 8-1 Super-
position of the low-
est stationary states
of an infinite square
well. (a) Potential
energy function for
the well, with the
two lowest energy
eigenvalues
shown. (b) Eigen-
functions for the
two lowest energy
states (broken lines)
and the superposi-
tion of these func-
tions (solid line), at
$t = 0$. (c) Proba-
bility density func-
tion at $t = 0$ for the
superposition
shown in (b). (d)
and (e) Plots corre-
sponding to (b) and
(c) for the later time
$t = h/(2E_1)$.

ing $E = 0$ at the bottom of the well)

$$E_n = \frac{n^2 h^2}{8mL^2}$$

Therefore,

$$\omega_2 = \frac{E_2}{\hbar} = 4\frac{E_1}{\hbar} = 4\omega_1$$

Thus one half-cycle for ω_1 occupies the same time as two
complete cycles for ω_2. This means that between $t = 0$ and
$t = h/2E_1$ the relative signs of the two components reverse, as

The time dependence of quantum states

indicated in Figure 8-1b and d. This reversal shifts the maximum value of the probability distribution from the left half to the right half of the box, as shown in Figures 8-1c and e. At later times, periodically, the original distribution is restored. In other words, we have constructed a probability distribution which "sloshes back and forth" within the well.

The above discussion is open to criticism, as you may have noticed, because it makes use of frequencies for the individual components, which in turn depend on the arbitrary choice of the zero of energy. The predicted physical results are correct, but to place the calculation on a sound footing, we use Eq. 8-3 to evaluate the probability function $|\Psi|^2$ as an explicit function of time. Writing the spatial factors as f_1 and f_2 for brevity, we have

$$\Psi(x, t) = f_1 e^{-i\omega_1 t} + f_2 e^{-i\omega_2 t}$$

Therefore,

$$|\Psi(x, t)|^2 = (f_1 e^{-i\omega_1 t} + f_2 e^{-i\omega_2 t})(f_1 e^{+i\omega_1 t} + f_2 e^{+i\omega_2 t})$$
$$= f_1^2 + f_2^2 + f_1 f_2 \left[e^{i(\omega_2 - \omega_1)t} + e^{-i(\omega_2 - \omega_1)t} \right]$$

that is,

$$|\Psi(x, t)|^2 = f_1^2 + f_2^2 + 2f_1 f_2 \cos \left(\frac{E_2 - E_1}{\hbar} t \right) \qquad (8\text{-}4)$$

This shows that the time dependence is governed by the *difference* between the two energies, $E_2 - E_1$, and *not* by their individual values. Hence no observable consequence depends on the choice of zero energy. Since the cosine factor varies periodically between $+1$ and -1, and since f_1 and f_2 are real, the extreme forms of $|\Psi|^2$ are $(f_1 + f_2)^2$ and $(f_1 - f_2)^2$. These correspond to the two probability distributions shown in Figures 8-1c and 8-1e. We now see, however, that the time to go from one extreme to the other is one half-cycle of the cosine factor of Eq. 8-4, which involves the energy difference. Since (again setting $E = 0$ at the bottom of the well) we have $E_2 - E_1 = 3E_1$, the distribution is given by $(f_1 - f_2)^2$ at times t defined by

$$\frac{3E_1}{\hbar} t = \pi, \ 3\pi, \ 5\pi, \text{ and so on}$$

Thus the time $t = h/2E_1$, marked on Figures 8-1d and 8-1e, does not give the first occurrence of the reversed distribution subsequent to $t = 0$ but rather the second occurrence.

In this example the motion of probability results from the superposition of two energy eigenstates. What does it mean to say that the particle has more than one energy? As mentioned in the preceding section, the question must be examined in terms of a possible experiment. If an energy-determining experiment is carried out on the system, what will be the result? The result will be *either* E_1, the lowest energy, *or* E_2, the second energy, never any other value. In this sense the particle energy is quantized, even in the superposition state. The explicit form of Eq. 8-3 is

$$\Psi(x, t) = \frac{1}{\sqrt{2}} \cdot \sqrt{\frac{2}{L}} \sin\left(\frac{\pi x}{L}\right) e^{-i\omega_1 t}$$

$$+ \frac{1}{\sqrt{2}} \cdot \sqrt{\frac{2}{L}} \sin\left(\frac{2\pi x}{L}\right) e^{-i\omega_2 t} \tag{8-5}$$

Here a factor $1/\sqrt{2}$ multiplies each of the normalized component eigenfunctions. This makes the integral of $|\Psi|^2$ over all x equal to unity and corresponds to equal probabilities (50 percent each) for finding the particle in the energy states E_1 and E_2. That is, if very large numbers of measurements of the energy are made on separate systems, each of which is prepared in the state described by Eq. 8-5, the results will be E_1 and E_2 in equal numbers (within the statistical fluctuations). In this sense one can associate an *average* energy $(E_1 + E_2)/2$ with the state, *but this average value will never be found in an individual measurement of the energy*; the sharp values E_1 and E_2 of the component pure states are the only values observed.

If, instead of measuring energy, one were to investigate the spatial location of the particle at a given time, the statistical result of many such measurements should approach the probability function $|\Psi(x, t)|^2$ at that time. One must be doubly careful in interpreting this function. First, at any given time t after preparation of the system, the probability function predicts the statistical result of many measurements of position of the particle. Second, verification of this prediction requires access to a large number of particles, all prepared in the given quantum state, each one allowed to evolve for the same time t, and each of which is subjected to a single

The time dependence of quantum states

measurement and then discarded. Such a procedure is the same as that required for a stationary state, with the added provision that an equal time interval t elapse after each system is prepared.[1]

If one were to sample the position distributions at *random* times after preparation, rather than at the same (or equivalent) times, the results would build up to the *time average* of $|\Psi(x, t)|^2$. It is interesting to note that in this time-averaging process the term involving the interference of the two component states (and which embodies the moving part of the probability distribution) would wash out, leaving only the sum of the probability distributions due to the component states separately. This is evident from inspection of Eq. 8-4.

The most important property of the spatial probability function $|\Psi(x, t)|^2$ for superposed energy states is that it changes with time. One may imagine constructing a series of movie stills of the probability function, each frame representing a different instant after preparation, each one requiring for verification an exhaustive set of experiments as described above for that single instant. Stringing these frames together in time sequence then yields a moving picture of the predicted time development of the spatial probability function of the particle. Figure 8-2 shows just such a sequence for a superposition of a number of eigenfunctions for a particle in a box. Because the frequencies of the different components are commensurable, they combine, after a certain repetition period, to reconstitute the original form of Ψ (compare the first and last pictures in this sequence).

8-4 PACKET STATES IN A SQUARE-WELL POTENTIAL

One goal of this chapter is to learn how to construct a wave function that describes a *free* particle that is spatially localized to some limited extent but that moves along the x axis. On the way to this goal we consider a particle trapped within a rigid box of very large but not infinite width. Extending the analysis of Section 8-3 to the superposition of more than two energy states, we shall see how to construct a wave function that describes a particle initially confined to an arbitrarily small range of positions within the box, and that

[1]See the film, *Individual Events in One-Dimensional Scattering*, available from the Education Development Center, Inc., Newton, Mass.

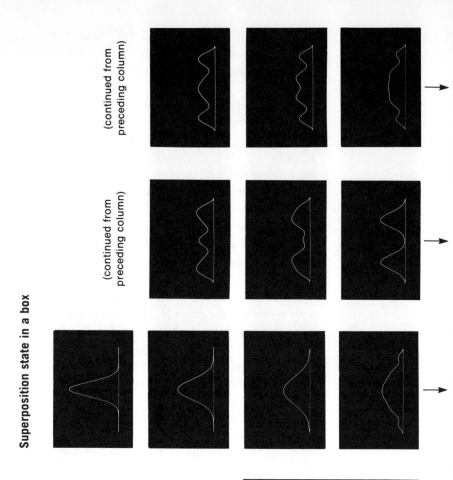

Superposition state in a box

(continued from preceding column)

(continued from preceding column)

Fig. 8-2   Stills from the film Particle in a Box, showing
the time development of the position probability distribution from a given initial configuration. (Reproduced with
permission of Education Development Center, Newton,
Mass.)

describes the subsequent motion of the particle probability
function. A moving probability that is limited spatially is called
a *packet* and the corresponding quantum-mechanical description is called a *packet state*.

A superposition of *all* of the energy eigenfunctions for a
rigid square well of width L has the form

$$\Psi(x, t) = \sum_{n=1}^{\infty} B_n \sin\left(\frac{n\pi x}{L}\right) e^{-iE_n t/\hbar} \qquad (8\text{-}6)$$

The time dependence of quantum states

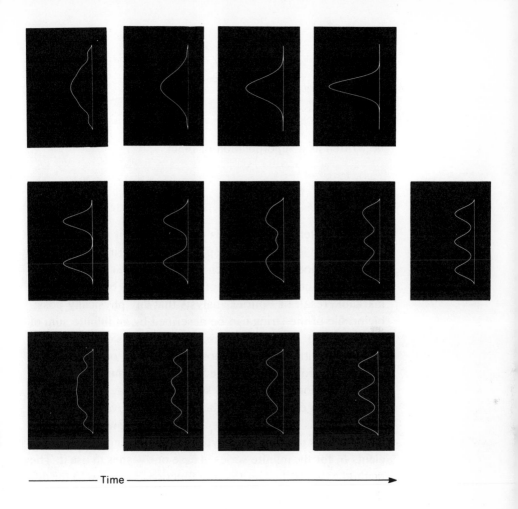

Time

At $t = 0$ this reduces to the form

$$\Psi(x,\, t = 0) = \psi(x) = \sum_{n=1}^{\infty} B_n \sin\left(\frac{n\pi x}{L}\right) \qquad (8\text{-}7a)$$

We need to find the values of the coefficients B_n that will yield this initial wave function.

An expression mathematically identical to that of Eq. 8-7a is encountered in the classical description of the initial profile of a plucked violin string. The violinist draws the string to one side with a finger near one end of the string (Figure 8-3). The shape of the displaced string does not correspond to the profile of any one of the normal modes, but can

8-4   Packet states in a square-well potential

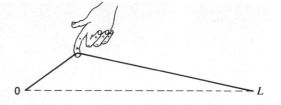

Fig. 8-3 *Profile of a violin string plucked aside at one point. (Transverse deflection greatly exaggerated.)*

be written as a superposition of them:

$$f(x) = \sum_{n=1}^{\infty} B_n \sin\left(\frac{n\pi x}{L}\right) \tag{8-7b}$$

There are alternative methods of plucking that yield an initial displacement localized to one portion of the string. The mathematical method by which such profiles are described in terms of normal modes is called Fourier analysis.[2] Now it is a remarkable result of Fourier analysis that any physically possible profile of the string can be described by an infinite sum of sinusoids (Eq. 8-7b) with signs and coefficients chosen to construct the initial pulse. In most applications one is satisfied with a superposition of a limited number of terms that conform to the function within some specified accuracy. The coefficients B_n can be chosen by trial and error to give a reasonable result for this limited number of modes. However, Fourier analysis does provide an analytic method for finding the coefficients B_n for the infinite series in case one wishes to use it. The method exploits the fact that the functions $\sin(n\pi x/L)$ for different n have the following mathematical property:

$$\int_0^L \sin\left(\frac{n_2\pi x}{L}\right) \sin\left(\frac{n_1\pi x}{L}\right) dx = 0 \qquad (n_1 \neq n_2) \tag{8-8}$$

This property is called *orthogonality*. Although it seems a far cry from orthogonality as defined for simple vectors or for polarization states (as discussed in Chapters 6 and 7), we shall see later (in Section 8-7) that there is a very basic equivalence involved.[3]

To find the values of the coefficients B_n for a given form of the function $f(x)$ we single out a particular value of n, say $n = j$,

---

[2]See, for example, Chapter 6 of the volume *Vibrations and Waves* in this series.

[3]See also the volume *Vibrations and Waves* in this series, pp. 195–196.

The time dependence of quantum states

by multiplying both sides of Eq. 8-7b by $\sin(j\pi x/L)$ and integrating from $x = 0$ to $x = L$. Using Eq. 8-8 we have

$$\int_0^L f(x) \sin\left(\frac{j\pi x}{L}\right) dx = \sum_{n=1}^{\infty} B_n \int_0^L \sin\left(\frac{n\pi x}{L}\right) \sin\left(\frac{j\pi x}{L}\right) dx$$

$$= B_j \int_0^L \sin^2\left(\frac{j\pi x}{L}\right) dx$$

Since, for any integral value of n,

$$\int_0^L \sin^2\left(\frac{n\pi x}{L}\right) dx = \frac{L}{2}$$

we find

$$B_n = \frac{2}{L} \int_0^L f(x) \sin\left(\frac{n\pi x}{L}\right) dx \qquad (8\text{-}9)$$

This equation gives us all the information to construct the Fourier series of Eq. 8-7b for any possible profile $f(x)$ of the string.

The corresponding quantum-mechanical analysis (Eq. 8-7a) proceeds identically, with the initial wave function $\psi(x)$ replacing the initial string profile $f(x)$. The result is the desired coefficients B_n that yield the initial function at $t = 0$.

If we now ask what happens as time goes on in the two cases (violin string and particle in box) both the mathematical analysis and the physical interpretation differ. For the violin string, each term in the Fourier series is multiplied by the time factor $\cos \omega_n t$, where ω_n is the characteristic angular frequency of the nth normal mode of the string. The resulting sum gives directly the profile of the string at any later time. In the quantum case the time factor for each term has the form $e^{-i\omega_n t}$ where $\omega_n = E_n/\hbar$ is determined by the energy of the state n. The result gives the wave function (Eq. 8-6) at any later time. The probability density for finding the particle at any position within the box at a later time t is then determined by taking the squared magnitude of the wave function and evaluating it at that position.

To give a simple example of the determination of the coefficients B_n, consider an initial wave function $\psi(x)$ that has the constant value A over a small range b centered at $L/2$ and is

8-4   Packet states in a square-well potential

zero everywhere else, as shown in Figure 8-4a. Then Eq. 8-9 becomes

$$B_n = \frac{2A}{L} \int_{(L-b)/2}^{(L+b)/2} \sin\left(\frac{n\pi x}{L}\right) dx$$

This gives

$$B_n = \frac{2A}{n\pi} \left[\cos\frac{n\pi}{2L}(L-b) - \cos\frac{n\pi}{2L}(L+b) \right]$$
$$= \frac{4A}{n\pi} \sin\left(\frac{n\pi b}{2L}\right) \sin\left(\frac{n\pi}{2}\right)$$

The factor $\sin(n\pi/2)$ makes $B_n = 0$ for all even values of n. [This is to be expected from the symmetry of $f(x)$. Why?] For successive odd values of n the coefficients B_n alternate in sign; they can be written as follows:

$$B_n = (-1)^{(n-1)/2} \frac{2Ab}{L} \frac{\sin(n\pi b/2L)}{(n\pi b/2L)} \qquad (n \text{ odd}) \qquad (8\text{-}10)$$

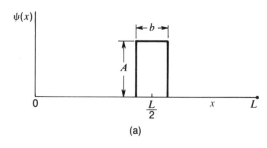

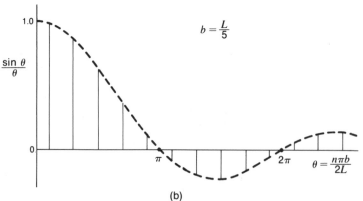

Fig. 8-4
(a) Localized wave function of rectangular shape within a one-dimensional box of width L. (b) Graph of (sin θ)/θ, where θ = nπb/2L. Ordinates are shown at points corresponding to n = 1, 3, 5 ... with b = L/5. Values of B_n are obtained by multiplying these ordinates by ±2Ab/L = ±2A/5, in accordance with Eq. 8-10.

The time dependence of quantum states

The function $(\sin \theta)/\theta$, where $\theta = n\pi b/2L$, is plotted as the dashed line in Figure 8-4b. The values of B_n depend on the relative sizes of b and L. The vertical bars in the figure are drawn for the ratio $b/L = \frac{1}{5}$.

8-5   THE POSITION-MOMENTUM UNCERTAINTY RELATION

In section 3-6 we mentioned that a particle confined to a region of space cannot have a unique single value of *momentum*. This result is in strict analogy to the classical wave result that a wave confined to a region of space cannot have a unique single value of *wavelength*. The connection between the two cases comes from the wave-like properties of particles as expressed in the de Broglie expression that relates momentum and wavelength. Using the background of the present chapter, we can make these notions more precise and formulate them in a well-known form called the *Heisenberg position-momentum uncertainty relation*, first enunciated by Werner Heisenberg in 1927.[4]

To begin with a simple case, consider again a rectangular profile, as in Figure 8-4, which is confined to a region of width b in the center of an infinite square well of width L. This could equally well represent an initial profile imposed on a string, or a localized wave function at $t = 0$ for a particle in a box. The Fourier analysis of this profile, as we have just seen, is a combination of sinusoids of the form $B_n \sin k_n x$, where $k_n = n\pi/L$. The distribution of values of B_n, as shown in Figure 8-4b, has as its envelope the function $(\sin \theta)/\theta$, with $\theta = n\pi b/2L$, and has its first zero at $\theta = \pi$. We can use the value $\theta = \pi$ to define a characteristic width for the function, even though this function has nonzero values for larger θ. Applied to Eq. 8-10, this width specifies a range of values of n for which the coefficients B_n make their most significant contribution to the Fourier series. Specifically, the range of values of n so represented extends from zero up to the value n_{max} defined by the condition

$$\frac{\pi b}{2L} n_{max} = \pi$$

We can express this result in more suggestive form by in-

---

[4]W. Heisenberg, Z. Phys. **43**, 172 (1927)

troducing the wave numbers k_n $(= n\pi/L)$, whose range Δk (from $n = 0$ up to n_{max}) is given, according to the above relation, by

$$\Delta k = \frac{2\pi}{b}$$

But b is the range of distance, Δx, within which the original profile is contained, and so we have the result[5]

$$\Delta x \cdot \Delta k \approx 2\pi \qquad\qquad (8\text{-}11)$$

We write this as an approximate equality for two reasons. First, it is clear that the value of Δk in the present example is based on an arbitrary criterion for the width of the k spectrum. But second, and more important, the relationship as represented by Eq. 8-11 expresses a *general* approximate connection between *any* spatial distribution confined mainly within some distance Δx and the spread Δk of component wave numbers contributing significantly to its Fourier analysis.

Let us now turn this discussion into the language of wave mechanics. If Figure 8-4a represents the initial wave function $\psi(x)$ of a particle, then at $t = 0$ we shall find the particle somewhere in the region of width b, but nowhere else. Many observations carried out on identical systems prepared in this state will yield a range of positions $\Delta x = b$. If, now, we take another set of systems prepared in this state and test them for values of *momentum p*, we shall find that most of the values lie within a range Δp equal to $\hbar \Delta k$, where Δk is related to Δx through Eq. 8-11. (This result of course involves the basic de Broglie relationship $p = \hbar k$.) Thus we have the condition

$$\Delta x \cdot \Delta p_x \approx h \qquad\qquad (8\text{-}12)$$

where we have added the subscript x to the momentum to remind us of its direction.

The result expressed by Eq. 8-12 is, as we implied above, much more general than the specific case considered here.

Notice that, as we confine the particle more and more by making $b = \Delta x$ smaller and smaller, the range of momenta

[5]The result in Eq. 8-11 is consistent with a more general statement about the product $\Delta x \cdot \Delta k$ as an inequality. See Eq. 8-14 below.

The time dependence of quantum states

must become wider and wider in accordance with Eq. 8-12. In the contrary case, in which the momentum spread is made smaller and smaller, the particle must be less and less confined. The limiting condition of this second case is provided by the uniform beam of unique momentum p. The wave function, as we saw in Chapter 3, is then proportional to $e^{ikx} = e^{ipx/\hbar}$. The squared magnitude of this wave function is independent of position: the particle is equally likely to be found everywhere along the x axis. Hence, when Δp approaches zero, Δx must approach infinity. Equation 8-12 provides a specific measure of spread for the intermediate cases.

Equation 8-12 describes a minimization of the product of the spreads in particle position and momentum. We can always do *worse* than this. For example, if we let time increase from zero, each energy eigenfunction in Eq. 8-6 will change its phase at the characteristic rate, different from that of the other eigenfunctions, and the position probability distribution will expand to fill the box. The momenta will not change, so the *range* of momenta will not change. However, Δx will now be equal to L, not the smaller value b. Hence the product $\Delta x \cdot \Delta p$ will be greater than h. When such considerations are taken into account, Eq. 8-12 takes on the form of an inequality:

$$\Delta x \cdot \Delta p_x \gtrsim h$$

This equation was set up in such a way that a large majority of position and momentum measurements will fall in the ranges Δx and Δp, respectively. The conventional measures of widths of statistical distributions (such as the *standard deviation*) are narrower than this and do not include within their boundaries so large a fraction of the observations. When these standard statistical measures of widths are applied to a wide range of cases of position-momentum uncertainty, the minimum product in Eq. 8-12 is somewhat less than that given above, in fact a factor of 2π less. This converts the right side from h to $\hbar$ and yields the relation

$$\Delta x \cdot \Delta p_x \gtrsim \hbar \qquad (8\text{-}13)$$

This is the Heisenberg position-momentum uncertainty relation (or principle). The corresponding classical wave expression relating position uncertainty and spread in wave number

8-5   The position-momentum uncertainty relation

is also stated most generally as an inequality with a minimum product smaller by a factor of 2π than that given in Eq. 8-11:

$$\Delta x \cdot \Delta k_x \gtrsim 1 \qquad (8\text{-}14)$$

8-6 THE UNCERTAINTY PRINCIPLE AND GROUND-STATE ENERGIES

The position-momentum uncertainty relation provides a simple and powerful way of estimating, at least in order of magnitude, the energy of the lowest bound state in a given potential.

The basis of the calculation is the assumption that the value of Δp is of the same order as that of the momentum itself, since the *magnitude* of p at any point is defined by E and the potential V at that point, but its *direction* is undefined (right or left).

As the simplest possible illustration of this approach, consider a particle of mass m in a one-dimensional square well of width L. For the value of Δx in this problem we take the value of L itself. Thus we have, from Eq. 8-13, that the smallest possible value of Δp is given approximately by the equation

$$\Delta p \approx \frac{\hbar}{L}$$

We deduce that the magnitude of p is also of about this size. It follows that the energy of the particle, which is equal to its kinetic energy if we set $V = 0$ at the bottom of the well, is given by

$$E = \frac{p^2}{2m} \approx \frac{\hbar^2}{2mL^2} = \frac{h^2}{8\pi^2 mL^2}$$

This can be compared to the exact result, $E = h^2/(8mL^2)$, for the infinitely deep square well. We see that our estimate is low by the factor π^2 (about 10), which is not too bad, especially if we remember that for wells of *finite* depth the ground-state energy may be substantially less than that for the infinitely deep well.

If we apply our approximate result to an electron confined within typical atomic dimensions ($L \approx 1$ Å), we have (with

330   The time dependence of quantum states

$\hbar \approx 10^{-27}$ erg-sec and $m \approx 10^{-27}$ g)

$$E \approx \frac{10^{-54}}{10^{-27} \times 10^{-16}} = 10^{-11} \text{ erg} \approx 10 \text{ eV}$$

We know that this is about right for the kinetic energy of an electron in the ground state of a hydrogen atom.

If, on the other hand, we consider a neutron or a proton ($m \approx 10^{-24}$ g) confined within a nucleus of diameter $L \approx 10^{-12}$ cm, we have

$$E \approx \frac{10^{-54}}{10^{-24} \times 10^{-24}} = 10^{-6} \text{ erg} \approx 1 \text{ MeV}$$

Kinetic energies of this order are indeed typical for individual nucleons in their lowest state in a medium-weight nucleus.

The uncertainty principle gives such a direct and simple approach to the estimation of ground-state energies that you may well wonder why we did not introduce it when we first discussed such matters in Chapter 3. The reason is that the theoretical basis for the uncertainty relations is far from trivial, and involves a clear understanding of the wave-particle duality and the statistical interpretation of quantum amplitudes. But once we have that, the uncertainty principle is a powerful tool, especially in cases where the binding potential is of such a form that an exact analytic solution of the Schrödinger equation is difficult or impossible (see the exercises).

8-7 FREE-PARTICLE PACKET STATES

Returning now to the analysis of localized probabilities described by packet states, we shall remove the conditions imposed by the boundaries of any box and describe a *free* particle which is, at a given instant, localized in a limited region of space. Such a state can be constructed from the complete set of eigenstates corresponding to all possible values of the momentum. A single-momentum particle (whose probability function has the same value everywhere along the x axis) has the space wave function

$$\psi_k(x) \sim e^{ikx} \tag{8-15}$$

where $p_x = \hbar k$. (Such momentum eigenfunctions, combined in pairs $\pm k$, yield sinusoidal functions such as we have used to construct packet states in a box.)

In order to obtain a localized packet state, we must superpose states of different momenta. For a free particle the wave number k is a continuous variable and the quantum summation analogous to Eq. 8-7a becomes an integral:

$$\psi(x) = \int B_k e^{ikx} \, dk \tag{8-16a}$$

The coefficients B_k are given by a process of Fourier analysis analogous to that which led to Eq. 8-9 in Section 8-4. The result is

$$B_k = \int \psi(x) e^{-ikx} \, dx \tag{8-17a}$$

which is analogous to Eq. 8-9.

Mathematically, Eqs. 8-16a and 8-17a together are an example of the use of *Fourier integral* analysis, the extension of ordinary Fourier analysis to situations in which the characteristic interval in x is expanded from some limited length L to the whole range of x between $-\infty$ and $+\infty$.[6] In the language of Fourier analysis, $\psi(x)$ and B_k are *Fourier transforms* of one another (apart from factors $1/2\pi$ or $1/\sqrt{2\pi}$ that do not concern us here). *Physically*, in the context of quantum mechanics, we can read Eqs. 8-16a and 8-17a as expressions of the analysis of a given state $|\psi\rangle$ in terms of eigenstates of either momentum or position—an example of *alternative representations*, analogous to (although more complicated than) the analysis of an arbitrary polarization state in terms of different sets of basic polarization states xy, $x'y'$, or RL (see Chapter 7, especially Eqs. 7-21 and 7-24).

In the present case, we take a given state, represented by a state vector $|\psi\rangle$, and expand it in terms of a complete set of position states $|x\rangle$ or momentum states $|k\rangle$:

$$|\psi\rangle = \sum_{\text{all } k} |k\rangle \langle k|\psi\rangle \equiv \sum_{\text{all } x} |x\rangle \langle x|\psi\rangle \tag{8-18}$$

The quantities $\langle k|\psi\rangle$ and $\langle x|\psi\rangle$ are the projection amplitudes: $\langle k|\psi\rangle$ is the amplitude B_k associated with any given wavenumber k, and $\langle x|\psi\rangle$ in the Schrödinger amplitude $\psi(x)$ associated with any given value of x.

If we go through the procedure of projecting the state

[6]For a presentation of the mathematical basis of this extension see for example, F. B. Hildebrand, *Advanced Calculus for Applications*, 2d ed., Prentice-Hall, Englewood Cliffs, N.J., 1976.

The time dependence of quantum states

vector $|\psi\rangle$, as given by Eq. 8-18, into a particular x or a particular k, we have

$$\langle x|\psi\rangle = \sum_{\text{all } k} \langle x|k\rangle\langle k|\psi\rangle$$

and

$$\langle k|\psi\rangle = \sum_{\text{all } x} \langle k|x\rangle\langle x|\psi\rangle$$

But putting

$$\langle x|\psi\rangle \equiv \psi(x) \quad \text{and} \quad \langle k|\psi\rangle \equiv B_k$$

these can be rewritten

$$\psi(x) = \sum_{\text{all } k} \langle x|k\rangle B_k \tag{8-16b}$$

$$B_k = \sum_{\text{all } x} \langle k|x\rangle\psi(x) \tag{8-17b}$$

A comparison of these equations with Eqs. 8-16a and 8-17a (and making due allowance for the need to replace sums by integrals when x and k are continuous variables) allows us to identify the exponential factors in the integrands of Eqs. 8-16a and 8-17a as the projection amplitudes $\langle x|k\rangle$ and $\langle k|x\rangle$—the former from a particular momentum state to any position state, and the latter from a particular position state to any momentum state:

$$\langle x|k\rangle = e^{+ikx}$$
$$\langle k|x\rangle = e^{-ikx}$$

We can also note the relationship

$$\langle k|x\rangle = \langle x|k\rangle^*$$

in conformity with the general result expressed by Eq. 7-17.

This formalism can be used to interpret the orthogonality condition between two square-well eigenfunctions given in Eq. 8-8. Using Eq. 8-18, let $|\psi_1\rangle$ be written

$$|\psi_1\rangle = \sum_{\text{all } x} |x\rangle\langle x|\psi_1\rangle$$

Then the integral on the left side of Eq. 8-8 can be replaced by a summation:

$$\int_0^L \sin\left(\frac{n_2\pi x}{L}\right) \sin\left(\frac{n_1\pi x}{L}\right) dx \rightarrow \sum_{\text{all } x} \langle \psi_2|x\rangle \langle x|\psi_1\rangle$$

This summation involves an infinite number of terms, since x is a continuous variable, but *formally* it corresponds precisely to the scalar product of the two state vectors $|\psi_1\rangle$ and $|\psi_2\rangle$, each expressed in terms of basic position state vectors $|x\rangle$:

$$|\psi_1\rangle = \sum_{\text{all } x_i} |x_i\rangle\langle x_i|\psi_1\rangle$$

$$|\psi_2\rangle = \sum_{\text{all } x_j} |x_j\rangle\langle x_j|\psi_2\rangle$$

Since the basis vectors $|x_i\rangle$ and $|x_j\rangle$ are orthogonal—$\langle x_i|x_j\rangle = 0$ for $x_i \neq x_j$—the result is a sum of the products of the amplitudes $\langle\psi_2|x\rangle$ and $\langle x|\psi_1\rangle$ at the same x, which is equivalent to the integral over x of the product $\psi_1(x)\psi_2(x)$. Thus the property represented by Eq. 8-8, that this integral vanishes identically, can be seen as equivalent to a statement of the orthogonality of the two state vectors in the hyperspace (of infinitely many dimensions) defined by the complete set of possible state vectors $|x\rangle$.

Now suppose, for example, that $\psi(x)$ (Figure 8-5a) is a rectangular function just like that in Figure 8-4a, extending from $x = -b/2$ to $x = +b/2$, and thus of width $\Delta x = b$. Then from Eq. 8-17a we have

$$B_k = \langle k|\psi\rangle \sim \int_{-b/2}^{b/2} e^{-ikx}\, dx$$

$$= \frac{1}{ik}(e^{ikb/2} - e^{-ikb/2}) \qquad (8\text{-}19a)$$

$$= \frac{\sin(kb/2)}{k/2} \qquad (8\text{-}19b)$$

This distribution of values of b_k is shown in Figure 8-5b.

Note the similarity in form between Eq. 8-19b for the coefficients B_k and Eq. 8-10 for the coefficients B_n for the packet state in a box. But Eq. 8-19a shows that the wave function includes contributions of equal magnitude from positive k and negative k. Therefore the *average* momentum of the particle is zero and the average position of the packet will move neither to the right nor to the left.

It is easy to modify the rectangular packet to yield a non-zero average momentum. We already know that a wave function $\psi(x) \sim e^{ikx}$ corresponds to a particle of momentum $\hbar k$, so it makes good sense to assume that if we multiply a wave function of zero average momentum by the factor $e^{ik_0 x}$ we shall generate a new wave function associated with the momentum

The time dependence of quantum states

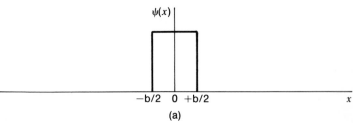

(a)

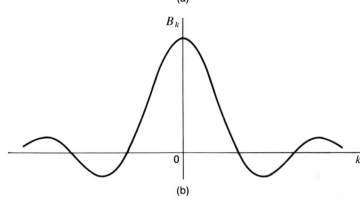

(b)

$\hbar k_o$. Let us therefore try a wave function given by

$$\psi(x) = 0 \qquad x < -\frac{b}{2}, \text{ and } x > +\frac{b}{2}$$

$$\psi(x) = e^{ik_o x} \qquad -\frac{b}{2} \leqslant x \leqslant +\frac{b}{2} \qquad (8\text{-}20)$$

This still represents a rectangular probability distribution, $|\psi(x)|^2 = \text{const.}$, contained between $x = -b/2$ and $x = +b/2$, but the complex phase factor makes a crucial difference. Substituting the new form of $\psi(x)$ into Eq. 8-17a one finds (check it!)

$$B_k \sim \frac{\sin\,[(k-k_o)b/2]}{(k-k_o)/2}$$

This is a curve of just the same form as Figure 8-5b, but centered on $k = k_o$, so that the packet does have a nonzero average momentum, namely $p = \hbar k_o$ as expected.

The free-particle packet states described in the above examples satisfy the position-momentum uncertainty relation. The width of the momentum distribution described by Eq. 8-19b can be characterized by putting $kb/2 = \pi$. The spread in position Δx is equal to b. Thus we have

8-7   Free-particle packet states

$$\Delta k = \frac{2\pi}{b} = \frac{2\pi}{\Delta x}$$

or

$$\Delta k \cdot \Delta x = 2\pi$$

Since $p = \hbar k$, this corresponds to

$$\Delta p \cdot \Delta x = 2\pi\hbar = h$$

which is greater than the minimum spread given by the uncertainty relation, Eq. 8-14. You can verify for yourself that the same equation describes the state with nonzero average momentum.

8-8 PACKET STATES FOR MOVING PARTICLES

The wave function given in Eq. 8-16a describes a localized particle. The right-hand side is a superposition of different momentum states whose relative contributions are determined by the factors $B_k(= \langle k|\psi \rangle)$ derived from the initial wave function using Eq. 8-17a.

We are now able to describe how this wave function will change with time, since each component momentum state simply increases its phase linearly with time according to the complex phase factor $e^{-i\omega t}$, so that Eq. 8-16a becomes

$$\Psi(x, t) \sim \int e^{i(kx - \omega t)} B_k \, dk \qquad (8\text{-}21)$$

In most cases this is a difficult integral to evaluate because not only does B_k depend on k but also the angular frequency ω is a continuous function of k according to the relationship for free particles:

$$\omega = \frac{E}{\hbar} = \frac{p^2}{2m\hbar} = \frac{\hbar k^2}{2m} \qquad (8\text{-}22)$$

If the function B_k is symmetrically peaked about some average value of k, say k_o, then the maximum of the spatial probability

The time dependence of quantum states

distribution moves with the corresponding group velocity given by Eq. 2-8 of Chapter 2:

$$v_{\text{group}} = \frac{d\omega}{dk}\bigg|_{k\,=\,k_o} = \frac{\hbar k}{m}\bigg|_{k_o} = \frac{p_o}{m} \qquad (8\text{-}23)$$

Equation 8-23 yields just the classical (Newtonian) expression for the velocity of a free particle of given momentum. But there is more to it than this. The probability packet described by Eq. 8-21 will in fact spread with time as it moves. The reason for this spreading can be found in the relation between ω ($= E/\hbar$) and k ($= p/\hbar$). The group velocity in Eq. 8-23 is proportional to k. The higher k components in the overall distribution of k values have a larger group velocity than the lower k components; a broadening of the spatial packet with time is an implicit consequence of the connection between ω and k as given in Eq. 8-22. The rate of spreading depends on the width of the packet in "momentum space." In principle if we could construct a packet of sufficiently small width Δx in position space, the minimum width Δp in momentum space would be set by the Heisenberg uncertainty limit $\Delta p = \hbar/\Delta x$. The narrower the packet in position space, the greater the spread of momenta, and thus the more rapid the broadening of the packet with time.[7] No one has yet made a free packet confined enough in space and momentum to approach the limits given by the Heisenberg relation. Most real pulses of particles are, in fact, so broad that the position-momentum uncertainty products are very much larger than the minimum set by the Heisenberg relation. In particular, the extension in position space is so great that the fundamental spreading is not observable. Why, then, have we talked about this unrealized case? There are two reasons: (1) It is implicit, as a limiting case, in any analysis of packet states of free particles. (2) It is the quantum-mechanically correct distillate of the fuzzy semiclassical idea of de Broglie wavelength.

In the preceding analysis we have entirely ignored the fact that the free-particle single-momentum wave function e^{ikx} cannot be normalized because it has magnitude unity over all space. But this difficulty is more mathematical than real. No

---

[7]See the film, *Free Wave Packets* (No. QP-5) available through the Education Development Center, Newton, Mass.

8-8   Packet states for moving particles

actual beam extends to infinity in both directions, and the wave functions for any finite beam *can* be normalized.

8-9 EXAMPLES OF MOVING PACKET STATES

We described in Section 8-7 how a wave function $\psi(x)$, representing a spatial distribution of quantum amplitudes at some instant, can be analyzed into momentum components, as given by Eqs. 8-16a and 8-17a. The amplitudes B_k for a free-particle state are constant—that is, *time-independent*—so that the development of the total wave function in time is given by Eq. 8-21 in which only the complex phase factor varies with t.

In order, then, to construct a moving packet state we do not necessarily have to begin with the momentum analysis of a wave function $\psi(x, 0)$ defined as a function of x at $t = 0$. We can, if we like, go directly to Eq. 8-21, insert in it any specific distribution of values of B_k, and integrate it out. Of course, we are not free to do this if the initial spatial probability distribution is previously specified, but if it is not, we can take some particularly simple or convenient distribution of B_k values and work from there.

To illustrate this approach, suppose that we have a *momentum spectrum* as shown in Figure 8-6a—a rectangular distribution in which B_k is constant for all values of k between $k_o - \Delta k$ and $k_o + \Delta k$, and zero for all other k values. Using Eq. 8-21 and putting $t = 0$, we can readily find what *initial spatial distribution* of probability this corresponds to

$$\Psi(x, 0) \sim \int_{k_0 - \Delta k}^{k_0 + \Delta k} e^{ikx} \, dx = \frac{1}{ix} \left[e^{ikx} \right]_{k_0 - \Delta k}^{k_0 + \Delta k}$$

$$= \frac{e^{ik_0 x}}{ix} (e^{ix\Delta k} - e^{-ix\Delta k})$$

$$= 2 e^{ik_0 x} \frac{\sin (x\Delta k)}{(x\Delta k)} \Delta k$$

Therefore

$$|\Psi(x, 0)|^2 \sim \left[\frac{\sin (x\Delta k)}{(x\Delta k)} \right]^2$$

This distribution is shown in Figure 8-6b.[8] It has its maximum

[8] The shapes of these $\psi(x)$ and B_k distributions are just the converse of those shown in Figure 8-5, in which we began with a rectangular form for $\psi(x)$.

The time dependence of quantum states

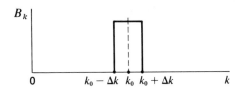

(a)

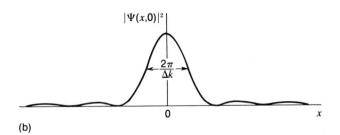

(b)

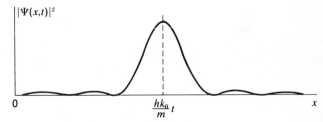

(c)

Fig. 8-6 (a) Continuous spectrum of amplitudes for a narrow, rectangular distribution of wave numbers constituting a wave packet. (b) Spatial probability distribution at t=0 for the wave packet derived from (a). (c) Spatial probability distribution for the same packet at a later time.

at $x = 0$, and is mostly contained between its first zeros at $x\Delta k = \pm\pi$, that is, within a range Δx equal to $2\pi/\Delta k$—another example of the uncertainty principle at work.

But now let us consider the more difficult problem of evaluating the probability distribution for some later time. We now have

$$\Psi(x, t) \sim \int_{k_0 - \Delta k}^{k_0 + \Delta k} e^{i(kx - \omega t)} \, dk$$

with $\omega = \hbar k^2/2m$.

8-9   Examples of moving packet states

To make things as simple as possible, we shall suppose $\Delta k \ll k_o$, and we shall put

$$k = k_o + k' \qquad (-\Delta k \leqslant k' \leqslant \Delta k)$$

Then

$$\omega = \frac{\hbar}{2m}(k_o + k')^2 \approx \frac{\hbar}{2m}(k_o^2 + 2k_o k')$$

Our expression for Ψ then becomes (approximately)

$$\Psi(x, t) \sim \int_{k'=-\Delta k}^{\Delta k} \exp i\left[(k_o + k')x - \frac{\hbar}{2m}(k_o^2 + 2k_o k')t\right] dk'$$

$$= \exp i\left(k_o x - \frac{\hbar k_o^2}{2m}t\right) \int_{-\Delta k}^{\Delta k} \exp\left[i\left(x - \frac{\hbar k_o}{m}t\right)k'\right] dk'$$

The integral can be evaluated directly and gives

$$\Psi(x, t) \sim \exp i\left(k_o x - \frac{\hbar k_o^2}{2m}t\right) \cdot \frac{\sin\left(x - \dfrac{\hbar k_o}{m}t\right)\Delta k}{\left(x - \dfrac{\hbar k_o}{m}t\right)\Delta k}$$

Therefore

$$|\Psi(x, t)|^2 \sim \left[\frac{\sin\left(x - \dfrac{\hbar k_o}{m}t\right)\Delta k}{\left(x - \dfrac{\hbar k_o}{m}t\right)\Delta k}\right]^2$$

This is a probability distribution just like that at $t = 0$ but shifted in the positive x direction by the distance $(\hbar k_o/m)\,t$—corresponding, exactly as we should expect, to the group velocity $\hbar k_o/m$ (Figure 8-6c). (Because $p_o = \hbar k_o$ according to the de Broglie relation, the corresponding classical result is $p_o/m = v_o$.)

Unfortunately the approximation that we have used here, in the interests of simplicity, suppresses the fact that the position probability distribution will also have broadened. The exact mathematics of this problem is rather cumbersome, and instead of going into it we shall simply quote the results of a different special case that is amenable to exact analysis. This is the so-called *Gaussian wave packet*, defined for all t by the momentum amplitude distribution

$$B_k \sim \exp\left[-\frac{(k - k_o)^2}{2(\Delta k)^2}\right] \tag{8-24}$$

The time dependence of quantum states

Such a momentum spectrum is shown in Figure 8-7a. It corresponds to a mean momentum $\hbar k_0$, and its characteristic width (between inflection points) is $2\Delta k$.

For such a wave packet, the space wave function is also Gaussian in form, but its evolution in time is given by the complicated equation

$$|\Psi(x, t)|^2 = \frac{\alpha^{1/2}}{(2\pi)^{1/2}\left[\alpha^2 + \left(\dfrac{\hbar t}{2m}\right)^2\right]^{1/2}} \exp\left[\frac{-\dfrac{\alpha}{2}\left(x - \dfrac{\hbar k_0}{m}t\right)^2}{\alpha^2 + \left(\dfrac{\hbar t}{2m}\right)^2}\right]$$

where $\alpha = [2(\Delta k)^2]^{-1}$.

This result [obtainable from direct integration of Eq. 8-21 with B_k given by Eq. 8-24] becomes less formidable if we look

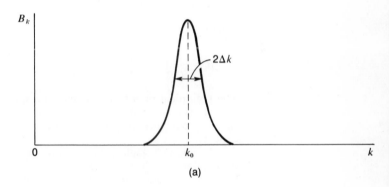

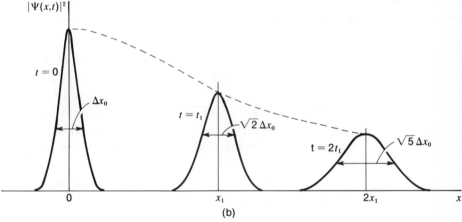

Fig. 8-7 (a) Momentum spectrum for a Gaussian wave packet of mean wavenumber k_0. (b) Spatial probability distribution for the wave packet derived from (a), as it might be observed at $t = 0$ and at two later times. Note the progressive spreading of the spatial distribution.

341   8-9   Examples of moving packet states

at its structure. At $t = 0$ it simplifies to

$$|\Psi(x, 0)|^2 = \frac{1}{\sqrt{2\pi\alpha}} \, e^{-x^2/2\alpha}$$

This describes a Gaussian probability distribution, centered at $x = 0$, and having a characteristic width, between its inflection points, given by

$$\Delta x_0 = 2\alpha^{1/2} = \frac{\sqrt{2}}{\Delta k}$$

For any t greater than zero, three things happen:

1. The center of the packet moves to $x = (\hbar k_0/m)t$, as expected.

2. The characteristic width of the distribution increases to a value given by

$$\Delta x(t) = \Delta x_0 \, (1 + \beta^2 t^2)^{1/2}$$

where

$$\beta = \frac{\hbar}{2m\alpha} = \frac{\hbar(\Delta k)^2}{m}$$

The reciprocal of β is a characteristic spreading time for the packet. The larger the value of Δk, the more rapidly does the spatial probability distribution broaden—again, as one would expect.

3. The peak value of $|\Psi|^2$, as given by the factor in front of the exponential, decreases as t increases, in such a fashion that the total area under the graph of $|\Psi|^2$ against x remains constant—the wave function automatically remains correctly normalized.

We saw that result 1 applied to the previous case. The spreading and decrease in peak value would have been obtained for that case too if we had been able to solve it without eliminating the term in $(k')^2$ from the expression for ω.

Figure 8-7b indicates how a Gaussian wave packet moves and changes shape in coordinate space, while in momentum space it is described throughout by the distribution of Figure 8-7a.

To conclude this section, let us re-emphasize how we are to interpret, physically, the picture of a moving wave packet. At any given instant, t_1, the spatial distribution of $|\Psi|^2$ describes what would be the result of position measurements on a host of identically prepared particles, the measurements being made at the time t_1 after the preparation of the state. Similarly, the distribution of $|\Psi|^2$ at some later time, t_2,

The time dependence of quantum states

describes the result of position measurements on *another* batch of exactly similar particles, also prepared at $t = 0$ but left unmolested until time t_2. The motion of the packet is only a description of the change with time of the probability distribution to which the particles would be found to conform.

Figure 8-8 illustrates this point. It is a computer simulation, based on random-number inputs, that displays the probability distribution for a certain packet state as it might be observed at two different times. Actual observations on such a packet state would be based on finite counts of particles with statistical fluctuations, but in the limit of arbitrarily large numbers the results would fit a smooth profile, spreading as its center moved steadily along, as illustrated in Figure 8-7b.

8-10 THE ENERGY-TIME UNCERTAINTY RELATION

In Section 3-6 we pointed out how a classical normal mode of a vibrating string is regarded as having a sharply defined frequency even though it cannot be said to have a unique wavelength. We alluded briefly to the fact that this difference in properties comes about because the normal mode is confined to a limited region of space but is thought of (or mathematically described) as if it extended indefinitely in time. The

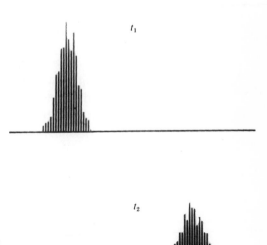

Fig. 8-8 *Computer-generated histograms showing the hypothetical results of position measurements made at two different times on finite-number samples of particles prepared at $t = 0$ in a certain packet state. (Reproduced from the film QP-10, "Individual Events in One-Dimensional Scattering," by permission of Education Development Center, Newton, Mass.)*

8-10   The energy-time uncertainty relation

limitation in spatial extent necessitates a spread in wave number according to the classical wave uncertainty relation [see Eq. 8-14]:

$$\Delta x \cdot \Delta k \gtrsim 1 \qquad\qquad\qquad (8\text{-}25)$$

In Sections 8-4 and 8-5 of the present chapter we developed in detail the quantum-mechanical analog of this result, as expressed in Heisenberg's position-momentum uncertainty relation, Eq. 8-13, which comes directly from Eq. 8-25 if we multiply through by $\hbar$ and substitute $p = \hbar k$:

$$\Delta x \cdot \Delta p \gtrsim \hbar$$

Our purpose in the present section is to point out that the same kind of limitation applies to the frequency of classical waves or the energy of particles in quantum mechanics because the time duration of waves or quantum states is never, in fact, unlimited and may indeed be extremely short. By way of introducing this discussion, let us consider a classical wave example in which the position and time measurements are closely connected.

If a radio transmitter is switched on for a short time interval Δt and then off again, the resulting radio waves lie within a range $\Delta x = c\Delta t$ of distances from the source. This spatial distribution may be converted back into a time sequence at the receiver. Any emission of progressive waves from a source has this same effect of transforming a certain time sequence within the source into a corresponding spatial sequence in the wave. Thus, the formation of a limited train of traveling waves can be described either in terms of a superposition of pure sinusoidal functions of position (corresponding to its extension in space) or in terms of a superposition of pure sinusoidal functions of time (corresponding to its duration). In the case of radio waves in free space, for which the phase velocity has the same value, c, for all frequencies, the uncertainty relation, Eq. 8-25, connecting the spatial extent of a wave train with its wavenumber spread can be immediately converted into another uncertainty relation connecting the duration of the signal (at source or receiver) and the associated spread of frequencies. We can put $k = \omega/c$, where ω is the angular frequency corresponding to a given k; and so $\Delta k = \Delta\omega/c$. Hence the product $\Delta x \cdot \Delta k$ is equal

The time dependence of quantum states

to $(c\Delta t) \cdot (\Delta\omega/c)$, and substituting this in Eq. 8-25 gives us

$$\Delta\omega \cdot \Delta t \geq 1 \qquad (8\text{-}26)$$

Although we have used a simple special way of arriving at this relationship in terms of the direct connection between frequency and wave number for progressive waves, we see that the wave speed c does not enter into the result. In fact, Eq. 8-26 expresses a general limitation on the related values of $\Delta\omega$ and Δt for *any* time-dependent signal or function. It is just a consequence of Fourier integral analysis applied to functions in which the connected variables are time and frequency instead of position and wavenumber. (Note that such analysis necessarily involves pairs of variables that are dimensionally inverse to one another.)

From Eq. 8-26 we can proceed at once to a quantum-mechanical *energy-time uncertainty relationship* by using the basic relation $\omega = E/\hbar$, and we have

$$\Delta E \cdot \Delta t \geq \hbar \qquad (8\text{-}27)$$

In any particular case, a state $|\Psi\rangle$ can be projected or Fourier-analyzed into a *spectrum* in time or frequency (energy). These spectra are Fourier transforms of one another, and are mathematically defined by a pair of equations exactly analogous to Eqs. 8-16a and 8-17a:

$$\Psi(t) = \int A_\omega e^{-i\omega t}\, d\omega \equiv \int A_E e^{-iEt/\hbar}\, dE \qquad (8\text{-}28)$$

where

$$A_\omega = \int \Psi(t) e^{i\omega t}\, dt \qquad (8\text{-}29\text{a})$$

or

$$A_E = \int \Psi(t) e^{iEt/\hbar}\, dt \qquad (8\text{-}29\text{b})$$

8-11   EXAMPLES OF THE ENERGY-TIME UNCERTAINTY RELATION

Probably the most important application of the energy-time uncertainty relation concerns the lifetimes of the excited states of atoms and nuclei. Suppose that, at time $t = 0$, a large number of identical atoms occupy an excited state. According to the classical picture, each of these excited atoms will radiate like a minature radio station (but typically in the visible region of the electromagnetic spectrum), emitting a signal whose intensity falls toward zero with some characteristic decay time τ. The corresponding spread in angular frequency $\Delta\omega$ of this radiation, according to Eq. 8-26, is given by $\Delta\omega \gtrsim 1/\tau$.

The *quantum* analysis is a little more subtle. In this picture (which conforms to experiment), individual atoms in the excited state drop back to the ground state as evidenced by the emission of photons. However, the *time* at which this transi-

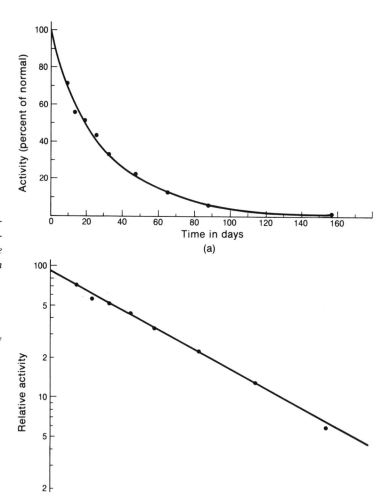

Fig. 8-9 Exponential decay of unstable nuclei. (a) The decay of uranium X_1 (an isotope of thorium) as observed by Rutherford in the early days of radioactivity research. (After a diagram in Rutherford's book, Radio-Activity, *Cambridge University Press, 1904.) (b) A semilogarithmic plot of the same data. The linearity of the graph shows the exponential character of the decay and indicates a half-life of about 22 days.*

The time dependence of quantum states

tion takes place for any one atom is unpredictable. All that one can say is that, between times t and $t + \Delta t$, a certain *fraction* of the atoms still in the excited state at time t will have dropped to the ground state.[9] If Δt is short, the number of atoms decaying during Δt is proportional to Δt itself, and we can put

$$\Delta N = -\gamma N(t) \Delta t$$

where the constant γ represents the *probability per unit time* that any one excited atom will decay. Integrating the above equation, we obtain the result

$$N(t) = N_o e^{-\gamma t} \tag{8-30}$$

This is the familiar law of exponential decay that describes all kinds of random decay processes—notably the radioactive decay of unstable atomic nuclei. Figure 8-9 shows a typical example.

We can relate Eq. 8-30 to a statement about the relative probability $|\Psi(t)|^2$ of finding an atom or nucleus in its excited state at any time t. [For $t < 0$, $|\Psi(t)|^2 = 0$.] Figure 8-10a displays this probability for a case in which a whole assemblage of atoms is raised into the excited state (for example, by electron bombardment or irradiation by light) at $t = 0$. The value of $|\Psi|^2 \Delta t$ is the probability of finding an atom in its excited state between t and $t + \Delta t$. Taking account of normalization, this requires

$$|\Psi(t)|^2 = \gamma e^{-\gamma t} \qquad (t \leq 0) \tag{8-31}$$

so that

$$\int_{-\infty}^{\infty} |\Psi(t)|^2 \, dt = 1$$

This is in strict analogy to $|\Psi|^2 \, dx$ being the probability of finding a particle within a range of position dx, with unit probability that it will be found *somewhere*: $\int |\Psi|^2 \, dx = 1$.

[The exact relation between Eqs. 8-30 and 8-31 is ob-

---

[9]Actually, since this is a random process, there are statistical fluctuations. If the *average* number of atoms decaying in a certain time is n, it is subject to a fluctuation of the order of $\pm\sqrt{n}$.

8-11   Examples of energy-time uncertainty

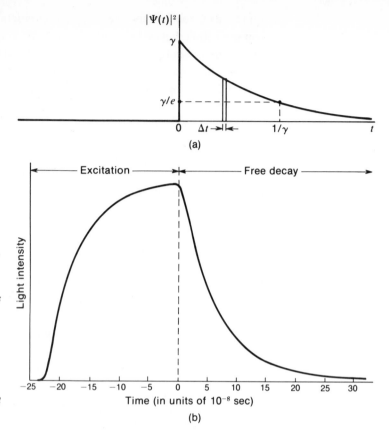

Fig. 8-10 (a) Theoretical graph of $|\psi|^2$ as a function of t for atoms excited from the ground state at $t = 0$. (b) Intensity of 3914 Å radiation by N_2^+ ions, excited by an electric discharge. For $t > 0$ the graph corresponds to (a) but, as can be seen, the excitation process is spread over a considerable interval prior to $t = 0$. [From data of R. G. Bennett and F. W. Dalby, J. Chem. Phys. 31, 434 (1959).]

tained by recognizing that $N(t)$ in Eq. 8-30 is the number of atoms, out of an initial number N_o, that have not decayed up to time t, and will thus be found in the excited state for times greater than t. This implies the relationship

$$\frac{N(t)}{N_o} = \int_t^{\infty} |\Psi(t)|^2 \, dt \qquad (t \geq 0)$$

Using the expression for $|\Psi(t)|^2$ from Eq. 8-31, it is easy to verify that this leads directly back to Eq. 8-30.]

A convenient measure of the width Δt in time of the distribution described by Eq. 8-31 is the value of $1/\gamma$ (see Figure 8-10a). The time τ equal to $1/\gamma$ is, in fact, the *mean lifetime* of atoms in the excited state, as defined by the equation

$$\tau = \int_{-\infty}^{\infty} t |\Psi(t)|^2 \, dt = \gamma \int_0^{\infty} t e^{-\gamma t} \, dt$$

The time dependence of quantum states

Expressing it in other words, τ is the *expectation value* of t for atoms in the excited state.

Having identified an appropriate value of Δt for this situation, we can use the energy-time uncertainty relation, Eq. 8-27, to infer the energy spread associated with the excited state. We have

$$\Delta E \approx \frac{\hbar}{\tau} = \hbar\gamma$$

Typical lifetimes for atomic transitions in the range of visible wavelengths are of the order of 10^{-8}–10^{-7} sec. Such lifetimes can be measured directly by suddenly exciting a collection of atoms in a discharge tube and observing the exponential decay of emitted intensity (Figure 8-10b). From the uncertainty relation we then have for the spread of energies ΔE in the excited state

$$\Delta E \gtrsim \hbar/10^{-7} \text{ sec} \approx 10^{-20} \text{ erg} \approx 10^{-8} \text{ eV}$$

Since the energies E_o of quanta of visible light are of the order of 2 eV, the ratios $\Delta E/E_o$ are typically about 10^{-8}. This means that the so-called *natural width* of a spectral line of, say, 5000 Å is only about 0.0001 Å. This natural width is usually masked by a far larger spreading due to Doppler effect of the moving gas atoms which are the source of light, but the natural width can be directly observed if proper precautions are taken.

By contrast, in nuclear physics the natural line width is often very apparent. For example, the yield of gamma radiation from nuclei bombarded by accelerated protons is found in many cases to exhibit resonances with characteristic widths of a few keV, as shown in Figure 8-11. This means that the excited nuclear states formed by such bombardments may typically have $\Delta E \approx 10^{-8}$ erg (1 keV $= 1.6 \times 10^{-9}$ erg). The corresponding lifetime is of the order of 10^{-19} sec. In such cases the lifetime is far too short to be directly measured, but once ΔE is known the lifetime may be confidently inferred with the help of the uncertainty relation.

Going still further, many of the so-called elementary particles produced by violent nuclear collisions at bombarding energies of many GeV (1 GeV $= 10^9$ eV) are so short-lived that there is a large and measurable spread in the *mass* of particles of a given type. This spread of mass reflects a corresponding uncertainty in the rest-energy E ($= mc^2$) of the par-

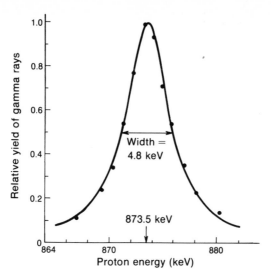

Fig. 8-11 Linewidth of a nuclear energy state excited by protons. Protons from an accelerator are incident on a sample of fluorine $^{19}_{9}F$ (superscript is atomic weight A; left-handed subscript is atomic number Z). An excited state of neon $^{20}_{10}Ne^*$, is produced (asterisk means excited state). This excited nucleus drops to the ground state emitting a gamma ray. The entire reaction may be written

$^1_1H + {}^{19}_9F \rightarrow {}^{20}_{10}Ne^* \rightarrow {}^{20}_{10}Ne + \gamma$

The number of gamma rays emitted depends crucially on the energy of the incident protons, as you can see from the plot above. From this "nuclear resonance curve" one can calculate the lifetime of the excited neon nucleus using the energy-time uncertainty relation (see the exercises). This lifetime is much too short to measure directly. [From data of R. G. Herb, S. C. Snowden, and O. Sala, Phys. Rev. **75**, 246 (1949).]

ticles. For example, Figure 8-12 shows an experimentally determined mass spectrum for particles of the type known as ρ^- (rho-minus) mesons. Their mean mass (measured in energy units) is about 770 MeV with an uncertainty (width of the mass spectrum at half-height) equal to about 170 MeV. From this value of ΔE ($\approx 2.4 \times 10^{-4}$ erg) one can infer a mean lifetime given by

$$\tau \approx \frac{\hbar}{\Delta E} \approx 4 \times 10^{-24} \text{ sec}$$

This is about as long (or short!) a time as it would take for a

The time dependence of quantum states

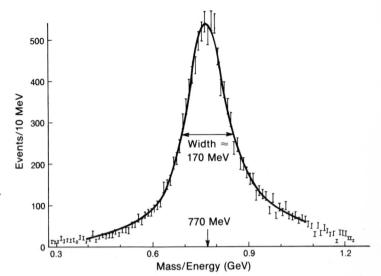

Fig. 8-12 Mass-
energy spectrum for
the ρ⁻ meson.
[After J. Pišút and
M. Roos, Nuc.
Phys. **B6**, 325
(1968).]

Events/10 MeV

Width ≈
170 MeV

770 MeV

Mass/Energy (GeV)

light signal to travel a distance equal to the diameter of a nucleon. Such a lifetime must be close to a lower limit for any particle that can meaningfully be said to exist.

The energy-time uncertainty relation is one of the great "fixed points" of quantum physics, fundamentally useful to both the theoretician and the experimentalist. As is true of all statements in quantum physics, one must be careful to employ this relation in a proper *statistical* manner. It does not limit in any way the accuracy with which either energy or time can be measured for a *single* event. Rather the energy-time uncertainty relation relates the minimum statistical spreads of energy and decay times of a large sample of atomic or nuclear systems identically prepared in a "factory" to be initially in a given excited state.

8-12 THE SHAPE AND WIDTH OF ENERGY LEVELS

In this section we shall illustrate how one can calculate a spectral line shape, in terms of energy, from a wave function appropriate to a decaying excited state of an atom or a nucleus. Let us suppose that the excited state in question is at an energy E_0 above a stable ground state taken to be at $E = 0$ (Figure 8-13). In Section 8-11 we made an argument that indicated that the wave function $\Psi(t)$ of the excited state was such that its

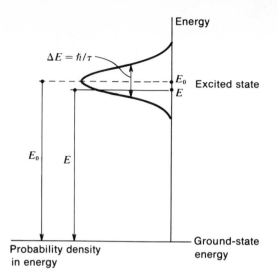

Fig. 8-13 *Relation between energy uncertainty in excited state and spread in energy in transitions to the ground state.*

squared modulus (Eq. 8-31) could be written

$$|\Psi(t)|^2 \sim e^{-\gamma t}$$

From this we infer

$$|\Psi(t)| \sim e^{-\gamma t/2}$$

However, if this "quasi-stationary" state has an energy E_0 above the ground state, we can ascribe to $\Psi(t)$ itself a complex phase factor equal to $e^{-iE_0 t/\hbar}$. Thus we shall put

$$\Psi(t) \sim e^{-\gamma t/2}\, e^{-iE_0 t/\hbar} \qquad (t \geq 0) \qquad (8\text{-}32)$$

Let us now use Eq. 8-29b to calculate the energy spectrum of photons or particles emitted when the system decays from this state to the ground state at $E = 0$. We have

$$A_E = \int \Psi(t)\, e^{iEt/\hbar}\, dt$$

$$\sim \int_0^\infty e^{-\gamma t/2}\, e^{-iE_0 t/\hbar}\, e^{-iEt/\hbar}\, dt$$

that is

$$A_E \sim \int_0^\infty \exp\left\{ \left[\frac{i(E - E_0)}{\hbar} - \frac{\gamma}{2} \right] t \right\} dt$$

The time dependence of quantum states

$$A_E \sim \frac{1}{\dfrac{i(E - E_o)}{\hbar} - \dfrac{\gamma}{2}} \exp\left\{\left[\frac{i(E - E_o)}{\hbar} - \frac{\gamma}{2}\right]t\right\}\Bigg|_{t=0}^{\infty}$$

This then gives the very simple result

$$A_E \sim \frac{1}{\dfrac{i(E - E_o)}{\hbar} - \dfrac{\gamma}{2}} \tag{8-33}$$

We see that A_E is a complex amplitude, but by forming the product $A_E A_E^* \ (= |A_E|^2)$ we can obtain a measure of the relative probability of finding the energy E for a photon or particle emitted from this state. We have, in fact,

$$|A_E|^2 \sim \frac{1}{\left(\dfrac{E - E_o}{\hbar}\right)^2 + \left(\dfrac{\gamma}{2}\right)^2}$$

$$\sim \frac{1}{(E - E_o)^2 + (\hbar\gamma/2)^2} \tag{8-34}$$

This energy spectrum (see Figure 8-13) has a peak at $E = E_o$, and falls to half-height at values of E given by

$$E - E_o = \pm\frac{\hbar\gamma}{2} = \pm\frac{\hbar}{2\tau}$$

The characteristic *width* ΔE of this energy spectrum is thus given by

$$\Delta E = \frac{\hbar}{\tau}$$

Where τ is the mean lifetime of the state. For this case, therefore, we have

$$\Delta E \cdot \Delta t = \hbar \tag{8-35}$$

which is the minimum uncertainty product, according to the Heisenberg relation Eq. 8-27.

The form of energy spectrum represented by Eq. 8-34 gives an extremely good fit to the observed shape of actual energy levels of nuclei, for example, the excited energy level of the Ne²⁰ nucleus already shown in Figure 8-11.

EXERCISES

8-1 *The simplest "sloshing" state of a particle in the infinite square-well potential: I.* A superposition of the two lowest energy states of the infinite square well is given in Eq. 8-3.

(a) Verify, for $t = 0$, that this superposition is correctly normalized by putting $A = 1/\sqrt{L}$, as in Eq. 8-5.

(b) Verify, further, that this normalization holds good for all later times.

(c) Calculate and sketch the probability distribution given by $|\Psi|^2$ (Eq. 8-4) for $t = \pi\hbar/[2(E_2 - E_1)]$; that is, after one quarter-cycle of the repetition period for this superposition.

(d) Find the (time-dependent) probability for the particle to be located in the left half of the well ($0 \leq x \leq L/2$).

(e) Find the expectation value $\langle x(t) \rangle$ of the position of a particle in this state, as defined by

$$\langle x(t) \rangle = \int_{\text{all } x} \Psi^* (x, t)\, x\, \Psi (x, t)\, dx$$

8-2 *The simplest "sloshing" state of a particle in the infinite square-well potential: II.* Again consider the superposition state described by Eq. 8-3.

(a) Show that the probability density $|\Psi|^2$ at $x = L/2$ is *independent* of time.

(b) Students A and B raise the following "paradoxes" concerning this sloshing state. Answer each of these as completely as you can.

(i) Student A asks: "How can probability slosh back and forth between the left and right halves of the well when the probability density at the center of the well does not vary with time?"

(ii) Student B asks: "How can $\langle x(t) \rangle$ vary with time when each state in the superposition is an equal blend of positive and negative momentum components?"

8-3 *Energy of a superposition state.* A particle in an infinite square well extending between $x = 0$ and $x = L$ has the wave function

$$\Psi(x, t) = A\ (2 \sin \frac{\pi x}{L}\ e^{-iE_1 t/\hbar} + \sin \frac{2\pi x}{L}\ e^{-iE_2 t/\hbar})$$

where $E_n = n^2 h^2 / 8mL^2$.

The time dependence of quantum states

(a) Putting $t = 0$ for simplicity, find the value of the normalization factor A.

(b) If a measurement of the energy is made, what are the possible results of the measurement, and what is the probability associated with each?

(c) Using the results of (b) deduce the average energy and express it as a multiple of the energy E_1 of the lowest eigenstate.

(d) The result of (c) is identical with the *expectation value* of E (denoted $\langle E \rangle$) for this state. A procedure for calculating such expectation values in general is based on the fact that, for a pure eigenstate of the energy, the wave function is of the form, $\Psi(x, t) = \Psi(x)e^{-iEt/\hbar}$, which yields the identity

$$ i\hbar\, \frac{\partial \Psi}{\partial t} = E\Psi $$

Clearly, in this case, $\int \Psi^* (i\hbar\, \partial/\partial t)\, \Psi\, dx = E \int \Psi^*\Psi\, dx = E$. An extension of this to a state involving an arbitrary superposition of energies suggests the following formula for calculating expectation values of E:

$$ \langle E \rangle = \int_{\text{all } x} \Psi^* \, (i\hbar\, \partial/\partial t)\, \Psi\, dx $$

This procedure is in fact correct. By applying it to the particular wave function of this exercise, verify that the value of $\langle E \rangle$ is identical with the average energy found in (c).

8-4 *Linear momentum in a superposition state.* In exercise 8-3 we indicated how one can calculate the expectation (average) value of the energy for a mixed-energy state. This exercise is concerned with an analogous procedure for linear momentum. We have seen that the spatial factor of a pure momentum state is given (Eq. 8-15) by $\psi(x) \sim e^{ikx}$. From this we have

$$ \frac{d\psi}{dx} = ik\psi = \frac{ip_x}{\hbar}\psi $$

which suggests the identity

$$ \frac{\hbar}{i}\, \frac{\partial \Psi}{\partial x} = p_x \Psi $$

We then calculate the expectation value of p_x for an *arbitrary* state by using the formula

$$\langle p_x \rangle = \int_{\text{all } x} \Psi^* \left(\frac{\hbar}{i} \frac{\partial}{\partial x} \right) \Psi \, dx$$

(a) Apply this to a pure eigenstate of a particle in an infinite square well, and verify that $\langle p_x \rangle = 0$.

(b) Show that the value of $\langle p_x \rangle$ for the superposition state of Exercise 8-3 oscillates sinusoidally at the angular frequency $\omega = (E_2 - E_1)/\hbar$. This result helps to illuminate the questions raised in Exercise 8-2, and gives some substance to the concept of a probability distribution moving back and forth in the well.

8-5 *A square-well packet state.* At a given instant of time, a particle has a constant wave function in the middle third of an infinite square well and vanishes elsewhere: $\psi(x) = \sqrt{3/L}$ for $L/3 \leqslant x \leqslant 2L/3$ and $\psi(x) = 0$ elsewhere.

(a) Use the results of Section 8-4 to obtain an expression for the coefficients B_n in the expansion of the wave function in energy eigenfunctions of the square well. (Don't work too hard; results are easily adapted from the example in that section.) Why is $B_n = 0$ for all even values of n?

(b) Use Eq. 8-10 to estimate roughly the most important values of n in the superposition. How many nonzero terms does this include?

(c) Evaluate B_1, B_3, and B_5.

(d) Construct a graph of the sum $B_1\sin(\pi x/L) + B_3\sin(3\pi x/L) + B_5\sin(5\pi x/L)$ as a function of x. (Use a computer if available.) Compare this graph with that of $\psi(x)$ itself. Note how good a "fit" is obtained with the use of only the first three nonzero terms in the superposition.

8-6 *Another square-well packet state.* A particle is initially in the lowest eigenstate of a one-dimensional infinite square well extending from $x = 0$ to $x = L/2$. Its space wave function, correctly normalized, is given by

$$\psi(x) = \frac{2}{\sqrt{L}} \sin \frac{2\pi x}{L}, \qquad 0 \leq x \leq \frac{L}{2}$$

(Check this.) Suddenly the right-hand wall of the well is moved to $x = L$.

(a) Using Eq. 8-9, calculate the probability that the particle is in the *second* ($n = 2$) state of the widened well. (Note that the wavelength within the well, and hence the energy, for this state is the same as for the initial state in the narrower well.)

The time dependence of quantum states

(b) What is the probability that the particle would be found in the *ground* state of the widened well?

(c) Would you expect energy to be conserved in the expansion? Discuss.

8-7 *A time-dependent state of the simple harmonic oscillator (SHO).* At time $t = 0$, a particle in the SHO potential $V(x) = \frac{1}{2}m\omega^2 x^2$ is described by the following wave function:

$$\Psi(x, 0) = A \sum_{n=0}^{\infty} \left(\frac{1}{\sqrt{2}}\right)^n \psi_n(x)$$

The $\psi_n(x)$ are the normalized SHO energy eigenfunctions, and $E_n = (n + \frac{1}{2})\hbar\omega$. Without invoking the explicit functional forms for the $\psi_n(x)$:

(a) Find the normalization constant A.

(b) Write an expression for $\Psi(x, t)$ for all $t > 0$.

(c) Show that $|\Psi(x, t)|^2$ is a periodic function of the time and find the (shortest) period T.

(d) Find the expectation value $\langle E \rangle$ of the particle's energy.

8-8 *The uncertainty principle and particle diffraction.* A well-collimated beam of particles of sharply defined momentum p falls on a screen in which is cut a slit of width d. On the far side of the slit the particles are traveling with a slight spread of directions, as described by the diffraction of their associated deBroglie waves. Show that the diffraction process can be described in terms of the uncertainty product $\Delta y \cdot \Delta p_y$ for the direction y transverse to the initial direction of the beam.

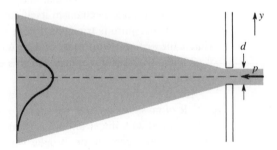

8-9 *The uncertainty principle and ground-state energies.* Consider a particle of mass m in a vee-shaped potential whose analytic form is

$$V(x) = -bx \qquad (x \leq 0)$$
$$V(x) = +bx \qquad (x \geq 0)$$

This provides a far better illustration of the use of the uncertainty principle than the square well discussed in the text (Section 8-6) because in that case an exact analytical solution of the Schrödinger equation was readily available, whereas here it is not. Show that the energy of the lowest state is of the order of $(\hbar^2 b^2/m)^{1/3}$. (Check that this somewhat strange-looking result is correct dimensionally.)

8-10   *A hydrogen atom in the earth's gravitational field.*
   (a) With the help of the uncertainty principle, estimate the lowest energy state of a hydrogen atom "resting" on a horizontal surface in the earth's gravitational field. (You should be able to take advantage of the result of the previous problem.)
   (b) According to your result, what is the typical elevation of the atom above the surface in this ground state?
   (c) For what temperature does kT equal the zero-point energy found in (a)?

8-11   *The uncertainty principle in the laboratory?*
   (a) A plant spore has a diameter of 1 micron ($= 10^{-4}$ cm) and a density of 1 g/cm³. What is its mass? (How many amu is this?)
   (b) Suppose that the spore is viewed through a microscope with which its horizontal position can be located to within about one wavelength of light (5×10^{-5} cm). An experimenter plans to measure its horizontal speed by timing its transit between markers 1 mm apart. What is the fractional error in the computed speed (assumed constant) due to the 0.5-micron uncertainties in the initial and final positions? (Treat the spore as a classical particle. Assume that the distance between markers is known to much better than 0.5 micron and that the errors in the timing mechanism itself can be neglected.)
   (c) If the spore is traveling with a speed of about 10^{-1} cm/sec, what is the *experimental* value of $\Delta p \cdot \Delta x$? What is this value in units of h? Does your result confirm that this experiment can be analyzed classically?
   (d) If you have not already done so, generalize your result: for a particle of mass m traveling with speed v through a region of length L and localized with uncertainty Δx, what is the *experimental* value of $\Delta p \cdot \Delta x$? Introduce the de Broglie wavelength $\lambda = h/mv$ and write your result as a dimensionless quantity times h.

8-12   *Free-particle packet states.* Below are seven assertions about a one-dimensional free-particle packet state for which the probability density differs significantly from zero only within a limited range of position Δx. Decide whether each assertion is true or false. If it is

The time dependence of quantum states

false, explain why it is false and then construct a true statement by altering only the italicized part(s).

(1) The packet state *can* be described as a superposition of *position states*.

(2) The packet state *can* be described as a superposition of *momentum states*.

(3) The average position of a particle in this state can change with time. Analytically this results from *the change in position of the probability distribution associated with each component momentum state*.

(4) As a packet propagates it also spreads with time. This is due to the fact that *the state is composed of states of different momentum*.

(5) The average position of a packet has a velocity *equal to c^2/v*, where v is the velocity of a classical particle with momentum equal to the mean momentum of the packet.

(6) In order to construct a packet state localized in a coordinate region Δx, one must use a range Δp of momenta *smaller* than or equal to (approximately) $\hbar/\Delta x$.

(7) The spatial probability distribution of a certain free-particle packet state spreads as time passes. The corresponding momentum probability distribution *becomes narrower* as time passes.

8-13 *The importance of relative phases in superpositions.* Consider a particle packet whose wave function is constant over a region centered on position x_0 but zero elsewhere:

$$\psi(x) = \sqrt{1/b} \quad \text{for } x_0 - b/2 < x < x_0 + b/2$$
$$\psi(x) = 0, \quad \text{elsewhere}$$

(a) Show that this packet state has momentum components given by the following expression:

$$B_k \equiv \langle k|\psi\rangle \sim e^{-ikx_0} \frac{\sin(kb/2)}{(kb/2)}$$

Compare with Eq. 8-19b for a similar packet centered at $x = 0$.

(b) What is the average momentum of the packet? [*Hint*: This need not involve computation.]

(c) In Section 8-7 we saw that if a wave function $\psi(x)$ describes a packet of zero average momentum, then $e^{ik_0x}\psi(x)$ describes a packet at the same location but with average momentum $\hbar k_0$. Obtain a similar rule for shifting the *position* of a packet without changing its average momentum.

8-14 *One-dimensional free packet states: I.* At $t = 0$, the wave function of a free particle of mass m is given by

$$\Psi(x, 0) = \frac{1}{\sqrt{x_2 - x_1}} e^{ik_0 x}, \quad x_1 \leq x \leq x_2$$
$$= 0, \quad x < x_1 \text{ and } x > x_2$$

(a) Calculate the expectation value $\langle x \rangle$ of the position of the particle at $t = 0$. (Common sense will tell you in advance what the answer should be.)

(b) Calculate the expectation value $\langle p \rangle$ of the linear momentum for this packet, using the recipe $\langle p \rangle = \int \Psi^* (-i\hbar \, \partial/\partial x) \Psi \, dx$.

(c) Calculate the expectation value $\langle K \rangle$ of the kinetic energy from the assumption that we can put

$$(K)_{op} = \frac{1}{2m} (p^2)_{op} = \left[\frac{1}{2m} (p)_{op} \right] (p)_{op}$$

8-15 *One-dimensional free packet states: II.* Consider a one-dimensional packet state $|\psi\rangle$ of a free particle of mass m, for which the *momentum* amplitudes $\langle k | \psi \rangle$ (alternatively denoted by B_k—see Section 8-7) are given by

$$\langle k | \psi \rangle = \frac{1}{\sqrt{k_2 - k_1}} e^{i\beta k}, \quad k_1 \leq k \leq k_2$$
$$= 0, \quad k < k_1 \text{ and } k > k_2$$

(a) Sketch the momentum probability distribution.

(b) Find the position amplitudes $\langle x | \psi \rangle$ [which collectively make up the space wave function $\psi(x)$ at $t = 0$]. Neglect normalization constants. (Use Eq. 8-16a.)

(c) Sketch the position probability distribution of this packet state at $t = 0$.

(d) Knowing that $p = \hbar k$, calculate the expectation value $\langle p \rangle$ of the linear momentum by direct use of the probability distribution obtained in (a).

(e) Calculate the expectation value $\langle E \rangle$ of the energy (= kinetic energy) of the particle. [*Hint:* If g is a function of k, then

$$\langle g \rangle = \int_{\text{all } k} g(k) \, |\langle k | \psi \rangle|^2 \, dk.]$$

8-16 *One-dimensional free packet states: III.* At $t = 0$, the wave function of a free particle of mass m is given by $\Psi(x, 0) = \sqrt{\alpha} \, e^{-\alpha |x|}$, where the constant α is real and positive.

(a) Sketch the position probability distribution of this packet state at $t = 0$.

(b) Using Eq. 8-17a, find the relative momentum amplitudes B_k and then normalize them. [Since the particle is free (that is $V(x) = 0$ everywhere) these momentum amplitudes do not change with time.]

(c) Sketch the momentum probability distribution.

(d) Using the results of (b) and (c) calculate the expectation value of momentum for this packet. [You should note that $\Psi(x, 0)$ has a discontinuity of slope at $x = 0$, so there is no obvious way of evaluating $\langle p \rangle$ by applying the operator $-i\hbar\, \partial/\partial x$ to the space wave function.]

(e) Calculate the expectation value $\langle E \rangle$ of the energy of the particle.

8-17 *Criterion for spreading of a wave packet.* We know that a packet of mean wavenumber k_o has a group velocity v_g equal to $\hbar k_o/m$. However, as pointed out in the text (Section 8-8) the finite spread Δk of wavenumbers in the packet means that there is a corresponding spread of group velocity, given in fact by $\hbar \Delta k/m$. It is this spread Δv_g of group velocity that leads to the progressive broadening of the spatial extent of the wave packet. After some time the spreading $t\Delta v_g$ is comparable to the initial width Δx of the packet.

(a) Show that significant spreading, defined by the above criterion, occurs after the packet has moved a distance x given, in order of magnitude, by the condition

$$ x \approx \frac{k_o}{\Delta k}\, \Delta x \approx \left(\frac{\Delta x}{\lambda_o}\right) \Delta x $$

where λ_o is the de Broglie wavelength corresponding to k_o. This means that the spreading distance can be expressed as the original width of the packet times the number of de Broglie wavelengths contained within that width.

(b) Verify that the mathematical expression for $|\Psi|^2$ for a Gaussian wave packet (Section 8-9) embodies a spreading factor that is consistent with this criterion.

8-18 *A packet describing thermionically emitted electrons.* Electrons are emitted from a heated cathode with a typical kinetic energy of about 0.1 eV and are then accelerated in the x direction through a potential difference of 10,000 V. Neglect the y and z motion and assume that the 0.1 eV of thermal energy at emission is the only source of uncertainty in the x momentum at the end of the acceleration.

(a) Calculate the mean wavenumber k_o and find the corresponding de Broglie wavelength λ_o, in Angstroms.

(b) What is the spread Δk in wave number? Therefore what is the approximate length of the wave packet describing one of the accelerated electrons? (Use the Heisenberg relation to make this estimate.)

(c) If the electrons are allowed to coast freely after traversing the accelerating region, after what time interval will quantum-mechanical spreading disperse the wave packet to a length of a millimeter ($= 10^7$ Å)? (Notice that after significant spreading has taken place then

$$\Delta k \frac{d(\text{group velocity})}{dk} = \Delta k \frac{d^2\omega(k)}{dk^2}$$

evaluated at k_o, is the approximate rate of spreading.) How far will the packet have traveled by this time?

8-19   *Links between quantum and classical physics.*

(a) In classical mechanics, from the definition of momentum, we can put $dx/dt = p_x/m$. In quantum mechanics, this is replaced by a corresponding relation between expectation values:

$$\frac{d}{dt}\langle x \rangle = \frac{\langle p_x \rangle}{m}$$

Verify this result with the help of the following outline:

(i)   Take the basic definition,

$$\langle x \rangle = \int_{\text{all } x} \Psi^*(x, t)\, x\, \Psi(x, t)\, dx$$

(Ψ will be the wave function of a moving wave packet, but we do not need to specify its precise form.)

(ii)   Taking the time derivative, we obtain

$$\frac{d}{dt}\langle x \rangle = \int_{\text{all } x} \frac{\partial \Psi^*}{\partial t}\, x\Psi\, dx + \int_{\text{all } x} \Psi^*\, x\frac{\partial \Psi}{\partial t}\, dx$$

(On the right, x is just the variable of integration and is not subject to the d/dt operation.)

(iii)   Replace $\partial\Psi/\partial t$ and $\partial\Psi^*/\partial t$, by using the time-dependent Schrödinger equation (Eq. 3-11) and its counterpart for Ψ^*:

$$-\frac{\hbar^2}{2m} \frac{\partial^2\Psi^*}{\partial x^2} + V(x)\Psi^* = -i\hbar\frac{\partial\Psi^*}{\partial t}$$

(iv)   Carry out the integrations over all x, taking advantage of

The time dependence of quantum states

the fact that Ψ vanishes for $x \to \pm\infty$. (Integration by parts is involved.)

(v) Use the relation $(p_x)_{op}\Psi = -i\hbar(\partial\Psi/\partial x)$

(b) See if, by means of a similar approach, you can obtain the quantum-mechanical counterpart of Newton's second law:

$$\frac{d}{dt}\langle p_x\rangle = \langle F_x\rangle = \left\langle -\frac{\partial V}{\partial x}\right\rangle$$

8-20 *Spectral line widths and atomic decay rates.* The atoms in a gas discharge tube are excited by a very brief burst of electrons at $t = 0$. The atoms subsequently fall back to the ground state, emitting visible light belonging to a single spectral line at 5500 Å. The intensity of this light falls off with time according to the law $I(t) = I_o e^{-\beta t}$ with $\beta = 5 \times 10^7$ sec^{-1}. Deduce the spread of wavelength of the spectral line, if the "natural" line width is the only source of broadening.

8-21 *Use of the energy-time uncertainty relation.* Construct a graph of log I versus t (for $t > 0$) for the decay of 3914 Å radiation from N_2^+, as shown in Figure 8-10b. Deduce the mean lifetime of the excited state and hence the width of the excited energy state (in electron-volts) and the natural width of the spectral line (in Angstroms).

8-22 *The energy-time uncertainty principle and a gravitational redshift experiment.* Relativity theory predicts that the quantum energy of photons emitted from a massive object will be progressively reduced as they move outward, because (crudely speaking) some of their kinetic energy must be transformed into gravitational potential energy. This manifests itself as a *gravitational redshift* of spectral lines from stars, compared to the same lines as observed from a laboratory source. In a famous experiment to measure the gravitational redshift, R. V. Pound and G. A. Rebka [Phys. Rev. Letters, **4**, 337 (1960)] determined the energy shift of 14-keV γ-ray photons ascending or descending through a distance l of only about 20 m in the gravitational field of the earth. The theoretical fractional change of frequency is given by $\Delta E/E_o = gl/c^2$. (See if you can justify this formula by means of a simple argument.)

(a) Pound and Rebka were able to claim a precision of about one part in 10^{16} for their energy measurement. How does this compare with the theoretical value of $\Delta E/E_o$?

(b) The lifetime of the excited nuclear state that radiates the $E = 14$-keV γ rays is about 10^{-7} sec. What is the natural width of this nuclear γ-ray line?

(c) Can you suggest how the experimental precision obtained by Pound and Rebka can be compatible with your answer to (b)?

8-23 *Natural widths in alpha decay.* The energies of alpha particles emitted from radioactive nuclei are all of the order of 5–10 MeV, but the mean lifetimes of the nuclei that emit them vary over an enormous range. Table 9-1 in the following chapter lists a number of examples. The longest-lived nucleus in that table is a thorium isotope (Th^{232}), which emits alpha particles of 4.05 MeV and has a half-life of 1.39×10^{10} years. The shortest-lived is a polonium isotope (Po^{212}), which emits alpha particles of 8.95 MeV and has a half-life of only 3.0×10^{-7} sec.

(a) Calculate the natural width ΔE, in electron-volts, associated with the finite lifetime of each of these nuclei.

(b) Deduce the values of $\Delta E/E$, and consider whether the fractional dispersion of energy would be detectable in a magnetic spectrometer in which the alpha particles, diverging from a slit 0.1 mm wide, are refocused into a sharp line image, also 0.1 mm wide, after traveling in a semicircular path of radius 0.5 m.

The time dependence of quantum states

In wave mechanics there are no impenetrable barriers, and as the British physicist R. H. Fowler put it after my lecture on that subject at the Royal Society of London . . . "Anyone at present in this room has a finite chance of leaving it without opening the door, or, of course, without being thrown out of the window."

GEORGE GAMOW, *My World Line* (1970)

9

Particle scattering and barrier penetration

9-1 SCATTERING PROCESSES IN TERMS OF WAVE PACKETS

Much of what we know about the structure of atoms and nuclei, and about the interactions between particles, has been derived from the results of *scattering* experiments. Particles from accelerators or radioactive sources, for example, are made to impinge on a target containing nuclei to be studied. The interaction between incident and target particles is inferred from the numbers and angular distribution of scattered particles and from their energy distribution. Such experiments are often used to test an assumed form of a potential well by comparing experimental results with the predictions of a quantum-mechanical analysis based on this assumed potential. The quantum-mechanical analysis has a great deal in common with the classical wave analysis of optical processes such as refraction, reflection, diffraction, etc.

In general, of course, a particle scattering process entails a change of direction. It was through the statistical study of such changes of direction that Rutherford (long before the advent of wave mechanics) inferred the properties of the nuclear Coulomb field inside an atom. Scattering in the most general case is a three-dimensional phenomenon, slightly too complicated for this book. However, some of the essential ideas can be more simply presented in terms of hypothetical one-dimensional scattering processes. Under these conditions, if one ignores the possibility of absorption at the scattering region, the description of the situation after the interaction has been completed must consist solely of wave packets traveling in the same direction as the incident packet, or else reversed in velocity.

It is important to recognize that a scattering region, as represented for example by a rectangular barrier, is a region where, in classical terms, forces are applied to the incident particles.[1] Thus we should expect that the mean momentum of a packet state may be significantly changed by the scattering process. Consider the case indicated in Figure 9-1, in which a probability packet is incident on a square barrier. The packet energy is actually a spread of energies in the neighborhood of some average value E_0 slightly less than the barrier height V_0. Initially the only significant momentum components are positive values close to $\sqrt{2mE_0}$. But after the interaction with the barrier has been completed, the final state is represented by two packets as shown. The packet on the right has a mean

Fig. 9-1 Results of the interaction of a probability packet of average energy E_0 with a barrier of slightly higher potential energy V_0. If this experiment were carried out with many particles, some would penetrate the barrier and the remainder would be reflected, the relative numbers being proportional to the respective areas under the "After" probability curves. Interaction with the barrier changes the average momentum of the state from positive to negative.

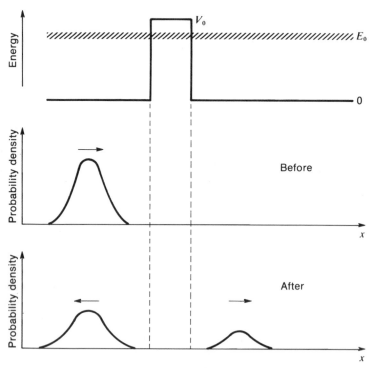

[1]We treat the scattering center as very massive, so that it experiences no recoil. By using the center-of-mass frame, the present analysis can be generalized.

Particle Scattering and Barrier Penetration

momentum equal to that of the incident packet. It corresponds to the interesting phenomenon of penetration through the potential barrier, as discussed in Section 9-5 below. The packet on the left corresponds to an elastic reflection; in classical particle mechanics this would, of course, be the only possible result. The areas under the graphs of probability density for the two separated packets measure the relative probabilities of transmission and reflection at the barrier. For the relative packet sizes indicated in the diagram, the *mean* momentum in the final state is clearly negative.

9-2 TIME-INDEPENDENT APPROACH TO SCATTERING PHENOMENA

A full analysis of scattering events must, in principle, embody the description of a localized free particle that travels toward a center of force and is scattered (deflected) by it. Quantum-mechanically, that implies a theoretical analysis in terms of packet states, with an explicit account of their change of position with time. The discussion in Chapter 8 (especially Sections 8-8 and 8-9) indicates that this is a difficult and laborious task. It is possible, however, to obtain many of the results of such scattering problems with the help of a time-*independent* treatment, using pure momentum states. The basis of the method can be appreciated by analogy with easily demonstrated water-wave scattering processes in a ripple tank.

Suppose that a short group of water waves with straight wave fronts is directed toward a small obstacle, as shown in Figure 9-2a. Partial scattering of the waves occurs, and at some later time there is seen a set of circular scattered wave fronts, plus the continuation of the remnant of the original straight waves on the far side of the obstacle (Figure 9-2b).

If, now, one maintains a *continuous* input of incident waves, the production of the scattered circular waves is likewise continuous, as shown in Figure 9-2c. For this steady-state situation one can consider the boundary conditions that must hold, independently of time, at the surface of the obstacle. From these one can calculate the amplitude of the outgoing scattered wave at any given direction and distance from the scatterer. Although at any given point, thus defined, there is a periodic variation of displacement of the water surface, the amplitude of this variation is constant and can be expressed as

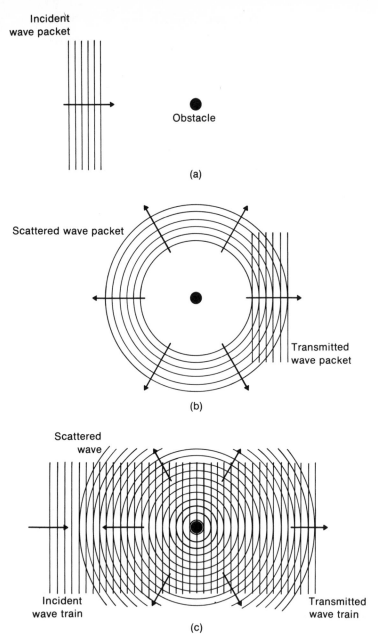

Fig. 9-2   Schematic drawing showing the wave fronts of water waves which encounter an obstacle.   (a) A short group or packet of waves is incident upon an obstacle.   (b) The pattern of waves which results from the encounter of the packet and the obstacle. (c) Steady-state pattern of water waves which arises when a very long train of waves encounters the obstacle. The theoretical analysis of the steady-state situation is usually easier than the analysis of the scattering of the isolated wave group. Similarly, in quantum mechanics the steady-state treatment of scattering (using long trains of de Broglie waves) avoids the explicit time dependence involved in the analysis of the scattering of de Broglie wave packets.

Incident wave packet

Obstacle

(a)

Scattered wave packet

Transmitted wave packet

(b)

Scattered wave

Incident wave train

Transmitted wave train

(c)

a fraction of the amplitude of the incident waves. Explicit consideration of the time is not necessary in such a steady-state analysis. A similar approach is used in the standard analysis of the partial reflection and refraction of light at a boundary between two media.

370   Particle Scattering and Barrier Penetration

The same stratagem of using a steady-state analysis greatly simplifies quantum-mechanical scattering theory in that, whereas the description of limited wave groups requires a whole spectrum of wave numbers, the steady-state analysis, with its assumption of infinitely extended wave trains, permits a calculation to be carried through for any unique value of the wavelength. In wave mechanics this means that one can fix attention on incident particles of sharply defined momentum and energy. The wave function can then be written without any mention of time dependence; for a one-dimensional problem one can put $\psi(x) \sim e^{ikx}$ simply, and obtain a steady-state solution with the aid of boundary conditions, just as is done for the truly time-independent bound-state energy eigenfunctions. And if ever one should want to reconstruct a description of the scattering process in terms of moving wave packets, this can be done by superposing the steady-state solutions for a range of energies (see Section 9-11).

To illustrate most simply how the steady-state method works, we consider the problem of scattering by a one-dimensional potential step as shown in Figure 9-3a. This potential step is not the same as a scattering *center* as described above,

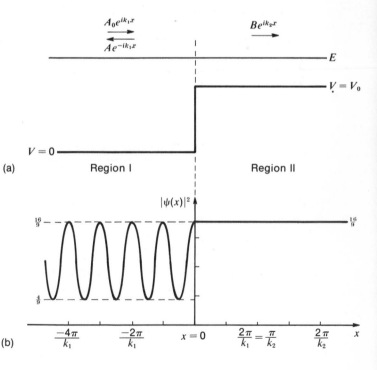

Fig. 9-3 (a) A beam of particles (each of energy E) is incident from the left on a potential-energy step of height $V_0 < E$. (b) The probability density $|\psi(x)|^2$ in the steady state, for the particular case $A_o = 1$, $k_2 = k_1/2$. Standing waves are set up by interference between the incident and reflected beams.

9-2   Time-independent approach to scattering

but is very much easier to analyze and will lead to models of scattering centers later on. Such a step change in potential might be experienced, for example, by an electron inside a metal and incident normally on its surface from within. In such a one-dimensional situation "scattering" is simply the partition of an incident beam into reflected and transmitted parts. We can assume that, just as classical propagating waves scatter only from abrupt changes in the medium, the de Broglie waves scatter only from abrupt changes in the potential.

Locate the coordinate system so that $x = 0$ at the position of the step and let V_o represent the step height. We suppose that a steady-state condition is set up corresponding to the incidence of particles of energy E ($> V_o$) from the region of negative x. (The case $E < V_o$ is considered in Section 9-5 below.) Divide the x axis into two regions:

Region I: $x \leq 0$
Region II: $x > 0$

We make the physical assumption that in region II we have only a wave traveling in the positive x direction, whereas in region I we have a superposition of the incident wave and a reflected wave. Our aim is to calculate the fractions of incident particles reflected by the step and transmitted across it. This means finding the *ratios* of the reflected and transmitted beams to the incident beam; absolute normalization is not needed. We therefore describe the incident beam by the wave function $A_0 e^{ik_1 x}$, where $|A_0|^2$ is a measure of the intensity of incoming particles, whose numerical value will not concern us. On this basis we write the following expressions for the form of the wave function in regions I and II:

$$\psi_{\text{I}}(x) = A_o e^{ik_1 x} + A e^{-ik_1 x}$$
$$\psi_{\text{II}}(x) = B e^{ik_2 x} \tag{9-1}$$

where $k_1 (= \sqrt{2mE}/\hbar)$ defines the magnitude of the momentum in region I and $k_2 (= \sqrt{2m(E - V_o)}/\hbar)$ gives this magnitude for region II. The value k_1 is used in region I for both the incident $(+k_1)$ and reflected $(-k_1)$ waves, since both travel in a region of zero potential.

The values of the constants A and B relative to A_o are found by applying the usual boundary conditions that both the

Particle Scattering and Barrier Penetration

wave function ψ and its spatial derivative $d\psi/dx$ be continuous across the potential step.[2] The derivatives of the expressions in Eq. 9-1 are

$$\frac{d\psi_{\mathrm{I}}}{dx} = ik_1 A_o e^{ik_1 x} - ik_1 A e^{-ik_1 x} \qquad (9\text{-}2)$$

$$\frac{d\psi_{\mathrm{II}}}{dx} = ik_2 B e^{ik_2 x}$$

Applying the boundary conditions at $x = 0$ and using Eqs. 9-1 and 9-2, we have

$$A_o + A = B$$
$$ik_1 A_o - ik_1 A = ik_2 B$$

From these two equations we easily obtain values for the coefficients A and B in terms of A_o:

$$A = \frac{k_1 - k_2}{k_1 + k_2} A_o$$
$$B = \frac{2k_1}{k_1 + k_2} A_o \qquad (9\text{-}3)$$

The fraction of incident particles reflected, the so-called *reflection coefficient R*, is the ratio of probability densities in the reflected and incident beams in region I:

$$R = \frac{|A|^2}{|A_o|^2} = \left(\frac{k_1 - k_2}{k_1 + k_2}\right)^2 \qquad (9\text{-}4)$$

The fraction transmitted, the *transmission coefficient T*, is that fraction not reflected, derivable from Eqs. 9-4 and 9-3:

$$T = 1 - R = \frac{4k_1 k_2}{(k_1 + k_2)^2} = \frac{k_2 |B|^2}{k_1 |A_o|^2} \qquad (9\text{-}5)$$

We have thus been able to predict the fractions of incident particles of given energy E reflected and transmitted at a potential step V_o, using $k_1 = \sqrt{2mE}/\hbar$ and $k_2 = \sqrt{2m(E - V_o)}/\hbar$.

---

[2]For a similar analysis of the partial reflection of a classical wave pulse at the junction between two strings, see, for example, the text *Vibrations and Waves* in this series, pp. 256 ff.

9-2   Time-independent approach to scattering

Notice how utterly different these results are from what one would have for classical particles approaching a potential step. In this case, if the particle energy were (as in the case analyzed above) greater than the step height, giving positive kinetic energy in region II, the transmission would be 100 percent. If, on the other hand, we had $E < V_o$ there would be 100 percent reflection. The results expressed by Eqs. 9-3 are, however, identical in form to what one has for partial reflection and transmission of sound or light waves falling at normal incidence on the boundary between two media.

9-3 PROBABILITY DENSITY AND PROBABILITY CURRENT

It is important to notice that in Eq. 9-5 the fraction of incident particles transmitted across the step is *not* equal to $|B|^2/|A_o|^2$, which is the ratio of transmitted to incident particle *densities*. The reason is that the transmitted (and reflected) fractions of the beam concern *probability currents* and not probability densities alone. Roughly speaking, the probability current is the product of a density and a characteristic velocity. The *reflected* beam in the problem considered above travels in the same potential as the incident beam, so its characteristic velocity is the same as that of the incident beam and thus cancels out in the reflection coefficient. In contrast, the transmitted beam in region II has lower kinetic energy ($K = E - V_o$) and thus a smaller characteristic velocity than the incident beam in region I, leading to a velocity-dependent factor $k_2/k_1 = p_2/p_1 = v_2/v_1$ in the expression for the transmission coefficient. To see this more clearly, consider the following analogy based on the flow of electric charge.

Suppose we have a steady flow of electric current in a circuit consisting of a single loop (Figure 9-4). For a steady flow, the same amount of charge must pass each point in the loop during any given time interval. However, this does not mean that the concentration of charge carriers (for example, conduction electrons) is necessarily the same at every point. A slow-moving high-density stream can deliver the same total charge across a surface in a given time interval as a fast-moving low-density stream. To express this in symbolic terms, and referring to Figure 9-4, let the linear charge density (charge per unit length along the direction of motion) be λ_1 in the vicinity of

Particle Scattering and Barrier Penetration

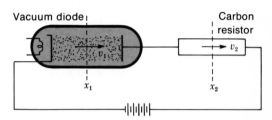

Fig. 9-4 A circuit in which a steady current is flowing. During any given time interval Δt the same amount of charge Δq must pass each point along the circuit. This same flow of charge can be maintained by a low density of fast-moving charges (as at x_1) or by a high density of slow-moving charges (as at x_2). The quantity which must not vary from place to place along the circuit is the product λv, where λ is the linear charge density and v is the local drift velocity.

point x_1, and call the local drift velocity v_1. Then during the time interval Δt an amount of charge $\Delta q = \lambda_1 \Delta x$ passes the point x_1, where $\Delta x = v_1 \Delta t$. This corresponds to a current $\Delta q/\Delta t = \lambda_1 v_1$ entering the region to the right of x_1. If there is at some other point x_2 a linear charge density λ_2 and a local drift velocity v_2, then a steady current flow in the loop requires that $\lambda_2 v_2 = \lambda_1 v_1$. This ensures that there is no net gain or loss of charge in the region between x_1 and x_2. In summary, a given steady current may be sustained at one point in the circuit by a large density of charge moving at low speed, and at another point by a low density of charge moving at high speed.

In applying the classical electric current analogy to quantum-mechanical scattering, one replaces the charge density function λ with the appropriate probability density and recognizes that the ratio of characteristic velocities in different regions of potential is equal to the ratio of wave numbers in those regions:

$$\frac{v_2}{v_1} = \frac{(p_2/m)}{(p_1/m)} = \frac{(\hbar k_2/m)}{(\hbar k_1/m)} = \frac{k_2}{k_1}$$

Using these substitutions one speaks of a *probability current*, given by the product of probability density and characteristic velocity. In applying this concept one may have to distinguish probability currents flowing in more than one direction in a given region of space. For example, in the scattering from a potential step discussed in the preceding section, region I contains both an incident beam (probability density $|A_0 e^{+ik_1 x}|^2 = |A_0|^2$) and a reflected beam (probability density $|A e^{ik_1 x}|^2 = |A|^2$). Since the characteristic speed $v_1 \sim k_1$ is

9-3   Probability density and probability current

the same for both directions in this region, the velocity factors cancel in the reflection coefficient

$$R = k_1 |A|^2 / k_1 |A_0|^2 = |A|^2 / |A_0|^2$$

In contrast, region II in this problem has a single rightward-moving beam with a velocity $v_2 \sim k_2$ *different* from the velocity in region I, and a current determined by the value of $v_2 |B|^2$. The transmission coefficient, defined as a ratio of probability currents, is then given by

$$T = \frac{v_2 |B|^2}{v_1 |A_0|^2} = \frac{k_2 |B|^2}{k_1 |A_0|^2}$$

We can see that this value of T agrees with the value of $1-R$ in Eqs. 9-4 and 9-5. Thus the use of probability currents automatically ensures the conservation of total probability.

The above discussion is clearly not very rigorous. A more general and convincing argument (given below) can be developed directly from the time-dependent Schrödinger equation. This calculation was first made in 1926 by Max Born; it played a central role in the development and acceptance of the probabilistic interpretation of the Schrödinger wave function. The general result is that the quantity $\Psi^*\Psi$, or $|\Psi|^2$, obeys a local conservation condition like that for electric charge. Any increase or decrease of net probability within a given region is exactly accountable in terms of a probability flow across the boundaries of that region.

In order to develop the relevant results, we make use of the time-dependent Schrödinger equations governing Ψ and its complex conjugate Ψ^*. These equations are as follows (one-dimensional only):

$$
\begin{aligned}
i\hbar \frac{\partial \Psi}{\partial t} &= -\frac{\hbar^2}{2m} \frac{\partial^2 \Psi}{\partial x^2} + V(x)\Psi \\
-i\hbar \frac{\partial \Psi^*}{\partial t} &= -\frac{\hbar^2}{2m} \frac{\partial^2 \Psi^*}{\partial x^2} + V(x)\Psi^*
\end{aligned}
\tag{9-6}
$$

[*Note:* Ψ^* can be used just as well as Ψ for the description of quantum states, but in its Schrödinger equation the $\partial/\partial t$ term is of opposite sign. The need for this sign reversal is plain if, for example, we consider the free-particle functions $\Psi \sim$

Particle Scattering and Barrier Penetration

exp $i(kx - Et/\hbar)$ and $\Psi^* \sim$ exp $-i(kx - Et/\hbar)$. The double differentiation $\partial^2/\partial x^2$ leads to the multiplier $-k^2$ in both cases, but the single differentiation $\partial/\partial t$ leads to $\mp iE/\hbar$.]

Now we ask about the time variation of total probability in a given one-dimensional region, say that between the boundaries x_1 and x_2. The total probability of finding a particle in this region is given by the integral

$$\int_{x_1}^{x_2} |\Psi(x, t)|^2 \, dx = \int_{x_1}^{x_2} \Psi^*(x, t) \cdot \Psi(x, t) \, dx$$

The time rate of change of this quantity is

$$\frac{\partial}{\partial t} \left[\int_{x_1}^{x_2} |\Psi|^2 \, dx \right] = \int_{x_1}^{x_2} \left[\frac{\partial \Psi^*}{\partial t} \cdot \Psi + \Psi^* \cdot \frac{\partial \Psi}{\partial t} \right] dx$$

Using the two variants of the Schrödinger equation from Eq. 9-6, we can change the right-hand side to the form

$$\frac{i}{\hbar} \int_{x_1}^{x_2} \left\{ \left[-\frac{\hbar^2}{2m} \frac{\partial^2 \Psi^*}{\partial x^2} + V(x)\Psi^* \right] \cdot \Psi - \Psi^* \cdot \left[-\frac{\hbar^2}{2m} \frac{\partial^2 \Psi}{\partial x^2} + V(x)\Psi \right] \right\} dx$$

The potential terms cancel and we have

$$\frac{\partial}{\partial t} \left[\int_{x_1}^{x_2} |\Psi|^2 \, dx \right] = \frac{i\hbar}{2m} \int_{x_1}^{x_2} \left(\Psi^* \cdot \frac{\partial^2 \Psi}{\partial x^2} - \Psi \cdot \frac{\partial^2 \Psi^*}{\partial x^2} \right) dx$$

But the integrand on the right can be made into a perfect differential:

$$\left(\Psi^* \cdot \frac{\partial^2 \Psi}{\partial x^2} - \Psi \cdot \frac{\partial^2 \Psi^*}{\partial x^2} \right) = \frac{\partial}{\partial x} \left(\Psi^* \cdot \frac{\partial \Psi}{\partial x} - \Psi \cdot \frac{\partial \Psi^*}{\partial x} \right)$$

so that the integral is immediately evaluated:

$$\frac{\partial}{\partial t} \left[\int_{x_1}^{x_2} |\Psi|^2 \, dx \right] = \frac{i\hbar}{2m} \left(\Psi^* \cdot \frac{\partial \Psi}{\partial x} - \Psi \cdot \frac{\partial \Psi^*}{\partial x} \right) \Bigg|_{x_1}^{x_2} \tag{9-7}$$

Equation 9-7 says that the rate of change of total probability in the region between x_1 and x_2 is equal to the difference between the two values of a single expression evaluated at the boundaries of the region. A consistent interpretation results if we in-

9-3   Probability density and probability current

terpret this expression as probability current. We can write Eq. 9-7 as

$$\left[\begin{array}{l}\text{rate of change of total probability} \\ \text{in the region between } x_1 \text{ and } x_2\end{array}\right] = J(x_1, t) - J(x_2, t)$$

where the probability current $J(x, t)$ is given by

$$J(x, t) = -\frac{i\hbar}{2m}\left(\Psi^* \cdot \frac{\partial \Psi}{\partial x} - \Psi \cdot \frac{\partial \Psi^*}{\partial x}\right) \tag{9-8a}$$

The signs have been arranged so that the probability current *into* the region at x_1 is *positive*. In one-dimensional problems the probability current has the dimension of reciprocal time (probability per unit time, with probability having no units).

For any state that has unique energy $\hbar\omega$ the wave function has the form

$$\Psi(x, t) = \psi(x)\, e^{-i\omega t}$$

In this special case, the time factors in Ψ and Ψ^* cancel in Eq. 9-8a and we can put

$$J = -\frac{i\hbar}{2m}\left(\psi^* \frac{d\psi}{dx} - \psi \frac{d\psi^*}{dx}\right) \qquad \text{(unique } E) \tag{9-8b}$$

The one-dimensional probability current takes on a particularly simple form for a monoenergetic free particle with the wave function of the form $\Psi = A e^{i(kx - \omega t)}$. Then the probability current is given by

$$J = -\frac{i\hbar}{2m}(ik + ik)(\Psi^* \cdot \Psi) = \frac{\hbar k}{m}(A^*A) = v|A|^2 \tag{9-9}$$

Here v is the classical velocity associated with a particle of momentum $p = \hbar k$. Thus the general expression for a one-dimensional probability current (Eq. 9-8a) confirms our assumption that, in the steady-state analysis of scattering problems, we can calculate transmission coefficients, etc., from the ratio of the values of $v|\Psi|^2$ in different regions.

When Eqs. 9-8 are extended to three dimensions, $|\Psi|^2$ becomes a probability per unit volume, normalized to unity when integrated over all space. The quantity **J** becomes a *vec-*

Particle Scattering and Barrier Penetration

tor probability current density with the units, for example, $cm^{-2} sec^{-1}$ (probability per square centimeter per second, with probability again having no units).

9-4 SCATTERING BY A ONE-DIMENSIONAL WELL

As a second example of scattering, let us consider the situation shown in Figure 9-5a. Particles of total energy E (relative to a zero-potential level represented by region I) encounter a potential "hole" of depth V_0 and width L. Partial reflection and transmission must be assumed to take place at both sides of the well. In regions I and III the wave numbers have the same

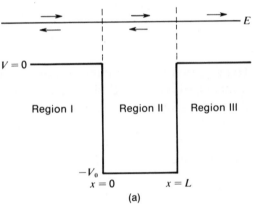

Region I   Region II   Region III

Fig. 9-5
(a) Steady-state situation for one-dimensional scattering of particles of unique energy E by a square well.
(b) Transmission coefficient T of square well as a function of incident particle energy, calculated for the dimensionless parameter $L\sqrt{2mV_0}/\hbar$ equal to $20.5\,\pi$. Note resonances giving 100 percent transmission at certain energies.

(a)

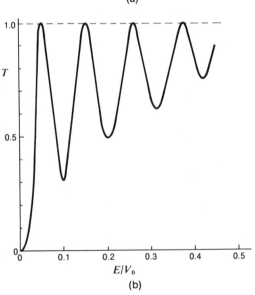

(b)

      9-4   Scattering by a one-dimensional well

value, $k_1 = \sqrt{2mE}/\hbar$; in region II there is a larger wave number, $k_2 = \sqrt{2m(E + V_0)}/\hbar$.

Appropriately extending the analysis of Section 9-2, we can write

$$\psi_{\mathrm{I}}(x) = A_0 e^{ik_1 x} + A e^{-ik_1 x}$$
$$\psi_{\mathrm{II}}(x) = B e^{ik_2 x} + C e^{-ik_2 x} \qquad (9\text{-}10)$$
$$\Psi_{\mathrm{III}}(x) = D e^{ik_1 x}$$

You can easily verify that the result of applying the continuity conditions on ψ and $d\psi/dx$ at $x = 0$ and $x = L$ is the following set of equations:

$$A_0 + A = B + C$$
$$ik_1 A_0 - ik_1 A = ik_2 B - ik_2 C$$
$$B e^{ik_2 L} + C e^{-ik_2 L} = D e^{ik_1 L} \qquad (9\text{-}11)$$
$$ik_2 B e^{ik_2 L} - ik_2 C e^{-ik_2 L} = ik_1 D e^{ik_1 L}$$

Here we have four equations relating five undetermined coefficients; this is enough information to obtain the values of A, B, C, and D as fractions of A_0. To calculate the transmission coefficient T of the well we need to find the value of D/A_0. The algebra of this is not difficult. From the first pair of Eqs. 9-11 we easily find

$$2k_1 A_0 = (k_2 + k_1) B - (k_2 - k_1) C$$

From the second pair of Eqs. 9-11 we can find B and C in terms of D:

$$B = \frac{k_2 + k_1}{2k_2} D e^{ik_1 L} e^{-ik_2 L}$$

$$C = \frac{k_2 - k_1}{2k_2} D e^{ik_1 L} e^{ik_2 L}$$

Substituting these expressions for B and C in the preceding equation then leads to the result

$$4k_1 k_2 A_0 = [(k_2 + k_1)^2 e^{-ik_2 L} - (k_2 - k_1)^2 e^{ik_2 L}] D e^{ik_1 L} \qquad (9\text{-}12)$$

The quantity $|D/A_0|^2 \, (= |D|^2/|A_0|^2)$ is the ratio of probability density in the transmitted beam to that in the beam incident on the well. Since, however, the potential energy is the

Particle Scattering and Barrier Penetration

same on both sides of the well (and hence k has the same value) the ratio $|D/A_0|^2$ as given by Eq. 9-12 is also the ratio of transmitted current to incident current. That is, the transmission coefficient T is equal to $|D/A_0|^2$; the form of its variation with particle energy is shown in Figure 9-5b.

Without evaluating the general result (done in exercise 9-11), we can identify certain properties of this scattering system:

1. For $k_1 \ll k_2$ (incident particle energy E much less than well depth V_0) we have

$$4k_1 k_2 A_0 \approx k_2{}^2 (e^{-ik_2 L} - e^{ik_2 L}) D e^{ik_1 L}$$
$$= -(2ik_2{}^2 \sin k_2 L) D e^{ik_1 L}$$

Therefore,

$$T \approx \frac{4k_1{}^2}{k_2{}^2 \sin^2 k_2 L}$$

Here, k_1 is proportional to $\sqrt{E}$, and k_2 $(= \sqrt{2m(V_0 + E)}/\hbar)$ is approximately constant as E is varied. Hence $T \sim E$—the transmission of the well rises linearly with incident particle energy.[3]

2. For $E \gg V_0$, we have $k_2 \approx k_1$, in which case

$$4k_1{}^2 A \approx [(2k_1)^2 e^{-ik_1 L}] D e^{ik_1 L} = 4k_1{}^2 D$$

Therefore

$$T \approx 1$$

Thus for incident particle energies much bigger than the well depth, the transmission approaches 100 percent.

3. For $k_2 L = n\pi$, we have a very interesting *resonance condition*. For values of k_2 satisfying this condition (n integral) we have

$$e^{ik_2 L} = e^{-ik_2 L} = +1 \quad (n \text{ even})$$
$$= -1 \quad (n \text{ odd})$$

---

[3]This result does not hold if $k_2 L \to n\pi$ as $E \to 0$. In that case $T \to 1$ as $E \to 0$, in the manner described in property 3 below.

9-4   Scattering by a one-dimensional well

Under these conditions, Eq. 9-8 gives us (exactly)

$$4k_1k_2A_0 = \pm[(k_2 + k_1)^2 - (k_2 - k_1)^2] \, De^{ik_1L}$$

Therefore

$$A_0 = \pm De^{ik_1L}$$

and

$$T = 1$$

Thus for all energies such that $k_2L = n\pi$ the well is completely transparent to the incident particles. The condition for this to happen ($k_2 = n\pi/L$) corresponds to the width L of the well being equal to an integral number of half-wavelengths λ_2 of the wave function ψ_{II} inside the well. We have $k_2 = 2\pi/\lambda_2$, and hence

$$2L = n\lambda_2$$

This behavior is closely analogous to the selective transmission of light of particular wavelengths by a thin layer of glass or dielectric—an effect that is exploited in optical interference filters, which by a careful choice of thickness transmit light within a narrow band of wavelengths with far less attenuation than occurs with normal colored filters (which work by selective absorption).

The wave-mechanical transparency of a potential well is observed in the scattering of electrons by noble-gas atoms, and is known as the Ramsauer effect.[4] It manifests itself as a minimum in the cross section (target area) presented by atoms to incident electrons at a certain value of the electron energy. If an atom of radius R could be regarded as a simple rectangular well of width $L = 2R$, the above analysis would imply a minimum in the cross section for $\lambda_2 = 4R$, corresponding to an electron kinetic energy *inside* the well equal to $h^2/2m\lambda_2^2$, or $h^2/32mR^2$. For $R \approx 1$ Å, this would give a value of about 10 eV. Actual experiments (Figure 9-6) show a minimum cross section for an *incident* electron energy of only about 1 eV, indicating that the effective well depth is of the order of 10 eV.

[4]More correctly, the Ramsauer-Townsend effect, discovered independently by C. W. Ramsauer [Ann. Phys. **64**, 513 (1921)] and J. S. Townsend [Phil. Mag. **43**, 593 (1922)].

Particle Scattering and Barrier Penetration

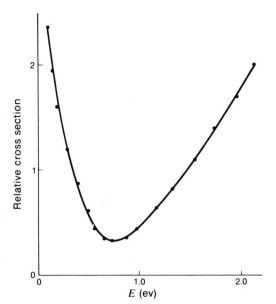

Relative cross section

E (ev)

Fig. 9-6 Ramsauer effect. Minimum in cross section for scattering of electrons by xenon at about 1 eV. [After S. G. Kukolich, Am. J. Phys. **36**, 701 (1968).]

9-5 BARRIER PENETRATION: TUNNELING

One of the most striking effects described by quantum mechanics is the transmission of incident particles across a potential barrier whose height is *greater* than the particle energy. Classically, these particles lack the energy to get over the barrier, yet for quantum-mechanical systems some fraction of incident particles are transmitted. Since from the energy standpoint (and certainly in spatial terms) the particle can be said to go *through* the barrier rather than over it, this effect is called *tunneling*. There is no doubt that tunneling takes place; the effect is used routinely in solid-state devices and is featured prominently in the theory of radioactive decay of nuclei. We develop here a simple theory of tunneling which we will employ later in this chapter (Sections 9-9 and 9-10) to obtain a crude but surprisingly accurate model of alpha decay for certain radioactive nuclei.

The fundamental idea of tunneling was already implicit in our treatment of the finite square well in one dimension (Section 3-8). There we saw that "outside the well" [that is, in the classically forbidden region where $E < V$—see Figure 9-7a] the energy eigenfunction has the form of a real exponential that decreases with distance from the well (Figure 9-7b). Quantum mechanics then tells us at the outset that particles can exist

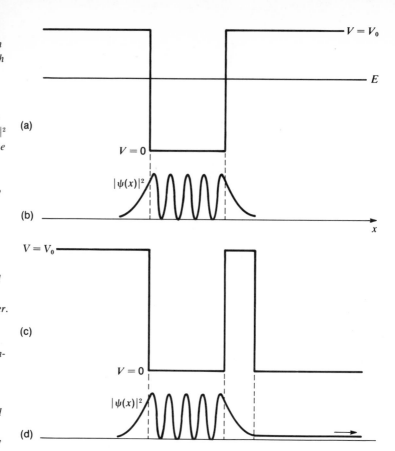

Fig. 9-7 (a) The finite square well in one dimension, with an energy eigenvalue indicated. (b) The quantum-mechanical probability density $|\psi(x)|^2$ for the energy value shown in (a). Notice that $|\psi(x)|^2$ is not identically zero in the classically forbidden region (where $E < V$). (c) A modified square well, in which the potential has been altered to produce a barrier. (d) The well shown in (c) has no bound eigenstates, and we can anticipate that a particle initially confined to the well will later exhibit a nonzero probability density to the right of the barrier. On physical grounds we can expect that the wave function to the right of the barrier will consist entirely of positive-momentum components, as indicated by the arrow.

outside the well. This existence might be verified if the potential on one side of the well were modified to be a barrier (Figure 9-7c). Then there would be some probability for the particle initially occupying the well to be found at the outside edge of the barrier, that is, to escape (Figure 9-7d). As usual, probability requires a statistical interpretation: out of many identical systems initially in such a bound state, some fraction will decay per unit time; in each decay a particle escapes from the bound condition. This model is very similar to the one we shall use to analyze alpha decay later (Section 9-9). But first we consider some details of the tunneling process.

To begin the analysis with the simplest possible example, consider a steady beam of monoenergetic particles incident on a potential step as examined in Section 9-2, but this time let the incident beam have an energy *less* than the height of the step (Figure 9-8a). Number the regions I and II as before. In region I we must expect both incident and reflected beams whose

Particle Scattering and Barrier Penetration

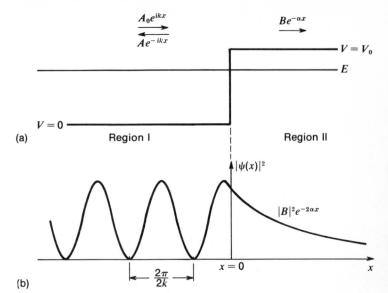

$A_0 e^{ikx}$ $\rightarrow$
$\leftarrow$
$A e^{-ikx}$

$B e^{-\alpha x}$ $\rightarrow$

$V = V_0$

E

$V = 0$

(a)   Region I | Region II

$|\psi(x)|^2$

$|B|^2 e^{-2\alpha x}$

$\dfrac{2\pi}{2k}$

$x = 0$ x

(b)

Fig. 9-8 (a) A beam of particles (each of energy E) is incident from the left on a potential-energy step of height $V_0 > E$. (b) The probability density $|\psi(x)|^2$ in the steady state.

wave functions have the usual complex exponential form. In region II, however, the wave function must be a decreasing exponential:

$$\psi_I = A_0 e^{ikx} + A e^{-ikx}$$
$$\psi_{II} = B e^{-\alpha x}$$

where the subscript has been dropped from k_1. The mathematically acceptable increasing exponential $e^{\alpha x}$ in region II is rejected because it would lead to an overwhelming relative probability of finding the particles deep inside the step (at large x), a result we know to be physically unacceptable. As discussed in the case of the finite square well, the positive real attenuation constant α is given by the expression

$$\alpha = \frac{\sqrt{2m(V_0 - E)}}{\hbar} \qquad (E < V_0)$$

The boundary conditions, as before, are continuity of ψ and $d\psi/dx$ across the step at $x = 0$. This leads to the pair of equations

$$A_0 + A = B$$
$$ikA_0 - ikA = -\alpha B$$

9-5 Barrier penetration: tunneling

which yield values of A and B in terms of A_o:

$$A = \frac{ik + \alpha}{ik - \alpha} A_o$$

$$B = \frac{2ik}{ik - \alpha} A_o$$

You can see that the magnitude $|A/A_o|$ has the value unity. The superposition of incident and reflected waves in region I creates a standing wave of probability (Figure 9-8b) which looks very much like that at the right-hand edge of a finite square well. Although particles may penetrate into the step, there is no way for them to get through it. Therefore it is not surprising that the reflection coefficient R for this case has the value unity:

$$R = \left| \frac{A}{A_o} \right|^2 = \left| \frac{ik + \alpha}{ik - \alpha} \right|^2 = \frac{k^2 + \alpha^2}{k^2 + \alpha^2} = 1$$

Once again probability is conserved, and particles that penetrate the step are reflected, not absorbed.[5]

In order to obtain transmission, we must have a barrier of finite width. The simplest such case, the so-called *square barrier*, is shown in Figure 9-9. Call the width of the barrier L. The x axis is now divided into three regions, with region III to the right of the barrier ($x > L$). Assuming as before that particles are incident from the left in the figure, we expect region III to contain only a transmitted beam with a wave function of the form e^{ikx}. Here the wave number k has the same value as in region I, since both have zero potential. The form of the wave function in region I is still a superposition of incident and reflected components characterized respectively by wave numbers $+k$ and $-k$.

The form of the wave function in region II requires a little thought. The decreasing exponential $e^{-\alpha x}$ will certainly be needed. For particles ($E < V_o$) incident on a potential step we rejected the addition of a term with a *rising* exponential $e^{+\alpha x}$ inside the step because it would lead to unlimited relative probability for large x. That argument does not hold for a finite bar-

---

[5]It is true that in the steady state there is a nonzero probability of finding the particle in region II. However, once it is established, this probability does not change with time.

　　Particle Scattering and Barrier Penetration

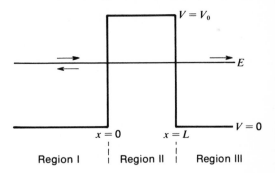

Fig. 9-9 *Steady-state scattering by a rectangular barrier of height V_o and width L.*

Region I ¦ Region II ¦ Region III

rier because $e^{+\alpha x}$ is limited to finite values in the region $x \leq L$. In addition we have the physical effect of the change in potential at the right-hand edge of the barrier, at the boundary between regions II and III. This discontinuity will cause some reflection back into region II, resulting in an exponential term that *decreases in the direction from right to left*, or *increases* in a direction from left to right. This is added reason to include a term $e^{+\alpha x}$ in the wave function for region II. Finally, it simply proves formally impossible to satisfy all boundary conditions without the rising exponential term in region II. For these reasons we use both decreasing and increasing exponential terms in the wave function for region II. The anticipated forms of $\psi(x)$ in the three distinct regions are then given by

$$\psi_I(x) = A_o e^{ikx} + A e^{-ikx}$$
$$\psi_{II}(x) = B e^{-\alpha x} + C e^{+\alpha x} \tag{9-13}$$
$$\psi_{III}(x) = D e^{ikx}$$

where $k = \sqrt{2mE}/\hbar$ and $\alpha = \sqrt{2m(V_o - E)}/\hbar$. Matching of the boundary conditions on ψ and $d\psi/dx$ then leads to the following two pairs of equations:

At $x = 0$:

$$A_0 + A = B + C$$
$$ikA_o - ikA = -\alpha B + \alpha C$$

$$\tag{9-14}$$

At $x = L$:

$$D e^{ikL} = B e^{-\alpha L} + C e^{\alpha L}$$
$$ikD e^{ikL} = -\alpha B e^{-\alpha L} + \alpha C e^{\alpha L}$$

The exact evaluation of the coefficients in the above equa-

387   9-5   Barrier penetration: tunneling

tions is tedious but straightforward. It follows precisely the same path as we indicated in Section 9-4 for scattering by a square well. The only coefficient we need is D, and this is given by the following equation, analogous to Eq. 9-12:

$$4ik\alpha A_o = [(\alpha + ik)^2 e^{-\alpha L} - (\alpha - ik)^2 e^{\alpha L}] D e^{ikL} \qquad (9\text{-}15)$$

As in the case of a scattering well with equal potentials on the two sides (see Section 9-4) the transmission coefficient T of the barrier is equal to $|D/A_o|^2$. Because the condition $E < V_o$ conjures up a picture of particles penetrating *through* the barrier, the transmission coefficient in this case is called the *penetrability*.

Equation 9-15 can be replaced by a simple approximate form for a "thick" barrier, that is, one for which the barrier width L is large compared with the characteristic length $1/\alpha$ of the exponential, so that $\alpha L \gg 1$. In this case the first term in the square bracket of Eq. 9-15 is much smaller than the second term and the equation becomes

$$\frac{D}{A_o} \approx -\frac{4ik\alpha e^{-(\alpha + ik)L}}{(\alpha - ik)^2} \qquad (\text{for } \alpha L \gg 1) \qquad (9\text{-}16)$$

The transmission coefficient (penetrability) $T = |D/A_o|^2$ is calculated by multiplying the expression for D/A_o by its complex conjugate. (Recall that the complex conjugate of an algebraic expression involving complex numbers is obtained simply by writing the same expression with each complex number conjugated.) The result is

$$T = \left|\frac{D}{A_o}\right|^2 \approx \frac{16\alpha^2 k^2 e^{-2\alpha L}}{(\alpha^2 + k^2)^2} \qquad (\alpha L \gg 1)$$

or

$$T \approx 16\left(\frac{E}{V_o}\right)\left(1 - \frac{E}{V_o}\right)e^{-2\alpha L} \qquad (\alpha L \gg 1) \qquad (9\text{-}17)$$

Now for many values of the ratio E/V_o, provided $\alpha L \gg 1$, the *order of magnitude* of the ratio $|D/A_o|^2$, given by Eq. 9-17, is dominated by the exponential factor alone. Thus we obtain

Particle Scattering and Barrier Penetration

the following very simplified expression for the penetrability:

$$T \approx e^{-2\alpha L} = \exp\left\{-\frac{2}{\hbar}[2m(V_0 - E)^{1/2}L]\right\} \qquad (\alpha L \gg 1) \quad (9\text{-}18)$$

As a numerical example, consider electrons of kinetic energy 3 eV incident on a rectangular barrier of height 10 eV and width 4 Å (such as might be presented by one or two atomic layers of oxide separating two sheets of the same metal). What is the approximate fraction of these electrons that will succeed in penetrating the barrier? In Eq. 9-18 we have

$$\alpha = \frac{[2m(V_0 - E)]^{1/2}}{\hbar}$$

$$= \frac{[2 \times 9.1 \cdot 10^{-31} \times 7 \times 1.6 \cdot 10^{-19}]^{1/2}}{1.05 \times 10^{-34}} \qquad \text{(SI units)}$$

$$\approx 1.35 \times 10^{10} \text{ m}^{-1}$$

Thus $2\alpha L \approx 10.8$

$$T \approx e^{-2\alpha L} \approx 2 \times 10^{-5}$$

[We may note that the more accurate approximation, as given by Eq. 9-17, yields a value of the penetrability about three times greater than the value just calculated.]

9-6 PROBABILITY CURRENT AND BARRIER PENETRATION PROBLEMS

The introduction of probability currents throws further light on barrier penetration problems as treated in the time-independent approach. Consider, for example, a one-dimensional problem in which the wave function describes particles incident from the left on a rectangular barrier as shown in Figure 9-10a. This, in contrast to the case considered in Section 9-5, represents a general type of rectangular barrier in which the potential levels are different on the two sides (regions I and III).

In region I, the space-dependent factor ψ is given by

$$\psi_1(x) = A_0 \, e^{ik_0x} + A e^{-ik_0x}$$

Then

$$\psi_1^*(x) = A_0^* \, e^{-ik_0x} + A^* e^{ik_0x}$$

9-6 Probability current and barrier penetration

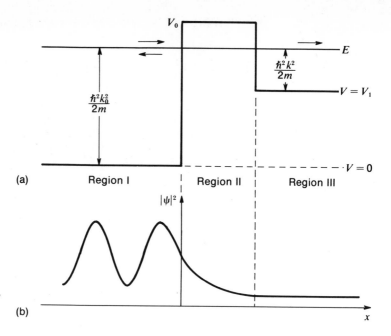

Fig. 9-10 (a) Potential barrier for study of probability current. (b) Resulting probability density.

(a)

(b)

Differentiating, we have

$$\frac{d\psi_1}{dx} = ik_oA_oe^{ik_ox} - ik_oAe^{-ik_ox}$$

$$\frac{d\psi_1{}^*}{dx} = -ik_oA_o{}^*e^{-ik_ox} + ik_oA^*e^{ik_ox}$$

The probability current J_I at any point in region I is now given (Eq. 9-8b) by

$$J_I = -\frac{i\hbar}{2m}\left(\psi_I{}^*\frac{d\psi_1}{dx} - \psi_I\frac{d\psi_1{}^*}{dx}\right)$$

Substituting the explicit expressions for ψ and $d\psi/dx$, one obtains

$$J_I = v_o\,(A_oA_o{}^*) - v_o(AA^*)$$

where $v_0 = \hbar k_o/m$. Thus, as we indicated previously in Section 9-3, the net probability current is simply the difference of two component flows in opposite directions.

In region III, we have a wave function of the form

$$\psi_{III} = De^{ikx}$$

Particle Scattering and Barrier Penetration

and an associated probability current given by

$$J_{\text{III}} = v(DD^*)$$

In a steady-state situation (that is, one in which all observable quantities are constant in time) continuity of probability current requires that we put $J_\text{I} = J_{\text{III}}$, that is

$$v_o(A_oA_o^*) - v_o(AA^*) = v(DD^*)$$

This describes an incident current, represented by $v_o (A_oA_o^*)$, being split into a reflected current and a transmitted current. This statement of conservation of probability is *automatically* satisfied by the values of D and A calculated from the continuity conditions on ψ and $d\psi/dx$ at the sides of the barrier.

Clearly there must also be continuity of the probability current across region II, even though here the wave function is made up of real exponentials. We have

$$\psi_\text{II} = Be^{-\alpha x} + Ce^{\alpha x}$$

Then

$$\psi_\text{II}^* = B^*e^{-\alpha x} + C^*e^{\alpha x}$$

and it is easy to verify, using Eq. 9-8b, that the probability current is given by the following expression:

$$J_\text{II} = -\frac{i\hbar\alpha}{m} (B^*C - BC^*)$$

A nonzero current thus exists if B and C are, in general, complex amplitudes with a phase difference between them. If we put

$$B = B_o e^{i\beta}, \qquad C = C_o e^{i\gamma}$$

then

$$B^*C - BC^* = 2iB_oC_o \sin(\gamma - \beta)$$

Note that the current within the barrier region can be nonzero only if both the positive and negative exponentials are present

9-6 Probability current and barrier penetration

in ψ. In the case of a semi-infinite potential step (Figure 9-8) for which $C = 0$ and only the term $e^{-\alpha x}$ remains, there is a non-zero *probability density* for all $x > 0$, but zero net *probability current* throughout this region. Particles incident from the left on such a step have a finite probability of penetrating beyond it, but are certain of being ultimately reflected.

9-7 AN APPROXIMATION FOR BARRIER PENETRATION CALCULATIONS

We saw in Section 9-5 (Eq. 9-18) that the ratio of transmitted wave function amplitudes for penetration through a *rectangular* barrier of height V_o and width L is given approximately by the equation

$$\frac{\psi(L)}{\psi(0)} \approx \exp\left\{-\frac{[2m\,(V_o - E)]^{1/2}}{\hbar} L\right\} \tag{9-19a}$$

We shall now give a simple analysis that indicates how this result can be extended to obtain an approximate value for the penetrability of *nonrectangular* barriers.

Consider a potential $V(x)$, as shown in Figure 9-11a, that rises and falls smoothly, but in any arbitrary way, as a function of x. We shall consider the penetration of this barrier by particles of some energy E incident from the left. The barrier width is a function of E, being defined by that range of x for which $V(x) > E$.

We shall assume that the only places at which reflections occur are the points x_1 and x_2, where $E - V(x)$ changes sign and the character of the solution to Schrödinger's equation changes abruptly. Furthermore, as we saw in Section 9-5, the effect of the leftward reflection at x_2 is small if the overall attenuation of $\psi(x)$ between x_1 and x_2 is large. In such a case, therefore, in which the reflection at x_1 is close to 100 percent, and the additional contribution from a leftward reflection at x_2 is negligible, we can approximate $\psi(x)$ within the barrier region by a single negative exponential function of the form $Ae^{-\alpha x}$, where both A and α are functions of x. Across the barrier, going from left to right, there is just some smooth continuous decrease in the amplitude of ψ.

To calculate this decrease, we now *imagine* that our arbi-

Particle Scattering and Barrier Penetration

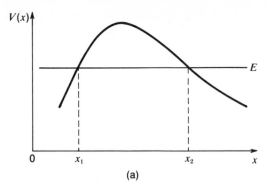

(a)

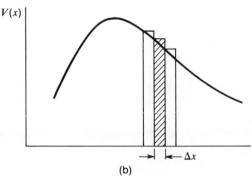

(b)

Fig. 9-11 (a) A
barrier of arbitrary
shape, classically
impenetrable by
particles of the en-
ergy E shown.
(b) Visualization of
the barrier in (a)
as a succession
of thin rectan-
gular barriers.

trary barrier is made up of a whole succession of thin rectangular barriers, as shown in Figure 9-11b, except that [because
the true barrier has no discontinuities in $V(x)$] there are no intermediate reflections at the edges of the slices—only a certain
exponential decrease in ψ. This decrease is easily found;
across any one slice we put

$$\frac{d^2\psi}{dx^2} = \frac{2m}{\hbar^2} [V(x) - E] \psi = [\alpha(x)]^2 \psi$$

where

$$\alpha(x) = \sqrt{\frac{2m}{\hbar^2} [V(x) - E]}$$

The attenuation of ψ across the slice is then described as in Eq.
9-19a, with L replaced by the small interval Δx:

$$\psi(x + \Delta x) = \psi(x) e^{-\alpha(x)\Delta x} \qquad (9\text{-}19b)$$

393   9-7   An approximation for barrier penetration

If we approximate $\psi(x)$ by the first two terms of its Taylor expansion, and likewise approximate the exponential on the right-hand side, we have

$$\psi(x) + \frac{d\psi}{dx} \Delta x \approx \psi(x) \left[1 - \alpha(x)\Delta x \right]$$

Therefore,

$$\frac{d\psi}{dx} \approx -\alpha(x)\psi$$

or

$$\frac{1}{\psi} \frac{d\psi}{dx} \approx -\alpha(x) \tag{9-20}$$

Now integrate across the whole width of the barrier, and we have

$$\ln \left[\frac{\psi(x_2)}{\psi(x_1)} \right] \approx -\int_{x_1}^{x_2} \alpha(x) \, dx = -\int_{x_1}^{x_2} \frac{\sqrt{2m[V(x) - E]}}{\hbar} \, dx$$

The penetrability of the barrier is then given by

$$T = \left[\frac{\psi(x_2)}{\psi(x_1)} \right]^2 \approx \exp \left\{ -2 \int_{x_1}^{x_2} \frac{\sqrt{2m[V(x) - E]}}{\hbar} \, dx \right\} \tag{9-21}$$

A more polished version of this approximation, leading to a slightly more complicated result, is generally known as the WKB method, after three theoreticians (Wentzel, Kramers, and Brillouin) who developed it in 1926.[6] But for all situations in which the penetrability is very small, the result is dominated by the simple exponential factor of Eq. 9-21, and we shall stop at that.

[6]The WKB approximation is described in many texts. See, for example, L. I. Schiff, *Quantum Mechanics*, 3rd ed., McGraw-Hill, New York, 1968, Chap. 8. Sometimes it is called the JWKB method, since the mathematical basis for it was published by Harold Jeffreys several years earlier (1923). But the method should really be credited to Lord Rayleigh, who discussed the analogous problem in optics (propagation through a medium of varying refractive index) as long ago as 1912.

Particle Scattering and Barrier Penetration

A nice application of the foregoing approximation for barrier penetration is the so-called *field emission* of the electrons from a metal.

The conduction electrons in a metal can be regarded as an example of particles confined within a potential box with sides of some finite height V_0, extending away to infinity. But if the metal is made strongly negative with respect to its surroundings, the resulting electric field modifies the potential energy diagram. If the external electric field can be considered as uniform, and of magnitude $\mathscr{E}$, then the potential seen by an electron at positions outside the surface is as shown in Figure 9-12a. Taking x as the direction normal to the surface, and putting $x = 0$ at the surface, we have

$$V(x) = 0 \qquad\qquad (x < 0)$$
$$\quad\;\; = V_0 - e\mathscr{E}x \qquad (x \geq 0)$$

An electron of energy $E\,(< V_0)$ then sees a triangular potential barrier extending from $x = 0$ to $x = L$, where L is defined by the condition

$$E = V(L) = V_0 - e\mathscr{E}L$$

Therefore

$$L = \frac{V_0 - E}{e\mathscr{E}}$$

An electron of energy E then has a certain probability of penetrating this barrier. The penetrability, according to Eq. 9-21, will be given approximately by

$$T = \exp\left\{ -\frac{2\sqrt{2m}}{\hbar} \int_0^L (V_0 - e\mathscr{E}x - E)^{1/2}\,dx \right\}$$

Let us take the case where E corresponds to the electrons of highest energy inside the metal. Then the difference $V_0 - E$ is the work function W, and the penetrability can be written

$$T \approx \exp\left\{ -\frac{2\sqrt{2me\mathscr{E}}}{\hbar} \int_0^L \sqrt{L - x}\,dx \right\}$$

where $L = W/e\mathscr{E}$.

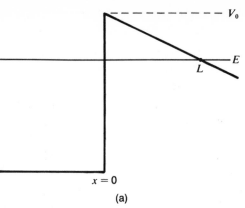

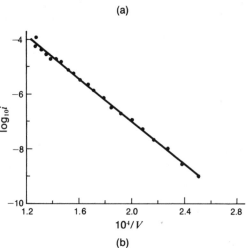

Fig. 9-12 (a) Po-
tential-energy dia-
gram appropriate to
field-emission pro-
cesses. (b) Verifi-
cation of the theo-
retical expression
(Eq. 9-22): the loga-
rithm of the field-
emission current
varies linearly with
the reciprocal of the
applied voltage.
[From data of R. A.
Millikan and C. C.
Lauritsen, Proc.
Natl. Acad. Sci. 14,
45 (1928).]

The integral can be directly evaluated and is equal to $2L^{3/2}/3$; therefore, we have

$$T \approx \exp\left\{-\frac{4}{3}\frac{\sqrt{2me\mathscr{E}}}{\hbar}\cdot\left(\frac{W}{e\mathscr{E}}\right)^{3/2}\right\}$$

that is

$$T \approx \exp\left[-\frac{4}{3}\frac{\sqrt{2m}}{\hbar}\frac{W^{3/2}}{e\mathscr{E}}\right] \qquad (9\text{-}22)$$

The strength of electric field needed to obtain a significant amount of field emission is extremely high—of the order of at least 10^8 V/m. We can understand this if we insert numerical

Particle Scattering and Barrier Penetration

values into Eq. 9-22. Let us assume $W = 4$ eV (a value approximately correct for many metals). Then in Eq. 9-22 we have

$W = 6.4 \times 10^{-19}$ J
$m = 0.91 \times 10^{-30}$ kg
$\hbar = 1.05 \times 10^{-34}$ J-sec
$e = 1.6 \ \times 10^{-19}$ C

Substituting these values, one finds

$$\frac{4}{3} \frac{\sqrt{2m}}{\hbar} \frac{W^{3/2}}{e} \approx 5.5 \times 10^{10} \text{ V/m}$$

This defines a field strength $\mathscr{E}_0$ characteristic of the system, and we can put

$$T \approx e^{-\mathscr{E}_0/\mathscr{E}}$$

We can see that T will be vanishingly small unless the ratio $\mathscr{E}_0/\mathscr{E}$ is less than something like 50 ($e^{-50} \approx 10^{-20}$). This suggests that field emission will be negligible for fields less than about 10^9 V/m. Fields of such magnitude can be achieved by forming the metal into a needle-point with a radius of curvature of a few thousand Angstroms and making it several hundred volts negative with respect to ground. The field at the tip is equal to the voltage divided by the radius, and would be exactly equal to 10^9 V/m for 500 V and a radius of 5000 Å $= 5 \times 10^{-7}$ m.

A simple criterion for the feasibility of field emission is given by the thickness L of the barrier as defined by the value of $W/e\mathscr{E}$. Taking $W = 4$ eV once again, and $\mathscr{E} = 10^9$ V/m, we have $L = 4 \times 10^{-9}$ m $= 40$ Å. Generally speaking, field emission is very improbable unless L is reduced to the order of 10 Å.

It may be seen from Eq. 9-22 that the barrier penetrability as a function of applied electric field should obey the relation

$$\log T = C_1 - \frac{C_2}{\mathscr{E}}$$

where C_1 and C_2 are constants. Figure 9-12b exhibits some

9-8   Field emission of electrons

data that confirm this relationship beautifully over a range of five powers of 10 in the current.

The phenomenon of field emission was of only academic interest for many years, but in 1937 Erwin Mueller used it as the basis of an entirely new form of microscopy in which electrons, traveling in straight lines after being extracted by field emission from a metal tip, form a greatly enlarged ($\times 10^5-10^6$) image of the tip. The image reveals the surface structure of the tip in terms of the relative intensity of emitted electrons from place to place over the surface (Figure 9-13a). Electrons are emitted most readily at corners or edges of the layers of atoms in the tip, where the electric field is greatest and hence the barrier thinnest.

Later, Mueller developed from this the still more sensitive technique of field ion microscopy, in which the electrons were replaced by helium atoms of far shorter de Broglie wavelength. This gave resolution so high that the image revealed the arrays of individual atomic sites on the surface (Figure 9-13b). However, since the helium atoms were first deposited on the surface from outside, the phenomenon is not, like field emission of electrons, a clear case of barrier penetration.

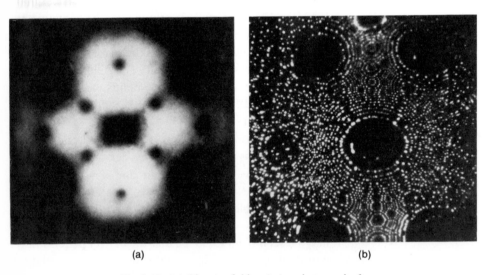

(a)　　　　　　　　　　　　　　　(b)

Fig. 9-13 (a) Electron field-emission photograph of a tungsten tip (radius $\approx$ 2000 Å). (b) Helium ion field-emission photograph of a tungsten tip (radius $\approx$ 600 Å) at 25 kV. (Both photographs courtesy of E. W. Mueller. Reproduced with permission of Springer-Verlag, New York, Inc.)

　　Particle Scattering and Barrier Penetration

The ideas of Section 9-6 can be fairly easily extended to three-dimensional systems with spherical symmetry. We saw in Chapter 5 (Eq. 5-16) that in such cases there is an equivalent one-dimensional Schrödinger equation:

$$-\frac{\hbar^2}{2m}\frac{d^2u}{dr^2} + V(r)u = Eu \qquad (9\text{-}23)$$

where $u(r) = r\psi(r)$.

Now the radial probability current *density* is given by

$$j(r) = -\frac{i\hbar}{2m}\left(\psi^* \cdot \frac{d\psi}{dr} - \psi \cdot \frac{d\psi^*}{dr}\right)$$

[In this three-dimensional case, the total probability *current J*, through the surface of a sphere of radius r, is equal to $4\pi r^2 j(r)$.]

Since $\psi = u/r$, we have

$$\psi^* \cdot \frac{d\psi}{dr} - \psi\frac{d\psi^*}{dr} = \frac{1}{r^2}\left(u^*\frac{du}{dr} - u\frac{du^*}{dr}\right)$$

Hence

$$j(r) = -\frac{i\hbar}{2mr^2}\left(u^*\frac{du}{dr} - u\frac{du^*}{dr}\right)$$

The net current flow $J(r)$, equal to $4\pi r^2 j(r)$, is thus given by

$$J = -\frac{2\pi i\hbar}{m}\left(u^*\frac{du}{dr} - u\frac{du^*}{dr}\right) \qquad (9\text{-}24)$$

Consider a spherical volume bounded by a surface at a given radius r_0. The integral of the probability density $\psi^*\psi$ within the volume is given by

$$P = \int_0^{r_0} (\psi^*\psi)\, 4\pi r^2 dr = 4\pi \int_0^{r_0} (u^*u)\, dr$$

9-9 Spherically symmetric probability currents

Conservation of probability requires that

$$\frac{\partial P}{\partial t} = - J(r_o) + J(0) \tag{9-25}$$

That is,

$$4\pi \frac{\partial}{\partial t} \int_0^{r_o} (u^*u)dr = 4\pi \frac{i\hbar}{2m} \left(u^* \frac{du}{dr} - u \frac{du^*}{dr} \right) \Big|_0^{r_o}$$

$$= - J(r_o) + J(0) \tag{9-26}$$

In this case there is a very important point that does not arise in one-dimensional discussions. The current $J(0)$ at $r = 0$ must be zero unless there is a source or a sink of particles at the origin. In the absence of any such source or sink we have two possibilities:

1. A time-independent situation, in which the probability current $J(r_o)$ vanishes for all r_o.
2. $J(r_o) \neq 0$, in which case then the integrated probability tween $r = 0$ and $r = r_o$ must change with time.

The second of the above possibilities can be used to describe (for example) alpha-particle emission from nuclei. The next section is devoted to a more sophisticated treatment of this problem, but for a first, rough-and-ready analysis we shall use the simplified potential shown in Figure 9-14. At any given instant, the amplitudes on the two sides of the barrier are related by the usual time-independent methods. Thus we put

$$u_1(r) = r\psi_1(r) = A_o e^{ik_o r} + A e^{-ik_o r}$$
$$u_{\text{III}}(r) = r\psi_{\text{III}}(r) = D e^{ikr}$$

Since the transmission coefficient is extremely small for all alpha-emitting nuclei, we have $|A| \approx |A_o|$. The integrated probability P within the sphere of nuclear radius R is then given by

$$P = 4\pi \int_0^R (u^*u) \, dr \approx 4\pi \, [2|A_o|^2 R]$$

(Consider for yourself the justifiability of this approximation.) The total probability current outside the nuclear barrier is

Particle Scattering and Barrier Penetration

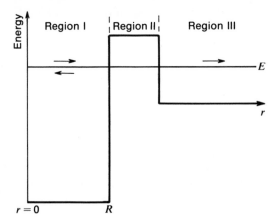

Fig. 9-14 Simple rectangular-barrier model for analysis of nuclear alpha-particle emission.

given by

$$J_{\mathrm{III}} = - 4\pi \, \frac{i\hbar}{2m} \, |D|^2 \, (2ik) = 4\pi v_{\mathrm{III}} \, |D|^2$$

where $v_{\mathrm{III}} \, (= \hbar k/m)$ is the velocity in the exterior region. By the conservation condition, Eq. 9-25, we then have

$$2R \, \frac{d}{dt} \, |A_o|^2 = - v_{\mathrm{III}} \, |D|^2$$

Dividing both sides of this equation by $|A_o|^2$, and making other rearrangements, we have

$$\frac{d|A_o|^2}{|A_o|^2} = - \frac{v_{\mathrm{III}}}{2R} \, \frac{|D|^2}{|A_o|^2} \, dt$$

Since the probability P of finding the alpha particle inside the nucleus is directly proportional to $|A_o|^2$, the last equation tells us how P varies with time:

$$\frac{dP}{P} = - \frac{v_{\mathrm{III}}}{2R} \, \frac{|D|^2}{|A_o|^2} \, dt \equiv - \gamma \, dt$$

that is,

$$P(t) = P(0) \, e^{-\gamma t}$$

401   9-9   Spherically symmetric probability currents

where the decay constant γ is given by

$$\gamma = \frac{v_{\mathrm{III}}}{2R} \frac{|D|^2}{|A_o|^2} \tag{9-27}$$

Thus we arrive once more at the characteristic exponential law of radioactive decay.

The above expression for γ can be cast into an interesting form if we introduce the barrier penetration coefficient. For this rectangular barrier, with different potential levels on the two sides, the transmission coefficient (the ratio of emergent to incident probability currents) is given by

$$T = \frac{v_{\mathrm{III}} |D|^2}{v_{\mathrm{I}} |A_o|^2} \tag{9-28}$$

Substituting this in Eq. 9-27 leads to the simple expression

$$\gamma = \frac{v_{\mathrm{I}}}{2R} T \tag{9-29}$$

If, now, we could picture an alpha particle as a classical point-particle, bouncing back and forth across the nucleus, the value of $v_{\mathrm{I}}/2R$ would be the number of times per second that it strikes the potential wall at $r = R$. And T represents the chance, for each such impact, that it will succeed in tunneling through the barrier to the outside (a purely wave-mechanical phenomenon!). Thus Eq. 9-29 expresses the decay constant γ as a product of Newtonian and wave-mechanical factors. But the picture of the alpha particle as a point object is, of course, scarcely defensible in this context; its de Broglie wavelength inside the nucleus would be comparable to the nuclear radius. Beyond this, however, the theory of nuclear structure suggests that alpha particles do not have a continuing existence as distinct objects inside a nucleus, but instead are formed occasionally from chance associations of two protons and two neutrons. This makes the observed rates of alpha decay much smaller than one would calculate from Eq. 9-29.

The major defect of the foregoing analysis is that the assumed shape of the barrier is completely different from the true (Coulomb, $1/r$) potential presented by a nucleus to a charged particle. In the next section we calculate how the decay rate

Particle Scattering and Barrier Penetration

might be expected to vary with alpha-particle energy when the correct barrier shape is used.

9-10 QUANTITATIVE THEORY OF ALPHA DECAY

The emission of alpha particles in natural radioactivity was the subject of one of the great triumphs of wave mechanics in its early days. Information obtained over several decades of the study of radioactivity had revealed a staggering range of mean lifetimes for alpha-emitting nuclei, from billions of years down to microseconds. It was also recognized that there is a strong correlation between lifetimes and the corresponding alpha-particle energies—the higher the energy, the shorter the lifetime. But whereas the lifetimes vary by a factor as huge as 10^{23}, the energies all lie within a factor of about 2—from 4 to 9 MeV approximately. In 1928 Gamow, Gurney, and Condon showed that this remarkable variation can be understood in terms of the quantum-mechanical theory of barrier penetration.[7]

The nuclear potential as seen by an alpha particle can be represented crudely by Figure 9-15a. An origin is placed at the center of the nucleus. Within a radial distance R from this center, the nucleus is regarded as providing a constant negative potential. The value of R defines the nuclear radius—the limit of the nuclear forces and the radius within which the nuclear particles are confined. Outside the range of the nuclear forces, an alpha particle experiences only the repulsive Coulomb potential, of the form $q_1 q_2 / r$. Putting $q_1 = (Z - 2)e$, $q_2 = 2e$, we thus have

$$V(r) = \frac{2(Z - 2)e^2}{r} \quad \text{(cgs units)} \quad (r > R) \quad \text{(9-30a)}$$

The maximum height of this barrier V_o is attained at $r = R$:

$$V_o = \frac{2(Z - 2)e^2}{R} \quad \text{(9-30b)}$$

As rough values we can put $Z - 2 \approx 90$, $R \approx 10^{-12}$ cm; we

[7]G. Gamow, Z. Phys. **51**, 204 (1928); R. W. Gurney and E. U. Condon, Nature **122**, 439 (1928).

9-10   Quantitative theory of alpha decay

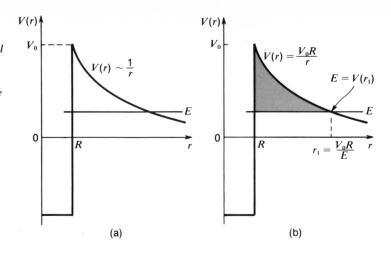

Fig. 9-15 Alpha-particle emission. (a) Potential energy of an alpha particle in the vicinity of a nucleus. The repulsive potential for r > R is the Coulomb potential. (b) The thickness of the classically forbidden region can be expressed in terms R, V_o, and the energy E of the alpha particle: $r_1 - R = \left(\dfrac{V_o - E}{E} \right) \cdot R$

The gray region in the diagram is the barrier through which an escaping alpha particle must tunnel.

then have (with $e = 4.8 \times 10^{-10}$ esu)

$$V_o \approx \frac{180 \times 2.3 \times 10^{-19}}{10^{-12}} \text{ erg} \approx 4 \times 10^{-5} \text{ erg}$$

Since 1 MeV = $1.6 \cdot 10^{-6}$ erg, we have

$$V_o \approx 25 \text{ MeV}$$

There is thus no question that the barrier height is much greater than the energies of the emitted alpha particles, which as we noted above range from 4 to 9 MeV.

The extent of the barrier is from the inner radius, $r = R$, to the radius r_1 at which an alpha particle of any given energy E breaks through into the region of positive kinetic energy (Figure 9-15b). This radius is clearly a function of E; it is defined by the condition

$$\frac{2(Z - 2)e^2}{r_1} = E$$

Since the maximum height of the barrier V_o is given by $V_o = 2(Z - 2)e^2/R$, we can express r_1 simply through the equation

$$r_1 = \frac{V_o}{E} R \tag{9-31}$$

Using Eqs. 9-30, we can also write the potential $V(r)$ at any

Particle Scattering and Barrier Penetration

point across the barrier in the form

$$V(r) = \frac{R}{r} V_o \qquad (9\text{-}32)$$

We now proceed to calculate the penetrability T of the barrier as given approximately by Eq. 9-21:

$$T \approx \exp\left\{-2 \int_R^r \frac{\sqrt{2m\left[V(r) - E\right]}}{\hbar} dr\right\}$$

$$= \exp\left\{-\frac{2\sqrt{2m}}{\hbar} \int_R^{r_1} \left(\frac{R}{r} V_o - E\right)^{1/2} dr\right\}$$

which can conveniently be rewritten as follows:

$$T \approx \exp\left\{-\frac{2\sqrt{2mE}}{\hbar} \int_R^{r_1} \left(\frac{r_1}{r} - 1\right)^{1/2} dr\right\} \qquad (9\text{-}33)$$

The integral in Eq. 9-33 is easily evaluated. Since $r \le r_1$ we can put $r = r_1 \sin^2\theta$, and we have

$$\int_R^{r_1} \left(\frac{r_1}{r} - 1\right)^{1/2} dr = \int_{r=R}^{r_1} (\operatorname{cosec}^2\theta - 1)^{1/2} \, d(r_1 \sin^2\theta)$$

$$= r_1 \int_{r=R}^{r_1} \cot\theta \cdot 2 \sin\theta \cos\theta \, d\theta$$

$$= r_1 \int_{r=R}^{r_1} 2\cos^2\theta \, d\theta$$

$$= r_1 \int_{r=R}^{r_1} (1 + \cos 2\theta) \, d\theta$$

$$= r_1 \left[\theta + \tfrac{1}{2}\sin 2\theta\right]_{r=R}^{r_1}$$

Now at $r = r_1$ we have

$$\theta = \frac{\pi}{2}, \qquad \sin 2\theta = 0$$

and at $r = R$ we have

$$\theta = \arcsin\left(\frac{R}{r_1}\right)^{1/2}, \quad \sin 2\theta = 2\left(\frac{R}{r_1}\right)^{1/2}\left(1 - \frac{R}{r_1}\right)^{1/2}$$

Therefore

$$\int_R^{r_1} \left(\frac{r_1}{r} - 1\right)^{1/2} dr$$

$$= r_1 \left[\frac{\pi}{2} - \arcsin\left(\frac{R}{r_1}\right)^{1/2} - \left(\frac{R}{r_1}\right)^{1/2}\left(1 - \frac{R}{r_1}\right)^{1/2}\right] \qquad (9\text{-}34)$$

9-10 Quantitative theory of alpha decay

For $R \ll r_1$ (which corresponds to $E \ll V_o$), the right-hand side of Eq. 9-34 can be simplified by the approximations

$$\text{arc sin } (R/r_1)^{1/2} \approx (R/r_1)^{1/2}, \text{ and } (1 - R/r_1)^{1/2} \approx 1.$$

We then find

$$\int_R^{r_1} \left(\frac{r_1}{r} - 1\right)^{1/2} dr \approx r_1 \left[\frac{\pi}{2} - 2 \left(\frac{R}{r_1}\right)^{1/2}\right] \tag{9-35}$$

Using Eq. 9-31, this gives

$$\int_R^{r_1} \left(\frac{r_1}{r} - 1\right)^{1/2} dr \approx \frac{\pi}{2} \frac{V_o}{E} R - 2 \left(\frac{V_o}{E}\right)^{1/2} R$$

Substituting this in Eq. 9-33 then gives

$$T \approx \exp\left\{-\frac{2 \sqrt{2mE}}{\hbar} R \left[\frac{\pi}{2} \frac{V_o}{E} - 2 \left(\frac{V_o}{E}\right)^{1/2}\right]\right\}$$

Multiplying this out, we have as the final result

$$T \approx \exp\left\{-\frac{\pi \sqrt{2m}}{\hbar} \cdot \frac{V_o R}{E^{1/2}} + \frac{4 \sqrt{2mV_o}}{\hbar} R\right\} \tag{9-36}$$

From Eq. 9-36, our approximation to the penetrability has the form

$$T(E) \approx A e^{-C/E^{1/2}} \tag{9-37}$$

where

$$C = \frac{\pi \sqrt{2m} V_o R}{\hbar} = \frac{\pi \sqrt{2m}}{\hbar} \cdot 2 (Z - 2)e^2$$

The discussions in Section 9-9 led to the result (Eq. 9-29) that the decay constant γ is given by

$$\gamma = \frac{v_I}{2R} T$$

where v_I is the alpha-particle velocity inside the nucleus and thus increases with E. However, the *variation* of γ with alpha-particle energy is almost completely dominated by the exponential factor in $T(E)$ because the exponent $-C/E^{1/2}$ is numerically very large. For alpha-particle emission from the

Particle Scattering and Barrier Penetration

heavy radioactive nuclei, such as radium and uranium, we have

$$Z - 2 \approx 90$$
$$m \approx 6.6 \times 10^{-24} \text{ g}$$

Substituting these and the values of e and $\hbar$ in the expression for C in Eq. 9-37 gives

$$C \approx 0.45 \text{ erg}^{1/2} \approx 360 \text{ MeV}^{1/2}$$

Our test of the success of the theory is to plot the logarithm of the decay constant against $E^{-1/2}$. If we assume $\gamma \sim T$, then from Eq. 9-37 we have

$$\log_{10} \gamma = \text{const} - \frac{C \log_{10} e}{E_{\text{MeV}}^{1/2}}$$

Substituting $C = 360 \text{ (MeV)}^{1/2}$, $\log_{10} e = 0.4343$, we have

$$\log_{10} \gamma = \text{const} - \frac{156}{E_{\text{MeV}}^{1/2}} \tag{9-38}$$

In Table 9-1 we list some data for alpha-particle emitters. The half-life $T_{1/2}$ is the time in which the number of undecayed atoms falls by a factor of 2. If $T_{1/2}$ is measured, γ can be in-

TABLE 9-1 Energies and Decay Constants for Some Alpha Emitters[a]

Parent Nucleus	E(MeV)	$T_{1/2}$	γ (sec^{-1})	$\log_{10}\gamma$	$1/E$	$1/\sqrt{E}$
Th^{232}	4.05	$1.39 \times 10^{10} y$	1.50×10^{-18}	-17.8	0.247	0.497
Ra^{226}	4.88	$1.62 \times 10^{3} y$	1.36×10^{-11}	-10.9	0.205	0.452
Th^{228}	5.52	$1.9 y$	1.16×10^{-8}	-7.9	0.181	0.425
Em^{222}(Rn)	5.59	$3.83 d$	2.10×10^{-6}	-5.7	0.179	0.423
Po^{218}	6.12	$3.05 m$	3.78×10^{-3}	-2.4	0.163	0.404
Po^{216}	6.89	0.16 sec	4.33	0.6	0.145	0.381
Po^{214}	7.83	1.5×10^{-4} sec	4.23×10^{3}	3.6	0.128	0.358
Po^{212}	8.95	3.0×10^{-7} sec	2.31×10^{6}	6.4	0.112	0.335

[a]Data from I. Kaplan, *Nuclear Physics*, Addison-Wesley, Reading, Mass., 1955.

9-10   Quantitative theory of alpha decay

ferred from the equation

$$e^{-\gamma T_{1/2}} = \frac{1}{2}$$

where

$$\gamma = \frac{\log_e 2}{T_{1/2}} = \frac{0.693}{T_{1/2}}$$

Figure 9-16 shows the graph of $\log_{10} \gamma$ against $E^{-1/2}$ for a range covering a factor of 10^{24} in γ. The empirical slope is within a few percent of that given by Eq. 9-38—a very impressive fact, since there are no adjustable parameters in the expression for the constant C. It is hard to imagine a more striking demonstration of the ability of wave mechanics to describe the subatomic world.

9-11 SCATTERING OF WAVE PACKETS

In Section 9-2 we pointed out that the description of a real scattering event should correspond to the experimental fact of

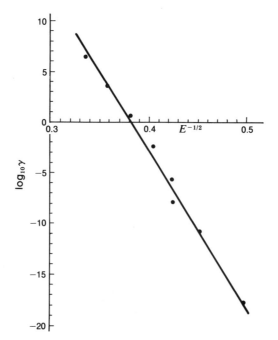

Fig. 9-16 Test of the theoretical relationship (Eq. 9-38) between decay constant and energy for alpha-particle emitters. γ is in sec^{-1} and E in MeV.

Particle Scattering and Barrier Penetration

localized particles approaching a scattering center and subsequently receding from it. In wave mechanics this means the construction of a packet state with some spectrum of energies and momenta, and the analysis of its evolution in time, rather than the use of a steady-state (unique energy) model.

The mathematical treatment of the scattering of wave packets is prohibitively difficult in most cases. However, the use of digital computers has brought such calculations into the realm of possibility. The results, when displayed visually in time sequence, provide a vivid picture of the scattering process. Figures 9-17 and 9-18 illustrate this technique, and contain a wealth of interesting detail.[8] For example, in the first few stills of Figure 9-17 we see an apparently featureless wave packet approaching a scattering center in the form of a square well. There is nothing to indicate that this wave packet is a superposition of plane waves with a rather well-defined average wavelength. But as the wave packet begins to interact with the well, the reflections at the two edges give rise to ephemeral standing-wave patterns that clearly reveal the existence and the magnitude of characteristic wavelengths outside and inside the well (the latter being shorter, as we should expect).

We can see how the amount of reflection or transmission varies with the mean energy of the packet; we can see also (Figure 9-18, columns printed white-on-black) how the amount of scattering (reflection) is greatly reduced by modifying the abruptness of the potential changes at the sides of the well or barrier. Above all, these films display very beautifully the whole course of the interaction, showing exactly how the final state evolves from the initial state during the time of interaction with the potential. This holds equally well for the picture of a scattering process from the viewpoint of "momentum space" (Figure 9-18, column printed black-on-gray). The effect of the potential in modifying the momentum distribution in the course of the interaction is clearly shown.

[8]These are from a set of computer-generated films (QP Series) entitled "Scattering in One Dimension." These films, devised by Judah Schwartz, Harry Schey, and Abraham Goldberg, were made by the Education Development Center, Newton, Mass., in 1968.

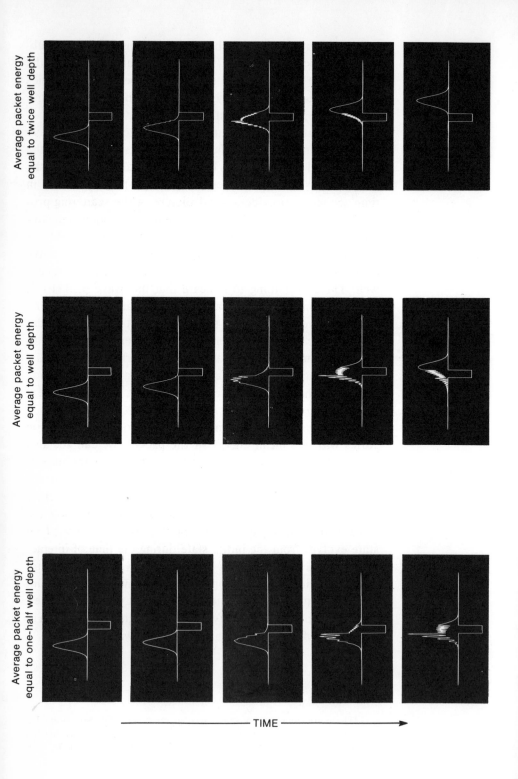

Average packet energy equal to twice well depth

Average packet energy equal to well depth

Average packet energy equal to one-half well depth

TIME

Particle Scattering and Barrier Penetration

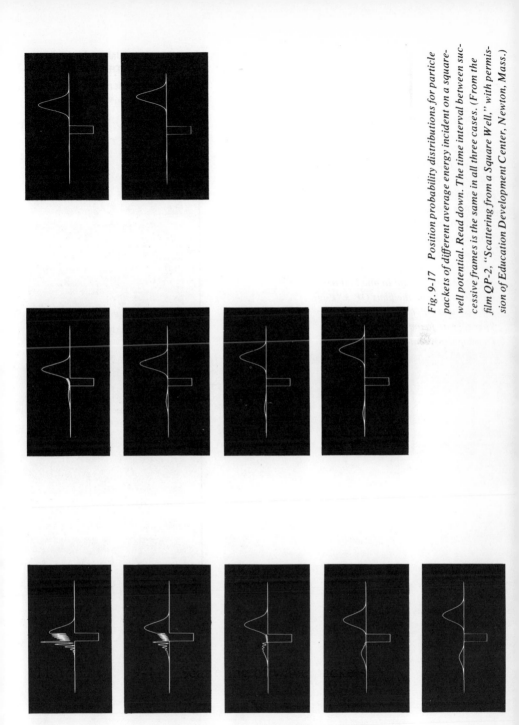

Fig. 9-17 Position probability distributions for particle packets of different average energy incident on a square-well potential. Read down. The time interval between successive frames is the same in all three cases. (From the film QP-2, "Scattering from a Square Well," with permission of Education Development Center, Newton, Mass.)

Fig. 9-18 Position probability distributions (white-on-black) and momentum probability distribution (black-on-gray) for particle packets incident on potential wells. Read down. The position probability distributions show reduced reflection when the edges of the potential are rounded. The momentum probability distribution shows changes in momentum during particle interaction with the well. In all cases, the average packet energy is equal to one-half the well depth. (From the films QP-3 "Edge Effects" and QP-4 "Momentum Space," with permission of Education Development Center, Newton, Mass.)

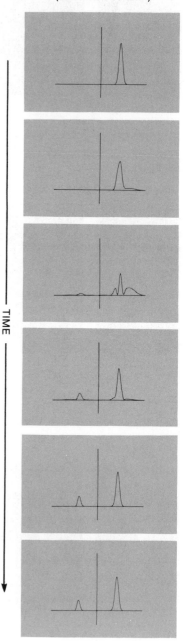

TIME

POSITION PROBABILITY DISTRIBUTIONS

| Zero surface thickness (square well) | Surface thickness approximately $1/8$ well width | Surface thickness approximately $1/4$ well width |

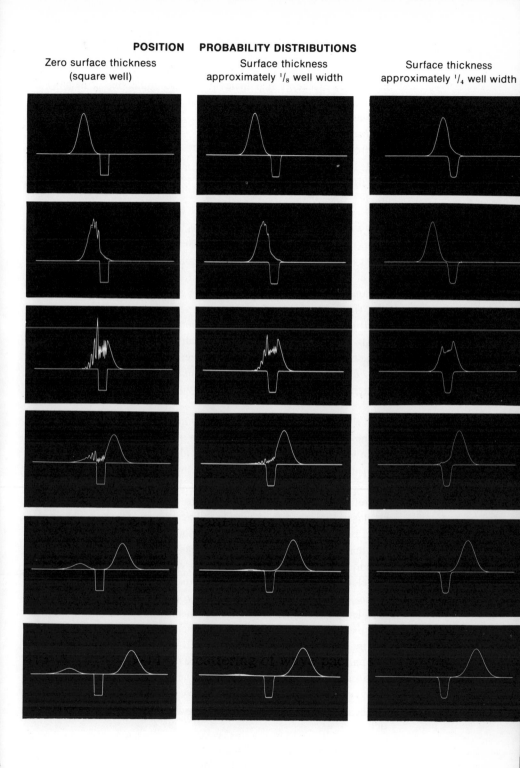

413   9-11   Scattering of wave packets

9-1 *Energy dependence of transmission at a potential step.* A beam of particles of energy E is incident on a potential step of height V_0. What are the reflection and transmission coefficients for the following incident energies?

(a) $E = 4V_0/3$

(b) $E = (1.001)V_0$

(c) $E = 2V_0$

(d) $E = 10V_0$

(e) $E = 10^3 V_0$

9-2 *Particle reflection at a down-step.*

(a) Derive expressions for the reflection and transmission coefficients for a particle beam of energy E incident on a potential down-step of height V_0 ($E > V_0$). Show that these coefficients are the same as those for the corresponding case of a particle incident on an up-step, treated in Section 9-2.

(b) In classical particle physics there would be no reflection at a down-step. (Marbles roll downstairs without hindrance.) In classical particle physics would there be reflection at an up-step of height $V_0 < E$? What are the classical *wave* analogs for particle reflection at a potential down-step and a potential up-step?

9-3 *Reflection at a "rounded" potential step.* Equations 9-4 and 9-5 for the reflection and transmission at a potential up-step can be written in terms of the momenta of the incident and transmitted particle. The same equations apply to a particle incident on a down-step (Exercise 9-2). *Planck's constant does not appear in these expressions.* Therefore they should apply equally to the large particles described by classical mechanics. A marble rolling off a stair-step should sometimes be reflected, but it is not. What is wrong with the quantum analysis that leads to this error in the classical limit? The answer has to do with the sharpness of the potential step compared with the de Broglie wavelength of the incident particle. If the distance over which the potential changes significantly is small compared with the de Broglie wavelength, there will be reflection; otherwise not. The ideal potential step used here has zero width, so is "sharp" compared with the wavelength of *all* particles. But for a classical particle, this relative sharpness is not physically realizable, as you will show below, so the quantum analysis does not apply. The remainder of the problem analyzes a "double step" that does have a characteristic width which can be compared with the de Broglie wavelength of the incident particle.

(a) Estimate the de Broglie wavelength of a marble rolling toward the edge of a stair-step. Is this wavelength long or short with

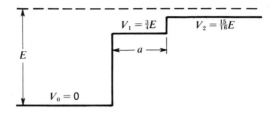

respect to the sharpness of the step? What do you conclude about the probability of reflection of this classical particle?

(b) As the analytically simplest way to make less sharp the ideal step used in the quantum analysis, consider the double up-step shown in the figure. The first step is of width a, yielding a characteristic dimension for the step as a whole. To begin with, ignore the relative heights of each step and write down the form of the wave function in each of the regions. Use k_1, k_2, and k_3 for wavenumbers in the three regions, respectively.

(c) Set up the boundary conditions at $x = 0$ and $x = a$. The result is four equations in four unknowns.

(d) The four equations of part (c) are difficult to solve. Most of the physical insight can be obtained with an even more special case. Set $k_1 = 2k_2 = 4k_3$. Because of the square-root relation between momentum and kinetic energy, this yields the heights of the two steps shown in the figure. Verify this and simplify the four boundary conditions using this added requirement.

(e) You may now solve the system of four boundary-condition equations to derive the reflection coefficient. Determinants are useful. Or you may accept the following simple result:

$$R = \left| \frac{3(1 + e^{ik_1a})}{9 + e^{ik_1a}} \right|^2$$

Verify that this (or your) result yields the single-step result derived in the text (Eq. 9-4) when the width $a = 0$ and for the values of k in this special case.

(f) Can the expression for R be written in terms of particle momenta without explicit or implicit use of Planck's constant? Can it therefore correspond to a classical particle analysis?

(g) Recall that $k_1 = 2\pi/\lambda$, where λ is the de Broglie wavelength of the incident particle. What value does R take on when the step width a has each of the values $\lambda/4$, $\lambda/2$, $3\lambda/4$, and λ? What is the *maximum* value of R for these k values? Are any of these reflection coefficients *greater* than for the one-step potential? (Typically, rounding the corners of a scattering potential reduces the amount of reflection.)

9-4 *Density and current in streams of classical particles.* The distinction between the number density of moving particles and the rate at which such particles pass a given point is as important classically as it is in quantum mechanics. The classical analysis illustrated by the following problem may help you feel more comfortable with the fact that a uniform current of particles can be associated with a nonuniform number density of particles. The corresponding quantum statement is that a uniform particle probability *flux* at different points in an apparatus can be associated with a nonuniform particle probability *density* at those points.

(a) Imagine a long line of bicyclists pedaling in single file up a hill at a constant speed of 400 ft/min (about 4.5 mph). Each bicyclist remains 20 ft behind the one ahead of him. For an observer standing on the uphill slope, how many cyclists are there per 1000 ft along the road? Also:

> How long after a cyclist passes him does the next cyclist pass?
>
> How many cyclists per minute pass him?

(b) As each cyclist reaches the crest of the hill and starts down the other side, he stops pedaling and coasts, quickly reaching a constant coasting speed of 2000 ft/min. For an observer standing on the downhill slope, how many cyclists are there per 1000 ft along the road? Also:

> How long after a cyclist passes him does the next cyclist pass?
>
> How many cyclists per minute pass him?

(c) Why are your answers to the two indented questions respectively the same in parts (a) and (b)?

9-5 *Properties of the probability current.*

(a) We have seen that a one-dimensional stationary-state wave function can be written as a real function $\psi(x)$ multiplied by an exponential time factor $e^{-iEt/\hbar}$. Show that for such functions the probability current $J(x, t)$ vanishes for all x.

(b) Show that the probability current J is always real. [*Hint*: A quantity is real if it equals its complex conjugate.]

9-6 *Probability current in the "sloshing" state of a particle in the infinite square well.*

(a) Calculate the probability current $J(x, t)$ at $x = L/2$ for the square-well sloshing state given in Eq. 8-5.

(b) This state of a particle in the square well was considered in Exercise 8-1. Verify directly that the rate at which probability disappears from the left half of the well (Exercise 8-1d) is equal to $J(x = L/2, t)$.

Particle Scattering and Barrier Penetration

9-7 *Normalization of an incident beam using flux.* The wave function in different regions for a particle of energy E incident on a potential step $V_o < E$ is given in Eq. 9-1, with relations among coefficients given in Eq. 9-2. Normalize the wave function so that it corresponds to unit flux (one particle per second) in the incident beam. In what way(s) would your result be different if the potential step were replaced by a square barrier? by a square well? by any other scattering center for which the potential has the value zero in the region of the incident beam?

9-8 *Conditions for a "zero" in probability current.* A wave function $\Psi(x, t)$ has an associated probability current $J(x, t)$. Each of the following conditions is proposed as sufficient to require $J(x, t)$ to be zero for a particular position $x = x_0$ and time $t = t_0$. Decide whether each is sufficient or not. If insufficient, give a counterexample.
   (a) $\Psi(x_0, t_0) = 0$
   (b) $|\Psi(x_0, t_0)|^2 = |\Psi(x_0, t)|^2$, all t
   (c) $|\Psi(x_0, t_0)|^2 = |\Psi(x, t_0)|^2$, all x
   (d) $\Psi(x_0 + b, t_0) = \Psi(x_0 - b, t_0)$, all $b > 0$
   (e) $\Psi(x = x_0, t = t_0)$ is real

9-9 *Probability current for a wave packet.* At time $t = 0$ a wave packet describing a free particle of mass m has the following (normalized) wave function:

$$\psi(x, 0) = e^{ik_0 x} f(x)$$

where $f(x)$ is a real function of x, centered on $x = x_0$, that vanishes outside the range $x = x_0 \pm a$.
   (a) Obtain an equation for the probability current $J(x)$ at $t = 0$.
   (b) Show that $\displaystyle\int_{-\infty}^{\infty} J(x)\, dx = \hbar k_0 / m$. Interpret this result.
   (c) Qualitatively, how would you expect $J(x)$ to vary with time at positions such that (i) $x \ll x_0 - a$, (ii) $x \gg x_0 + a$?

9-10 *A footnote concerning probability currents.* Exercise 8-19a described a way of developing the connection between the expectation values of momentum and position for a moving wave packet.
   (a) Re-examine that analysis, and show that it also embodies the following result involving the one-dimensional probability current:

$$\frac{d}{dt}\langle x \rangle = \int_{\text{all } x} J(x)\, dx$$

   (b) Discuss the physical interpretation of this equation.

9-11 *Transmission across a rectangular well.* Equation 9-12 relates the incident and transmitted *amplitudes* for particles scattered by a one-dimensional rectangular well.

(a) Deduce that the general expression for the transmission coefficient T can be written

$$T = \frac{4k_1^2 k_2^2}{4k_1^2 k_2^2 + (k_2^2 - k_1^2)^2 \sin^2 k_2 L}$$

(b) Substituting $k_1^2 = 2mE/\hbar^2$, $k_2^2 = 2m(E + V_o)/\hbar^2$, and putting $E/V_o = \epsilon$, show that the above expression leads to the result

$$T(\epsilon) = \frac{4\epsilon(1 + \epsilon)}{4\epsilon(1 + \epsilon) + \sin^2 (\beta\sqrt{1 + \epsilon})}$$

where $\beta = L\sqrt{2mV_o}/\hbar$ is a dimensionless parameter characterizing the scattering well. In this form the equation for T is suited for direct computations of the transmission coefficient as a function of particle energy.

(c) Calculate and graph the variation of T with the ratio E/V_o for $\beta = 10.5\pi$. Compare with Figure 9-5b, which gives the results of a similar calculation for $\beta = 20.5\pi$.

(d) Explore the behavior of T for low energies ($E/V_o \to 0$) for the case $\beta = 10\pi$ exactly. The difference between this and (c) may surprise you. Can you give a simple physical interpretation of it?

9-12 *Flying over a rectangular hurdle.* Consider a particle beam of energy E incident on a rectangular barrier of height V_o for the case $E > V_o$.

(a) Show that the analysis of the scattering by a one-dimensional *well* in Section 9-4 leads to equations identical in form to those of the present case, so that the general result (Eq. 9-12) applies to the barrier also.

(b) For the special cases $E \gg V_o$ and $k_2 L = n\pi$ treated in Section 9-4, show that the results are the same as for the present case. The case $k_1 \ll k_2$ for the well cannot apply to the barrier. Instead consider the case $k_2 \to 0$ and show that the resulting transmission coefficient is the same as that calculated in the next exercise ("Skimming a barrier").

9-13 *Skimming a barrier.* Find the fraction of incident particles transmitted by a rectangular potential barrier in the very special case that the energy E of the incident particles is exactly equal to the barrier height V_o. Let k stand for the wave number of the incident particles and L stand for the barrier width.

(a) Starting from the Schrödinger equation, show directly that the form of the wave function inside the barrier is linear:

$$\psi_{II}(x) = Bx + C$$

(In general, B and C may be complex numbers, so it is not strictly correct to say that ψ_{II} represents a "straight line" in the usual sense.)

(b) Set up and solve the boundary condition equations to obtain an expression for the transmission coefficient. Does this expression have the expected limiting values for $L = 0$ and $L \rightarrow \infty$?

(c) For what value of L/λ (where λ is the de Broglie wavelength) is the transmission fraction equal to $\frac{1}{2}$?

9-14 *Burrowing through walls.* A beam of electrons of energy 2 eV is incident on a rectangular potential barrier of height 4 eV and thickness 10 Å. What is the transmission coefficient?

9-15 *Barrier penetration Olympics.* Proton and deuteron beams, each with kinetic energy of 4 MeV, are incident on a rectangular barrier of height 10 MeV and thickness 10^{-12} cm.

(a) From general physical principles predict which type of particle has the greater probability of penetrating the barrier.

(b) Evaluate the transmission coefficient of each beam.

9-16 *Through versus over.*

(a) Show that the transmission *through* a rectangular barrier $(E < V_o)$ can be formally analyzed in the same way as transmission *across* a rectangular barrier $(E > V_o)$. In particular, show that Eq. 9-15 can be obtained from Eq. 9-12 by the simple substitution $i\alpha = k_2$ (and $k = k_1$).

(b) In the case $E > V_o$ (hurdle) or $E > 0$ (ditch), the transmission coefficient is a periodic function of the length of the scattering region. Is the same feature present for transmission through a potential barrier? Why or why not?

9-17 *Tunneling through a threshold.* A one-dimensional potential barrier has the shape shown in the figure. Find the transmission coefficient for particles of mass m coming from the left with energy E such that $V_1 < E < V_o$.

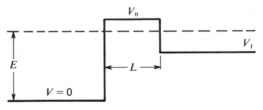

9-18 *Probability current within a barrier.* In Section 9-6 it is stated that the probability current within a rectangular barrier is given by

$$J = \frac{i\hbar\alpha}{m}(B^*C - BC^*)$$

where the steady-state wave function in this region is given by

$$\psi(x) = Be^{-\alpha x} + Ce^{\alpha x} \qquad [\alpha = \sqrt{2m(V_0 - E)}/\hbar]$$

(a) Verify that this equation for J follows from the basic equation for probability current (Eq. 9-8b).

(b) Using the continuity conditions on ψ and $d\psi/dx$ at $x = L$ for this problem (Eqs. 9-14), show that the current within the barrier is equal to $(\hbar k/m)|D|^2$, which is equal to the current of particles that have managed to penetrate completely through the barrier.

9-19 *Penetration through a thick barrier.* A certain potential barrier rises parabolically $[V(x) \sim x^2]$ from $x = 0$ to a height of V_0 at $x = L$ (see the figure).

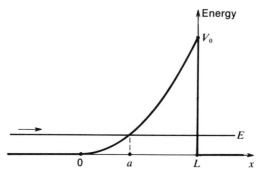

(a) Using Eq. 9-21 obtain an equation for the penetrability of this barrier for particles of mass m and energy E. You may find it convenient to put $V(x) = \frac{1}{2}Cx^2$, and to note that the barrier, for particles of energy E, extends from $x = a$ to $x = L$, where $E = \frac{1}{2}Ca^2$, $V_0 = \frac{1}{2}CL^2$.

(b) Show that, for $E \ll V_0$, the result of (a) can be approximated by the relatively simple expression

$$T(E) \approx \exp\left\{\frac{-L\sqrt{2mV_0}}{\hbar}\left[1 - \frac{E}{V_0}\log_e\left(2\sqrt{\frac{V_0}{E}}\right)\right]\right\}$$

(c) Show that, for $E \to 0$, the penetrability of this barrier is equal to the *square root* of the penetrability of a *rectangular* barrier of height V_0 and width L—which means that the parabolic barrier is vastly more penetrable than the rectangular one. (For instance, if the rectangular barrier had a penetrability of only 10^{-6}, the parabolic barrier would have a penetrability of 10^{-3}, a thousand times larger.)

Particle Scattering and Barrier Penetration

9-20 *Field emission of electrons.* A typical arrangement for obtaining field-emission data such as shown in Figure 9-12b is to apply a high voltage V between a hollow metal cylinder and a thin wire mounted along its axis.

(a) If the radii of cylinder and wire are R and r, respectively, verify that the electric field at the surface of the wire is given by $E = V/[r \log_e(R/r)]$.

(b) Suppose $r = 10^{-3}$ cm and $R = 1$ cm. Calculate the theoretical slope of a graph of $\log_{10} i$ versus $1/V$, where i is the field-emission current, assuming a work function W equal to 4 eV. (Refer to Eq. 9-22.) Compare your result with the slope of the graph of Figure 9-12b.

9-21 *Spherically symmetric probability currents.* The space wave function describing a certain steady state is given by

$$\psi(r) = A \frac{e^{-ikr} + b e^{ikr}}{r} \qquad (r > R)$$

This can be interpreted as the wave function corresponding to a scattering process due to a center of force located in a spherical volume of radius less than R.

(a) Find an expression for the net radial probability current.

(b) Interpret the result in terms of separate incoming and outgoing currents.

(c) What physical circumstances would correspond to the conditions $|b|^2 = 1$? $|b|^2 < 1$? $|b|^2 > 1$?

9-22 *Nuclear Coulomb barrier penetration.* Equation 9-36 gives the approximate penetrability of a nuclear Coulomb barrier in terms of the nuclear radius R, the height V_o of the barrier at $r = R$, and the energy E and mass m of the particles penetrating the barrier. This formula was calculated for particles escaping from inside the nucleus to the outside world, but it applies equally well to particles approaching a nucleus from outside.

(a) Suppose that a nucleus of atomic number Z and radius R is bombarded with protons, of mass m_o and charge e, and alpha particles of mass $4m_o$ and charge $2e$. Sketch graphs of log T versus $E^{-1/2}$ for both types of bombarding particles. (Remember that the barrier height V_o is different for the two.) What are the relative slopes of these lines?

(b) On the basis of Eq. 9-36, calculate the probability that incident protons of kinetic energy 2MeV will get through to the nucleus of an atom of aluminum ($Z = 13$). Assume a nuclear radius of about 4.5×10^{-13} cm.

9-23 *The steady-state description of scattering processes.* Consider a wave function given by

$$\psi(r, z) \sim e^{ikz} + A\frac{e^{ikr}}{r}$$

This can be interpreted as the wave function corresponding to a scattering process in which particles, initially traveling along the z direction, are scattered by a spherically symmetric center of force located at the origin ($r = 0$). It is a mathematical description of the situation illustrated in Figure 9-2c but extended from two to three dimensions.

(a) Calculate the probability current in the z direction at a large distance from the scattering center.

(b) Calculate the net *radial* probability current at large r.

(c) The coefficient A has the dimension of length. Can you give a physical interpretation to the value of $|A|^2$?

9-24 *Effect of scattering on the momentum distribution of a wave packet.* Sketched in the figure are two momentum probability distributions for a packet state, one of them *before* the packet enters a scattering region (that is, a region in which the potential V changes with x) and the other one *after* it has left the scattering region. Before the scattering begins and after it ends, the momentum probability distribution does not change with time.

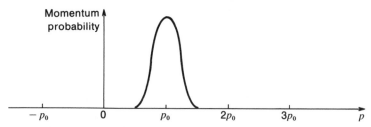

(a) Momentum probability distribution *before* scattering

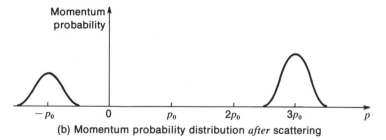

(b) Momentum probability distribution *after* scattering

(a) Sketch a simple potential that could lead to the change shown in the momentum distribution *and* locate the average energy of the packet on this plot.

Particle Scattering and Barrier Penetration

(b) Sketch the *spatial* probability distribution for this packet state at three different times: before the collision, during the collision, and after the collision.

9-25   *Scattering of a wave packet.* A wave packet of mean energy E_o is incident on a potential square well of depth $V = -2E_0$. Figures (a)-(c) are stills from the film WELLS (QP-2) showing three successive stages in the interaction. In a few words, answer each of the following questions about these stills.

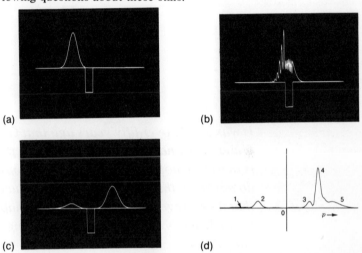

(a)   (b)

(c)   (d)

(a) True or false? The incident packet in Figure (a) describes many particles, each in a different momentum state. If false, tell what the incident pulse *does* describe.

(b) In Figure (b), what is the reason for the fine structure of maxima and minima, both inside and outside the well?

(c) Why are adjacent maxima more closely spaced inside than outside the well in Figure (b)?

(d) True or false? During the interaction (Figure (b)), each probability maximum will oscillate up and down with an average angular frequency $\omega_o = E_o/\hbar$, where E_o is the mean kinetic energy of the incident packet. If false, say what the time dependence of these fine-structure maxima will be.

(e) True or false? The reflected and transmitted probability maxima in Figure (c) describe a particle in a single state consisting of a superposition of positive and negative momentum states. If false, tell what the two maxima *do* represent.

(f) Figure (d) shows the *momentum* probability distribution during the interaction [at the same time as Figure (b)]. Account for the peaks numbered 1, 2 and 4, 5. [For a discussion of this momentum distribution, including peak number 3, see R. H. Good, "Momentum Space Film Loops," Am. J. Phys. **40**, 343 (1972).]

It was a little over fifty years ago that George Uhlenbeck and I introduced the concept of spin . . . It is therefore not surprising that most young physicists do not know that spin had to be introduced. They think that it was revealed in Genesis or perhaps postulated by Sir Isaac Newton, which most young physicists consider to be about simultaneous.

SAMUEL A. GOUDSMIT, *address to American Physical Society* (February 1976)

10

Angular momentum[1]

[1]The authors gratefully acknowledge the work of Charles P. Friedman in developing parts of this chapter.

10-1 INTRODUCTION

We know that in the classical analysis of the motion of objects under the action of a central force, the angular momentum of the motion is a very important feature. Thus, for example, the size and shape of the orbit of a planet about the sun is completely specified by a knowledge of the total energy and the total orbital angular momentum. Angular momentum plays a similarly important role in atomic systems, such as an electron subject to the central force provided by the Coulomb field of a nucleus. In this case, as in the classical planetary system, the angular momentum is an important constant of the motion.

As you probably already know, the comparison between an orbiting planet and an orbiting electron can be drawn even closer by virtue of the fact that an electron has an *intrinsic* angular momentum (usually called *spin*) just as a planet has rotational angular momentum in addition to its orbital angular momentum. It will be our purpose in this chapter to introduce the properties of angular momentum, both orbital and spin, in the quantum-mechanical scheme of things.

Angular Momentum of Atoms.

In atomic systems, the most notable characteristic of angular momentum is the fact that it is *quantized*. We encountered this feature in a naive way in our original discussion of the Bohr atom in Chapter 1. Now we shall verify and examine this quantization from the more realistic standpoint of wave mechanics.

The angular momentum of atomic systems is inseparable from the magnetic properties of atoms. This is easy to understand in general terms because electrons and protons are electrically charged, and we have learned from classical physics that rotating or circulating charges act like magnets in giving rise to magnetic fields. The existence and the quantized magnitudes of angular momentum are, in fact, often revealed by energy shifts due to magnetic interactions. The so-called *fine structure* in atomic spectra results in part from such energy shifts, the most famous example being the pair of closely spaced components (the *D* lines) in the spectrum of the orange light from sodium vapor[2] (Figure 10-1). The energy difference (splitting) between these lines is actually quite small, corresponding to only about 1/1000 of the photon energy for either line. Thus, in quantitative terms, it is relatively unimportant for a description of the energy level structure of the atom. However, the various permitted values of the quantized angular momentum play a central role in determining the number and variety of possible quantum states in atomic systems. It is this aspect that will be our main concern, although we shall at various points give consideration to the modifications and splittings of energy levels due to atomic magnetism.

Gyromagnetic Effect

Probably the most direct link between magnetic properties and angular momentum in atoms is provided by an effect whose possibility was first suggested by O. W. Richardson in 1908.[3] The effect is as follows: When an initially unmagnetized iron rod is suddenly magnetized along its length, it tends to twist about this axis. The picture of what is happening is that in the initial unmagnetized state the elementary magnetic dipoles (actually individual spinning electrons) in the iron are all ran-

---

[2]Sodium vapor street lamps are in common use and may be recognized by their yellow color. When one of these is viewed through an inexpensive spectrometer, such as a hand-held piece of plastic diffraction grating, the yellow *D* line is the most prominent color in the spectrum. It takes a somewhat more expensive instrument to resolve the *D* line into its two separate components.

[3]O. W. Richardson, Phys. Rev. **26**, 248 (1908).

Angular momentum

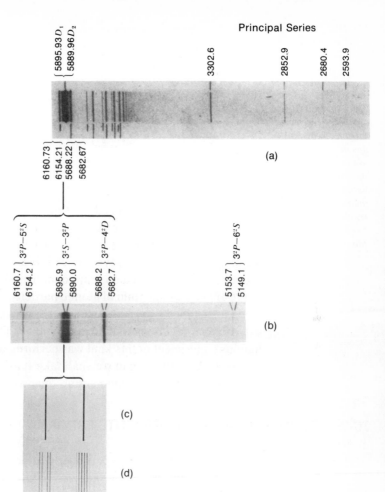

Fig. 10-1 The closely spaced "D lines" in the spectrum of sodium. (a) through (c): The D lines seen under different dispersions. (d) The further splitting of the D lines under the influence of a magnetic field. (Spectra taken from G. Herzberg, Atomic Spectra and Atomic Structure, Dover Publications, Inc., New York 1944. Reprinted through the permission of the publisher.)

domly oriented (see Figure 10-2a). When a magnetic field is applied along the axis of the rod, the dipoles align themselves in this direction, bringing with them their intrinsic angular momentum (Figure 10-2b). Since the total angular momentum of the rod must remain zero (there being no source of external torques), the body of the rod must develop a rotational angular momentum equal and opposite to that of the aligned electron spins. Thus if the rod is suspended vertically from a delicate torsion fiber, a small angular impulse can be observed upon magnetization. The effect is generally known as the *Einstein-de Haas effect*, since the existence of the phenomenon was first demonstrated by them (in 1915) using a resonance

10-1   Introduction

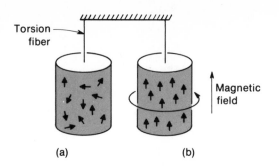

Fig. 10-2 Schematic diagram of the Einstein-deHaas effect. (a) Initially unmagnetized iron rod has elementary magnetic dipoles oriented at random. (b) Sudden magnetization of the rod lines up the dipoles; conservation of angular momentum then requires that the rod as a whole rotate.

Torsion fiber

Magnetic field

(a)   (b)

method. The effect is exceedingly tiny (see the exercises) but has been measured with considerable accuracy.[4]

The Einstein-de Haas effect clearly links magnetism with angular momentum. However, since it is a macroscopic effect, it does not by itself demonstrate quantization—the limitation of the angular momentum of atomic particles to discrete values along a specified axis. For such a verification we must turn to experiments involving individual particles. The first and most famous experiment of this kind was performed by O. Stern and W. Gerlach in 1922,[5] and we shall take it as the starting point of our study of angular momentum quantization.

10-2   STERN-GERLACH EXPERIMENT: THEORY

The Stern-Gerlach experiment was deliberately designed to test whether or not the angular momentum of individual *neutral* atoms is quantized in a magnetic field. To do this, a beam of silver atoms, traveling in a well-defined direction, was passed through a nonuniform magnetic field perpendicular to the beam direction. The role of the magnetic field was to supply magnetic forces to deflect the beam. Deflection of the atoms by the magnetic field was studied using a detector located far from the region of deflection.

Force on a Magnetic Dipole

Since the heart of the experiment is the effect of the magnetic field, we shall consider that first and describe the other

[4]A. P. Chattock and L. F. Bates, Phil. Trans. Roy. Soc. A **223**, 257 (1923). Some earlier measurements of fair accuracy were made by J. Q. Stewart, Phys. Rev. **11**, 100 (1918).

[5]W. Stern and O. Gerlach, Z. Phys. **8**, 110 (1921).

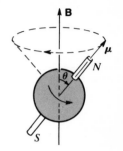

Fig. 10-3 *Classical model of the precession of an atomic dipole moment about an applied magnetic field.*

details later. Suppose that an atom can be pictured (rather like the earth) as a rotating sphere with a magnetic dipole moment $\boldsymbol{\mu}$. On a naive view this magnetic moment is a vector parallel to the angular momentum vector (pointing in the same direction if a positive charge is rotating and in the opposite direction if the rotating charge is negative). Now if such a dipole finds itself in a magnetic field **B**, it does not (as a compass needle would) simply oscillate about the field direction and (because of friction) finally line up with it. Instead, by virtue of its rotational angular momentum the dipole will *precess* about the field direction in such a way that the angle between $\boldsymbol{\mu}$ and **B** remains constant, as shown in Figure 10-3. For purposes of analysis one can think of the magnetic dipole being composed of equal and opposite magnetic "poles" $\pm q_m$ separated by a distance l, so that $\mu = q_m l$. Then an applied magnetic field creates forces acting on the north and south poles of the magnetic dipole and hence a torque **M** ($= \boldsymbol{\mu} \times \mathbf{B}$) pointing always perpendicular to the angular momentum vector. This is the condition for steady precession.[6] Thus for any initial angle θ between $\boldsymbol{\mu}$ and the field direction, which we shall take to lie along the z axis, the component $\mu_z (= \mu \cos \theta)$ of the magnetic moment along z is constant.

If the field **B** is uniform, the magnetic dipole will precess but will experience *no net force*. But consider what happens if the magnetic field strength varies along the z direction, the direction transverse to the beam. This variation is characterized by a nonzero field gradient $\partial B_z / \partial z$. The existence of this gradient leads to a net force in the z direction. The simplest way to calculate this force is again to use the magnetic pole model in which pole strength q_m is defined so that the force due to an

[6]See, for example, the volume *Newtonian Mechanics* in this series, p. 686.

10-2   Stern-gerlach experiment: theory

external field $\mathbf{B}$ is $q_m\mathbf{B}$. Then for a uniform magnetic field, the downward force $-q_m B_z$ (in the z direction) on one end of the dipole is balanced by the upward force $+q_m B_z$ on the other end. In an inhomogeneous field, however, the field at one end of the dipole is stronger than the field at the other end. A dipole of length l that makes an angle θ with the z direction has its poles separated by a distance $l\cos\theta$ along z. Thus if the field gradient along z is $\partial B_z/\partial z$, the net force on the dipole is

$$F_z = -q_m B_z + q_m \left(B_z + l\cos\theta\frac{\partial B_z}{\partial z} \right)$$

$$= q_m l\cos\theta\frac{\partial B_z}{\partial z}$$

But $q_m l\cos\theta$ is just the component μ_z of the magnetic dipole vector $\boldsymbol{\mu}$ along the z direction. Thus[7]

$$F_z = \mu_z\frac{\partial B_z}{\partial z} \tag{10-1}$$

Magnetic Deflection of an Atomic Beam

Suppose now that an atom with a magnetic moment $\boldsymbol{\mu}$, traveling with speed v in the x direction, moves for a distance d through a transverse inhomogeneous magnetic field of the type described above (Figure 10-4). Its time of transit t is d/v, during which it experiences a constant acceleration F_z/M, where M is the mass of the atom. The transverse displacement of the

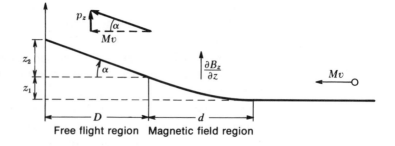

Fig. 10-4 Trajectory of an atomic dipole through the Stern-Gerlach apparatus.

Free flight region   Magnetic field region

[7]The same result for the somewhat more realistic "current loop model" of the magnetic dipole is treated in the exercises.

430   Angular momentum

atom during this time will be

$$z_1 = \frac{1}{2} at^2 = \frac{1}{2}\left(\frac{F_z}{M}\right)\left(\frac{d}{v}\right)^2 = \frac{d^2}{2Mv^2}\mu_z\frac{\partial B_z}{\partial z} \qquad (10\text{-}2)$$

In the process the atom will acquire a transverse linear momentum p_z given by

$$p_z = F_z t = \frac{d}{v}\mu_z\frac{\partial B_z}{\partial z}$$

The direction of the motion is turned through an angle α given by

$$\tan\alpha = \frac{p_z}{Mv} = \frac{d}{Mv^2}\mu_z\frac{\partial B_z}{\partial z} \qquad (10\text{-}3)$$

After leaving the field, the atom travels in a straight line. While moving an additional distance D in the x direction, it undergoes an additional transverse displacement z_2:

$$z_2 = D\tan\alpha = \frac{dD}{Mv^2}\mu_z\frac{\partial B_z}{\partial z} \qquad (10\text{-}4)$$

Thus the total transverse deflection z is given (Eqs. 10-2 and 10-4) by

$$z = z_1 + z_2 = \frac{d}{Mv^2}\left(D + \frac{d}{2}\right)\mu_z\frac{\partial B_z}{\partial z} \qquad (10\text{-}5)$$

The above calculation closely parallels that of the deflection of an electron beam by a transverse *electric* field in a cathode-ray tube.[8] In Section 10-4 we shall use Eq. 10-5 to calculate the approximate value of an atomic magnetic moment.

The central feature of the result expressed by Eq. 10-5 is that if the magnetic moment μ of the atom, and the associated angular momentum, can point in any direction then their z components take on a continuous range of values. Then the final deflections z for a large sample of atoms should also be continuously distributed. On the other hand, if the values of μ_z are limited to certain discrete, quantized values, then the cor-

---

[8]See, for example, the volume *Newtonian Mechanics* in this series, p. 195.

10-2   Stern-Gerlach experiment: theory

responding deflections z will also be discrete. As described in the following section, the Stern-Gerlach experiment showed convincingly that, for the atoms tested, the value of μ_z, and hence of the z component of angular momentum, is indeed quantized.

We have kept this discussion simple by naively modeling the magnetic atom in classical terms (except for the possibility of a discrete set of orientations). As we discuss the quantum-mechanical description of angular momentum in more detail, we shall retreat from the view of angular momentum vectors pointing in specific directions.

10-3 STERN-GERLACH EXPERIMENT: DESCRIPTIVE

We pointed out earlier that the Stern-Gerlach experiment was performed with a beam of neutral silver atoms. Since it is very difficult to produce, steer, and detect a beam of *neutral* atoms, one might wonder why the quantization of angular momentum should not be explored with free electrons or charged atomic ions, which are easily produced and detected and may be focused and controlled by electric and magnetic fields. The main reason is that electric and magnetic forces on *charged* particles are typically so much larger than forces on atomic magnetic dipoles due to magnetic field gradients that the deflections we seek to study would be swamped by deflections due to electric charge (see the exercises). However, in the case of electrons there is also a general argument that proves the impossibility of splitting a beam into separate magnetic components.[9]

Stern-Gerlach apparatus

Figure 10-5a shows a schematic diagram of an atomic beam apparatus, and Figure 10-5b is a photograph of a modern apparatus used to demonstrate the Stern-Gerlach experiment.[10] A basic requirement for atomic beam experiments is a good vacuum, since at atmospheric pressure atoms have free paths of only about 10^{-4} cm between collisions. By maintain-

---

[9]See, for example, O. Klemperer, *Electron Physics*, 2d. ed., Butterworths, London, 1972.

[10]From the film, *The Stern-Gerlach Experiment*, by Jerrold R. Zacharias, Education Development Center, Newton, Mass., 1967.

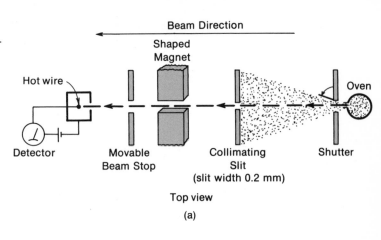

Fig. 10-5 The Stern-Gerlach experiment. (a) Schematic diagram. (b) Photograph of demonstration apparatus. (Reproduced from the film "Stern-Gerlach Experiment" with permission of the Education Development Center, Newton, Mass.)

Beam Direction

Shaped Magnet

Hot wire

Oven

Detector

Movable Beam Stop

Collimating Slit
(slit width 0.2 mm)

Shutter

Top view

(a)

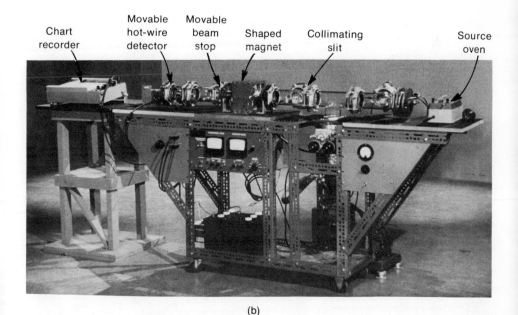

Chart recorder

Movable hot-wire detector

Movable beam stop

Shaped magnet

Collimating slit

Source oven

(b)

ing a low pressure of about 10^{-8} atm, the experimenter can ensure that atoms in an atomic beam will travel through the apparatus (a total distance of 1 or 2 m) with negligible chance of collision with a molecule of the residual gas.

Referring to Figure 10-5a, the atoms under study (cesium in the experiment now to be described) emerge from an "oven" shown at the extreme right of the figure. They are simply vaporized by warming a sample of cesium metal contained in

10-3   Stern-Gerlach experiment: descriptive

the oven. A small fraction of the atoms, selected (collimated) by a narrow slit, enter the magnetic field region as a narrow and almost parallel beam. After leaving the magnetic field region (about 10 cm long), the atoms travel freely for about 50 cm before reaching the detector.

The detector consists of a niobium wire heated to incandescence and connected to an electrometer which is, in effect, a very sensitive galvanometer (ammeter) capable of measuring currents of the order of 10^{-12} A. This is a highly efficient detector for alkali atoms such as cesium, which are ionized when they strike the hot wire and can be drawn off by an auxiliary electrode. Oriented parallel to the collimating slit, the wire acts as a narrow probe which can be moved along the z direction to reveal the distribution of transverse deflections of the atoms in the beam.

The inhomogeneous magnetic field is provided by an elec-

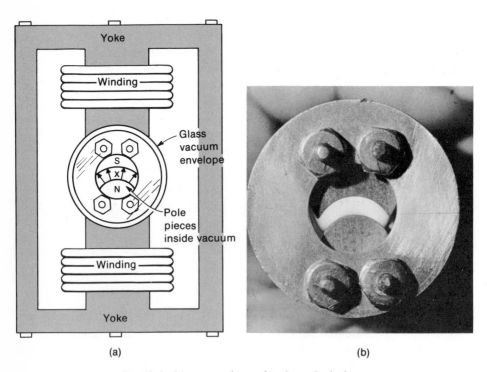

Fig. 10-6   Magnet used in modern Stern-Gerlach apparatus.   (a) Schematic diagram of entire magnet (viewed along beam direction).   (b) Photograph of the pole pieces, the part of the magnet enclosed in the vacuum system.   (See Figure 10-5 for source.)

434   Angular momentum

tromagnet with pole faces shaped like portions of cylinders, as shown in Figure 10-6. This particular geometry provides a nearly constant value of the gradient $\partial B_z/\partial z$ across the width of the gap (several millimeters). The shaped pole pieces are mounted inside the vacuum system, with the magnet yoke and field coils outside (see Figure 10-6). This greatly simplifies the practical problems of making electrical connections and keeping the vacuum chamber clean and small.

Quantization of Angular Momentum

Figure 10-7 shows the results of a typical run with the apparatus. In the absence of a magnetic field, the beam is quite

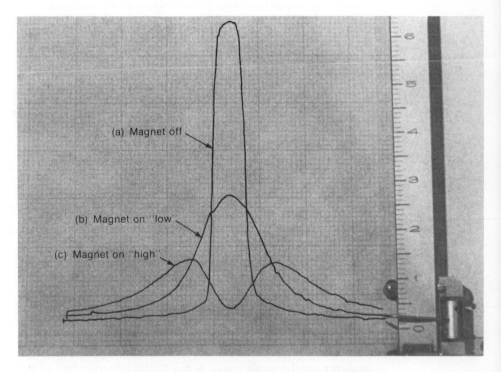

(a) Magnet off

(b) Magnet on "low"

(c) Magnet on "high"

Fig. 10-7   Beam profiles obtained for cesium beam using the equipment of Figure 10-5. Curve (a) shows the undeflected beam with the analyzer magnet turned off. Curve (b) shows the spreading of the beam with a low magnetic field. Gradient is not great enough to cause separation of components. Curve (c) shows separation in high field gradient. Notice the range of deflections, due principally to spread of velocities of cesium atoms from oven. (Redrawn. See Figure 10-5 for source.)

10-3   Stern-Gerlach experiment: descriptive

narrow (about 1 mm), producing a single high peak of detector current. With a moderate magnetic field gradient the beam broadens symmetrically about $z = 0$. At this stage there is no indication of quantization. However, with a larger value of the magnetic field gradient $\partial B_z / \partial z$, one sees quite plainly the splitting of the beam into two main components, with equal and opposite deflections along z. The pattern is smeared because of the wide range of speeds of atoms in the beam, resulting from the distribution of atomic speeds in the cesium vapor in the source oven.[11] The broadening of the beam (no observed splitting) at low magnetic fields is also due to this spread of atomic speeds. When allowance is made for this and for the finite widths of the collimating slits and the detector, the results are completely consistent with the quantization of the atoms into two sharply defined states with equal and opposite values of μ_z.

Figure 10-8 shows the results of the original Stern-Gerlach experiment, in which the silver atoms of the beam were detected by allowing them to build up a visible deposit on a glass plate. Only for the central section of the pattern are the field gradients sufficiently large to show the splitting. The figure is a reproduction of a postcard sent by Stern and Gerlach to Niels Bohr congratulating him on his prediction of the splitting that they had verified.

Quantization of the z component of angular momentum is the central feature of the Stern-Gerlach experiment in both its historical and modern forms. In some important ways the analyzing of cesium atoms into angular momentum states using an inhomogeneous magnetic field is similar to analyzing photons into polarization states using calcite and other birefringent materials (Chapter 6). In both experiments the separation is discrete, not continuous (exhibiting *quantization*). In both experiments every incident particle appears in one or another output beam (exhibiting *completeness*). In both cases, an analyzer can be made into a projector by covering one of the output beams, and the different states into which an incident beam is analyzed are *orthogonal* in the sense defined for photons. In both cases other properties (for example, energy) *not* sorted by the analyzer must be specified in a complete description of a particle's quantum state.

[11] O. Stern, Z. Phys. **2**, 49 (1920); J. F. Zartman, Phys. Rev. **37**, 383 (1931).

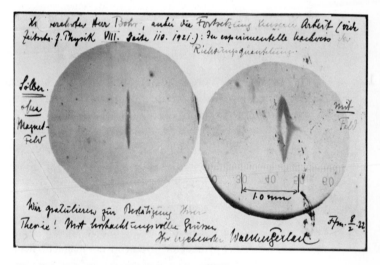

Fig. 10-8   The front and back of a postcard from
Walter Gerlach to Niels Bohr. The left-hand image
shows the beam profile without magnetic field, the
right-hand image with the field. Only in the center of
the image was the field gradient sufficient to cause splitting. Translation of post card message: "My esteemed
Herr Bohr, attached [is] the continuation of our work
(vide Zeitschr. f. Phys. **8**, 110 (1921)): the experimental proof of directional quantization. We congrat-
ulate [you] on the confirmation of your theory! With
respectful greetings, Your most humble Walter Gerlach." (Reproduced with permission of Aage Bohr and
the late Walter Gerlach.)

These similarities cannot mask the differences between the two cases: The types of "particles" under study are utterly different (cesium atoms are not photons!); there are always only two orthogonal photon polarization states in a complete set, whereas experiments have discovered different atoms (and even excited cesium atoms) that can be analyzed into two, three, four or more discrete and orthogonal angular momentum states. And, of course, the Stern-Gerlach experiment is *much* more difficult to carry out than the photon polarization experiments.

Despite the obvious differences, the similarities are striking and result from the unified nature of quantum descriptions. We have seen and will continue to discover that many of the terms and formalisms used to describe photon polarization states have their analogies in descriptions of quantum states of atomic systems.

10-4 MAGNITUDES OF ATOMIC DIPOLE MOMENTS

Measured Value of μ_z

We can estimate the value of the magnetic moment of cesium using values from the filmed Stern-Gerlach experiment. In that experiment

$2z =$ separation of peaks in detector plane $= 0.37$ cm

$kT =$ typical kinetic energy of atoms emerging from oven $= 3.4 \times 10^{-2}$ eV $= 5.5 \times 10^{-14}$ erg

$\dfrac{\partial B_z}{\partial z} = z$ component of field gradient at beam $= 10^4$ gauss/cm

$d =$ length of magnetic field region $= 12.5$ cm

$D =$ detector distance from magnetic field region $= 50$ cm

A careful analysis using these values and Eq. 10-5 yields the following numerical estimate for μ_z of cesium:

$\mu_z = \pm 8.7 \times 10^{-21}$ erg/gauss

(The $\pm$ is written to emphasize the presence of two quantized values of the z component of angular momentum.)

Theoretical Value for μ

How does the value for μ_z obtained above compare with

Angular momentum

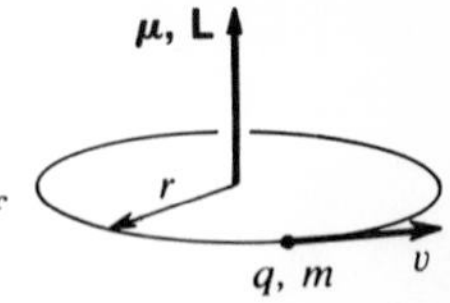

Fig. 10-9   *Angular momentum and dipole moment of an orbiting charge.*

the values one expects to find for atomic magnetic moments? We make a rough estimate using the Bohr theory and attributing the magnetic moment to the *orbital* motion of the electron. (The equally important spin magnetic moment of the electron itself will be considered later.) So consider an electric charge q traveling with speed v in a circular orbit of radius r (Figure 10-9). Since the charge passes a given point on the circle once in every period $T = 2\pi r/v$, it is equivalent to a current I given by

$$I = \frac{q}{T} = \frac{qv}{2\pi r}$$

Now it is a well-known result of classical electromagnetism[12] that a current I flowing in a circuit that encloses an area A constitutes a magnetic dipole moment μ whose magnitude is

$$\mu = \frac{IA}{c} \text{ (cgs)} \quad [IA \text{ (SI)}]$$

In the present case, with $I = qv/2\pi r$, this gives

$$\mu = \frac{qvr}{2c} \text{ (cgs)} \quad \left[\frac{qvr}{2} \text{ (SI)}\right] \tag{10-6}$$

However, the orbital angular momentum of the charged particle of mass M is given by

$$L = Mvr$$

Thus the magnetic moment is proportional to the angular momentum; combining the last two equations, we have

$$\mu = \frac{q}{2Mc} L \text{ (cgs)} \quad \left[\frac{q}{2M} L \text{ (SI)}\right] \tag{10-7}$$

[12]See, for example, E. M. Purcell, *Electricity and Magnetism*, Berkeley Physics Course, Vol. II, McGraw-Hill, New York, 1965, p. 364.

439   10-4   Magnitudes of atomic dipole moments

The constant of proportionality between L and μ is called the *gyromagnetic ratio*. Now we have a broad hint from the Bohr model, and we shall verify it more rigorously in the present chapter, that the basic unit of angular momentum in atomic systems is equal to $\hbar$ $(= h/2\pi)$. Using this hint and Eq. 10-7, we can propose a natural atomic unit for magnetic dipole moment with $q = e$ and $L = \hbar$ and M equal to the mass m_e of the electron. This unit is called the *Bohr magneton* μ_B:

$$\mu_B = \frac{e\hbar}{2m_e c} \text{ (cgs)} \qquad \left[\frac{e\hbar}{2m_e} \text{ (SI)} \right] \qquad \text{(10-8a)}$$

The numerical values are

$$\mu_B = 9.27 \times 10^{-21} \text{ erg/gauss (cgs)}$$
$$= 9.27 \times 10^{-24} \text{ J/tesla (SI)} \qquad \text{(10-8b)}$$

Notice that this is of the same order of magnitude as μ_z calculated earlier from the results of the demonstration Stern-Gerlach experiment.

Different atomic systems have different values of magnetic moment, of which the Bohr magneton is a natural unit of measure. For any system the magnetic moment can be expressed as a certain multiple g of the Bohr magneton:

$$\mu = g\mu_B$$

This factor g is called the *Landé g factor*.[13] For magnetic moments due only to orbital motion of electrons the above Bohr analysis gives the correct result and the g factor is unity. For other systems, the g factor is not unity but nevertheless is useful as a dimensionless measure of magnetic moment.

In summary, we have derived a natural unit for atomic magnetic moments and have found from experiment that ground-state cesium has a z component of magnetic moment comparable in value to this natural unit.

Electron Spin

The above estimate of the value of an atomic magnetic moment assumes that these moments result solely from the or-

---

[13]Named after Alfred Landé, who introduced it in 1923 in connection with the analysis of atomic spectra.

bital motion of electrons about the nucleus. The truth of the matter is that many experimental results, *including those quoted above for atomic beams*, can be accounted for only by assuming that electrons also exhibit an intrinsic or *spin* magnetic moment. From evidence to be cited later, the angular momentum connected with this spin is $\hbar/2$ instead of $\hbar$. Although the angular momentum is only $\hbar/2$, nevertheless the magnetic moment associated with the spin is approximately one Bohr magneton as well. One way of expressing this is to say that the g factor for electron spin magnetism is 2. This perplexing double magnetism of the electron spin has no classical explanation. The gyromagnetic ratio of free electrons has been directly measured by observing their precession in a magnetic field.[14] The results verified a g factor very close to 2. (Later, however, more refined experiments have confirmed theoretical predictions that the g factor for electron spin is *not exactly* 2; see Section 11-4.)

Magnetic Moments of Other Atoms

Well over half of the different atomic species in the periodic table in their ground states have nonzero magnetic moments of the order of the Bohr magneton. Speaking approximately and rather incorrectly, we can say that in many-electron atoms the electrons pair off so as to cancel spin and orbital angular momenta. A net atomic magnetic moment results when one or more electrons are left out of this pairing process. A more exact statement can be made after the analysis in Chapter 13. Despite the especially strong magnetism that is exhibited by ferromagnetic elements in bulk (iron, nickel, cobalt), the individual atoms of these elements (including iron itself) do *not* have larger magnetic moments than, for example, the individual cesium atoms used in the Stern-Gerlach experiments (see the exercises).[15] Those atoms that do *not* have magnetic moments of the order of one Bohr magneton (such as the noble

---

[14]W. H. Louisell, R. W. Pidd, and H. R. Crane, Phys. Rev. **94**, 7–16 (1954).

[15]The large-scale magnetization of so-called magnetic materials, such as iron in bulk, is due to cooperative effects which influence the separate atomic magnetic moments to line up with one another. In "nonmagnetic" materials, such as solid cesium, these cooperative tendencies are weaker or lacking, even though the magnetic moments of the individual atoms are of much the same magnitude as that of iron atoms.

10-4   Magnitudes of atomic dipole moments

gases helium, neon, and argon in their ground states) can be recognized as those whose electron structures have zero net angular momentum. Many of these atoms do have tiny net magnetic moments which are associated with the angular momentum of the atomic nucleus. Assuming the same kind of connection between magnetism and angular momentum that holds for electrons, the order of magnitude of μ for nuclei is given by replacing the electron mass in Eq. 10-8 by the mass of a nucleon (proton or neutron). The resulting quantity is called a *nuclear magneton*. It is smaller by a factor of about 1840 than the Bohr magneton.

We have cited experimental evidence that angular momentum is quantized, and have presented a classical analysis of the relation between orbital angular momentum and magnetism to account for the approximate magnitudes of atomic magnetic moments. Now we proceed to the quantum-mechanical theory of angular momentum per se.

10-5   ORBITAL ANGULAR MOMENTUM OPERATORS

Our earlier analysis of the quantized energy of bound states was based on the Schrödinger eigenvalue equation, which made formal use of an energy operator $(E)_{op}$ whose specific form depends on the system being described (Section 5-3):

$$(E)_{op}\psi_n = E_n\psi_n$$

Here E_n is a quantized energy value and ψ_n is the corresponding energy eigenfunction. A similar analysis of the quantized orbital angular momentum of bound states depends on eigenvalue equations that make formal use of operators for angular momentum. As in the development of the Schrödinger equation, classical physics gives some hints of the correct forms of these angular momentum operators.

Classical Angular Momentum

Classically, the orbital angular momentum **L** of a particle with respect to any chosen center is defined by the cross-product of the position vector **r** (measured from the given

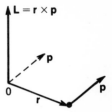

Fig. 10-10   Angular momentum of a particle in arbitrary motion.

center) and the linear momentum vector $\mathbf{p}$ (Figure 10-10):

$$\mathbf{L} = \mathbf{r} \times \mathbf{p} \tag{10-9}$$

In order to explore the quantization of orbital angular momentum, we shall seek to convert $\mathbf{L}$ into an operator analogous to the energy operator in the Schrödinger equation. A preliminary step is to identify operators corresponding to *linear momentum*, which can then be incorporated into a quantum-mechanical operator form of Eq. 10-9.

Linear Momentum Operators

We saw in Chapter 3 that free particles with a unique value of linear momentum, $p_x = \hbar k_x$, are described (ignoring the time factor) by a wave function ψ of the form $e^{ik_x x}$. This free-particle wave function is a *linear momentum eigenfunction*. If we are to have a linear momentum operator, $(p_x)_{op}$, it must be such as to yield the relation

$$(p_x)_{op}\psi = p_x\psi = \hbar k_x\psi$$

where $\psi = e^{ik_x x}$. It is easy to verify that this requirement is met if we put

$$(p_x)_{op} = \frac{\hbar}{i}\frac{\partial}{\partial x} \tag{10-10a}$$

for then we have

$$(p_x)_{op}\psi = \frac{\hbar}{i}\frac{\partial}{\partial x}(e^{ik_x x}) = \frac{i\hbar k_x}{i}e^{ik_x x} = p_x e^{ik_x x} = p_x\psi$$

By analogy we can immediately write down the operators

for the y and z components of linear momentum:

$$(p_y)_{op} = \frac{\hbar}{i}\frac{\partial}{\partial y} \tag{10-10b}$$

$$(p_z)_{op} = \frac{\hbar}{i}\frac{\partial}{\partial z} \tag{10-10c}$$

We can then show that the form of the linear momentum operator $(p_x)_{op}$ is consistent with the form of the kinetic energy operator, $(-\hbar^2/2m)(\partial^2/\partial x^2)$, used in the one-dimensional Schrödinger equation. The classical expression for kinetic energy of particle, $p_x^2/2m$, takes on this form if p_x is replaced by $(p_x)_{op}$, provided the symbol $(p_x^2)_{op}$ is given its standard meaning: the application of $(p_x)_{op}$ twice in succession, yielding in this case a second derivative

$$(p_x^2)_{op}\,\psi = (p_x)_{op}\,\frac{\hbar}{i}\frac{\partial\psi}{\partial x} = -\hbar^2\frac{\partial^2\psi}{\partial x^2}$$

Therefore

$$(KE)_{op}\,\psi = \frac{(p_x^2)_{op}}{2m}\,\psi = \frac{-\hbar^2}{2m}\frac{\partial^2\psi}{\partial x^2}$$

Operators for Angular Momentum

Using the linear momentum operators we can now construct an operator for orbital angular momentum based on the classical expression for $\mathbf{L}$ in Eq. 10-9. The classical expression can conveniently be written as a determinant:

$$\mathbf{L} = \mathbf{r} \times \mathbf{p} = \begin{vmatrix} \hat{x} & \hat{y} & \hat{z} \\ x & y & z \\ p_x & p_y & p_z \end{vmatrix} \tag{10-11}$$

Here, $\hat{x}$, $\hat{y}$, $\hat{z}$ are unit vectors along the respective directions. The z component of angular momentum is thus

$$L_z = (\mathbf{r} \times \mathbf{p})_z = xp_y - yp_x$$

Using the quantum-mechanical operators corresponding to p_x and p_y, we obtain

$$(L_z)_{op} = \frac{\hbar}{i}\left(x\frac{\partial}{\partial y} - y\frac{\partial}{\partial x}\right) \qquad (10\text{-}12a)$$

The corresponding operator expressions for the x and y components of $\mathbf{L}$ are

$$(L_x)_{op} = \frac{\hbar}{i}\left(y\frac{\partial}{\partial z} - z\frac{\partial}{\partial y}\right) \qquad (10\text{-}12b)$$

$$(L_y)_{op} = \frac{\hbar}{i}\left(z\frac{\partial}{\partial x} - x\frac{\partial}{\partial z}\right) \qquad (10\text{-}12c)$$

From these component operators we can construct a vector operator $(\mathbf{L})_{op}$ and try to use it in a vector eigenvalue equation for angular momentum. In Chapter 11 we shall see that the operator $(L^2)_{op} = (\mathbf{L})_{op} \cdot (\mathbf{L})_{op}$ proves to be more useful. An eigenvalue equation using the operator $(L^2)_{op}$ yields eigenvalues of the *squared magnitude* of the orbital angular momentum. For the present we limit ourselves to analyzing the quantization of a single component of orbital angular momentum, such as that along the magnetic field gradient in the Stern-Gerlach experiment. It is customary to choose the z axis to lie along the axis of quantization selected by the physical conditions in any given experiment. Therefore we concentrate on the quantization of the z component of orbital angular momentum L_z.

10-6   EIGENVALUES OF L_z

A particle moving in a plane has an angular momentum that, according to the classical analysis, points perpendicular to that plane. If the xy plane is chosen to be the plane of motion of the particle, then classically the angular momentum lies along the z axis.

The analogous situation in quantum mechanics requires a particle described by a two-dimensional wave function $\psi(x, y)$. If the particle has a unique value L_z (an eigenvalue) of the z component of angular momentum, then the wave function $\psi(x, y)$ is an eigenfunction of the operator $(L_z)_{op}$ according to the equation:

$$(L_z)_{op}\,\psi\,(x, y) = L_z\,\psi\,(x, y) \qquad (10\text{-}13)$$

$(L_z)_{op}$ in Polar Coordinates

The operator $(L_z)_{op}$ takes on a particularly simple form if, instead of x and y, we use the plane polar coordinates r and ϕ as shown in Figure 10-11:

$$x = r \cos \phi$$
$$y = r \sin \phi$$

We then find that $(L_z)_{op}$ can be written as follows:

$$(L_z)_{op} = \frac{\hbar}{i} \frac{\partial}{\partial \phi} \qquad (10\text{-}14)$$

This is most easily verified by starting with Eq. 10-14 and working backward. By the chain rule of implicit differentiation, we have

$$\frac{\partial}{\partial \phi} = \frac{\partial y}{\partial \phi} \frac{\partial}{\partial y} + \frac{\partial x}{\partial \phi} \frac{\partial}{\partial x}$$

But $\partial y / \partial \phi = r \cos \phi = x$, and $\partial x / \partial \phi = -r \sin \phi = -y$. Hence

$$\frac{\partial}{\partial \phi} = x \frac{\partial}{\partial y} - y \frac{\partial}{\partial x}$$

which leads directly from Eq. 10-14 back to the original expression for $(L_z)_{op}$ (Eq. 10-12a). Actually, the form of $(L_z)_{op}$ given in Eq. 10-14 remains valid in a three-dimensional system described by spherical polar coordinates (see the exercises). Thus the two-dimensional analysis we are carrying out here will be directly useful in the more general three-dimensional analysis to follow (Chapter 11).

Quantized Eigenvalues of L_z

If the wave function ψ satisfies the eigenvalue equation for

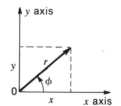

Fig. 10-11   Conversion from Cartesian to plane polar coordinates.

Angular momentum

the z component of angular momentum, we have, using Eq. 10-14,

$$\frac{\hbar}{i}\frac{\partial \psi}{\partial \phi} = L_z\psi \tag{10-15}$$

where L_z is a constant. Note that this is a *partial* differential equation. If we limit ourselves to a two-dimensional form of ψ, described by the coordinates r and ϕ only, then Eq. 10-15 tells us that the ϕ dependence of the solution is an exponential. Including the r dependence as a product function, we have

$$\psi(r, \phi) = R(r)e^{i(L_z\phi/\hbar)} \tag{10-16}$$

where $R(r)$ is some function of the radius r alone. This form of the wave function leads directly to the theoretical description of quantized angular momentum observed experimentally. The quantization of L_z emerges (as in all eigenvalue problems) from the imposition of a physically required boundary condition. In this case we make the physically reasonable demand that ψ be single-valued at all points on the xy plane.[16] After all, the point specified by the polar coordinates r and ϕ is exactly the *same* point on the plane as that specified by r and $\phi + 2\pi$. So the value of the wave function for $(r, \phi + 2\pi)$ should equal its value for (r, ϕ). Imposing this requirement, we have

$$e^{i(L_z/\hbar)(\phi+2\pi)} = e^{i(L_z/\hbar)\phi}$$

or

$$e^{i(2\pi L_z/\hbar)} = 1$$

---

[16]One might argue that only the squared magnitude $|\psi|^2$, which represents a measurable probability density, has to be single-valued. This weaker requirement would permit *any* real value of m whatever, even noninteger, since it occurs in a phase factor. But for nonintegral values of m the derivative $\partial \psi / \partial \phi$ is not single-valued and the wave function cannot be an acceptable solution of the eigenvalue equation (Eq. 10-15). Hence we exclude nonintegral values of m for *orbital* angular momentum. However, we have already referred to the fact that the spin angular momentum of the electron is $\frac{1}{2}\hbar$, as contrasted with integral multiples of $\hbar$ for orbital angular momentum. This case of half-integer spin is simply not covered by the eigenvalue equation (10-15). For further discussion see E. Merzbacher, "Single Valuedness of Wave Functions," Am. J. Phys. **30**, 237–247 (1962).

10-6   Eigenvalues of L_z

This equation is satisfied only if the quantity $2\pi L_z/\hbar$ is zero or an integral multiple of 2π. Denote this integer by the symbol m. Thus we deduce

$$\frac{2\pi L_z}{\hbar} = 2\pi m$$

so that

$$L_z = m\hbar \qquad (m = 0, \pm 1, \pm 2, \ldots) \qquad (10\text{-}17)$$

with

$$\psi(r, \phi) = R(r)e^{im\phi} \qquad (10\text{-}18)$$

The quantization of the z component of orbital angular momentum is a general and extremely powerful result. All observations confirm that, for *any* atomic system, *every* measurement made of one component of its *orbital* angular momentum yields a positive or negative integral multiple of $\hbar$, or zero, as implied by Eq. 10-17, and *never* any other value.

Not quite so general is the particular form of the wave function given in Eq. 10-18. The product of functions in the separate coordinates r and ϕ forms an acceptable wave function only if the particle is in a *central* potential, that is, a potential that is a function only of the distance r from the force center and is independent of direction from that center. The argument that leads to this conclusion is given in Chapter 11.

It is evident from the basic symmetry of Eqs. 10-12 for the separate quantum operators for L_x, L_y, L_z that we could just as well have chosen the x or the y axis as the axis of quantization, yielding quantized values for L_x or L_y corresponding exactly to the results for L_z expressed in Eq. 10-17. However, it turns out that, as in the old adage, "You pay your money and you take your choice." For when we consider these other possibilities, we find that if we have an eigenfunction of $(L_z)_{op}$, the same function is *not* an eigenfunction of either $(L_x)_{op}$ or $(L_y)_{op}$. In terms of a hypothetical atomic beam experiment, if a beam of particles, all prepared in an eigenstate of L_z, is tested for, say, L_x (through the application of a field gradient $\partial B_x/\partial x$), then every particle in the beam will yield one of the quantized values for L_x *but not all the same value.* A particle cannot simultaneously be in an eigenstate of each of two perpendicular components of angular momentum. This is just one example of the general problem of *simultaneous eigenvalues.* The next section discusses this problem.

10-7 SIMULTANEOUS EIGENVALUES

We pose the question: Under what conditions can two or more observable properties of a quantum system have unique eigenvalues for a given quantum state? Fortunately, this question can be considered in a way that does not entail examining individually every possible state of a system. A very simple argument suggests a criterion that can be applied to the quantum operators that correspond to the observable quantities.

Commuting Operators

Consider an operator $(P)_{op}$ that represents an observable quantity (for instance, linear momentum). Let us suppose that in a certain state ψ the quantity has a particular quantized value p. This means that the operator $(P)_{op}$, operating on ψ, yields the same ψ multiplied by the eigenvalue p:

$$(P)_{op} \psi = p\psi \qquad (10\text{-}19a)$$

Suppose now that a particle in the *same* state also has a specific value q of another quantity (for instance, energy) for which the operator is $(Q)_{op}$. Then we have a second eigenvalue equation for the *same* wave function:

$$(Q)_{op} \psi = q \psi \qquad (10\text{-}19b)$$

In the circumstance just described, p and q are called *simultaneous eigenvalues.* What are the *conditions* under which simultaneous eigenvalues can be found for two different observables, so that 10-19a and 10-19b are both valid?

Apply the operator $(Q)_{op}$ to both sides of the first equation and the operator $(P)_{op}$ to both sides of the second. Then we get

$$(Q)_{op} (P)_{op} \psi = (Q)_{op} (p \psi) = p (Q)_{op} \psi = pq \psi \qquad (10\text{-}20a)$$

and

$$(P)_{op} (Q)_{op} \psi = (P)_{op} (q|\psi) = q (P)_{op} \psi = qp \psi \qquad (10\text{-}20b)$$

Since p and q are just numbers (with associated physical dimensions, of course), the products pq and qp are identical. Therefore, by subtracting both sides of Eq. 10-20a from the corresponding sides of Eq. 10-20b, we obtain

$$[(P)_{op} (Q)_{op} - (Q)_{op} (P)_{op}] \psi = (qp - pq) \psi = 0$$

This suggests that the chance for simultaneous eigenvalues to occur is best if the operators $(P)_{op}$ and $(Q)_{op}$ *commute*—that is, if

$$(P)_{op} (Q)_{op} - (Q)_{op} (P)_{op} = 0 \qquad (10\text{-}21)$$

The expression on the left-hand side of Eq. 10-21 is called the *commutator* of the two operators. Notice that Eq. 10-21 does not include the wave function ψ at all, so the possibility of simultaneous eigenstates can be examined without knowing the eigenfunctions of either operator. This generality has its price. If two operators $(P)_{op}$ and $(Q)_{op}$ commute, it *does not* follow that *all* possible eigenstates of one observable are also eigenstates of the other observable. But what *does* follow is that a set of states *can* be found, each of which is an eigenstate of both P and Q.[17] In contrast, if $(P)_{op}$ and $(Q)_{op}$ do *not* commute then no such complete set of states can be found even though an occasional exceptional state may be an eigenstate of both P and Q.

As an example, consider the linear momentum and energy of a *free* particle in one dimension. The momentum operator is $(\hbar/i) (d/dx)$. The energy of a free particle is all kinetic, so the energy operator in this case is $(-\hbar^2/2m)(d^2/dx^2)$. The commutator of these two operators is zero:

$$\left(\frac{\hbar}{i} \frac{d}{dx}\right)\left(\frac{-\hbar^2}{2m} \frac{d^2}{dx^2}\right) - \left(\frac{-\hbar^2}{2m} \frac{d^2}{dx^2}\right)\left(\frac{\hbar}{i} \frac{d}{dx}\right)$$
$$= -\frac{\hbar^3}{2mi}\left(\frac{d^3}{dx^3} - \frac{d^3}{dx^3}\right) = 0$$

[17]We have not proved this assertion, but it can be proved. See Albert Messiah, *Quantum Mechanics* Vol. I, John Wiley, New York, 1961, p. 199.

Angular momentum

Therefore a free particle state *can* be characterized by unique values of both linear momentum and energy. This does *not* mean that *all possible* energy eigenstates of a free particle must also be momentum eigenstates. For example, $\psi \sim (e^{ikx} + e^{-ikx})$ is the wave function of an energy state $E = \hbar^2 k^2 / 2m$ that is clearly *not* a momentum eigenstate.

As another example, consider the linear momentum and energy operators of a particle *bound* in a one-dimensional potential $V(x)$. The potential operator $V(x)$ must be added to the kinetic energy operator to give the total energy operator $(E)_{\text{op}}$. In evaluating the commutator of linear momentum and energy, we note that the linear momentum and kinetic energy operators still commute, but the potential energy operator does not commute with the linear momentum operator. You should show that the commutator of $(p)_{\text{op}}$ and $(E)_{\text{op}}$ has the value $(\hbar/i)(dV/dx)$. Since this is not everywhere equal to zero, we conclude that, in general, bound particle states do not have simultaneous eigenvalues of energy and linear momentum. This conclusion is not new to us: a bound particle with a unique energy cannot have a unique linear momentum, as first discussed in Chapter 3 (Section 3-6 on "Unique Energy without Unique Momentum."). In the next section we consider other pairs of noncommuting operators.

Commutators and Uncertainty Relations

Typically, when two operators do not commute, the corresponding observables exhibit an uncertainty relation. A good case in point is that of the components of momentum and displacement along the same direction, say the x axis. We have already seen that these quantities obey an uncertainty relation $\Delta x \cdot \Delta p_x \gtrsim \hbar$. We know the operators for this pair of observables to be $(\hbar/i)(\partial/\partial x)$ and x itself. Thus we have (using a dummy wave function ψ to keep the differentiations straight)

$$(x)_{\text{op}} (p_x)_{\text{op}} \psi = x \left[\frac{\hbar}{i} \frac{\partial \psi}{\partial x} \right] = \left[\frac{\hbar}{i} x \frac{\partial}{\partial x} \right] \psi$$

That is,

$$(x)_{\text{op}} (p_x)_{\text{op}} = \frac{\hbar}{i} x \frac{\partial}{\partial x}$$

10-7 Simultaneous eigenvalues

But, taking the operators in reverse order, we have

$$(p_x)_{\text{op}}(x)_{\text{op}}\,\psi = \frac{\hbar}{i}\frac{\partial}{\partial x}(x\,\psi) = \frac{\hbar}{i}\,\psi + \frac{\hbar}{i}\,x\,\frac{\partial\psi}{\partial x}$$

that is,

$$(p_x)_{\text{op}}(x)_{\text{op}} = \frac{\hbar}{i} + \frac{\hbar}{i}\,x\,\frac{\partial}{\partial x}$$

Hence

$$(x)_{\text{op}}(p_x)_{\text{op}} - (p_x)_{\text{op}}(x)_{\text{op}} = -\frac{\hbar}{i} = i\hbar \neq 0 \tag{10-22}$$

This commutation relation embodies the same message as the uncertainty principle: One cannot find a single state for which both position and momentum have precise eigenvalues.

Simultaneous Angular Momentum Eigenvalues

Now we shall use the commutation criterion to justify our assertion at the end of Section 10-6 that, in general, separate components of angular momentum cannot be simultaneously specified. Consider the expressions for $(L_x)_{\text{op}}$ and $(L_y)_{\text{op}}$ in Eqs. 10-12b and 10-12c. If these are applied in succession to a wave function, these two results follow:

$$
\begin{aligned}
(L_x)_{\text{op}}(L_y)_{\text{op}}\,\psi &= -\hbar^2\left(y\frac{\partial}{\partial z} - z\frac{\partial}{\partial y}\right)\left(z\frac{\partial\psi}{\partial x} - x\frac{\partial\psi}{\partial z}\right) \\
&= -\hbar^2\left(y\frac{\partial\psi}{\partial x} + yz\frac{\partial^2\psi}{\partial z\,\partial x} - xy\frac{\partial^2\psi}{\partial z^2}\right. \\
&\qquad \left. - z^2\frac{\partial^2\psi}{\partial x\,\partial y} + zx\frac{\partial^2\psi}{\partial y\,\partial z}\right)
\end{aligned}
$$

and

$$
\begin{aligned}
(L_y)_{\text{op}}(L_x)_{\text{op}}\,\psi &= -\hbar^2\left(z\frac{\partial}{\partial x} - x\frac{\partial}{\partial z}\right)\left(y\frac{\partial\psi}{\partial z} - z\frac{\partial\psi}{\partial y}\right) \\
&= -\hbar^2\left(yz\frac{\partial^2\psi}{\partial z\,\partial x} - z^2\frac{\partial^2\psi}{\partial x\,\partial y} - xy\frac{\partial^2\psi}{\partial z^2}\right. \\
&\qquad \left. + x\frac{\partial\psi}{\partial y} + zx\frac{\partial^2\psi}{\partial y\,\partial z}\right)
\end{aligned}
$$

Taking the difference, we then find

$$(L_x)_{\text{op}}(L_y)_{\text{op}} - (L_y)_{\text{op}}(L_x)_{\text{op}} = \hbar^2 \left(x \frac{\partial}{\partial y} - y \frac{\partial}{\partial x} \right) \tag{10-23}$$

$$= i\hbar (L_z)_{\text{op}}$$

In summary, $(L_x)_{\text{op}}$ and $(L_y)_{\text{op}}$ do *not* commute and hence there is not a set of states for which *both* L_x and L_y (or any other pair of angular momentum components) have simultaneous eigenvalues. It follows that the *vector* angular momentum **L** generally cannot have a unique value (that is, both a unique magnitude *and* direction in space), since its three components cannot usually be simultaneously specified.

The reason we keep saying "generally" and "usually" is that there is at least one exception to the statements above: there *is* a state for which all three orbital angular momentum components L_x, L_y, and L_z are equal to zero. In this state the vector angular momentum also has the unique magnitude zero (although, of course, no definable direction). This does not dilute the general conclusion that no complete *set* of states can be found with unique values of the vector orbital angular momentum.

The operators for the three components of angular momentum do not commute with one another. With what operators *do* they commute, and what does this imply for results of experiments? In Chapter 11 we shall examine the operator for the squared magnitude of the angular momentum $(L^2)_{\text{op}}$. We shall find that it commutes with *each* of the components of angular momentum. (Since L^2 is a scalar, it fixes the magnitude of the total orbital angular momentum without determining a direction.) In the remainder of this chapter and in later chapters we deal with particles in *central* potentials, for which the *energy* of the state can be specified simultaneously with one component of angular momentum and the squared magnitude of the angular momentum. For each of these central potentials it is easy to show that the energy operator in the Schrödinger equation commutes with both $(L_z)_{\text{op}}$, for example, *and* with $(L^2)_{\text{op}}$.

The next section discusses a particular example of this—the two-dimensional simple harmonic oscillator—in a way that helps to carry our analysis of quantized angular momentum a stage further. First, however, we preview the answer to a question that will arise repeatedly: For a given system, what is the *maximum number* of observables which can have simultaneous eigenvalues? That is, how many physical quantities can be simultaneously specified? Of course, the answer depends on the system under consideration. But the general requirement is that the operator corresponding to each observable in such a set must commute with *every other* operator in the set. For a particle in a central field (such as the electron in the hydrogen atom or a three-dimensional simple harmonic oscillator), one may specify energy, the squared angular momentum, *and* the z component of angular momentum. The general result (not proved here[18]) is that for every system one may identify at least one *complete set of commuting observables*. Specifying a permissible eigenvalue for each observable in the complete set then specifies a quantum state of the system uniquely. For a one-particle system in three dimensions the number of commuting observables is three. (If the particle has spin, a fourth must be added.)

10-8 QUANTUM STATES OF A TWO-DIMENSIONAL HARMONIC OSCILLATOR[19]

Our ultimate goal is to analyze angular momentum in *three* dimensions, as exhibited by real atoms and molecules. However, by exploring the *two*-dimensional case further, we can expose some features of both two- and three-dimensional systems that are not displayed in one-dimensional systems. These are:

1. More than one quantum state can have the same total energy; in such cases the total energy alone cannot specify the quantum state completely (code phrase for this circumstance: "energy degeneracy").

2. For many systems, information about angular momentum, when added to information about energy, *can* specify a quantum state completely.

3. In such systems, complete specification of quantum states using energy and angular momentum reveals limits on the values of angular momentum associated with a given energy.

[18]Albert Messiah, *op. cit.*, p. 202.
[19]C. P. Friedman and E. F. Taylor, Am. J. Phys. **39**, 1073 (1971).

Rather than deal with these features in the abstract, we illustrate them with a particular system, that of the two-dimensional simple harmonic oscillator. The simple harmonic oscillator (abbreviated SHO in what follows) is an important model in its own right, useful in the approximate analysis of almost every system characterized by a potential minimum. The SHO is also easy to analyze and, most important for our purposes, the angular momentum properties of the SHO can be applied directly to *any* central potential, such as the Coulomb (inverse square) potential due to the atomic nucleus. Thus our discussion of the two-dimensional SHO gives us a running start on atomic systems, which are three-dimensional and therefore more complicated.

Review of the One-Dimensional SHO

We have already estimated the lowest energy of the one-dimensional SHO, drawn qualitative plots of its wave functions (exercises of Chapter 3), and studied some of its analytic solutions (Chapter 4). The SHO potential is parabolic:

$$V(x) = \tfrac{1}{2}Cx^2 = \tfrac{1}{2}M\omega^2 x^2$$

where C is the "spring constant" and $\omega = (C/M)^{1/2}$ is a constant equal to the classical angular frequency of oscillation. The Schrödinger equation for this SHO is

$$-\frac{\hbar^2}{2M}\frac{d^2\psi}{dx^2} + \frac{1}{2}M\omega^2 x^2 \psi = E\psi \tag{10-24}$$

Its solutions are all of the form

$$\psi_n(x) \sim H_n\!\left(\frac{x}{a}\right) \cdot e^{-x^2/2a^2} \tag{10-25}$$

where $H_n(x)$ is a polynomial (Hermite polynomial) in either even or odd powers of x, and $a^2 = \hbar/(CM)^{1/2}$. The eigenvalues of the energy are given by

$$E_n = (n + \tfrac{1}{2})\hbar\omega \qquad (n = 0, 1, 2, \ldots) \tag{10-26}$$

(Some complete normalized solutions are presented in Table 4-1 and in the accompanying discussion, but we do not need all

10-8   Two-dimensional harmonic oscillator

the details here.) Equation 10-26 is an important reminder that the energy levels are equally spaced and that, by convention, the state of lowest energy is numbered zero.

The Two-Dimensional SHO

Suppose now that a bound particle moves in the xy plane under the influence of a two-dimensional parabolic potential $V = \frac{1}{2}C(x^2 + y^2) = \frac{1}{2}M\omega^2(x^2 + y^2)$. Think, for example, of an atom in a crystal of simple cubic structure and limit attention to motion of the atom within a densest plane of atoms. If the surrounding atoms are considered fixed in position, this atom acts very much like a two-dimensional SHO provided its displacements from equilibrium are small.

The two-dimensional Schrödinger equation corresponding to this potential is (recall Section 4-5 and see Figure 10-12 for a plot of the potential):

$$-\frac{\hbar^2}{2M}\left(\frac{\partial^2\psi}{\partial x^2} + \frac{\partial^2\psi}{\partial y^2}\right) + \frac{1}{2}M\omega^2(x^2 + y^2)\psi = E\psi \qquad (10\text{-}27)$$

This equation has as its solutions wave functions $\psi(x, y)$ that are functions of both x and y. The squared magnitude $|\psi(x, y)|^2$ is the *probability per unit area* of finding the particle in a small area about the point (x, y).

Now we shall verify that a set of possible solutions to Eq.

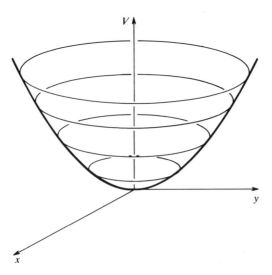

Fig. 10-12 The paraboloidal potential well of a two-dimensional simple harmonic oscillator.

Angular momentum

10-27 can be constructed from *products* of the individual solutions for one-dimensional oscillators along x and y:

$$\psi(x, y) = f(x) \cdot g(y)$$

That is, we construct wave functions for the two-dimensional SHO by multiplying any one of the eigenfunctions $f(x)$ of Eq. 10-25 by a function $g(y)$ that is also a solution of a Schrödinger equation of the form of Eq. 10-24 but with y in place of x. We then have

$$\frac{\partial^2 \psi}{\partial x^2} = g(y)\frac{d^2 f}{dx^2}$$
$$\frac{\partial^2 \psi}{\partial y^2} = f(x)\frac{d^2 g}{dy^2}$$

Substituting these and the wave function $\psi(x, y) = f(x) \cdot g(y)$ into Eq. 10-27 yields

$$g(y)\left[\frac{-\hbar^2}{2M}\frac{d^2 f}{dx^2} + \frac{1}{2}M\omega^2 x^2 f\right] + f(x)\left[\frac{-\hbar^2}{2M}\frac{d^2 g}{dy^2} + \frac{1}{2}M\omega^2 y^2 g\right]$$
$$= Ef(x)g(y)$$

This looks complicated, but since we chose f and g to be eigenfunctions of the one-dimensional SHO, the bracketed expressions can be written

$$\left[-\frac{\hbar^2}{2M}\frac{d^2 f}{dx^2} + \frac{1}{2}M\omega^2 x^2 f\right] = E_{n_x}f(x) \qquad (n_x = 0, 1, 2, \ldots)$$

and

$$\left[-\frac{\hbar^2}{2M}\frac{d^2 g}{dy^2} + \frac{1}{2}M\omega^2 y^2 g\right] = E_{n_y}g(y) \qquad (n_y = 0, 1, 2, \ldots)$$

Substituting these relations for the bracketed expressions in the preceding equation, we then find

$$(E_{n_x} + E_{n_y})f(x)g(y) = Ef(x)g(y)$$

Hence if we put $E = E_{n_x} + E_{n_y}$ we have found an energy eigenstate for the two-dimensional oscillator. Its total energy,

10-8   Two-Dimensional harmonic oscillator

in terms of the individual quantum numbers n_x and n_y, is given (Eq. 10-26) by

$$E = (n_x + n_y + 1)\hbar\omega \qquad n_x = 0, 1, 2, \ldots \qquad (10\text{-}28)$$
$$n_y = 0, 1, 2, \ldots$$

and its wave function is given by the product of one-dimensional solutions (Eq. 10-25):

$$\psi_{n_x, n_y}(x, y) \sim H_{n_x}\left(\frac{x}{a}\right) H_{n_y}\left(\frac{y}{a}\right) e^{-(x^2 + y^2)/2a^2} \qquad (10\text{-}29)$$

Equation 10-29 specifies a complete set of energy eigenfunctions for the two-dimensional harmonic oscillator.

Energy Degeneracy

If we define $n = n_x + n_y$ then it follows at once from Eq. 10-28 that the energy levels of the two-dimensional SHO are given by

$$E_n = (n + 1)\hbar\omega \qquad (n = 0, 1, 2, \ldots)$$

We then see that, except for the state of lowest energy ($n_x = n_y = 0$), more than one quantum state indexed by n can have the same energy. For example, the second energy $E = 2\hbar\omega$ with $n = 1$ is common to *two* quantum states: $n_x = 1$, $n_y = 0$ and $n_x = 0$, $n_y = 1$. We say that these two states are *degenerate* with respect to energy. The $n = 1$ energy level is said to be "twofold degenerate" since there are two states of the same total energy with $n = 1$. Similarly, for $E = 3\hbar\omega$ and $n = 2$ there is a threefold degeneracy, since the states $(n_x, n_y) = (2,0)$ and $(1,1)$ and $(0,2)$ all have this same energy. Using the energy quantum number $n = n_x + n_y$ we can generalize by saying that for the two-dimensional SHO the energy level indexed by n is $(n + 1)$-fold degenerate. (You should verify this.)

Energy degeneracy means that the total-energy quantum number n is not, by itself, sufficient to specify the quantum state of the system. The energy degeneracy in this two-dimensional model is in marked contrast to one-dimensional models, for which a particular energy corresponds to a unique quantum state. In two-dimensional systems we need something more than energy alone to specify a quantum state completely.

Our next step will be to show how, from the degenerate states of a given energy E, we can construct eigenstates of the z component of angular momentum L_z. Then values of E and L_z, taken together, specify the eigenstates completely and constitute a complete set of quantum numbers alternative to the set n_x, n_y.

To simplify the following analysis, we describe the two-dimensional SHO states using the Dirac notation, employed in Chapter 7 to describe photon polarization states. In this notation $|n_x, n_y\rangle$ symbolizes the state with quantum numbers n_x and n_y. So, for example, the two degenerate states with $E = 2\hbar\omega$ are symbolized by the state vectors $|1, 0\rangle$ and $|0, 1\rangle$. Formally the projection of the state $|n_x, n_y\rangle$ onto all possible positions x, y yields the wave function ψ of Eq. 10-29:

$$\langle x, y | n_x, n_y \rangle \equiv \psi_{n_x, n_y}(x, y)$$

Our goal in what follows is to replace the set of states $|n_x, n_y\rangle$ with a set of states described by energy, indexed by n, and one other observable. We shall find this other observable to be the z component L_z of orbital angular momentum, indexed by the quantum number m. The resulting set of states is symbolized by the Dirac ket vector $|n, m\rangle$.

Angular Momentum and the Two-Dimensional Oscillator

The quantization of L_z is defined by the eigenvalue equation

$$(L_z)_{\text{op}} \psi = L_z \psi$$

We have seen that for a two-dimensional central field system any wave function which is an eigenfunction of L_z is of the form stated in Eq. 10-18:

$$\psi(r, \phi) = R(r)e^{im\phi} \tag{10-18}$$

The associated value of L_z is $m\hbar$ where m is an integer or zero. We have also seen that energy eigenfunctions of the two-dimensional SHO can be constructed from one-dimensional solutions indexed by quantum numbers n_x and n_y:

$$\psi_{n_x,n_y}(x, y) \sim H_{n_x}\left(\frac{x}{a}\right) \cdot H_{n_y}\left(\frac{y}{a}\right) e^{-(x^2 + y^2)/2a^2} \qquad (10\text{-}29)$$

We now show that, by superposing $|n_x, n_y\rangle$ states of the same energy (that is, for which $n_x + n_y = n = $ const), it is possible to generate energy eigenstates that are eigenfunctions of L_z. The limitation of superposed states to those of the same energy is very important! States of the same total energy E_n have a common time factor exp $(-iE_n t/\hbar)$, which factors out of each term to become an *overall* time factor for the superposition. The resulting superposition is thus *stationary* in the usual sense that the spatial probability distribution described by $|\Psi|^2$ does not change with time. If, in contrast, we should superpose states of different energy, then the time factor for each term would include a different value of E_n, the time could not be segregated in an overall time factor, and the resulting state would not be stationary.

Once we limit attention to superposition of states with equal energy, the common time factor does not enter any further calculations and so will be omitted in what follows.

Angular Momentum Eigenfunctions

In mathematical terms, our task is to find linear combinations of the wave functions of Eq. 10-29 (indexed by n_x and n_y) such that the resultant functions are of the form of Eq. 10-18 (indexed by n and m). We can make a good start by noting that the exponential factor in Eq. 10-29 is simply equal to $e^{-r^2/2a^2}$ for all of the states in question. This factor has no angular dependence, yields zero when operated on by $(L_z)_{op}$, and will therefore be included in the radial function $R(r)$ of Eq. 10-18. Thus our only real concern is with the products $H_{n_x}(x/a) \cdot H_{n_y}(y/a)$ in Eq. 10-29.

In Table 10-1 we show the first few wave functions in the n_x, n_y representation, organized in terms of the quantum number n that defines the total energy. (Normalization of the functions is unnecessary for our purpose and is ignored.) We now consider how to construct the eigenfunctions of L_z for each value of n.

$n = 0$: The only contributing state is $|0,0\rangle$ in the $|n_x, n_y\rangle$

TABLE 10-1 Some Two-Dimensional SHO Wave Functions in the (n_x, n_y) Representation (Unnormalized)

n	n_x	n_y	$\langle x, y \vert n_x, n_y \rangle$
0	0	0	$e^{-r^2/2a^2}$
1	1	0	$\dfrac{2x}{a} e^{-r^2/2a^2}$
	0	1	$\dfrac{2y}{a} e^{-r^2/2a^2}$
2	2	0	$\left(\dfrac{4x^2}{a^2} - 2\right) e^{-r^2/2a^2}$
	1	1	$\dfrac{4xy}{a^2} e^{-r^2/2a^2}$
	0	2	$\left(\dfrac{4y^2}{a^2} - 2\right) e^{-r^2/2a^2}$

representation. Since the product $H_o(x/a) \cdot H_o(y/a)$ is unity, there is no angular dependence at all. Thus $e^{im\phi} = $ const for all ψ and hence $m = 0$ and hence also $L_z = 0$. No other values of L_z are possible in this lowest energy eigenstate.

$n = 1$: Since this energy level is twofold degenerate, we need to construct two independent functions of the form of Eq. 10-18, by superposing the states $\vert 1,0 \rangle$ and $\vert 0,1 \rangle$ of the $\vert n_x, n_y \rangle$ representation. This is easily done. The relevant functions in Table 10-1 are proportional to x and y, which are related to polar coordinates by the equations

$$x = r \cos \phi \qquad y = r \sin \phi$$

Superposing x and y with the relative phase factors $+i$ and $-i$, we have

$$x \pm iy = r(\cos \phi \pm i \sin \phi) = re^{\pm i\phi}$$

These correspond to states with $m = \pm 1$ and $L_z = \pm \hbar$. This particular case is reminiscent of the *classical* two-dimensional oscillator, which executes right- or left-handed circular motion if its x and y displacements are equal in amplitude and 90° out of phase (represented here by the complex phase factors $\pm i = e^{\pm i\pi/2}$).

$n = 2$: For this, the third energy level, we need to construct three eigenfunctions in what we may call the $\vert E, L_z \rangle$ or $\vert n, m \rangle$ representation from the three possibilities for $n = 2$ shown in Table 10-1 in the $\vert n_x, n_y \rangle$ representation. Knowing that we are looking for functions of the form $e^{im\phi}$, we can take our cue from the $n = 1$ case by noting that

$$(x \pm iy)^2 = x^2 \pm 2ixy - y^2 = r^2 e^{\pm 2i\phi}$$

10-8   Two-dimensional harmonic oscillator

By consulting Table 10-1, we can obtain two such functions from the following combinations of the $|n_x, n_y\rangle$ states:

$$|\psi\rangle = |2, 0\rangle \pm 2i|1, 1\rangle - |0, 2\rangle$$

That gives us two eigenfunctions having $m = \pm 2$, $L_z = \pm 2\hbar$, respectively. The third function we need is provided by a simple addition of the eigenfunctions $\langle x, y|2, 0\rangle$ and $\langle x, y|0, 2\rangle$:

$$\langle x, y|2, 0\rangle + \langle x, y|0, 2\rangle \sim \frac{4(x^2 + y^2)}{a^2} - 4 \sim \left(\frac{r^2}{a^2} - 1\right)$$

Since this is a function of r alone, it corresponds to a state with $m = 0$, $L_z = 0$.

Table 10-2 summarizes these results for the first three energy levels.

A pattern is emerging from this analysis that can be verified for higher energy levels. For each energy level, indexed by n, there are $n + 1$ states with different eigenvalues of L_z, corresponding to the $(n + 1)$-fold degeneracy of that energy level. For a given energy quantum number n, the permitted values of m (the quantum number for the z component of angular momentum) are

$$m = n, n - 2, n - 4, \ldots, -(n - 4), -(n - 2), -n$$

This every-other-integer value of m is a somewhat misleading peculiarity of the two-dimensional SHO system. In *three-dimensional* systems with central potentials, the quantum

TABLE 10-2 Some Two-Dimensional SHO Wave Functions in the (n, m) Representation (Unnormalized)

n	m	$\langle x, y|n, m\rangle$
0	0	1
1	1	$e^{+i\phi}re^{-r^2/2a^2}$
	−1	$e^{-i\phi}re^{-r^2/2a^2}$
2	2	$e^{+2i\phi}r^2e^{-r^2/2a^2}$
	0	$\left(\dfrac{r^2}{a^2} - 1\right)e^{-r^2/2a^2}$
	−2	$e^{-2i\phi}r^2e^{-r^2/2a^2}$

Angular momentum

number m typically takes on *all* positive and negative integer values up to some maximum absolute value that depends on the energy level under consideration.

Although we have developed these properties of angular momentum states in the context of the two-dimensional harmonic oscillator, the fact is that the $|n, m\rangle$ representation, once we have it, is of wide usefulness. For any quantum-mechanical system the total energy—or a change in the total energy—is a primary feature of the interactions of the system with the outside world. Moreover, in many situations, changes in angular momentum play a role comparable in importance to the changes of energy. Indeed, energy and angular momentum are the physical quantities on which the fundamental classification scheme of the periodic table of the elements is based. Hence, of the two representations $|n_x, n_y\rangle$ and $|n, m\rangle$ for the two-dimensional SHO, the $|n, m\rangle$ representation, with its eigenvalues of energy and angular momentum, has the greater physical interest. Our next task will be to extend this description to include a characteristic quantization associated with the *total* orbital angular momentum of a three-dimensional system.

We have come a very long way in this chapter and have now in hand some extremely powerful tools for analyzing atomic systems. Starting with an atomic beam experiment that demonstrated the quantization of orientation of atomic magnetic moments in a magnetic field, we surmised the corresponding quantization of angular momentum. By looking at two-dimensional central field systems, we derived the quantization of the z component of orbital angular momentum. Applying this analysis to the two-dimensional simple harmonic oscillator, we discovered how to specify quantum states by total energy and the z component of orbital angular momentum. The result showed also that the z component of angular momentum is not only quantized but has limits on its maximum value that depend on the total energy of the system. Along the way we examined a powerful mathematical device, the commutator, for determining which observables may be simultaneously specified for the quantum states of a system.

EXERCISES

10-1 *Slow music*. Suppose that a 12-in. phonograph record has one billion (10^9) units of angular momentum $\hbar$. How many days will it take to complete one revolution? Will you live that long? (Estimate the moment of inertia $I = \frac{1}{2}MR^2$ for the disk and set angular momentum equal to $I\omega$.)

10-2   *The size of the gyromagnetic effect.* In this exercise we consider a possible experimental arrangement for observing the gyromagnetic effect, and we estimate the response of the experimental equipment (See text, Sec. 10-1).

(a) Consider a solid cylinder of density ρ, radius R, and height H. Its mass M is $\rho(\pi R^2 H)$ and its moment of inertia about its axis is $I = MR^2/2$. The cylinder is suspended by a torsion fiber which exerts a restoring torque $-K\phi$, where ϕ is the angular displacement of the cylinder (see Figure 10-2a). What is the angular frequency ω_o of torsional oscillations of the system?

(b) Each atom in the cylinder has mass A amu and net angular momentum s_a, but initially the atomic spins are randomly oriented. Suppose that a strong magnetic field is suddenly applied parallel to the axis of the cylinder, causing the atomic spins to become aligned (see Figure 10-2b). What is the consequent angular momentum L of the solid cylinder? What is the amplitude ϕ_o of the resulting torsional oscillation of the system?

(c) Your answer to part (b) should indicate that to obtain a measurable angular motion ϕ_o, K should be as small as possible. It can be shown[20] that the torsional constant K of a fiber of radius r and length l is given by $K = \pi n r^4/2l$, where n (called the *shear modulus*) is a property of the material. But the maximum safe mass that can be suspended from the same fiber is approximately $10^{-2} n(\pi r^2)$. Given these facts and an assortment of variously sized cylinders (together with the thinnest fibers that will support each of them), which cylinder should be used to maximize the amplitude ϕ_o of the angular motion? (Assume that the cylinders are geometrically similar, but that all the fibers are of the same given length.)

(d) Suppose that this experiment is to be carried out using an iron ($\rho = 7.8$ g/cm³; $A = 56$) cylinder with $R = H = 0.5$ cm and using a glass fiber of length 10 cm and of the smallest "safe" radius. If $n = 2.5 \times 10^{11}$ dyne/cm² for glass, what is the appropriate fiber radius r? What is the corresponding natural period, T? If $s_a = \hbar/2$, what is the amplitude ϕ_o of the oscillation? Express your answer both in radians and in degrees of arc. Compare with the rms amplitude ϕ_{th} due to thermal motion at 300°K. [*Note:* $\phi_{th} \approx (kT/K)^{1/2}$.]

10-3   *Magnetic moment of cesium.* Using the data from the filmed Stern-Gerlach experiment (as given in Section 10-4), employ Eq. 10-5

[20]See, for example, the volume *Vibrations and Waves* in this series, pp. 54–57.

to compute a numerical value for the magnetic moment μ_z of the cesium atom in its ground state. [*Note*: The result you will obtain is only approximate. The main reason for the discrepancy is the fact that the speeds of the atoms that form the two observed peaks are statistically distributed and do not have the single value given by $Mv^2/2 = kT$. The accurate analysis of the experiment, of course, takes account of this.]

10-4 *Stern-Gerlach analysis of a beam of charged particles?* In this exercise you will verify that the Stern-Gerlach experiment must be carried out on neutral atoms. The following comparisons show that even tiny inhomogeneities and uncertainties in the electric and magnetic forces on charged particles would overwhelm the Stern-Gerlach effect.

(a) Compare (in order of magnitude) the force $\mu_z(\partial B_z/\partial z)$ with the magnetic force evB/c on an ion of charge $\pm e$, magnetic dipole moment μ, and speed v. Use numerical values from Section 10-4. For such an ion, to what fractional precision must the magnetic field B be known in order to extract the effect of the $\mu_z(\partial B_z/\partial z)$ force with an accuracy of 1 percent? (Considering that the path of the charge through the field region is not precisely known, even this estimate is optimistic!)

(b) Compare the force $\mu_z(\partial B_z/\partial z)$ with the fluctuating electric force on one charged particle in the beam due to a nearby charge in the beam. Take this force to be $e^2/r^2 \approx e^2 n^{2/3}$ where r is the distance between a given charge and its nearest neighbor and n is the number of particles per unit volume in the beam. Find n from the following information: The beam current is 10^{12} particles/sec, the beam cross section is 10^{-2}cm^2, and the speed v of the charges is 2×10^4 cm/sec.

10-5 *Classical prediction of the beam profile in the Stern-Gerlach experiment.* Consider once again the Stern-Gerlach experiment.

(a) Find the beam intensity as a function of z in the detector plane on the assumption that each atom has a magnetic moment of given magnitude μ_o (and associated angular momentum) which can have *any* orientation with respect to the magnetic field. (Assume that the distribution of orientations is random and that each atom has the same initial speed v.)

(b) To a good approximation, the source and collimator produce a beam in which all atoms initially travel in the same direction (the $+x$ direction, say). However, the atoms have a wide range of initial speeds v. Without introducing a specific speed distribution, describe how the prediction obtained in (a) will be modified. Let $(dN/dz)\,dz$ be the rate of arrival of atoms between z and $z + dz$ in the detector plane. Give an argument to show that no matter how large the field gradient

$\partial B_z/\partial z$ is, the distribution dN/dz is symmetric with a single maximum at $z = 0$.

10-6 *The magnetic field in the Stern-Gerlach experiment.* As you may know from previous study of electromagnetism, the magnetic field obeys the law div $\mathbf{B} = 0$.

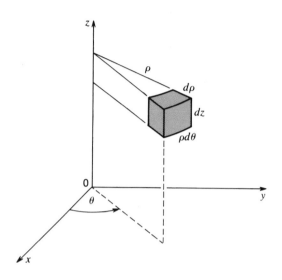

(a) The text (Sections 10-2, 10-3) implicitly assumes that in the Stern-Gerlach experiment there is a region of space throughout which $\partial B_z/\partial z \neq 0$ but that $B_x = B_y = 0$. Can this be strictly true? Explain.

(b) Apply the definition of divergence (Exercise 5-5) to show that in cylindrical coordinates (see Fig.)

$$\text{div } \mathbf{B} = \frac{1}{\rho}\frac{\partial}{\partial \rho}(\rho B_\rho) + \frac{1}{\rho}\frac{\partial B_\theta}{\partial \theta} + \frac{\partial B_z}{\partial z}$$

(c) Consider a cylindrically symmetric field $B = \hat{z}B_z(\rho,\ z) + \hat{\rho}B_\rho(\rho,\ z)$. [Note that cylindrical symmetry implies that $B_\rho(0,\ z) = 0$.] Show that for small ρ we have $B_\rho(\rho,\ z) = -\frac{1}{2}\rho(\partial/\partial z)B_z(0,\ z)$ For a field $\mathbf{B}$ with $B_z(0,\ 0) = 10^5$ gauss and $(\partial/\partial z)B_z(0,\ 0) = -10^4$ gauss/cm, estimate the radius within which $B_\rho(\rho,\ 0)/B_z(\rho,\ 0) < 0.1$.

10-7 *Magnetic moment of an iron atom.* The atomic weight of iron is 55.85, and solid iron has a density of 7.8 g/cm³. When a strong magnetic field is applied to a sample of iron, the magnetization (or magnetic moment density) reaches a maximum value of 1.83×10^3 erg/gauss-cm³.

(a) Assuming that the magnetic moments of all the atoms are parallel, find the magnetic moment of an iron atom.

Angular momentum

(b) If the alignment is *not* complete, is your answer to (a) an underestimate or an overestimate of the true value?

10-8   *The current-loop model of atomic magnetic moments*. In the text, the effect of an external magnetic field on an atom is computed using a pair of separated magnetic "poles" $\pm q_m$ to represent the magnetic moment of the atom. Although no classical picture is strictly correct, a more physical description uses a tiny loop of circulating electric charge (Section 10-4). Consider such a circular current loop located in a uniform external magnetic field.

(a) Add up the forces $d\mathbf{F}$ on the elements $d\mathbf{l}$ of the loop to show that the net force $\mathbf{F}$ on the loop is zero. [*Hints*: The magnetic force $d\mathbf{F}$ on a loop element carrying current I is $d\mathbf{F} = (I d\mathbf{l} \times \mathbf{B})/c$. If you carefully examine the expression for the net force $\mathbf{F}$, you can avoid doing the calculation explicitly.]

(b) Add up the torques $d\mathbf{N}$ on the various elements of the loop to show that the net torque on the loop is $\mathbf{N} = \boldsymbol{\mu} \times \mathbf{B}$. [*Hints*: $d\mathbf{N} = \mathbf{r} \times d\mathbf{F}$, where $\mathbf{r}$ is the radius vector from a chosen center to the loop element. When $\mathbf{F} = \Sigma \, d\mathbf{F}$ vanishes, the torque $\mathbf{N}$ is independent of the choice of center, but an obvious and convenient choice is the center of the loop.]

(c) The current loop possesses an angular momentum $\mathbf{L}$ proportional to $\boldsymbol{\mu}$. Classically, what is the effect of the torque found in (b)?

(d) (Optional) If you are ambitious, you may wish to show that if a tiny current loop is located in a *non*uniform external field, it experiences a net force $\mathbf{F} = (\boldsymbol{\mu} \cdot \boldsymbol{\nabla})\mathbf{B}$, and a torque $\mathbf{N} = \boldsymbol{\mu} \times \mathbf{B}_o$, where $\mathbf{B}_o$ is the value of the external field at the center of the loop. (You will need to use div $\mathbf{B} = 0$ and you will also need to use curl $\mathbf{B} = 0$, an equation satisfied by a time-independent external field. Retain only first-order terms in the expansion of $\mathbf{B}$.)

10-9   *Precession of the electron spin*. A beam of electrons is polarized with spin direction perpendicular to the direction of motion. The beam enters a region of uniform magnetic field whose direction is perpendicular to both the beam direction and the spin orientation. The beam emerges from the field moving in a direction 90° from its initial direction. Assuming $g = 2$ for an electron, show that in the emerging beam the electron spin is still perpendicular to the direction of motion. [*Note*: A classical analysis of the problem is sufficient since the time variation of the expectation value of the electron spin is identical to what one would calculate from the corresponding classical dynamical equation.]

10-10   *Classical magnetic moment*. The net angular momentum of any distribution of mass, referred to the center of mass, is given by

$$L = \int \rho_{mass} (\mathbf{r}) [\mathbf{r} \times \mathbf{v(r)}] \, dV$$

Let the total mass be M. Show that if such a system also has a net charge Q distributed at every point proportional to mass, that is, $\rho_{charge}(\mathbf{r}) = (Q/M)\rho_{mass}(\mathbf{r})$, then the magnetic moment of the system is given by $\boldsymbol{\mu} = (Q/2Mc)\mathbf{L}$ (cgs units). This indicates that the result given in Eq. 10-7 for an electron orbiting an infinitely massive nucleus is a special case of a quite general result.

10-11   *g factors in classical magnetic moments.* If a classical system does not have a constant charge-to-mass ratio throughout the system, the magnetic moment can be written

$$\boldsymbol{\mu} = g \frac{Q}{2Mc} \mathbf{L} \qquad \text{(cgs units)}$$

where Q is the total charge, M the mass, and $g \neq 1$.
   (a) Show that $g = 2$ for a solid cylinder of uniform mass density ($I = \frac{1}{2}MR^2$) that spins about its axis and has a uniform surface charge density on the cylindrical surface.
   (b) Find g for a solid sphere ($I = \frac{2}{5}MR^2$) that has a uniform ring of charge at the equator.
   (c) Find g for a solid sphere that has a uniform surface charge density over its whole surface.

10-12   *The operator $(L_z)_{op}$ in three dimensions.* In Section 10-6 the operator $(L_z)_{op}$ is shown to be equal to $(\hbar/i)(\partial/\partial\phi)$ for a particle confined to the xy plane. Show that the same expression is also valid in three dimensions, where ϕ is the azimuthal angle in spherical polar coordinates: $x = r \sin \theta \cos \phi, y = r \sin \theta \sin \phi, z = r \cos \theta$. [*Hint*: Start with the spherical polar coordinate expression for $(L_z)_{op}$ and transform to the basic Cartesian definition.]

10-13   *Eigenstates of L_z.* A particle moves in a certain potential in two dimensions. For a certain energy E, there are two possible independent wave functions, as follows:

$$\psi_1(x, y) = xf(r),$$

$$\psi_2(x, y) = yf(r),$$

where $r = (x^2 + y^2)^{1/2}$

In general, an eigenfunction of energy E is a linear combination of these. [*Note*: The exact form of the function $f(r)$ is not needed.]

(a) Find the particular combination(s) of ψ_1 and ψ_2 which correspond to a quantized value of the z component of the angular momentum.

(b) Find the permitted quantized value(s) of the z component of angular momentum for the state(s) you obtain in part (a).

10-14 *Properties of the energy eigenstates of the two-dimensional simple harmonic oscillator (2D SHO).* Investigate the following properties of the energy eigenstates of the two-dimensional simple harmonic oscillator.

(a) Verify that the nth energy level $E_n = (n + 1)\hbar\omega$ is $(n + 1)$-fold degenerate. (*Hint:* Use Eq. 10-28: $n = n_x + n_y$.)

(b) Show that $\psi_{n_x n_y}(-x, -y) = (-1)^{n_x + n_y} \psi_{n_x n_y}(x, y)$.

(c) The functions $\psi_{nm}(r, \phi) \equiv \langle r, \phi | n, m \rangle$ are linear combinations of the $(n + 1)$ functions $\psi_{n_x n_y}$ for which $n_x + n_y = n$. Since ψ_{nm} is an eigenfunction of $(L_z)_{op}$, we know that $\psi_{nm} = R_{nm}(r)e^{im\phi}$. Use this equation, together with the result of part (b), to show that $(m - n)$ must be an even integer. [*Hint:* If (r, ϕ) corresponds to (x, y), what corresponds to $(-x, -y)$?]

(d) Use the fact that

$$x = \frac{1}{2}r(e^{i\phi} + e^{-i\phi}) \text{ and } y = -\frac{i}{2}r(e^{i\phi} - e^{-i\phi})$$

to show that the m values associated with energy E_n are $n, n - 2, \ldots -(n - 2), -n$ as stated in Section 10-8. [*Hint:* Consider the function $H_{n_x}(x/a) \cdot H_{n_y}(y/a)$. What happens if you substitute the given expressions for x and y into this product of Hermite polynomials?]

10-15 *Counting the states of the 2D SHO.* It is often important to know how many distinct energy eigenstates a given system possesses within a given energy interval or less than a given energy. The 2D SHO provides a simple system on which to practice "counting states". How many distinct states (n_x, n_y) or (n, m) does the 2D SHO possess with $E \leq N\hbar\omega$?

10-16 *Projection amplitudes for SHO states.* Given the two normalized 2D SHO wave functions $\langle x, y | m_x, n_y \rangle$ for the second energy level $n = n_x + n_y = 1$ in the n_x, n_y representation:

$$\langle x, y | 1, 0 \rangle = \sqrt{\frac{2}{\pi}} \, x e^{-(x^2 + y^2)/2}$$

$$\langle x, y | 0, 1 \rangle = \sqrt{\frac{2}{\pi}} \, y e^{-(x^2 + y^2)/2}$$

and in the alternative n, m representation $\langle x, y | n, m \rangle$:

$$\langle x, y | 1, +1 \rangle = \frac{1}{\sqrt{\pi}} (x + iy) e^{-(x^2 + y^2)/2}$$

$$\langle x\, y | 1, -1 \rangle = \frac{1}{\sqrt{\pi}} (x - iy) e^{-(x^2 + y^2)/2}$$

(a) Construct a table of projection amplitudes between these two representations similar to Table 7-1 for photon polarization states.

(b) Take *any one* entry in the table and explain carefully, using examples, what predictive value the entry has for experiment.

(c) How many entries will the corresponding table have for the energy level $n = n_x + n_y = 3$?

10-17 *A nonisotropic two-dimensional SHO.*

(a) Write the Schrödinger equation for a particle of mass m moving in two dimensions under the influence of the potential $V(x, y) = \frac{1}{2}m\omega_x^2 x^2 + \frac{1}{2}m\omega_y^2 y^2$, where ω_x and ω_y are incommensurable.

(b) Show that the equation allows product solutions and describe these solutions. What are the energy levels of this oscillator? Are these energy levels degenerate?

(c) In the corresponding classical problem, the particle experiences a force $\mathbf{F} = -\nabla V(x, y)$. Is this force directed toward the center (origin)? What does this imply about the angular momentum $L_z = x p_y - y p_x$ of the particle?

(d) Rewrite $V(x, y)$ in terms of polar coordinates r and ϕ. Is $V(r, \phi)$ a function of r only?

(e) The energy operator is

$$(E)_{\mathrm{op}} = -\frac{\hbar^2}{2m} \nabla^2 + V$$

It can be shown that the kinetic-energy term commutes with the operator $(L_z)_{\mathrm{op}} = (\hbar/i)(\partial/\partial\phi)$. Therefore, $(L_z)_{\mathrm{op}}(E)_{\mathrm{op}} - (E)_{\mathrm{op}}(L_z)_{\mathrm{op}} = (L_z)_{\mathrm{op}}V - V(L_z)_{\mathrm{op}}$. Find the value of this commutator using $V(r, \phi)$ from part (d). Does $(L_z)_{\mathrm{op}}$ commute with $(E)_{\mathrm{op}}$? Does there exist a set of energy eigenstates which are also eigenstates of angular momentum?

(f) Contrast the system treated in this exercise with that of the *isotropic* two-dimensional simple harmonic oscillator analyzed in Section 10-8.

Perhaps to the student there is no part of elementary mathematics so repulsive as spherical trigonometry.

P. G. TAIT, *Article on Quaternions, Encyclopaedia Britannica* (1911)

11

Angular momentum of atomic systems

11-1 INTRODUCTION

In Chapter 10 we described experimental evidence that atomic angular momentum is quantized, estimated the value of atomic magnetic moments, began a formal analysis of orbital angular momentum, and applied this analysis to the orbital angular momentum of a two-dimensional simple harmonic oscillator. In the present chapter we extend the analysis of orbital angular momentum to three dimensions and introduce the spin angular momentum of the electron. With these tools we can then "return with power" to describe real atoms and molecules.

11-2 TOTAL ORBITAL ANGULAR MOMENTUM IN CENTRAL FIELDS

The Classical Equation for Central Force Motion

We now extend to three dimensions the analysis of orbital angular momentum in a central force field. We are familiar with the fact that the time-independent Schrödinger equation is closely related to the classical equation for the total energy of a particle in a given potential. Therefore it is helpful to begin by considering what classical mechanics can tell us about angular momentum in three dimensions. Now classically, if the potential is central, it is advantageous to analyze the motion at any

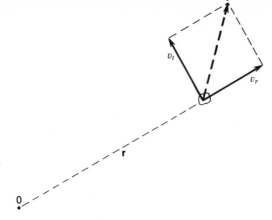

Fig. 11-1 Resolution of a velocity vector into radial and transverse components with respect to a center of attraction at O.

instant into components along the position vector **r** and transverse to this direction (Figure 11-1). If the radial and transverse velocity components are labeled v_r and v_t respectively, the classical energy equation is[1]

$$\tfrac{1}{2}Mv_r{}^2 + \tfrac{1}{2}Mv_t{}^2 + V(r) = E$$

However, we know that the total angular momentum **L** is conserved for central potentials. Its constant magnitude is

$$L = Mrv_t$$

Substituting $v_t = L/Mr$ and also $v_r = p_r/M$, where p_r is the radial component of linear momentum, we have

$$\frac{p_r{}^2}{2M} + \frac{L^2}{2Mr^2} + V(r) = E \tag{11-1}$$

The Schrödinger Equation for Central Force Motion

Now we ask: What is the quantum-mechanical equivalent of this classical equation?

Unfortunately, there is no really simple way to tackle this question. There are, however, two slightly different approaches to it that end up at the same point:

1. Instead of starting with Eq. 11-1, we can go back to a

---

[1]See, for example, the volume *Newtonian Mechanics* in this series, Chapter 13. We use capital M for mass, reserving lower case m to indicate azimuthal quantum number.

Angular momentum of atomic systems

more basic statement of the Schrödinger equation in three dimensions:

$$-\frac{\hbar^2}{2M}\nabla^2\psi + V(r)\psi = E\psi \qquad (11\text{-}2)$$

We have already considered (in Section 5-5) a special case of solutions to this equation for which ψ is spherically symmetric. We quoted there the result

$$\nabla^2\psi = \frac{1}{r}\frac{d^2}{dr^2}(r\psi) \qquad \text{(spherically symmetric } \psi \text{ only)}$$

As we noted in Chapter 5, the fact that the potential $V(r)$ is spherically symmetric does *not* make the wave function spherically symmetric any more than the spherically symmetric gravitational field of the sun, for example, makes the orbits of the comets circular; comet orbits can be highly eccentric. If we admit the possibility that the function ψ varies with the angles θ, ϕ in a spherical polar coordinate system, then we must employ the full form of the Laplacian operator ∇^2 in terms of r, θ, and ϕ. When we do this, we find that Eq. 11-2 can be written

$$\left[-\frac{\hbar^2}{2M}\frac{\partial^2}{\partial r^2} + \frac{1}{2Mr^2}\,\text{Op}(\theta,\phi) + V(r)\right](r\psi) = E\cdot(r\psi) \qquad (11\text{-}3)$$

where $\text{Op}(\theta,\phi)$ is a combination of mathematical operators involving the coordinates θ and ϕ only. Then, by comparison of Eqs. 11-1 and 11-3, we can identify $\text{Op}(\theta,\phi)$ with the scalar operator $(L^2)_{\text{op}}$. This approach is carried out in the Appendix to this chapter.

2. Alternatively, we can lean more directly on Eq. 11-1 and accept the hint it carries that the scalar quantity L^2—the square of the total orbital angular momentum—is of basic importance. Typically, a conserved quantity in the classical analysis can have a unique quantized value in the quantum analysis. In order to convert Eq. 11-1 into a Schrödinger equation that is a function of r alone, we must construct the operator $(L^2)_{\text{op}}$ and determine whether there are eigenfunctions ψ such that

$$(L^2)_{\text{op}}\,\psi = L^2\,\psi$$

11-2   Total orbital angular momentum

where L^2 is the value of the square of the orbital angular momentum. We saw in Section 10-7 that the individual components L_x, L_y, and L_z cannot have simultaneous eigenvalues. This means that the *direction* of **L** cannot be specified quantum-mechanically. It does not preclude the possibility, anticipated in Section 10-7 and verified below, that L^2 has unique eigenvalues corresponding to specific *magnitudes* of **L**.

We shall adopt procedure number 2, taking as our starting point the operators for the individual components of **L**, as given in Eqs. 10-12.

Eigenvalue Equation for L^2

We begin by restating the operators for the angular momentum components in Cartesian coordinates:

$$(L_x)_{op} = \frac{\hbar}{i}\left(y\frac{\partial}{\partial z} - z\frac{\partial}{\partial y}\right)$$

$$(L_y)_{op} = \frac{\hbar}{i}\left(z\frac{\partial}{\partial x} - x\frac{\partial}{\partial z}\right) \qquad (11\text{-}4)$$

$$(L_z)_{op} = \frac{\hbar}{i}\left(x\frac{\partial}{\partial y} - y\frac{\partial}{\partial x}\right)$$

Our plan is to express these in terms of spherical polar coordinates and then to evaluate the operator corresponding to L^2:

$$(L^2)_{op} = (L_x{}^2)_{op} + (L_y{}^2)_{op} + (L_z{}^2)_{op}$$

As in the case of $(p^2)_{op}$, we shall assume that $(L_x{}^2)_{op}\,\psi$, for example, means that $(L_x)_{op}$ is applied twice in succession:

$$(L_x{}^2)_{op}\,\psi = (L_x)_{op}\left[(L_x)_{op}\,\psi\right]$$

The important manipulations are summarized in Box 11-1. The expression for $(L^2)_{op}$ derived there is

$$(L^2)_{op} = -\hbar^2\left(\frac{\partial^2}{\partial\theta^2} + \cot\theta\,\frac{\partial}{\partial\theta} + \frac{1}{\sin^2\theta}\frac{\partial^2}{\partial\phi^2}\right)$$

Now we wish to look for eigenfunctions of this operator. Note the important fact that the radial component r appears nowhere in the expression for $(L^2)_{op}$. This suggests that we can

Angular momentum of atomic systems

separate out a radial function in the eigenfunction ψ:

$$\psi(r, \theta, \phi) = R(r)\, F(\theta, \phi) \qquad (11\text{-}8)$$

Box 11-1. Derivation of the Operator $(L^2)_{op}$.

What we are faced with here is a calculation similar to (although more complicated than) the one used to obtain the L_z operator. The relations between the two coordinate systems (Figure 11-2) are

$$x = r \sin \theta \cos \phi, \quad y = r \sin \theta \sin \phi, \quad z = r \cos \theta \qquad (11\text{-}5a)$$

or

$$r = (x^2 + y^2 + z^2)^{1/2}, \qquad \cos \theta = \frac{z}{(x^2 + y^2 + z^2)^{1/2}},$$

$$\tan \phi = \frac{y}{x} \qquad (11\text{-}5b)$$

We must employ these relationships together with the rules of implicit differentiation. Take $(L_x)_{op}$ as an example. First consider the operator $\partial/\partial z$ in that expression. We have

$$\frac{\partial}{\partial z} = \frac{\partial r}{\partial z} \frac{\partial}{\partial r} + \frac{\partial \theta}{\partial z} \frac{\partial}{\partial \theta} + \frac{\partial \phi}{\partial z} \frac{\partial}{\partial \phi}$$

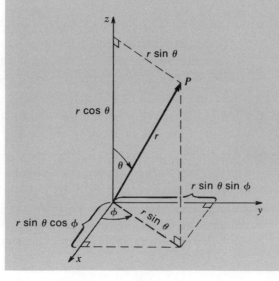

Fig. 11-2 Relation between the Cartesian coordinates (x, y, z) and spherical polar coordinates (r, θ, φ) for a point P.

11-2   Total orbital angular momentum

Now use Eqs. 11-5 to find

$$\frac{\partial r}{\partial z} = \cos\theta, \qquad \frac{\partial \theta}{\partial z} = -\frac{1}{r}\sin\theta, \qquad \frac{\partial \phi}{\partial z} = 0$$

Hence we have

$$\frac{\partial}{\partial z} = \cos\theta\frac{\partial}{\partial r} - \frac{\sin\theta}{r}\frac{\partial}{\partial \theta}$$

In similar fashion,

$$\frac{\partial}{\partial y} = \sin\theta\sin\phi\frac{\partial}{\partial r} + \frac{\cos\theta\sin\phi}{r}\frac{\partial}{\partial \theta} + \frac{\cos\phi}{r\sin\theta}\frac{\partial}{\partial \phi}$$

Substituting these, together with the relations $z = r\cos\theta$ and $y = r\sin\theta\sin\phi$, in the expression for $(L_x)_{op}$, one finds

$$(L_x)_{op} = \frac{\hbar}{i}\left(-\sin\phi\frac{\partial}{\partial \theta} - \cot\theta\cos\phi\frac{\partial}{\partial \phi}\right) \qquad (11\text{-}6a)$$

You should check the foregoing and then use similar methods to verify the following:

$$(L_y)_{op} = \frac{\hbar}{i}\left(\cos\phi\frac{\partial}{\partial \theta} - \cot\theta\sin\phi\frac{\partial}{\partial \phi}\right) \qquad (11\text{-}6b)$$

$$(L_z)_{op} = \frac{\hbar}{i}\frac{\partial}{\partial \phi} \qquad (11\text{-}6c)$$

The expression for $(L^2)_{op}$ then follows by applying each of these operators twice in succession. That is, if F is any function, we put

$$(L^2)_{op}F = (L_x)_{op}[(L_x)_{op}F] + (L_y)_{op}[(L_y)_{op}F] \\ + (L_z)_{op}[(L_z)_{op}F]$$

In evaluating this expression, care must be taken to apply each differential operator to every relevant factor to the right of it. For example, in the calculation of $(L_x{}^2)_{op}$, the second term in the parenthesis of Eq. 11-6a will involve calculation of the term shown below when it is applied twice in succession:

$$\left(-\cot\theta\cos\phi\,\frac{\partial}{\partial\phi}\right)\left(-\cot\theta\cos\phi\,\frac{\partial F}{\partial\phi}\right)$$

$$=\cot^2\theta\cos\phi\,\frac{\partial}{\partial\phi}\left(\cos\phi\,\frac{\partial F}{\partial\phi}\right)$$

$$=\cot^2\theta\cos\phi\left(-\sin\phi\,\frac{\partial F}{\partial\phi}+\cos\phi\,\frac{\partial^2 F}{\partial\phi^2}\right)$$

The complete calculation is cumbersome but not difficult. One is rewarded by a considerable simplification through cancellations and combination of different terms, and the final result is as follows:

$$(L^2)_{\text{op}}=-\hbar^2\left(\frac{\partial^2}{\partial\theta^2}+\cot\theta\,\frac{\partial}{\partial\theta}+\frac{1}{\sin^2\theta}\,\frac{\partial^2}{\partial\phi^2}\right) \qquad (11\text{-}7)$$

A similar separation of variables was used in the analysis of the two-dimensional simple harmonic oscillator (Section 10-8). This separation of variables pays off in simplicity, as we now show. The eigenvalue equation for the squared angular momentum

$$(L^2)_{\text{op}}\,\psi=L^2\,\psi$$

becomes

$$(L^2)_{\text{op}}\left[R(r)\,F(\theta,\,\phi)\right]=L^2\left[R(r)\cdot F(\theta,\,\phi)\right]$$

Now, L^2 on the right is simply a number and $(L^2)_{\text{op}}$ on the left (which involves the variables θ and ϕ only) does not act on any function of r alone. Therefore $R(r)$ simply "passes through" these operators and can be cancelled from both sides:

$$R(r)\cdot(L^2)_{\text{op}}\,F(\theta,\,\phi)=R(r)\cdot L^2 F(\theta,\,\phi)$$

or

$$(L^2)_{\text{op}}\,F(\theta,\,\phi)=L^2 F(\theta,\,\phi)$$

Then substituting the expression for $(L^2)_{\text{op}}$ from Eq. 11-7 gives us an eigenvalue equation for squared angular momentum that is a function of θ and ϕ but not r:

$$-\hbar^2\left(\frac{\partial^2}{\partial\theta^2} + \cot\theta\,\frac{\partial}{\partial\theta} + \frac{1}{\sin^2\theta}\frac{\partial^2}{\partial\phi^2}\right)F = L^2 F \qquad (11\text{-}9)$$

There is another consideration that can be used to guide our attack on this problem. The operator for L^2 commutes with the individual operators (Eqs. 11-6) for each component of $\mathbf{L}$. This is very easily verified in the case of the mathematically simple operator[2] L_z ($\sim \partial/\partial\phi$). (It is this mathematical simplicity, rather than any physical reason, that makes us choose our coordinate system so that the z axis lies along the direction of any one component of $\mathbf{L}$ we wish to analyze.) By symmetry L_x and L_y must also commute with L^2, although the use of polar coordinates gives them a complicated form that prevents this from being mathematically obvious. The fact that $(L^2)_{op}$ and $(L_z)_{op}$ commute means that we can find states that are eigenstates of both L^2 and one component of $\mathbf{L}$, which we choose to be the z component L_z. But we have already studied the eigenfunctions of $(L_z)_{op}$ (Section 10-6); they are of the form

$$\psi \sim e^{im\phi}$$

with associated quantized values of L_z equal to $m\hbar$ with m an integer. This suggests that we seek solutions of Eq. 11-9 in which the function $F(\theta, \phi)$ has the form

$$F(\theta, \phi) = P(\theta)\,e^{im\phi} \qquad (11\text{-}10)$$

so that in Eq. 11-9 we can put

$$\frac{\partial^2 F}{\partial\phi^2} = -m^2 F$$

Substituting this into Eq. 11-9 and dividing out the common factor $e^{im\phi}$ we arrive at a differential equation in θ alone:

$$(L^2)_{op}P = -\hbar^2\left[\frac{d^2P}{d\theta^2} + \frac{\cos\theta}{\sin\theta}\frac{dP}{d\theta} - \frac{m^2}{\sin^2\theta}P\right] = L^2 P$$

[2]Commutation follows from the fact that ϕ appears in $(L^2)_{op}$ only in a second partial derivative, which commutes with the first partial derivative in $(L_z)_{op}$.

Angular momentum of atomic systems

Since the eigenvalue L^2 is the square of an angular momentum, we write it as $\lambda \hbar^2$, where λ is a pure number. The preceding equation can then be written in a form ready for solution:

$$\frac{d^2 P}{d\theta^2} + \frac{\cos \theta}{\sin \theta}\frac{dP}{d\theta} + \left(\lambda - \frac{m^2}{\sin^2 \theta}\right)P = 0 \qquad (11\text{-}11)$$

The solutions of Eq. 11-11 will give us those functions $P(\theta)$ and those values of the squared angular momentum $L^2 (= \lambda \hbar^2)$ that are consistent with the eigenvalue $m\hbar$ of L_z. The product $P(\theta)e^{im\phi}$ then describes the complete angular dependence of an eigenfunction ψ that can be expressed in the form

$$\psi(r, \theta, \phi) = R(r)P(\theta)e^{im\phi} \qquad (11\text{-}12)$$

Eigenfunctions of $(L^2)_{op}$

Now if one is faced with an unfamiliar and rather complicated equation like Eq. 11-11, it is instructive to explore it via individual cases. The sines and cosines in Eq. 11-11 suggest that the solutions $P(\theta)$ themselves are various combinations of sine and cosine functions, although there is one even simpler solution: $P(\theta) = $ const. Let us indicate how one can begin discovering the form of a few of the simplest solutions.[3]

a. $P = const.$

If this is taken as a trial solution, Eq. 11-11 requires

$$\lambda - \frac{m^2}{\sin^2 \theta} = 0$$

Since this condition must be satisfied for every value of θ, the only possible choice is $\lambda = 0$ with $m = 0$.

b. $P = sin\ \theta.$

In this case we find

$$-\sin \theta + \frac{\cos^2 \theta}{\sin \theta} + \left(\lambda - \frac{m^2}{\sin^2 \theta}\right)\sin \theta = 0$$

---

[3]Some readers may recognize Eq. 11-11 as the differential equation for Legendre functions. It is discussed, for example, in F. B. Hildebrand, *Advanced Calculus for Applications*, 2d ed., Prentice-Hall, Englewood Cliffs, N.J., 1976. All trial-and-error methods of solving differential equations are "legal," since all alleged solutions are easily checked by direct substitution into the original equation.

11-2   Total orbital angular momentum

or

$$-2 \sin \theta + \frac{1}{\sin \theta} + \lambda \sin \theta - \frac{m^2}{\sin \theta} = 0$$

This requires $\lambda = 2$ and $m^2 = 1$ and so $m = \pm 1$.

c. $P = \cos \theta$.
This leads to the condition

$$-2 \cos \theta + \left(\lambda - \frac{m^2}{\sin^2 \theta} \right) \cos \theta = 0$$

which is satisfied only if $\lambda = 2$ and $m = 0$.

d. $P = \sin^2 \theta$.
In this case,

$$4 - 6 \sin^2 \theta + \lambda \sin^2 \theta - m^2 = 0$$

This requires $\lambda = 6$ and $m^2 = 4$ so $m = \pm 2$.

e. $P = \cos^2 \theta$. (?)
If you try this as a solution, you will find

$$2 - 6 \cos^2 \theta + \left(\lambda - \frac{m^2}{\sin^2 \theta} \right) \cos^2 \theta = 0$$

Putting $m = 0$ and $\lambda = 6$, all terms cancel except for the 2 at the beginning. However, by putting $P = \cos^2 \theta - \frac{1}{3}$ instead of just $\cos^2 \theta$, we introduce a constant -2 in the last term, and we then have a solution with $\lambda = 6$ and $m = 0$.

f. $P = \sin \theta \cos \theta$.
This function leads to the condition

$$-6 \sin \theta \cos \theta + \frac{\cos \theta}{\sin \theta} + \lambda \sin \theta \cos \theta - m^2 \frac{\cos \theta}{\sin \theta} = 0$$

which is satisfied for $\lambda = 6$ and $m^2 = 1$ so $m = \pm 1$.

Looking at the above few solutions, we can discern the beginnings of a pattern. Solution (a) stands by itself, with $\lambda = 0$ and $m = 0$. But solutions (b) and (c) both have $\lambda = 2$ and, taken together, give $m = 0$ and ± 1. Solution (d), together with (e) (revised) and (f), all have $\lambda = 6$, with $m = 0$, ± 1, ± 2. Table

Angular momentum of atomic systems

11-1 groups these solutions according to the values of λ (that is, according to the different values of L^2).[4]

Eigenvalues for L^2

Exploration of progressively more complicated solutions beyond those given in Table 11-1 shows that the constant λ has the following set of values:

$$\lambda = 0, 2, 6, 12, 20, 30, \ldots$$

For the first time in our study we encounter a sequence of quantum numbers that have no immediately obvious pattern of progression. However, the mathematical prescription that generates them is quite simple. Let l be a positive integer or zero. Then all the values of λ are given by the expression

$$\lambda = l(l + 1) \qquad l = 0, 1, 2, 3, \ldots$$

The number l is called the *orbital angular momentum quantum number*. The quantity of *physical* importance is λ because it measures the possible values of the square of the orbital angular momentum (in units of $\hbar^2$). However, the corresponding value of the quantum number l is most often used in *describing* the state because it takes on simple integer values.

Referring to Table 11-1, we observe that for each l there are $(2l + 1)$ values of the quantum number m ranging in integer steps from $-l$ to $+l$. Thus the simultaneous quantization of L^2

TABLE 11-1 Solutions $P(\theta)$ of Eq. 11-11 for Low Values of $L^2 = \lambda \hbar^2$

$\lambda = l(l + 1)$	$P(\theta)$	m
0	1	0
2	$\cos \theta$	0
	$\sin \theta$	± 1
6	$\cos^2 \theta - \frac{1}{3}$	0
	$\sin \theta \cos \theta$	± 1
	$\sin^2 \theta$	± 2

[4]For a more formal analytical treatment, see Hildebrand, *loc. cit.*

and L_z is expressed by the following pair of relationships:

$$L^2 = l(l + 1)\, \hbar^2 \qquad (l = 0, 1, 2, 3, \ldots)$$
$$L_z = m\hbar \qquad (|m| \leq l; \; m \text{ an integer})$$

(11-13)

Properties of the Eigenfunctions

The functions $P(\theta)$ in Table 11-1 are different for different values of l and m (actually for different values of $|m|$). For a given l, the form of $P(\theta)$ is different for different values of $|m|$, even though $P(\theta)$ is not a function of the angle ϕ most closely associated with the quantum number m. The importance of l and m in determining the form of $P(\theta)$ is indicated by using these quantum numbers as subscripts. Then the simultaneous eigenfunctions of $(L^2)_{op}$ and $(L_z)_{op}$ can be expressed in the following way:

$$F(\theta, \phi) = P_{l, m}(\theta)\, e^{im\phi} \equiv Y_{l, m}(\theta, \phi)$$

(11-14)

There is an easy way to recognize the value of l associated with any given $P_{l,m}(\theta)$: the sum of the powers of $\sin\theta$ and $\cos\theta$ in the highest term of $P_{l,m}(\theta)$ is equal to l.

The functions $P_{l,m}(\theta)$ are known as the *associated Legendre functions*. They are discussed in many mathematics and physics texts.[5] The combination of an associated Legendre function with its exponential factor, as in Eq. 11-14, when normalized, is called a *spherical harmonic* and is denoted $Y_{l,m}(\theta, \phi)$. [The normalization consists of making the integral of $|Y|^2$ over all solid angles $d\Omega$ $(d\Omega = \sin\theta\, d\theta\, d\phi)$ equal to unity.] Table 11-2 lists a few normalized spherical harmonics. We shall make use of spherical harmonics in later chapters.

The spherical harmonics have an important symmetry property that leads directly to conclusions about the *parity*—the even or odd character—of the complete eigenfunctions ψ as given by Eq. 11-8:

$$\psi(r, \theta, \phi) = R(r)F(\theta, \phi) = R(r)Y_{l,m}(\theta, \phi)$$

In one-dimensional systems the parity property (see the sub-

[5] See, for example, L. Pauling and E. B. Wilson, *Introduction to Quantum Mechanics with Applications to Chemistry*, McGraw-Hill, New York, 1935, pp. 125–136.

Angular momentum of atomic systems

TABLE 11-2 Some Normalized Spherical Harmonics $Y_{l,m}(\theta, \phi)$

l	m	$Y_{l,m}(\theta, \phi)$
0	0	$\dfrac{1}{\sqrt{4\pi}}$
1	0	$\sqrt{\dfrac{3}{4\pi}}\cos\theta$
	± 1	$\mp\sqrt{\dfrac{3}{8\pi}}\sin\theta\, e^{\pm i\phi}$
2	0	$\sqrt{\dfrac{5}{16\pi}}(3\cos^2\theta - 1)$
	± 1	$\mp\sqrt{\dfrac{15}{8\pi}}\cos\theta\sin\theta\, e^{\pm i\phi}$
	± 2	$\sqrt{\dfrac{15}{32\pi}}\sin^2\theta\, e^{\pm 2i\phi}$

Note: These are *normalized* in the sense that the integral of the squared magnitude of $Y_{l,m}(\theta, \phi)$ over all solid angles equals unity:

$$\int_{\phi=0}^{2\pi}\int_{\theta=0}^{\pi}|Y_{l,m}|^2\sin\theta\, d\theta d\phi = 1$$

section entitled "Symmetry Considerations" in Section 3-11) is that, for a symmetric potential $[V(-x) = V(x)]$ the value of $\psi(-x)$, where ψ is an energy eigenfunction, is equal to either $\psi(x)$ (even parity) or $-\psi(x)$ (odd parity). We are now dealing with a spherically symmetric potential $V(r)$, and the parity operation consists in replacing the vector **r** by the vector $-$**r**. In terms of the spherical polar coordinates, this is achieved by the following operations:

$r \to r$ (r is, by definition, a positive scalar)
$\theta \to \theta + \pi$
$\phi \to \phi + 2\pi$

Thus, under the parity operation, the factor $R(r)$ in ψ remains unchanged, so the symmetry of ψ itself is that of the angular factor alone. It happens that this angular factor, in turn, depends only on the value of l. All the spherical harmonics of

11-2   Total orbital angular momentum

different m for a given value of l are even if l is even, and are odd if l is odd. That is, we have formally

$$Y_{l,m}(\theta + \pi, \phi + 2\pi) = (-)^l Y_{l,m}(\theta, \phi)$$

and hence

$$\psi(-\mathbf{r}) = (-)^l \psi(\mathbf{r})$$

The vector model for orbital angular momentum.

The results expressed by Eqs. 11-13 suggest a simple and widely used pictorial representation of quantized angular momentum. It is possible to visualize the orbital angular momentum as a vector $\mathbf{L}$ of length $\sqrt{l(l+1)}\hbar$ whose projection on the z axis has a value $m\hbar$ between $-l\hbar$ and $+l\hbar$ (Figure 11-3). We can make no statement, however, about the azimuthal orientation ϕ of $\mathbf{L}$; $\mathbf{L}$ can, in this picture, lie anywhere along a cone whose semiangle α is given by $\cos \alpha = m/\sqrt{l(l+1)}$. This complete lack of definition of the azimuthal angle ϕ can be regarded as another example of Heisenberg's uncertainty principle: when the angular momentum about the z axis is sharply defined ($L_z = m\hbar$), then the

Fig. 11-3
(a) Semiclassical pictorial representation of space quantization for the case $l = 2$. The angular momentum vector $\mathbf{L}$ takes on only those orientations for which the z component L_z has values given by integer multiples of $\hbar$. (b) Refinement of pictorial representation for the case $l = 2$ and $m = +2$. For a unique value of L_z the azimuthal angle ϕ is completely undetermined.

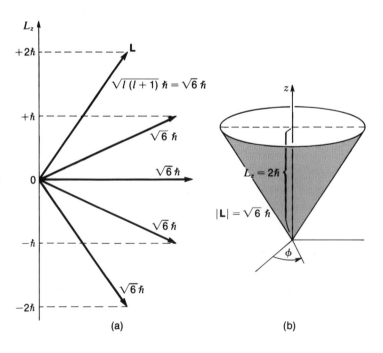

(a) (b)

486 Angular momentum of atomic systems

angular position around z is completely undefined. [Recalling Section 10-7, you can relate this uncertainty relation to the fact that $(L_z)_{op}$ does not commute with $(L_x)_{op}$ or $(L_y)_{op}$. More simply, note that the angular position operator ϕ does not commute with the angular momentum operation $(\hbar/i)(\partial/\partial\phi)$.]

This pictorial representation of the angular momentum vector is sometimes called *space quantization*, although it is quantization not of space but of the "orientation" of **L** *in* space. Like all classical representations of quantum-mechanical results, the visualization can be misleading and must be used with caution. For example, it may lead to experimentally meaningless questions such as, "In what direction is the angular momentum vector *really* pointing at such-and-such an instant?" However, if the limitations of the model are kept in mind, it is a helpful aid that is used by most physicists.

The description we have developed here in terms of angular wave functions is far removed from the classical picture of an orbiting particle with which we began this analysis. In the wave function we have the basis of a probability distribution for a large number of identical systems in a given angular momentum state. The description of a localized particle in orbit requires a superposition of eigenfunctions analogous to the packet state that describes motion in one dimension.

11-3 ROTATIONAL STATES OF MOLECULES

Before we can apply our knowledge of quantized angular momentum to the structure of individual atoms (Chapters 12 and 13), we will need to take account of the angular momentum due to the spin of the electron itself, as described in Section 11-4 below. However, using only our present knowledge of orbital angular momentum we can make a simple and powerful analysis of the rotation of slightly larger systems: molecules. To a good approximation, discussed more fully below, molecules can be treated as *rigid rotators* that revolve without changing the separations or relative positions of their constituent atoms. In the simplest case, that of a diatomic molecule, one has the classical picture of the two atoms of the molecule revolving with fixed separation at opposite ends of a line that passes through their common center of mass (Figure 11-4). In what follows we recall a previous treatment of molecular *vibrations* in Chapter 4 and proceed to show that, to a good ap-

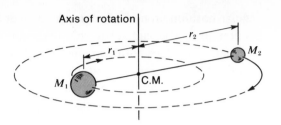

proximation, vibrational and rotational motion can be treated separately. Real molecules at not-too-low temperatures rotate and vibrate at the same time.

Rotation of a Rigid Rotator about a Fixed Center

We need not, for our present purposes, be concerned with the exact details of the forces between the two atoms in a diatomic molecule. Indeed, the nature of the forces will be different in a polar molecule (that is, a molecule with an electric dipole moment) such as HCl, and a covalent molecule such as N_2. The only essential feature, in dealing with rotational states, is the fact that the forces are such that the nuclei of the two atoms are maintained at a more or less well-defined equilibrium separation, r_0. For any smaller separation a strong repulsion develops, and for separations greater than r_0 the force becomes attractive (but decreasing in magnitude as r increases). The resulting potential-energy curve is of the form shown in Figure 11-5.

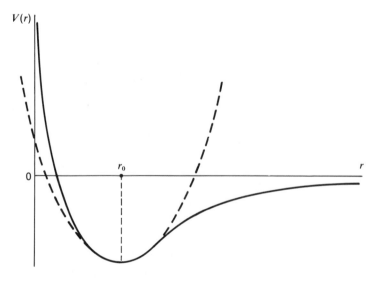

Fig. 11-5 Schematic diagram of the potential energy curve for a diatomic molecule (solid curve). Near the equilibrium separation r_0 the potential energy curve may be approximated by a parabola (dashed curve), which is the potential for a simple harmonic oscillator.

Angular momentum of atomic systems

Around the equilibrium point, $r = r_0$, we may approximate the potential by a parabola (dashed curve in Figure 11-5):

$$V(r) \approx \tfrac{1}{2}M\omega_o^2(r - r_0)^2 = \tfrac{1}{2}M\omega_o^2x^2 \qquad (x = r - r_0 \ll r_0)$$

We can then find the energy states for vibrational (radial) motion of the system by using our knowledge of the one-dimensional simple harmonic oscillator. We have already discussed this problem in Chapter 4; the present question is how the situation is modified when we take into account the rotation as well. The discussion is greatly simplified by the fact that, for all diatomic molecules, the energy levels associated with rotations of the molecule are closely spaced when compared with those associated with vibrations; this is known from observational evidence (mainly spectroscopic). Moreover, the increase of potential energy with displacement in the neighborhood of the equilibrium separation r_0 is so great that, in any given vibrational state, the amplitude of vibrational motion is only a small fraction of r_0 itself. Or, speaking quantum mechanically, the probability function for the radial separation of the two nuclei is sharply peaked in the neighborhood of $r = r_0$. This is the correct physical interpretation of the classical term *rigid*. (In the HCl molecule, for example, the mean amplitude of vibration is about 6 percent of the mean internuclear distance.) These features allow us, with high accuracy, to treat the rotational motions separately from the vibrational. Hence we consider a spectrum of rotational states, all of which are associated with a single vibrational state. This model is called the *rigid rotator*.

If one of the atoms in a diatomic molecule is very much more massive than the other, we can to some approximation regard the more massive atom as constituting a fixed center about which the other atom moves. A good example of this is the hydrogen chloride (HCl) molecule, in which the chlorine atom has 35 times the mass of the hydrogen (or 37 times for the second, less abundant chlorine isotope). We then treat the entire hydrogen atom as a satellite particle in a central potential. If the distance between molecules is fixed, all the energy *changes* are rotational kinetic energy changes, and *the quantization of total angular momentum gives rise to discrete rotational energy levels* (Figure 11-6). If angular momentum were not quantized, we would observe a continuous, not discrete, rotational energy spectrum for HCl and other diatomic molecules.

11-3   Rotational states of molecules

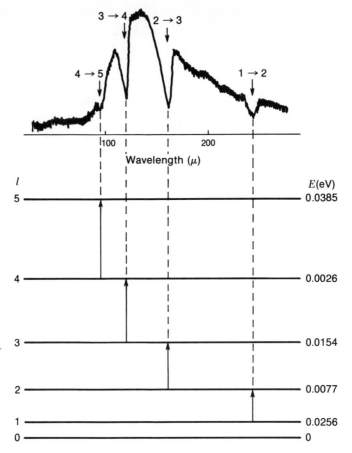

Fig. 11-6   Rotational energy spectrum of HCl (top) by absorption and the transitions that account for it (bottom). [Spectrum from D. Bloor et al., Proc. Roy. Soc. A **260**, 510 (1961) Reprinted with permission of the Royal Society.]

The analytic tools developed for angular momentum make short work of the rigid rotator problem. The *classical* equation for the rotational kinetic energy is

$$E = \frac{1}{2} M v_t^2 = \frac{L^2}{2Mr_o^2}$$

where v_t is the transverse velocity (with $v_r = 0$ for circular motion) and $L = M v_t r_o$. This can be converted to the equivalent Schrödinger operator equation

$$\frac{(L^2)_{op}}{2Mr_o^2} \psi = E_{rot} \psi$$

or

$$(L^2)_{op} \psi = 2Mr_o^2 E_{rot} \psi \qquad (11\text{-}15)$$

Angular momentum of atomic systems

Inspection shows this equation to be an eigenvalue equation for $(L^2)_{op}$. Solutions to this equation are the spherical harmonics $Y_{l,m}(\theta, \phi)$ and the eigenvalues of $(L^2)_{op}$ are $l(l+1)\hbar^2$:

$$(L^2)_{op} Y_{l,m} = l(l+1)\hbar^2 Y_{l,m} = 2Mr_0^2 E_{rot} Y_{l,m}$$

Dividing through by $Y_{l,m}$, we obtain

$$2Mr_0^2 E_{rot} = l(l+1)\hbar^2$$

or

$$E_{rot} = \frac{l(l+1)\hbar^2}{2Mr_0^2} \tag{11-16}$$

Equation 11-16 is a marvelously compact statement that will be easily generalized below to a much larger class of molecules.

Equation 11-16 tells us that the rotational levels have energies proportional to the numbers $0, 2, 6, 12, 20, \ldots$, and the spacing between adjacent levels (now using the subscript l) is given by

$$\Delta E_l = E_l - E_{l-1} = \frac{l(l+1)\hbar^2}{2Mr_0^2} - \frac{(l-1)l\hbar^2}{2Mr_0^2} = \frac{l\hbar^2}{Mr_0^2} \tag{11-17}$$

This systematic increase of rotational energy level spacing with l is to be contrasted with the theoretically equal spacing (Eq. 4-22) of the *vibrational* levels (to the approximation that these are modeled by a one-dimensional harmonic oscillator).

Equation 11-16 predicts the energies of the rotational states with no adjustable parameters if one assumes that the distance r_0 is known. Actually, in research, the result is usually applied the other way round, by using spectroscopic data on the rotational level spacings to infer the value of r_0 (the distance between the nuclei in the diatomic molecule). We may make use of the knowledge that all such distances are of the order of 1 Å (about twice an atomic radius) and use Eq. 11-17 to obtain an order-of-magnitude estimate of the level spacings. For the HCl molecule, putting $r_0 = 1$ Å and $M = 1.67 \times 10^{-24}$ g (the mass of the proton) we find

$$E_l - E_{l-1} \approx (6 \times 10^{-15} \text{ erg})l \approx (4 \times 10^{-3} \text{ eV})l$$

11-3   Rotational states of molecules

Figure 11-6 shows the results of some actual measurements on the rotational levels of the HCl molecule. The observed transition energies are of the order of 10^{-2} eV, in agreement with the rough estimate above. As is most usual in molecular spectroscopy, the experiment was performed by passing "white" (continuous) infrared radiation through a gaseous sample and finding the wavelengths at which the radiation was strongly absorbed. When pure rotational states are involved, this corresponds to quantum jumps in which the molecule is raised from each of a set of rotational states to the next higher state, with quantum energies given by the energy spacings of Eq. 11-17. The incident radiation in this case had a mean wavelength of about 150 micron ($= 0.15$ mm), corresponding to a frequency of about 2×10^{12} Hz. This is in the region of very long infrared wavelengths or extremely short microwaves. The source of the primary continuous spectrum in this experiment was a white light source (high-pressure mercury lamp) drastically filtered to pass radiation only in the region of wavelengths of interest.

Figure 11-6 relates the observed absorption wavelengths to transitions between the rotational levels of the HCl molecule, and in Table 11-3 the data are analyzed in such a way as to show how the quantum jumps between the successive levels ($l - 1$ to l) have a frequency (or energy) proportional to l, just as Eq. 11-17 requires.

The energy separation of rotational energy levels is of the same order of magnitude as the (magnetic) fine-structure splitting of electronic states in atoms as discussed at the beginning of Chapter 10. The spacing of *vibrational* levels of diatomic molecules is typically of the order of 50 times greater than this,

TABLE 11-3 Rotational Transitions in HCl (Absorption Spectrum)

Transition $l-1 \rightarrow l$	λ (microns)	$\nu = c/\lambda$ (10^9 Hz)	ν/l (10^9 Hz)	λl (cm)	$h\nu$ (eV)
$(0 \rightarrow 1)$[a]	(479)	(626)	(626)	(0.0479)	(0.0026)
$1 \rightarrow 2$	243	1235	618	0.0486	0.0051
$2 \rightarrow 3$	162	1852	617	0.0486	0.0077
$3 \rightarrow 4$	121	2479	620	0.0484	0.0103
$4 \rightarrow 5$	96	3125	625	0.0480	0.0129

[a]This transition not shown in Figure 11-6.

although still far smaller than the energy change corresponding to emission of a photon in the visible spectrum (resulting from a quantum jump by an electron between major energy levels in an atom).

Using the quantitative data on rotational energy levels, we can deduce a better value of the internuclear distance r_0 in HCl. For the transition $l - 1$ to l we have, by Eq. 11-17,

$$\Delta E_l = h\nu = \frac{l\hbar^2}{Mr_0^2}$$

Therefore,

$$r_0^2 = \frac{\hbar^2}{hcM}\lambda l = \frac{\hbar}{2\pi cM}\lambda l \qquad (\lambda = \text{wavelength} = c/\nu)$$

Putting in the values of the fundamental constants, this gives

$$r_0^2 = (3.36 \times 10^{-15}\lambda l) \text{ cm}^2$$

But from Table 11-3, we have

$$\lambda l = 4.83 \times 10^{-2}$$

Hence

$$r_0^2 = 1.62 \times 10^{-16} \text{ cm}^2$$
$$r_0 = 1.27 \text{ Å}$$

Thus the precise results of experiment give a value for r_0 of the expected order of magnitude. As pointed out above, experiments such as these are of great use in the study of molecular structure and dimensions.

Rotation of a Diatomic Molecule

The main shortcoming of the preceding analysis is its limitation to those diatomic molecules in which one of the atoms is so massive that it can be regarded as stationary. This is never strictly justifiable, and in any case we need to be able to apply the theory to molecules in which the atoms are of comparable or even equal mass. In short, we need to treat the motions of the atoms in a molecule relative to their center of mass.[6]

We treat the center of mass as a fixed origin for the purpose of analyzing internal motions such as rotation and vibration. In fact, we have already considered two-body vibrational motion in these terms in Chapter 4. Consider, then, a diatomic molecule composed of atoms of masses M_1, M_2 at distances r_1, r_2, respectively, from the center of mass (Figure 11-4). Treating this molecule as a rigid rotator we then have

$$r_0 = r_1 + r_2$$
$$L = M_1 v_{1t} r_1 + M_2 v_{2t} r_2$$
$$E = \tfrac{1}{2} M_1 v_{1t}{}^2 + \tfrac{1}{2} M_2 v_{2t}{}^2$$

where v_{1t}, v_{2t} are the (transverse) velocities of the individual masses with respect to the center of mass. But by the definition of the center of mass,

$$M_1 r_1 = M_2 r_2$$

and

$$M_1 v_{1t} = M_2 v_{2t}$$

Using these equations we deduce the following result (which you should verify for yourself):

$$E = \frac{M_1 + M_2}{2 M_1 M_2} L^2 = \frac{L^2}{2 \mu r_0{}^2}$$

where $\mu = M_1 M_2 / (M_1 + M_2)$ is the *reduced mass* of the two-particle system (see Section 4-4). We have thus obtained an equation identical in form to the single-particle equations of the preceding section. It can quite easily be shown (see the exercises) that the separation of internal and overall motions,

---

[6]More fundamentally, we are concerned with a two-body system whose motion, just as in classical mechanics, should be separable into (a) motion of the two parts relative to the center of mass, and (b) translational motion of the center of mass. In the absence of a net external force, the center of mass moves with uniform velocity, so we may analyze the internal motion from an inertial frame with respect to which the center of mass is stationary. See, for example, the book *Newtonian Mechanics* in this series, p. 629.

Angular momentum of atomic systems

leading to the above description of the internal motions, is completely valid in the quantum-mechanical analysis.

Imposing the quantization condition on L^2, we have

$$E = \frac{l(l+1)\hbar^2}{2\mu r_0^2} \tag{11-18}$$

In this form the result can be applied to the rotational energy-level system of *any diatomic molecule*, regardless of the relative masses of the two atoms.

The expression for the moment of inertia I of the diatomic rotator about its center of mass appears in the denominator on the right side of Eq. 11-18:

$$I = M_1 r_1^2 + M_2 r_2^2 = \mu r_0^2$$

Thus in place of Eq. 11-18, we can put

$$E = \frac{l(l+1)\hbar^2}{2I} \tag{11-19}$$

By using the appropriate generalized form of the moment of inertia, Eq. 11-19 can be applied to *any rigid molecule whatever*.

The Rotation-Vibration Spectrum

Actually, the data on HCl presented in Figure 11-6 and Table 11-3 are not the kind most easily obtained experimentally. Pure rotational transitions for diatomic molecules are difficult to observe because the radiation associated with such quantum jumps falls in an awkward range of the electromagnetic spectrum—the frequencies are too high to be easily accessible to microwave techniques and yet are too low for optical (infrared) spectroscopy. As a result, most of our knowledge of rotational levels comes from observations of transitions involving a change of vibrational level as well as rotational. As we have said, vibrational levels are relatively widely spaced. (It was this very fact that permitted meaningful separation of rotational motion from vibrations.) Transitions between adjacent vibrational states correspond to photon energies that fall in the part of the infrared region which is spectroscopically accessible. When rotational transitions are superposed on

11-3   Rotational states of molecules

these, the result is a combined *rotation-vibration* spectrum. The energies of the states involved in these combined transitions are given, to first approximation, by the simple sum of the vibrational energy and the rotational energy; that is,

$$E_{n,J} \approx (n + \tfrac{1}{2})\hbar\omega_o + \frac{J(J+1)\hbar^2}{2\mu r_o^2} \tag{11-20}$$

For the rotational quantum number, we have here replaced l by the symbol J, this being the customary notation in molecular spectroscopy.

The most commonly observed rotation-vibration spectra arise from transitions between the vibrational states $n = 0$ and $n = 1$. Each observed absorption line represents a transition between one rotational level associated with the $n = 0$ state and another rotational level associated with the $n = 1$ state. In practice, there are, however, so-called *selection rules* (discussed further in Chapter 14) which limit the possible transitions in this case to those for which J changes by ± 1. Hence the observed spectrum is made up of lines at the following frequencies (see Figure 11-7):

$$\nu_{J,J-1} = \nu_0 \pm \frac{h}{4\pi^2 \mu r_o^2} J \quad (J = 1, 2, 3, \ldots) \tag{11-21}$$

where $\nu_0 \ (= \omega_0/2\pi)$ is the frequency that would characterize a pure vibrational transition. A complete rotation-vibration frequency spectrum is thus a set of equally spaced lines, centered at ν_0 (although there is no line at ν_0 itself). In the laboratory these spectra are almost always studied by the absorption methods already described in connection with the pure rotation spectrum of the HCl molecule. Figure 11-8 shows a nice example of results so obtained.

Note that, by Eq. 11-21, the *spacing* between adjacent lines in the rotation-vibration spectrum is given by

$$\Delta\nu = \nu_{J+1,J} - \nu_{J,J-1}| = \frac{h}{4\pi^2 \mu r_o^2} \tag{11-22}$$

Table 11-4 presents some data on the values of $\Delta\nu$ obtained from studies of rotation-vibration spectra of several molecules, together with corresponding values of the internuclear distance r_o as inferred from Eq. 11-22.

Angular momentum of atomic systems

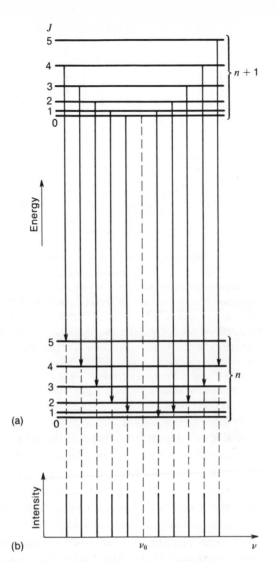

Fig. 11-7
(a) Transitions
in the rotation-
vibration spectrum
of a diatomic mole-
cule. (b) Schemat-
ic diagram of the
resulting spectrum
(see Figure 11-8).

Corrections to the Rotation-Vibration Spectrum

The simple account above ignores various features that have to be considered in the precise theory and analysis of rotation-vibration states. The most important of these is the fact that the potential energy curve which governs the vibrational motion (Figure 11-5) is not the perfectly symmetrical parabolic potential of a harmonic oscillator but rises from the equilibrium position more steeply for decreased separation of the atoms than for increased separation. Because of this, the vibra-

11-3   Rotational states of molecules

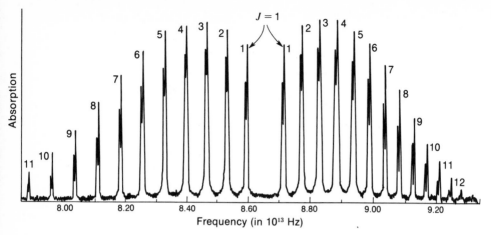

Fig. 11-8 *Rotation-vibration spectrum for HCl. This is an absorption spectrum that measures the radiation after it has passed through the gas rather than the emission spectrum of radiation given off by the gas (Figure 11-7b). Note that each peak is double, because of the presence of two different types of HCl molecule—one type containing the more abundant isotope of chlorine (Cl^{35}) and the other the less abundant isotope (Cl^{37}). [Data of T. Faulkner and T. Nestrick, as reproduced in P. A. Tipler,* Foundations of Modern Physics, *Worth Publishers, Inc., New York, 1969, p. 375.*]

tional levels depart from those of the simple harmonic oscillator, and the average distance between the two nuclei (which we took to be equal to the equilibrium distance r_0) varies with the vibrational state. Specifically, the effective separation, which we now write as r_n to indicate its dependence on the vibrational state, increases as n increases. As a result, the rotational energies are affected, since we must replace r_0 by r_n in Eqs. 11-16 and 11-21. Thus, when calculating the energies of

TABLE 11-4 Internuclear Distances from Rotation-Vibration Spectra

Molecule	Reduced Mass (amu)	$\Delta\nu\,(=\nu_{0\to1})\,Hz$	$r_0\,(\text{Å})$
CO	6.85	1.16×10^{11}	1.13
NO	7.47	1.02×10^{11}	1.15
HF	0.95	12.6×10^{11}	0.92
HCl	0.97	6.5×10^{11}	1.27
NaCl	13.9	0.13×10^{11}	2.36

Angular momentum of atomic systems

TABLE 11-5 Centrifugal Stretching in OCS Molecule (Absorption Spectrum)

Transition $(J \rightarrow J + 1)$	ν (in 10^{10} Hz)	I (in 10^{-38} g-cm^2)
$1 \rightarrow 2$	2.432592	1.379450
$2 \rightarrow 3$	3.648882	1.379454
$3 \rightarrow 4$	4.865164	1.379459
$4 \rightarrow 5$	6.081408	1.379468

Data courtesy of M. W. P. Strandberg and M. Lipsett.

quantum jumps in a rotation-vibration spectrum, one has to use different values of r_n for different n in Eq. 11-21.

We shall conclude this section by mentioning one subtle but very beautiful feature of pure rotational states. In a classical system of two masses connected by a spring, the spring stretches as the rotation rate is increased. A diatomic molecule, although it is not amenable to being described classically, is a completely analogous system. The "spring" (the interatomic force in the molecule), is exceedingly rigid, but not infinitely so, and very sensitive measurements of pure rotational spectra do reveal a tiny centrifugal stretching. That is, for a given vibrational state (n) the interatomic separation r increases with the rotational quantum number J. As one might expect, such stretching is not limited to diatomic molecules, and some excellent data on it have been obtained with the linear triatomic molecule OCS (carbonyl sulfide). The OCS molecule is linear in geometry, so it is dynamically similar to a diatomic molecule. Table 11-5 shows how the moment of inertia I of this molecule depends on the rotational transition from which its value is inferred.[7] It is rather pleasing to find, in the midst of all this language of eigenstates and wave functions, that such specific and picturesque concepts as centrifugal force still have their place.

You now have the basic equipment necessary to analyze rotational energy states of more complicated molecules. Comparison with experiment gives results rich in information about these physical systems. We break off the analysis at this point, however, in order to return to the study of the atom. Having treated orbital angular momentum of the electron in the atom,

[7]Since each transition in Table 11-5 involves the values of I for two different rotational states (initial and final) the value listed in each case is only a kind of average between the two states. Nevertheless, the general trend of I with J is correctly indicated.

11-3   Rotational states of molecules

we now look at the part played by electron spin in the total angular momentum of this system.

11-4 SPIN ANGULAR MOMENTUM

Orbital Angular Momentum Alone Insufficient to Describe Atoms

The analysis of orbital angular momentum is a powerful tool for describing molecular rotation. However, orbital angular momentum alone does not provide a complete description of angular momentum in *atoms*. In particular it cannot account for the most obvious feature of the Stern-Gerlach experiment with cesium: splitting into *two* beams in the presence of a transverse magnetic field gradient.

We have seen that for any given value of the orbital angular momentum quantum number l there are $(2l + 1)$ compatible eigenvalues of L_z—that is, $(2l + 1)$ orientations of $\mathbf{L}$ with respect to a given direction. Since l is integral (Section 11-2) this number of different orientations is always *odd*: $1, 3, 5, 7, \ldots$. But in the Stern-Gerlach experiment definite evidence was obtained that a beam of cesium atoms (or of silver atoms in Stern and Gerlach's original experiment) is split into only *two* components by the action of an inhomogeneous magnetic field. That implies only two possible orientations of the magnetic dipoles with respect to the z axis, and hence only two distinct quantized values of the z component of angular momentum. This is inconsistent with the prediction of an odd number of orientations arising from the integer values of l for orbital angular momentum alone.

Electron Spin

To account for situations in which even numbers of z states appear, the bold suggestion was made that the z component of angular momentum can exist also in *half*-integer multiples of $\hbar$ (as mentioned in Section 10-4). Specifically, it was postulated in 1925 by Uhlenbeck and Goudsmit[8] (when both were graduate students) that the intrinsic *spin* of an electron is characterized by a spin quantum number s equal to $\frac{1}{2}$. Uhlenbeck and Goudsmit recognized that, with the single assump-

---

[8]G. E. Uhlenbeck and S. Goudsmit, Naturwiss. **13**, 953 (1925); Nature **117**, 264 (1926).

Angular momentum of atomic systems

tion $s = \frac{1}{2}$, electron spin could be described by analogy to orbital angular momentum. Then the value of $2s + 1$, defining the number of possible projections along some z axis, is equal to 2. Moreover, in analogy to $L^2 = l(l + 1)\hbar^2$ and $L_z = m_l\hbar$, we write[9]

$$S^2 = s(s + 1)\hbar^2 = \tfrac{1}{2}(\tfrac{1}{2} + 1) = \tfrac{3}{4}\hbar^2$$
$$S_z = m_s\hbar = \pm\tfrac{1}{2}\hbar$$

(11-23)

In contrast to multiple values of the angular momentum quantum number l, the spin quantum number s for an individual electron has the single value $s = \frac{1}{2}$ because this measures an intrinsic property of the electron. In consequence, the z component quantum number m_s takes on only two values: $m_s = \pm\frac{1}{2}$.

Electron Spin and the Stern-Gerlach Experiment

The results of the Stern-Gerlach experiment can now be accounted for if we assume that, in atoms of silver and cesium, there is *zero* net *orbital* angular momentum, and that the observed magnetic moment and angular momentum are due solely to one electron spin. Both silver ($Z = 47$) and cesium ($Z = 55$) have odd values of atomic number and hence odd numbers of electrons. The angular momentum properties of these atoms are nicely accounted for on the following assumption: In each of them the electrons pair off as far as possible with equal and opposite orbital and spin angular momenta, and the last unpaired electron (the valence electron) is in an orbital angular momentum state with $l = 0$. In this case the atom as a whole has the net angular momentum and magnetic moment associated with the spin of a single electron.[10] We shall consider states of many-electron atoms more fully in Chapter 13.

Spin Magnetic Moment of the Electron

The Stern-Gerlach experiment and many other lines of evidence show that the measured magnetic moment associated

---

[9]From this point on we use the subscripts in m_l and m_s to distinguish between these quantum numbers for orbital and for spin angular momentum, respectively.

[10]We are ignoring here the contribution due to the nucleus. This may supply further angular momentum equal to a multiple of $\hbar/2$, but scarcely affects the magnetic properties, since (as we noted in Section 10-4) nuclear magnetic moments are so small.

11-4   Spin angular momentum

with an electron spin is almost exactly one Bohr magneton. We demonstrated in Section 10-4 that the Bohr magneton ($\mu_B = e\hbar/2M_ec$) is the amount of magnetic moment that we can associate semiclassically with an amount $\hbar$ of orbital angular momentum. We concluded that the ratio of magnetic moment to angular momentum (the gyromagnetic ratio) is about twice as large for the electron spin as for orbital magnetism. We are now in a position to discuss these results more explicitly. An electron with a z component of orbital angular momentum characterized by the quantum number m_l has a corresponding z component of magnetic moment given by

$$\mu_z = -m_l\mu_B \quad \text{(orbital)} \tag{11-24}$$

(The minus sign comes from the fact that the electron charge is negative.) But for the z component of magnetic moment due to spin, we have[11]

$$\mu_z = -g_s m_s \mu_B \quad \text{(spin)} \tag{11-25}$$

where the "g factor" g_s is very close to 2.

11-5 SPIN-ORBIT COUPLING ENERGY

As we have discussed in the previous section, the splitting of a cesium beam into two components by a transverse magnetic field gradient (Section 10-3) tells us that the total angular momentum of a cesium atom in the ground state is that of its valence electron, with spin $\frac{1}{2}$ and zero orbital angular momentum. The orbital and spin angular momenta of all the other electrons (an even number) pair off and so their magnetic moments cancel.

When orbital angular momentum and spin angular mo-

[11]Electron spin, presented as a useful postulate by Uhlenbeck and Goudsmit in 1925, was described theoretically by Dirac in 1928 [P. A. M. Dirac, Proc. Roy. Soc. **A 117**, 610 (1928)]. For a readable modern account and references, see Chalmers W. Sherwin, *Introduction to Quantum Mechanics*, Henry Holt and Co., New York, 1959, Chap. 11. The original Dirac theory predicted the value of the g factor g_s to be exactly 2. More recent experimental and theoretical developments give the value $g_s = 2.00231911 \pm 0.00000006$. [See D. T. Wilkinson and H. R. Crane, Phys. Rev. **130**, 852 (1963) and A. Rich, Phys. Rev. Letters **20**, 967 (1968); **20**, 1221 (1968).]

Angular momentum of atomic systems

mentum both exist in the same atom, the magnetic moments that result from these two angular momenta interact to cause a splitting in the corresponding energy level. Roughly speaking, the energy of the system is slightly higher when the orbital and spin magnetic moments are "parallel" than when they are "antiparallel." Since the interaction that leads to this energy difference is between the spin and the orbital electronic structure, this interaction is called *spin-orbit coupling* or *L-S coupling*.

A simple semiclassical calculation of the order of magnitude of the energy of the spin-orbit coupling for a single electron can be made using the Bohr model. Suppose, to be specific, that an electron is orbiting a proton with one unit of angular momentum. (Recall that in the Bohr theory this is the smallest possible orbital angular momentum.) We then have

$$M_e v r = \hbar$$

Now, if we imagine ourselves sitting on the electron, we see the *proton* describing an orbit of radius r with speed v. This represents a circulating current I equal to $ev/2\pi r$ which generates a magnetic field at the position of the electron. The strength of the magnetic field given by this classical calculation is

$$B = \frac{2\pi I}{cr} = \frac{ev}{cr^2} \text{ (cgs)} \qquad \left[\frac{\mu_0 I}{2r} = \frac{\mu_0}{4\pi} \frac{ev}{r^2} \text{ (SI)} \right]$$

Substituting for v the Bohr value $\hbar/M_e r$ (Section 1-7), we have

$$B = \frac{e\hbar}{M_e c r^3} = \frac{2\mu_B}{r^3} \text{ (cgs)} \qquad \left[\frac{\mu_0}{4\pi} \frac{e\hbar}{M_e r^3} = \frac{\mu_0}{2\pi} \frac{\mu_B}{r^3} \text{ (SI)} \right] \qquad (11\text{-}26)$$

We know that r is of the order of 1 Å, and if we use this value, together with μ_B as given in Eq. 10-8, we find

$$B \approx 2 \times 10^4 \text{ gauss} = 2 \text{ tesla}$$

Now the electron spin magnetic moment will have an energy which depends on its orientation in this rather strong magnetic field. This spin-orbit coupling energy can be expressed quite simply in terms of the potential energy of the electron spin magnetic moment in the field B due to the orbital momentum.

11-5   Spin-orbit coupling energy

This potential energy is

$$\Delta E = \pm\mu_B B$$

(The $+$ and $-$ signs refer to the two possible orientations of the spin "parallel" or "antiparallel" to **B**.) With $B \approx 2 \times 10^4$ gauss, as above, this gives

$$\Delta E \approx \pm 2 \times 10^{-16} \, \text{erg} \approx 10^{-4} \, \text{eV} \qquad (11\text{-}27)$$

This energy difference is very much smaller than the energy difference (a few electron-volts) separating the excited states from ground states in atoms. Spin-orbit coupling is an important contribution to what is called *fine structure* in atomic spectra.

To compare the prediction of Eq. 11-27 with experiment, we need an experimental measure of the fine-structure energy separation between a pair of one-electron states, both having unit orbital angular momentum but opposite orientations of the spin magnetic moment. The simplest case should be hydrogen. Here we run into an interesting difficulty in comparing the results of a Bohr-theory analysis with reality. According to the Bohr theory, the lowest energy state of hydrogen has unit orbital angular momentum. In Section 11-2 we found that the orbital angular momentum in a central field can take on the value zero and in the following chapter we shall see that the lowest energy state for hydrogen indeed has zero orbital angular momentum. Thus there can be no fine-structure energy difference for the lowest state of hydrogen, since the spin magnetic moment of the electron has nothing to interact with. The lowest energy state that can have orbital angular momentum is the first excited state. The fine structure separation for the first excited state of hydrogen is 0.905×10^{-4} eV. This is the same order of magnitude as the estimate made above. (In actuality, the fine structure of hydrogen also includes contributions from at least two relativistic effects comparable in magnitude to the *L-S* coupling.)

Another example of spin-orbit coupling is the famous pair of *D* lines of the sodium spectrum (Section 10-1 and Figure 10-1). In sodium a single valence electron lies largely outside a set of completed shells (more on this in Chapter 13) and so behaves somewhat like an electron in hydrogen. Here again,

Angular momentum of atomic systems

the lowest state has $l = 0$ while the first excited state can have $l = 1$. The fine structure in this case yields a separation in energy of 21.4×10^{-4} eV between the two initial $l = 1$ states. This is a factor of 10 larger than that predicted above.

11-6 FORMALISM FOR TOTAL ANGULAR MOMENTUM

When an atom has both orbital angular momentum and spin angular momentum, the spin-orbit interaction couples the two into a resultant *total* angular momentum. This can be described very simply using a vector model similar to the model for orbital angular momentum alone.[12] According to this model, the total angular momentum of an electron is characterized by a quantum number j. For any given nonzero value of l, the possible values of j are given by

$$j = l \pm \tfrac{1}{2}$$

(Since j must always be positive, for $l = 0$ we have $j = \tfrac{1}{2}$ only.) Note that the same value of j can always be obtained from two successive values of l. For example, $j = \tfrac{3}{2}$ can result from $j = 1 + \tfrac{1}{2}$ or $j = 2 - \tfrac{1}{2}$.

The squared magnitude of the total angular momentum is given by the relation

$$J^2 = j(j + 1)\hbar^2 \tag{11-28}$$

The component of $\mathbf{J}$ along a prescribed z axis is given by

$$J_z = m_j \hbar \tag{11-29}$$

where m_j can take on any value from $-j$ to $+j$ in steps of unity. Thus, for example, if $l = 1$, we can have $j = \tfrac{3}{2}$ or $j = \tfrac{1}{2}$. For $j = \tfrac{3}{2}$ the possible values of m_j are $-\tfrac{3}{2}, -\tfrac{1}{2}, +\tfrac{1}{2}, +\tfrac{3}{2}$. For $j = \tfrac{1}{2}$ the values of m_j are $\pm\tfrac{1}{2}$ only.

For atomic structures in which more than one unpaired electron must be included in the analysis, both the theory and

---

[12]We are stating the properties of total angular momentum without proof. For justification of the results (which closely parallel those for orbital momentum alone) see, for example, E. Feenberg and G. E. Pake, *Notes on the Quantum Theory of Angular Momentum*, Addison-Wesley, Reading, Mass., 1953.

11-6   Formalism for total angular momentum

the experimental results are more complicated than those analyzed here. Rather than treat such cases, we return, in the following chapter, to the complete treatment of the Coulomb model of hydrogen. Basic to this treatment will be the permissible values of $L^2 = l(l + 1)\hbar^2$ and $L_z = m_l\hbar$ summarized in Eq. 11-13 for an electron bound in a central potential. Indeed, the angular dependence of the hydrogen atom wave functions has already been found in the spherical harmonics (Eq. 11-14). The energies of hydrogen—and the full wave functions—will then follow from the radial dependence of the wave function (Eq. 11-12).

Appendix: The Schrödinger Equation in Spherical Coordinates

The three-dimensional time-independent Schrödinger equation in Cartesian coordinates is

$$-\frac{\hbar^2}{2m}\left(\frac{\partial^2}{\partial x^2} + \frac{\partial^2}{\partial y^2} + \frac{\partial^2}{\partial z^2}\right)\psi + V\psi = E\psi$$

A coordinate-independent expression of this equation makes use of the Laplacian ∇^2:

$$-\frac{\hbar^2}{2m}\nabla^2\psi + V\psi = E\psi \tag{11A-1}$$

Now, in Cartesian coordinates, the gradient operator ∇ is defined by the equation

$$\nabla f = \hat{x}\frac{\partial f}{\partial x} + \hat{y}\frac{\partial f}{\partial y} + \hat{z}\frac{\partial f}{\partial z}$$

where f is any scalar function. The Laplacian operator ∇^2 is defined by

$$\nabla^2 f = \frac{\partial^2 f}{\partial x^2} + \frac{\partial^2 f}{\partial y^2} + \frac{\partial^2 f}{\partial z^2}$$

and is expressible as the scalar product of the operator ∇ with

Angular momentum of atomic systems

itself:

$$\nabla^2 f = (\nabla \cdot \nabla)f \qquad (11A\text{-}2)$$

In spherical coordinates we can begin with the coordinate-free expression of Eq. 11A-2, plus the fact that the gradient operator in spherical coordinates is given by

$$\nabla = \hat{r} \frac{\partial}{\partial r} + \hat{\theta} \frac{1}{r} \frac{\partial}{\partial \theta} + \hat{\phi} \frac{1}{r \sin \theta} \frac{\partial}{\partial \phi} \qquad (11A\text{-}3)$$

where $\hat{r}, \hat{\theta}, \hat{\phi}$ are orthogonal unit vectors in the directions of increasing r, θ, ϕ, respectively (Figure 11A-1). Equation 11A-3 arises directly from the fact that the elements of displacement in the $\hat{r}, \hat{\theta},$ and $\hat{\phi}$ directions are $dr, rd\theta,$ and $r \sin \theta \, d\phi$, respectively. The Laplacian in spherical coordinates is thus given by

$$\nabla^2 = \left(\hat{r} \frac{\partial}{\partial r} + \hat{\theta} \frac{1}{r} \frac{\partial}{\partial \theta} + \hat{\phi} \frac{1}{r \sin \theta} \frac{\partial}{\partial \phi} \right)$$
$$\left(\hat{r} \frac{\partial}{\partial r} + \hat{\theta} \frac{1}{r} \frac{\partial}{\partial \theta} + \hat{\phi} \frac{1}{r \sin \theta} \frac{\partial}{\partial \phi} \right) \qquad (11A\text{-}4)$$

In evaluating this we have to take account of the fact that, in contrast to the Cartesian coordinate analysis, the unit vectors of the spherical polar system are defined only locally, and their directions change with the angles θ and ϕ. As a preliminary step we must analyze this dependence.

First, consider the unit vector $\hat{r}$. If we keep θ and ϕ constant and vary r, there is no change in $\hat{r}$. But if we keep r and ϕ

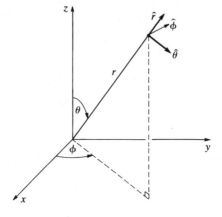

Fig. 11A-1   The variables r, θ, ϕ and the unit vectors $\hat{r}, \hat{\theta}, \hat{\phi}$ in a spherical polar coordinate system.

Appendix

constant and change θ to $\theta + d\theta$, the vector $\hat{r}$ is turned through an angle $d\theta$. The change in it, expressed as a vector, is equal to $|\hat{r}| d\theta$ in the direction $\hat{\theta}$. Since $|\hat{r}| = 1$, we have simply

$$\frac{\partial \hat{r}}{\partial \theta} = \hat{\theta}$$

Similarly, if we keep r and θ constant and change ϕ to $\phi + d\phi$, the tip of the $\hat{r}$ vector moves through a distance $\sin \theta\, d\phi$ in the direction $\hat{\phi}$. Hence

$$\frac{\partial \hat{r}}{\partial \phi} = \hat{\phi} \sin \theta$$

By considering how the other unit vectors change as r, θ, and ϕ are varied, it is quite straightforward to arrive at the following set of results:

$$
\begin{array}{lll}
\dfrac{\partial \hat{r}}{\partial r} = 0 & \dfrac{\partial \hat{r}}{\partial \theta} = \hat{\theta}, & \dfrac{\partial \hat{r}}{\partial \phi} = \hat{\phi} \sin \theta \\[2ex]
\dfrac{\partial \hat{\theta}}{\partial r} = 0 & \dfrac{\partial \hat{\theta}}{\partial \theta} = -\hat{r}, & \dfrac{\partial \hat{\theta}}{\partial \phi} = \hat{\phi} \cos \theta \\[2ex]
\dfrac{\partial \hat{\phi}}{\partial r} = 0 & \dfrac{\partial \hat{\phi}}{\partial \theta} = 0, & \dfrac{\partial \hat{\phi}}{\partial \phi} = -(\hat{r} \sin \theta + \hat{\theta} \cos \theta)
\end{array}
\qquad (11\text{A-}5)
$$

We now apply these results in Eq. 11A-4. In doing so, we use the fact that the derivative operators in the left-hand factor of Eq. 11A-4 operate on everything in the right-hand factor. Thus, for example, the term $\hat{r}(\partial/\partial r)$ in the left-hand factor, when applied to the right-hand factor, gives

$$\hat{r} \cdot \frac{\partial}{\partial r}\left(\hat{r}\frac{\partial}{\partial r} + \hat{\theta}\frac{1}{r}\frac{\partial}{\partial \theta} + \hat{\phi}\frac{1}{r \sin \theta}\frac{\partial}{\partial \phi}\right)$$

Making use of Eqs. 11A-5, together with the fact that $\hat{r}$, $\hat{\theta}$, and $\hat{\phi}$ are mutually orthogonal, you should not have much trouble in convincing yourself that Eq. 11A-4 reduces to the following:

$$
\begin{aligned}
\nabla^2 = {}&(\hat{r} \cdot \hat{r})\frac{\partial^2}{\partial r^2} + (\hat{\theta} \cdot \hat{\theta})\left[\frac{1}{r}\frac{\partial}{\partial r} + \frac{1}{r^2}\frac{\partial^2}{\partial \theta^2}\right] \\
&+ (\hat{\phi} \cdot \hat{\phi})\left[\frac{1}{r}\frac{\partial}{\partial r} + \frac{1}{r^2}\cot \theta \frac{\partial}{\partial \theta} + \frac{1}{r^2 \sin^2 \theta}\frac{\partial^2}{\partial \phi^2}\right]
\end{aligned}
$$

Angular momentum of atomic systems

Collecting terms, and putting $\hat{r} \cdot \hat{r} = \hat{\theta} \cdot \hat{\theta} = \hat{\phi} \cdot \hat{\phi} = 1$, this gives

$$\nabla^2 = \left(\frac{\partial^2}{\partial r^2} + \frac{2}{r} \frac{\partial}{\partial r} \right) + \frac{1}{r^2} \left(\frac{\partial^2}{\partial \theta^2} + \cot \theta \frac{\partial}{\partial \theta} \right)$$
$$+ \frac{1}{r^2 \sin^2 \theta} \frac{\partial^2}{\partial \phi^2} \qquad (11A\text{-}6)$$

The Schrödinger equation 11A-1 can then be written

$$-\frac{\hbar^2}{2m} \left(\frac{\partial^2 \psi}{\partial r^2} + \frac{2}{r} \frac{\partial \psi}{\partial r} \right) + \frac{\left[-\hbar^2 \left(\frac{\partial^2 \psi}{\partial \theta^2} + \cot \theta \frac{\partial \psi}{\partial \theta} + \frac{1}{\sin^2 \theta} \frac{\partial^2 \psi}{\partial \phi^2} \right) \right]}{2mr^2}$$
$$+ V\psi = E\psi$$

We can multiply through by r, thus putting the equation in the form of Eq. 11-3:

$$-\frac{\hbar^2}{2m} \frac{\partial^2 (r\psi)}{\partial r^2} + \frac{(L^2)_{\mathrm{op}} (r\psi)}{2mr^2} + V(r\psi) = E(r\psi) \qquad (11A\text{-}7)$$

Here $(L^2)_{\mathrm{op}}$ is the operator for squared angular momentum. By the above derivation we have shown that this operator is identical to Op (θ, ϕ) in Eqs. 11-3 and 11-7:

$$\mathrm{Op}\,(\theta, \phi) = (L^2)_{\mathrm{op}}$$
$$= -\hbar^2 \left(\frac{\partial^2}{\partial \theta^2} + \cot \theta \frac{\partial}{\partial \theta} + \frac{1}{\sin^2 \theta} \frac{\partial^2}{\partial \phi^2} \right) \qquad (11A\text{-}8)$$

EXERCISES

11-1 *Angular position-momentum uncertainty relation.* Consider a particle moving in a circle with tangential momentum p_t and angular momentum $L = rp_t$. Let Δs be the statistical spread of its position along the circle and ϕ be the angular position of the particle.

(a) Show that the relation $\Delta s \cdot \Delta p_t \gtrsim \hbar$ can be rewritten as $\Delta \phi \cdot \Delta L \gtrsim \hbar$.

(b) For an electron in a given Bohr orbit of hydrogen, what does this relation imply about locating the angular position of the electron?

(c) In terms of wave mechanics, estimate the range of angular momenta of wave functions that must be superposed in order that the angular position of the electron be known statistically to one-sixth of a radian.

11-2 *Normalization of spherical harmonics.*

(a) Verify for the cases $l = 0$ and $l = 1$ that the spherical harmonic functions given in Table 11-2 are normalized. [*Hint:* Use cos θ as a variable of integration so that sin θ $d\theta = - d(\cos \theta)$.]

(b) Suppose that a bound-state energy eigenfunction has the form

$$\psi(r, \theta, \phi) = R(r) \, Y_{l, m} \, (\theta, \phi)$$

where $Y_{l,m}(\theta, \phi)$ is a normalized spherical harmonic. What is the normalization condition on $R(r)$? [*Hint:* The element of volume in spherical coordinates is $dV = r^2 dr \sin \theta \, d\theta \, d\phi$.]

11-3 *Orthogonality of the spherical harmonics.* In the preceding exercise you verified the normalization of some of the spherical harmonics in Table 11-2. The spherical harmonics are also *orthogonal* in the sense that

$$\iint Y^*_{lm} \, Y_{l'm'} \, \sin \theta \, d\theta \, d\phi = 0 \text{ unless } both \ l = l' \ and \ m = m'$$

(a) Verify the orthogonality of the spherical harmonics for the following three cases:

(i) $l = 0, \ m = 0; \quad l' = 1, \ m' = 0$
(ii) $l = 0, \ m = 0; \quad l' = 1, \ m' = 1$
(iii) $l = 1, \ m = 0; \quad l' = 1, \ m' = -1$

[*Hint:* Again, it is helpul to use cos θ as a variable of integration: sin θ $d\theta = -d(\cos \theta)$.]

(b) Consider the orthogonality of the following two energy eigenfunctions in the same binding potential:

$$\psi_1 = R_1(r) \, Y_{lm} \, (\theta, \phi)$$
$$\psi_2 = R_2(r) \, Y_{l'm'} \, (\theta, \phi)$$

(i) If either $l \neq l'$ or $m \neq m'$, then the two wave functions are orthogonal no matter what forms of R_1 and R_2. True or false?

(ii) If $l = l'$ and $m = m'$, what is the orthogonality condition on R_1 and R_2?

11-4 *The radial function in a central potential.*

(a) As we have seen, for a particle in a central potential, the Schrödinger equation admits product solutions $\psi(\mathbf{r}) = R(r) \, Y_{lm}(\theta, \phi)$. (See Eqs. 11-8, 11-12, and 11-14.) Show that the equation for the radial function $u(r) = rR(r)$ depends on l but does *not* depend on m.

(b) Consider a classical particle of energy E and angular momentum $\mathbf{L}$ moving in a central potential. The size and shape of the orbit

Angular momentum of atomic systems

depend on E and on $\mathbf{L}$. Does the size or shape of the orbit also depend on the direction of $\mathbf{L}$?

(c) Consider once more the quantum-mechanical problem. Assume that l is given. How many independent eigenfunctions correspond to *each* solution $u(r) = rR(r)$ of the radial equation? Note that this "m degeneracy" exists for *all* central potentials.

11-5 *Commutation of* $(L^2)_{op}$ *with* $(L_z)_{op}$. It was shown in Chapter 10 that the operators $(L_x)_{op}$ and $(L_y)_{op}$ do not commute but satisfy the relation (Eq. 10-23)

$$(L_x)_{op}(L_y)_{op} - (L_y)_{op}(L_x)_{op} = i\hbar(L_z)_{op}$$

(a) Use this relation and the two similar equations obtained by cycling the coordinate labels to show that $(L^2)_{op}(L_z)_{op} = (L_z)_{op}(L^2)_{op}$, that is, these two operators commute. [*Hint:* You do *not* need to introduce the differential formulas for the operators, Eqs. 10-12. Use the fact that $(AB)C = A(BC)$ where A, B, and C are operators.]

(b) What does the result obtained in (a) tell you about possible simultaneous eigenfunctions of squared angular momentum and the z component of angular momentum in a central potential?

(c) Without carrying it through, do you expect that the operators $(L^2)_{op}$ and $(L_x)_{op}$ commute? Can the three quantities L^2, L_z, and L_x have simultaneous eigenvalues in a central potential?

11-6 *Simultaneous eigenvalues of angular momentum.* Roland claims that L^2, L_x, L_y, and L_z can *all* be specified simultaneously for an atomic system. His argument goes as follows:

(1) $(L^2)_{op}$ and $(L_z)_{op}$ commute so that L^2 and L_z can have simultaneous eigenvalues.
(2) $(L_x)_{op}(L_y)_{op} - (L_y)_{op}(L_x)_{op} = i\hbar(L_z)_{op}$
(3) There exist states for which $(L_z)_{op} = 0$.
(4) Therefore the commutator in (2) equals zero.
(5) Therefore L_x and L_y can have simultaneous eigenvalues.
(6) But x, y, and z are arbitrarily chosen directions, so that, for example, L_z and L_x must also have simultaneous eigenvalues.
(7) Therefore L^2, L_x, L_y, L_z can all be specified simultaneously. Is Roland right or wrong? If right, what is the error in the conclusions of Section 10-7? If wrong, what is his error?

11-7 *Internuclear distance in a diatomic molecule.* Carbon monoxide absorbs energy in the microwave region of the spectrum at 1.153×10^{11} Hz. This can be attributed to a transition between rotational states $l = 0$ and $l = 1$. Deduce the internuclear distance for this molecule. Use the atomic masses of the isotopes C^{12} and O^{16} for your calculation. Compare your result with the entry in Table 11-4.

11-8 *Semiclassical model of rotating diatomic molecule.*

(a) At what angular frequency ω must a classical rotator with moment of inertia $I = \mu r_o^2$ rotate in order to have angular momentum $L = \sqrt{l(l+1)}\,\hbar$? What is the ratio of $\hbar\omega$ to the energy spacing $(E_l - E_{l-1})$ obtained by a quantum-mechanical analysis?

(b) If a molecular rotator consists of two atoms separated by a distance r_o and with masses m_1 and m_2 [so that $\mu = m_1 m_2/(m_1 + m_2)$], what are the speeds v_1 and v_2 of the atoms in a semiclassical picture of the rotational state $l = 1$? Evaluate your result for H and for Cl atoms in the HCl molecule, using the value for r_o obtained in Section 11-3.

11-9 *Centrifugal stretching of the OCS molecule.* Consider the data on the OCS molecule presented in Table 11-5. Using the isotopes O^{16}, C^{12}, and S^{32}, and assuming that the three atoms are equidistant along a line, calculate the interatomic distance between adjacent atoms from the value of the moment of inertia given in the table. Compare the result with the value for carbon monoxide given in Table 11-4. How much do these bonds stretch under the centrifugal forces reflected in the figures of Table 11-5? Give your answer in fermi (1 fermi $= 10^{-13}$ cm, comparable to the size of a nucleus).

11-10 *Hydrogen atom as a diatomic molecule?* Why does the energy-level equation (11-18) for a diatomic molecule with atoms of unequal mass not apply to the electron-proton system composing the hydrogen atom?

11-11 *Center-of-mass separation in quantum mechanics.* Two particles of masses m_1 and m_2 at positions $\mathbf{r}_1$ and $\mathbf{r}_2$ interact with each other but are free from external forces. The time-independent Schrödinger equation for this system is

$$\left[\left(-\frac{\hbar^2}{2m_1}\nabla_1^2 - \frac{\hbar^2}{2m_2}\nabla_2^2\right) + V(\mathbf{r}_2 - \mathbf{r}_1)\right]\psi(\mathbf{r}_1, \mathbf{r}_2) = E\psi(\mathbf{r}_1, \mathbf{r}_2)$$

Here $|\nabla_1^2$ means that derivatives are taken only with respect to the position coordinates of particle one (for example).

(a) Show that the kinetic-energy operators in the above equation can be written in the form

$$-\frac{\hbar^2}{2M}\nabla_R^2 - \frac{\hbar^2}{2\mu}\nabla_r^2$$

where M is the total mass, μ is the reduced mass, $\mathbf{R}$ is the position of the center of mass, and $\mathbf{r} = \mathbf{r}_2 - \mathbf{r}_1$ is the relative position of the two particles. [*Hint:* Carry out the differentiations, using the chain rule,

Angular momentum of atomic systems

for x coordinates only and invoke symmetry to obtain the results for y and z.]

(b) Write the Schrödinger equation using the coordinates R and r, and show that it admits product solutions $\psi(\mathbf{r}, \mathbf{R}) = \psi_R(\mathbf{R}) \psi_r(\mathbf{r})$. Remember that $V = V(\mathbf{r})$. Write down the two equations into which the overall equation separates, using E_R to denote the eigenvalue in the R equation and E_r to denote the eigenvalue in the r equation. How are E, E_R, and E_r related?

(c) Interpret the two separated equations found in part (b) in terms of the description of two (fictitious) single particles.

11-12 *Analysis of a rotation-vibration spectrum.* According to Eq. 11-21 (Section 11-3) the frequencies of the lines of a molecular rotation-vibration spectrum should be given by

$$\nu_{J,J-1} = \nu_0 \pm \frac{h}{4\pi^2 \mu r_0^2} J$$

The $+$ sign refers to transitions in which J is the rotational quantum number associated with the higher vibrational level, and the $-$ sign to transitions in which J is the rotational quantum number associated with the lower vibrational level. Refer to the experimental spectrum for HCl shown in Figure 11-8.

(a) Read off the wavelengths for the two lines marked $J = 1$. According to the above formula, their frequency difference is equal to $h/(2\pi^2 \mu r_0^2)$. Use this to deduce a value for the internuclear distance r_0 (using a value for the reduced mass μ calculated from the known masses of the H and Cl atoms). Compare the result with the value 1.27 Å deduced in Section 11-3 from the pure rotational spectrum shown in Figure 11-6.

(b) From the complete spectrum of Figure 11-8, construct a graph of $1/\lambda$ versus J for all of the lines shown. How well does it fit the theoretical prediction?

(c) It is reasonable to expect that the mean internuclear distance r_0 will be different for the rotational states associated with the different vibrational states. (Why?) Go back to Eq. 11-20 and derive a revised expression for $\nu_{J,J-1}$ under this assumption. Test whether this leads to a more satisfactory analysis of the experimental data. Or does your graph in (b) suggest some other modification of the simple theory?

11-13 *The electron as a spinning Newtonian ball?* All Newtonian models of the electron designed to account for electron spin are disqualified not only because of quantum effects, but also because they fail to incorporate special relativity. Show that relativity must play an important role by calculating the peripheral speed of a spinning Newtonian ball (of radius equal to the classical electron radius e^2/mc^2 and uniform mass density so that $I = \frac{2}{5}mr^2$) if it is to possess a spin angular momentum of order $\hbar/2$. Compare the necessary peripheral speed with the speed of light.

11-14 *Rates of electron spin precession.* We have seen that the electron has an intrinsic angular momentum of $\hbar/2$ and a magnetic moment of approximately one Bohr magneton ($\mu_B = e\hbar/2M_e c$).

(a) Calculate what the rate of steady precession of the electron spin would be, according to the classical dynamics equation $\boldsymbol{\mu} \times \mathbf{B} = d\mathbf{L}/dt$ (magnetic torque = rate of change of angular momentum) in a magnetic field of the order of the earth's field—that is, about 1 gauss ($= 10^{-4}$ tesla).

(b) On the same basis, calculate what the angular velocity of precession of the electron spin would be in a magnetic field equal to $2\mu_B/a_0^3$, where a_0 is the Bohr radius. (We saw in Section 11-5 that this is the order of magnitude of field to be expected in the interior of an atom.)

(c) Compare the result of (b) with the orbital angular frequency of the first orbit in the Bohr model for hydrogen.

11-15 *Spin-orbit coupling and spectroscopic fine structure.* The potential energy of a magnetic dipole $\boldsymbol{\mu}$ in a magnetic field $\mathbf{B}$ is equal to $-\boldsymbol{\mu} \cdot \mathbf{B}$. For an electron in an atom, this energy comes from the interaction of the electron spin magnetic moment with the average atomic magnetic field to which it is exposed.

(a) From the fact that the electron's spin angular momentum is quantized ($\pm\frac{1}{2}\hbar$) along the direction of $\mathbf{B}$, show that the angular frequency defined by $(\boldsymbol{\mu} \cdot \mathbf{B})/\hbar$ is equal to half the precessional frequency calculated classically, as in the previous exercise. (The factor of $\frac{1}{2}$ is a relativistic effect; the result obtained here is the correct one.)

(b) The frequency difference between the D lines of the sodium spectrum (Figure 10-1) arises from the energy difference between two excited states having the two different possible orientations of the electron spin with respect to the internal atomic field. From the measured wavelengths of the D lines (5889.96 and 5895.93 Å) deduce the strength of the atomic magnetic field that causes this splitting.

11-16 *States of the total angular momentum, J.* The more advanced theory of quantum mechanics justifies the use of a very simple vector model (Section 11-6) for the total angular momentum $\mathbf{J}$ and its components along a specified axis, strictly analogous to what we found for orbital momentum alone. The figure indicates how the vector $\mathbf{J}$, of

Angular momentum of atomic systems

length $\sqrt{j(j+1)}\hbar$, can be thought of as formed from the individual vectors $\mathbf{L}$, of length $\sqrt{l(l+1)}\hbar$, and $\mathbf{S}$, of length $\sqrt{s(s+1)}\hbar = \sqrt{3}\hbar/2$.

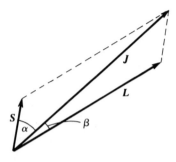

(a) Construct such a vector diagram for the specific case $l = 2$, $j = \frac{3}{2}$.

(b) Orient $\mathbf{J}$ so that it has the component $+\frac{1}{2}\hbar$ along a specified axis. What can you say about the components of $\mathbf{L}$ and $\mathbf{S}$ along this same axis?

11-17 *Magnetic moments for combined orbital and spin states.* The magnetic interaction between the electron spin magnetic moment and the magnetic field associated with the orbital motion can be described alternatively as an interaction between spin and orbital magnetic moments, $\boldsymbol{\mu}_s$ and $\boldsymbol{\mu}_l$. In this interaction, both $\boldsymbol{\mu}_l$ and $\boldsymbol{\mu}_s$ precess about an axis defined by the direction of the resultant angular momentum, $\mathbf{J}$. The figure shows this situation. The angles α and β are the same as those shown in the figure for Problem 11-16, but because of the extra factor 2 in the gyromagnetic ratio for spin, the magnetic moment vectors are given by

$$\mu_l = \sqrt{l(l+1)}\,\mu_B, \qquad \mu_s = 2\sqrt{s(s+1)}\,\mu_B$$

where μ_B is the Bohr magneton.

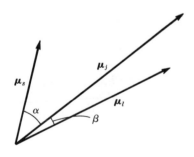

(a) Using values of cos α and cos β obtained from the vector diagram for **L**, **S**, and **J**, show that the resultant magnetic moment along the direction of **J** is given by

$$\mu_J = g_j \sqrt{j(j + 1)}\, \mu_B$$

where

$$g_j = 1 + \frac{j(j + 1) + s(s + 1) - l(l + 1)}{2j(j + 1)}$$

(g_j is called the Landé g-factor, after A. Landé who introduced it in 1921.)

(b) Obtain simplified general formulas for g_j, given that $s = \frac{1}{2}$ and $j = l \pm \frac{1}{2}$ for single-electron states. Evaluate for the case $l = 2$, $j = \frac{3}{2}$.

(c) Because of the rapid precession of μ_s and μ_l (see Exercises 11-14 and 11-15) about the direction of **J**, their components perpendicular to **J** average out to zero. Hence the energy shift of the electron in an *external* magnetic field is given simply by $W = -\mu_j \cdot \mathbf{B}$. Using the result of (a), show that this leads to the simple formula

$$W = -m_j g_j \mu_B B$$

where m_j can range between $-j$ and $+j$ in integral steps.

Where is the good, old-fashioned solid matter that obeys precise, compelling mathematical laws? The stone that Dr. Johnson once kicked to demonstrate the reality of matter has become dissipated in a diffuse distribution of mathematical probabilities.

MORRIS KLINE, *Mathematics in Western Culture* (1953)

12

Quantum states of three-dimensional systems

12-1 INTRODUCTION

We turn now to the simplest real system of atomic physics—a single electron in the field of a central charge. For definiteness we take the central charge to be a proton, so that the system under study is a hydrogen atom. The hydrogen atom is the most accurately and completely described system in all of physics. Its energy levels are known experimentally (and are described theoretically) with a precision limited only by observational uncertainties in our knowledge of the fundamental physical constants. The components of the hydrogen atom are so few that discrepancies between experiment and theory may reflect an error in our fundamental understanding of nature and therefore command the immediate attention of the scientific community.

12-2 THE COULOMB MODEL

The interaction of overwhelming importance in hydrogen is the Coulomb potential between electron and proton. Although a complete dynamical account of the hydrogen atom must recognize other effects—in particular relativistic corrections and magnetic interactions—these affect the Coulomb-model energy levels by only about one part in ten thousand.

519

Under these circumstances, standard procedure is: (1) find the exact wave functions for the Coulomb interaction alone. Despite its incomplete nature, this yields results of impressive accuracy. Then (2) the relativistic and magnetic interactions are approximated as small corrections to the Coulomb model.

Schrödinger Equation for the Coulomb Model

The time-independent Schrödinger equation in three dimensions is

$$-\frac{\hbar^2}{2M}\nabla^2\psi + V(x, y, z)\psi = E\psi$$

[We first stated this equation in Chapter 5, Eq. 5-1b.] In Chapter 11 (Section 11-2, Eq. 11-3) we stated the form to which this Schrödinger equation changes if one uses the spherical polar coordinates appropriate to central potentials:

$$\left[-\frac{\hbar^2}{2M}\frac{\partial^2}{\partial r^2} + \frac{\mathrm{Op}(\theta, \phi)}{2Mr^2} + V(r)\right](r\psi) = E(r\psi) \qquad (12\text{-}1)$$

(Refer to the Appendix to Chapter 11 for the detailed derivation of this result.)

We now begin the task of calculating complete energy eigenfunctions and associated energies for states that are also eigenstates of L^2 and L_z, in the particular case that the potential $V(r)$ is the Coulomb potential in hydrogen:

$$V(r) = -\frac{e^2}{r} \text{ (cgs)} \qquad \left[-\frac{e^2}{4\pi\epsilon_0 r} \text{ (SI)}\right] \qquad (12\text{-}2)$$

From Chapter 11, we know that the eigenfunctions of both E and L^2 are product functions:

$$\psi(r, \theta, \phi) = R(r) \cdot F(\theta, \phi) \qquad (12\text{-}3)$$

where

$$(L^2)_{\mathrm{op}} F(\theta, \phi) = l(l + 1)\hbar^2 F(\theta, \phi)$$

By substituting this form of ψ in Eq. 12-1, together with the

Quantum states of three-dimensional systems

eigenvalue condition on L^2, we obtain the basic radial equation for an electron (mass M) in the hydrogen atom[1]:

$$\left[-\frac{\hbar^2}{2M} \frac{d^2}{dr^2} + \frac{l(l+1)\hbar^2}{2Mr^2} - \frac{e^2}{r} \right](rR) = E(rR) \tag{12-4}$$

In Chapter 5 we found spherically symmetric solutions for hydrogen, starting from an equation (Eq. 5-19) very similar to Eq. 12-4 except that the term in l was missing. For $l = 0$ this term would disappear from Eq. 12-4. This gives a hint of a result we shall obtain shortly, namely that the spherically symmetric solutions for the hydrogen atom (or indeed for any spherically symmetric potential) belong to $l = 0$.

In that earlier treatment we reduced the radial equation to an equivalent one-dimensional equation by the substitution $u(r) = r\psi(r)$ where $\psi(r)$, being spherically symmetric in that case, was a function of r alone. In the present case, even though the complete function $\psi(r, \theta, \phi)$ may not be spherically symmetric, we look for a similar simplification by writing

$$u(r) = rR(r)$$

With this substitution, Eq. 12-4 takes the form of a one-dimensional equation in the single coordinate r:

$$\left[-\frac{\hbar^2}{2M} \frac{d^2}{dr^2} + \frac{l(l+1)\hbar^2}{2Mr^2} - \frac{e^2}{r} \right]u = Eu \tag{12-5a}$$

In solving Eq. 12-5a it helps to simplify it by converting to "natural" units. Let ρ be radial distance measured in units of Bohr radius a_0 and let ϵ be the energy measured in units of the Rydberg energy E_R:

$$r = a_0\rho = \frac{\hbar^2}{Me^2}\rho$$

$$E = E_R\epsilon = \frac{Me^4}{2\hbar^2}\epsilon$$

---

[1]We shall shortly be changing to dimensionless variables. To avoid unnecessary complication, we shall restrict ourselves to cgs units in all dimensional equations and formulas in the rest of this chapter. The conversion to SI units can easily be made with the help of Eq. 12-2.

With this substitution, Eq. 12-5a becomes

$$-\frac{d^2u}{d\rho^2} + \frac{l(l+1)}{\rho^2}\,u - \frac{2}{\rho}\,u = \epsilon u \tag{12-5b}$$

This is of the form

$$-\frac{d^2u}{d\rho^2} + V_{\text{eff}}u = \epsilon\theta \tag{12-6}$$

where V_{eff} is an effective potential energy defined by

$$V_{\text{eff}}(\rho) = \frac{l(l+1)}{\rho^2} - \frac{2}{\rho} \qquad \text{(dimensionless form)}$$

or $\tag{12-7}$

$$V_{\text{eff}}(r) = \frac{l(l+1)\hbar^2}{2Mr^2} - \frac{e^2}{r} \qquad \text{(normal form—}V\text{ in ergs)}$$

The first term on the right of Eq. 12-7 is called the *centrifugal potential*. A corresponding potential enters classical equations of satellite motion.[2] Physically the centrifugal potential represents the price we pay for limiting our attention to the radial dimension: it expresses the effect of angular momentum when the motion is referred to a rotating reference frame whose angular velocity keeps step with that of the orbiting particle. The form of V_{eff} for various values of l is shown in Figure 12-1. Clearly the effective potential energy for $l = 0$ is qualitatively different from all others; it has no minimum, and it provides an infinite attraction at $\rho = 0$, whereas those for $l \neq 0$ provide an infinite repulsion at the origin. It is not surprising that this difference in form leads to an important difference in the properties of the radial wave functions $R(\rho)$: those for $l \neq 0$ go to zero at $\rho = 0$, but for $l = 0$ there is a nonzero value of $R(0)$, corresponding to a finite probability density for the electron at the position of the proton. This, however, is looking ahead. For now, let us consider how to tackle the solution of Eq. 12-5 for an arbitrary nonzero value of l.

[2]See Eq. 11-1 and, for more detail, *Newtonian Mechanics* in this series, p. 564.

Quantum states of three-dimensional systems

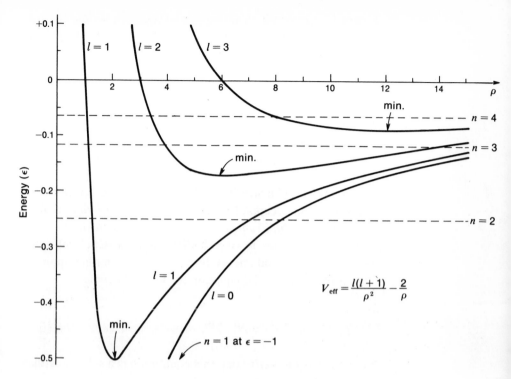

Fig. 12-1   The effective potential for hydrogen for
various values of the total angular momentum. The
energy levels for the Coulomb model are also indicated.
The figure is drawn to scale in natural units (radius in
Bohr radii and energy in Rydberg energy units).

12-3   GENERAL FEATURES OF THE RADIAL WAVE FUNCTIONS FOR HYDROGEN

Radial Solutions in Limiting Cases

To begin, we can gain valuable information by the technique of looking at the simpler approximate forms to which Eq. 12-5 reduces for $\rho \to \infty$ and $\rho \to 0$.

For very large radius ($\rho \to \infty$), both the Coulomb energy $(-2/\rho)$ and the centrifugal potential $[l(l + 1)/\rho^2]$ tend to zero, so that the equation becomes

$$\frac{d^2u}{d\rho^2} \approx -\epsilon u \qquad (\rho \to \infty)$$

Figure 12-1 makes it clear that bound states exist only for negative values of the total energy. Therefore, the quantity $-\epsilon$ is positive and an acceptable solution is

$$u(\rho) \sim e^{-b\rho}$$

where

$$\epsilon = -b^2 \qquad (12\text{-}8)$$

and we take b to be positive real. The alternative solution $e^{+b\rho}$ is physically unacceptable because it diverges for large radius.

For very small radius ($\rho \to 0$), it is the centrifugal term ($\sim 1/\rho^2$) that comes to overshadow both the Coulomb term ($\sim 1/\rho$) and the total energy term. Thus the limiting approximate form of the radial equation for $r \to 0$ becomes

$$\frac{d^2u}{d\rho^2} \approx \frac{l(l+1)}{\rho^2} u \qquad (\rho \to 0)$$

Now it is easy to verify that this equation has a solution either of the form

$$u(\rho) \sim \rho^{l+1}$$

or else of the form

$$u(\rho) \sim \rho^{-l}$$

We reject the latter possibility because it would (except for $l = 0$) make $u(\rho)$ tend to infinity at $\rho = 0$. And even for $l = 0$, the radial factor $R(\rho)$ in ψ, being given by $u(\rho)/\rho$, would go to infinity at $\rho = 0$ if the second form of $u(\rho)$ were used. Thus we assume $u(\rho) \sim \rho^{l+1}$ at small radius, for *all* l.

In view of these limiting forms for large and small radial distances, we can hope to find solutions embodying both of the limiting forms as factors[3]:

$$u(\rho) \sim \rho^{l+1}e^{-b\rho} \qquad (12\text{-}9)$$

---

[3]Strictly speaking, this does not have the form $e^{-b\rho}$ as ρ becomes very large. But see the long footnote in Section 4-3.

Quantum states of three-dimensional systems

We shall see that in some cases Eq. 12-9, just as it stands, does satisfy the full form of the Schrödinger equation (Eq. 12-5) for all radii. For other cases, it is necessary to introduce a further ρ-dependent factor in the equation for $u(\rho)$.

Permitted Energies for Hydrogen

Before taking the mathematical analysis any further, let us look at what we have learned up to this point. Figure 12-2a shows the effective potential energy diagram for some nonzero l. Figure 12-2b indicates the approximate form of $u(\rho)$ for small and large radii. We arbitrarily assume that $u(\rho)$ is positive near $\rho = 0$, but then we must allow the exponential func-

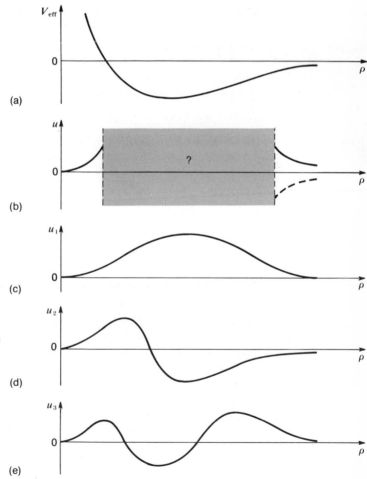

Fig. 12-2 Radial wave functions for hydrogen. (a) Form of the effective potential for $l \neq 0$. (b) Limiting forms of the radial wave function for small and large radii. (c) to (e) Qualitative plots.

12-3   General features of radial wave functions

tion for large ρ to approach the axis from either the positive or the negative side. Our task now is to find acceptable ways of bridging the gap between these regions.

As we have seen, the Schrödinger equation that we are trying to solve (Eq. 12-5) is essentially one-dimensional, and we can draw on our knowledge of the qualitative form of energy eigenfunctions, as developed in Chapter 3. We can guess that possible eigenfunctions of successively higher energy are of the forms indicated in lower portions of Figure 12-2, with 0, 1, 2, ..., nodes. How can we find the analytic forms of these functions, using Eq. 12-5 as a guide?

It turns out that one solution is already available. The form of $u(\rho)$ given by Eq. 12-9, as it stands, satisfies the Schrödinger equation (Eq. 12-5) *exactly*, and describes a nodeless radial function of the type shown in Figure 12-2c. You can easily verify, by substituting Eq. 12-9 into Eq. 12-5b and matching coefficients, that the following two conditions must be satisfied:

$$\epsilon = -b^2; \qquad b = \frac{1}{(l + 1)}$$

The first of these duplicates Eq. 12-8, and therefore provides no new information; but the second, by fixing the value of b (for a given l), specifies the exact quantized value of the energy:

$$\epsilon = -\frac{1}{(l + 1)^2}$$

Now, in the natural (atomic) units we are using here, the quantized energies of the Bohr model (Section 1-7)—or of the spherically symmetric states studied in Section 5-5—are given by $\epsilon_n = -1/n^2$. Hence the energy ϵ in the above equation is just one of these Bohr energies with n equal to $l + 1$. Turning this around, we can identify the solution we have here as belonging to a *principal quantum number n*, with $l = n - 1$.

In order to generate functions $u(\rho)$ that have one or more nodes, we introduce on the right of Eq. 12-9 an extra factor $f(\rho)$ having one or more roots. A factor with the desired properties is a polynomial in ρ with coefficients that alternate

Quantum states of three-dimensional systems

in sign.[4] To take the simplest case, we obtain a radial function with *one* node by putting

$$u(\rho) = (c_0 - c_1\rho)\rho^{l+1}e^{-b\rho}$$

You can verify that this can be made to satisfy Eq. 12-5 if

$$b = \frac{1}{l+2}$$

thus defining an energy given by

$$\epsilon = -\frac{1}{(l+2)^2}$$

This can be the *same* energy as for our previous (nodeless) radial function if we use the same principal quantum number n and now take $l = n - 2$. On the other hand, if we restrict ourselves to a particular value of l, then the two solutions that we have obtained define states of *different* energies, corresponding to principal quantum numbers $n_1 = l + 1$ and $n_2 = l + 2$, respectively. Both sets of possibilities exist; it is just a question of whether we classify the states according to n or according to l. However, as further examples are tried, an important general result emerges which is illustrated in the cases already treated. All possible quantized values of the energy are given by $\epsilon = -1/(\text{integer})^2$ and are thus the same set as those found for the Bohr atom (Section 1-7) and for the spherically symmetric states using the quantum analysis (Section 4-7). It is for this reason that we reintroduce the principal quantum number n and use it to index the energy states.

Limits on the Value of l

As more radial solutions are found, we shall discover that, for a given value of l, the possible values of n are restricted; or, alternatively, for a given n the values of l are restricted. This can be seen most easily by looking at Figure 12-1, on which

---

[4]The alternation in sign for sequential powers of ρ is important. The number of changes in sign in such a sequence gives the number of nodes for positive ρ. (Nodes for negative ρ are physically meaningless since $\rho \geq 0$ for spherical coordinates.)

12-3   General features of radial wave functions

the permitted energies have been drawn and indexed by n. You can see that for a given effective potential (for a given value of l) not all Bohr energies (all values of n) are permitted, because no solution is possible for an energy that lies *below* the bottom of the potential well. Since different effective potentials have different depths, the permitted values of n will differ for different values of l. Alternatively, for a given energy level (a given value of n), only certain effective potentials (indexed by certain values of l) have a minimum that lies below that energy.

To make these considerations more specific, we ask how deep is the effective potential well for a given value of l. From Eq. 12-7 the effective potential is

$$V_{eff} = \frac{l(l+1)}{\rho^2} - \frac{2}{\rho}$$

Set the derivative of this expression equal to zero to find the radius of the potential minimum ρ_{min}:

$$\rho_{min} = l(l+1)$$

Substitute this value back into the effective potential to find the value of the minimum:

$$V_{min} = -\frac{1}{l(l+1)}$$

Because of the $(l+1)$ in the denominator, this well is slightly *shallower* than the energy $-1/n^2$ for $n = l$. Therefore, for a given l, n is limited to values $l+1$, $l+2$, $l+3$, and so forth. Alternatively, for a given n, acceptable solutions can be found only for the values of l given by the expression

$$l \leq n - 1, \qquad l = 0, 1, 2, \ldots \tag{12-10}$$

12-4 EXACT RADIAL WAVE FUNCTIONS FOR HYDROGEN

The general radial solution is found by multiplying the "limiting cases" function, Eq. 12-9, by a polynomial in ρ:

$$u(\rho) = \rho^{l+1} e^{-b\rho} \sum_{i=0}^{i_{max}} (-1)^i c_i \rho^i$$

Quantum states of three-dimensional systems

Here the energy is determined by the value of b ($\epsilon = -b^2 = -1/n^2$). The signs in front of the coefficients c_i alternate with i in order to yield $i_{\max}$ nodes in the radial wave function. However, we know that in a given one-dimensional potential the wave function for the lowest energy state has no nodes, the wave function for the second energy has one node, and so forth. Looking again at Figure 12-1, we can see that, for a given effective potential (a given l), the lowest permissible energy has $n_1 = l + 1$, the second has $n_2 = l + 2$, and, in general, the ith permissible energy is indexed by $n_i = l + i$. But in general the wave function for the ith state has $i - 1$ nodes. This number of nodes for any n and l is therefore $n - l - 1$. Hence this is the value of $i_{\max}$ in the summation above. The general radial solution is then

$$u_{n,l}(\rho) = \rho^{l+1} e^{-\rho/n} \sum_{i=0}^{n-l-1} (-1)^i c_i \rho^i \qquad (12\text{-}11)$$

where we have substituted $b = 1/n$. Also we have added the subscripts n and l to $u(\rho)$ since each value of n and l requires a different solution. The coefficients c_i may be found by substituting Eq. 12-11 back into the radial Schrödinger equation, Eq. 12-5b, and equating coefficients of common powers of ρ. Only when this is done can one verify all the generalizations about which we have been speaking in the paragraphs above. Table 12-1 shows the first few radial wave functions to be used in the total wave function Eq. 12-3. See Figure 12-3, which shows the actual radial functions $R_{n,l}$ for $n = 1, 2, 3$:

$$R_{n,l}(\rho) = \frac{u_{n,l}(\rho)}{\rho} \qquad (12\text{-}12)$$

To summarize, we have seen that the energy states of the electron in a hydrogen atom are defined by the different integer

TABLE 12-1 Radial Wave Functions $R_{n,l}(\rho)$ of Hydrogen (Not Normalized)

n	$l = 0$	$l = 1$	$l = 2$
1	$e^{-\rho}$	No solution	No solution
2	$(1 - \tfrac{1}{2}\rho)e^{-\rho/2}$	$\rho e^{-\rho/2}$	No solution
3	$(1 - \tfrac{2}{3}\rho + \tfrac{2}{27}\rho^2)\,e^{-\rho/3}$	$\rho(1 - \tfrac{1}{6}\rho)e^{-\rho/3}$	$\rho^2 e^{-\rho/3}$

12-4   Exact radial wave functions for hydrogen

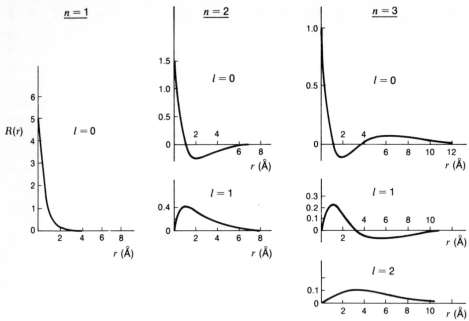

Fig. 12-3   The radial functions R(r) for hydrogen.
[Note: *The radius is measured here in angstroms rather
than in Bohr radii.*] *The units of R are* 10^{12} $cm.^{-3/2}$

values of the *principal quantum number n* ($n \geq 1$). For a given
value of *n* the *orbital angular momentum* number *l* can take on
positive integer values $l \leq n - 1$. (And the results of Chapter
11 tell us that for each value of *l* there are $2l + 1$ values of the
azimuthal quantum number m_l and all states have either
$m_s = +\frac{1}{2}$ or $m_s = -\frac{1}{2}$.) Although the functional form of the wave
equation depends on the values of *l* and m_l, these values turn
out to be irrelevant, in a sense, to the determination of the
energy, once we have introduced *n* and given it a specific
value. We have here, in fact, another example of *energy
degeneracy* (Sections 5-4 and 10-8), since the radial wave
functions of all different *l* and *m* (from $l = 0$ up to $l = n - 1$,
each with its values of m_l) for a given value of *n* have exactly
the same energy. The degeneracy with respect to *l* holds only
because the interaction between proton and electron is assumed to be *exactly* of the Coulomb form. Any departure from
a $1/r$ dependence of $V(r)$ will remove the degeneracy, so that
the energy will depend explicitly on the value of *l*. This occurs
in multielectron atoms: the inner electrons partially shield the

Quantum states of three-dimensional systems

outer electrons from the nuclear charge, giving a potential for the outer electrons that does *not* have a $1/r$ dependence, so that the energy *does* depend on l. We shall consider this in more detail in Chapter 13.

12-5 COMPLETE COULOMB WAVE FUNCTIONS

We are now in a position to bring together the results of the previous section [the radial functions $R(\rho)$ in a Coulomb potential] and the results of Chapter 11 (the angular momentum eigenfunctions in any central potential). The combination gives us the total wave function for an electron (treated simply as a point particle of mass M and charge $-e$) in a stationary state characterized by three quantum numbers n, l, and m:

$$\psi_{n,l,m}(\rho, \theta, \phi) = R_{n,l}(\rho) \, P_{l,m}(\theta) \, e^{im\phi} \tag{12-13}$$

where

$$R_{n,l}(\rho) = \frac{u_{n,l}(\rho)}{\rho} = \rho^l e^{-\rho/n} \sum_{i=0}^{n-l-1} (-1)^i \, c_i \rho^i$$

and

$$P_{l,m}(\theta) = \text{Associated Legendre Function}$$

The full details of these solutions are not important to us at the moment, but it is worth considering how the probability density for the electron varies with position. We can note, first, that the probability distribution does not depend on ϕ, since $|e^{im\phi}|^2 = 1$ everywhere. Thus, for a particular quantum state n, l, m the probability density has rotational symmetry about the z axis; it is given explicitly by

$$|\psi|^2 = [R_{n,l}(\rho)]^2 [P_{l,m}(\theta)]^2 \tag{12-14}$$

We may obtain the probability $dp(\rho, \theta, \phi)$ of finding the electron somewhere within an incremental volume by multiplying $|\psi|^2$ by the volume element $d\tau$ (Figure 12-4):

$$d\tau = (d\rho)(\rho d\theta)(\rho \sin \theta \, d\phi) = \rho^2 \, d\rho \sin \theta \, d\theta \, d\phi$$

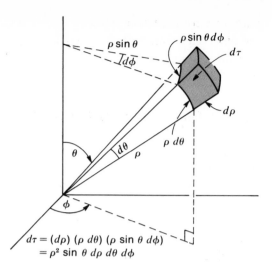

Fig. 12-4
Spherical coor-
dinates volume ele-
ment dτ.

$$d\tau = (d\rho)\,(\rho\,d\theta)\,(\rho\,\sin\theta\,d\phi)$$
$$= \rho^2 \sin\theta\,d\rho\,d\theta\,d\phi$$

Thus

$$dp(\rho, \theta, \phi) = [R(\rho)]^2\,\rho^2\,d\rho\,[P(\theta)]^2\,\sin\theta\,d\theta\,d\phi \qquad (12\text{-}15)$$

If we are interested only in the probability of finding the electron at a certain *distance* from the center of the atom (that is,
the probability of finding it somewhere within a spherical shell
radius ρ and thickness $d\rho$), then we integrate Eq. 12-15 over θ
(from zero to π) and ϕ (from zero to 2π).
This gives

$$dp(\rho) = 4\pi[R(\rho)]^2\rho^2\,d\rho \qquad (12\text{-}16a)$$

The probability of finding the electron between ρ and $\rho + d\rho$ is
equal to $w(\rho)d\rho$, where

$$w(\rho) = \frac{dp}{d\rho} = 4\pi\rho^2\,[R(\rho)]^2 = 4\pi[u(\rho)]^2 \qquad (12\text{-}16b)$$

These radial probability distributions have maxima that
shift to larger values of ρ as the principal quantum number n
increases. Figure 12-5 illustrates this for the states listed in
Table 12-1, whose radial functions $R(\rho)$ are shown in Figure
12-3. This trend may be compared with the results of the original Bohr theory, in which the radii of the permitted circular
orbits are equal to n^2 in units of a_o, the Bohr radius, and n is the
single quantum number of the Bohr theory.

Quantum states of three-dimensional systems

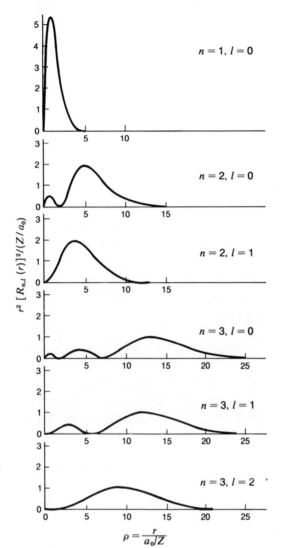

$$\rho = \frac{r}{a_0/Z}$$

Fig. 12-5 Radial probability distribution for several states of hydrogen. [Note: *the horizontal axis is in Bohr radii for* Z = 1.]

In the wave-mechanical model there is, of course, no unique value of orbit radius. However, an interesting comparison with the results of the Bohr theory can be made in the case of those states which, for a given value of the principal quantum number n, have the largest possible angular momentum. This is not surprising, since the circular Bohr orbit for a given n has angular momentum $n\hbar$. Although there is no solution of the Schrödinger equation corresponding exactly to this, the states of highest l ($l = n - 1$) yield the most "Bohr-like" radial probability function, with a single peak and no nodes. Further,

12-5 Complete coulomb wave functions

if we take the particular states having $|m| = l$, the probability distributions are concentrated toward the xy plane. This last result is expressed mathematically in a property of the associated Legendre functions $P_{l,m}(\theta)$. For $|m| = l$,

$$P_{l,l}(\theta) \sim (\sin \theta)^l$$

Hence for $l = n - 1$, the θ dependence will be as

$$|P_{l,l}|^2 \sim (\sin \theta)^{2(n-1)}$$

For large n this is very sharply peaked at $\sin \theta = 1$, that is, at $\theta = \pi/2$. Thus, for states having $|m| = l = n - 1$, the probability distribution is concentrated around a circle in the xy plane. The radius of this circle can be defined as corresponding to that value of r for which the radial probability density function $w(r)$ (Eq. 12-16b) is largest. For $l = n - 1$ we have (see Table 12-1)

$$w(\rho) \sim (u_{n,n-1})^2 = \rho^2(R_{n,n-1})^2 \sim \rho^{2n}e^{-2\rho/n}$$

By setting the derivative of this function (that is, $dw/d\rho$) equal to zero we find the radius for which the radial probability is maximum. You can quickly verify that this occurs for

$$\rho_n = n^2$$

or   (12-17)

$$r_n = n^2 a_o$$

which agrees exactly with the Bohr orbit radii. Figure 12-6 shows the probability function for such a "Bohr-like" state with $n = 5$, $l = m = 4$.

If one is interested in the dependence of the probability distribution on angle as well as on radial distance, then one can refer to diagrams such as those shown in Figure 12-7.[5] It is the total picture of such probability distributions that is commonly referred to as the *charge cloud* of an atom. One must

[5]A similar set of beautiful analog models was constructed by H. E. White many years ago. See Phys. Rev. **37**, 416 (1931) and also H. E. White *Introduction to Atomic Spectra*, McGraw-Hill, New York, 1934.

Quantum states of three-dimensional systems

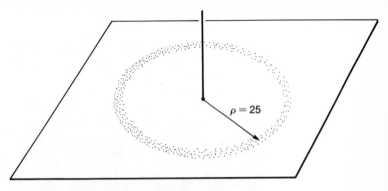

Fig. 12-6 *Visualization of the probability density function (stippled region) for the hydrogen atom state $n = 5$, $l = m = 4$. The radial density function*

$$|u|^2 \sim \rho^{10} e^{-\rho/5} \sin^8 \theta$$

is a maximum in the xy plane and at a radius equal to that of the Bohr orbit for n = 5.

never forget, however, that experimentally the charge cloud picture has a purely statistical meaning. A single observation on a single atom will discover either no electron or an *entire* electron at a given location. It is only as the result of many observations on an assemblage of identical atoms (for example by x-ray scattering) that one can build up the picture of the complete distribution of possible locations. As an alternative picture, one may imagine the electron in an individual atom occupying all possible locations in sequence, spending more time in regions of high probability. This moving electron picture is also deficient since it implies some kind of track or orbit which the electron follows in time. But we know that the probability function that describes states of unique energy does not change with time—it is stationary, so no electron orbit can be observed for such states. In reality, both the charge cloud picture and the moving electron picture are attempts to represent quantum results with classical models. It is natural to use such classical mental models, but we must be aware of their deficiencies. (See also the comments at the end of Section 11-2.)

It is important to realize that a probability distribution as described by Eqs. 12-15 and 12-16, or by any one of the charge cloud pictures of Figure 12-7, belongs to a unique state characterized by all three of the quantum numbers n, l, and m. How-

12-5   Complete coulomb wave functions

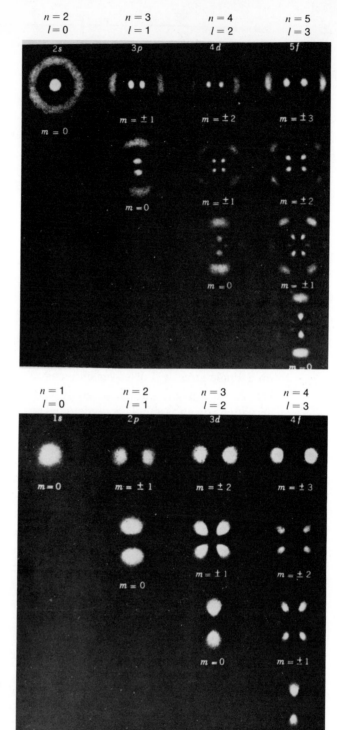

Fig. 12-7
Probability distributions for different states of hydrogen. Probability density is greater in regions of lighter shading. (From R. B. Leighton, Principles of Modern Physics, McGraw-Hill, New York, 1959. Reproduced with permission of the publisher.)

Quantum states of three-dimensional systems

ever, our ability to specify a particular value of m depends on having a physically real z axis of quantization, provided by the external environment (such as the magnetic field direction in a Stern-Gerlach experiment), that will allow us to define and to select an individual state from among the $2l + 1$ "sub-states" associated with a given value of l. Often no such axis exists naturally, and one has to take special steps (notably via external magnetic fields) in order to provide it. In the absence of such a preferred z direction, the probability distribution is in effect averaged over all the m values for a given n and l—and the result, as one might expect, is a spherically symmetric probability distribution.

12-6 CLASSIFICATION OF ENERGY EIGENSTATES IN HYDROGEN

We have now seen how the quantum state of an electron in the Coulomb model of hydrogen (ignoring spin) is completely specified by the three quantum numbers n, l, and m. We have also seen that, for a given value of n, all of the various possible states of different l and m belong to the same energy E_n. It is a simple matter to calculate how many of these energy-degenerate states exist for a given value of n. For each value of l we have $2l + 1$ different m states, and for a given n the permitted values of l range from zero to $n - 1$. Thus we have

$$\text{(total states belonging to } E_n) = \sum_{l=0}^{n-1} (2l + 1)$$
$$= n + 2 \sum_{l=0}^{n-1} l \qquad (12\text{-}18)$$
$$= n + n(n - 1)$$
$$= n^2$$

This is a remarkably simple result. (No doubt you have noticed repeated examples of this characteristic of quantum physics, that a simple question leads through a complicated manipulation to a simple conclusion!) When we add to this picture the fact of electron spin, there are *two* possible states for each triad of the quantum numbers n, l, m. To identify a particular state we must give the value of $m_s(= \pm\frac{1}{2})$ that labels the z component of the spin. Thus, the unique characterization of a state, including electron spin, actually requires *four* quantum numbers. Writing m_l instead of just m to identify unam-

12-6 Classifications of energy eigenstates

biguously the z component of $\mathbf{L}$, we can represent a given state by the ket $|n, l, m_l, m_s\rangle$. If we ignore the small energy splittings due to spin-orbit interaction and relativistic effects, we can then say that *for any given value of n there are $2n^2$ different quantum states of essentially the same energy for an electron in a hydrogen atom.*

n	Total Number of States
1	2
2	8
3	18
4	32
5	50

The first few entries in this tabulation contain numbers that are identical with the lengths of some of the periods in the periodic table of elements. This is no accident, although we shall defer the full discussion of many-electron atoms until Chapter 13.

We noted at the end of Chapter 11 that the existence of the spin-orbit interaction makes the *total* angular momentum $\mathbf{J}$ ($= \mathbf{L} + \mathbf{S}$) an important constant of the motion, and that the z component of the angular momentum is characterized by m_j rather than by m_l and m_s separately. In fact, the quantum numbers j and m_j are part of a different *representation* for the possible states of an electron in hydrogen. This other representation still requires four distinct quantum numbers; in ket notation, the basis states are written $|n, l, j, m_j\rangle$ with $2j + 1$ values of m_j for each possible j. This does not change our count of the total number of basis states associated with a given value of n, for we have

$$(\text{total states belonging to } n) = \sum_{l=0}^{n-1} (2j_+ + 1) + \sum_{l=1}^{n-1} (2j_- + 1)$$

where $j_+ = l + \frac{1}{2}$ and $j_- = l - \frac{1}{2}$.

One can easily verify that the combined result of the above summations is $2n^2$ once again. This is necessarily so, because the number of distinct states is independent of the particular choice of basis.

Why discuss alternative sets of energy eigenstates when, in the Coulomb model, all basis states of a given n have the same energy, no matter which representation is used? The answer is that in the next stage of approximation, in which the magnetic spin-orbit interactions are taken into account, the energy degeneracy disappears; that is, states with different val-

Quantum states of three-dimensional systems

ues of j take on slightly different energies, as described in a later section of this chapter. Then the $|n, l, j, m_j\rangle$ representation becomes the preferred one for describing hydrogen.

12-7 SPECTROSCOPIC NOTATION

It is convenient at this point to introduce the conventional notation by which the quantum states of a single electron are specified. We have seen that, in the absence of an external magnetic field, a given energy level can be characterized by the three quantum numbers n, l, and j. It is customary to identify the value of l by letter code, as follows:

l:	0	1	2	3	4	5	...
Designation:	s	p	d	f	g	h	...

(For higher l, the normal alphabetical sequence is followed, as indicated in the above tabulation.) The logic of the above sequence is not at all obvious; it is mainly a legacy from empirical spectroscopy. The first four letters (s, p, d, f) were introduced to label certain notable series of spectral lines observed from the alkali metals (lithium, sodium, potassium, etc.). These four series, which were named *sharp*, *principal*, *diffuse*, and *fundamental*, respectively, were later recognized to involve sets of energy levels having particular values of l. More detail on this is given in Chapter 13.

The state of an individual electron (except for its z component of angular momentum) is then specified by writing the value of the principal quantum number n, followed by the letter identifying l—for example, $3p$ for $n = 3, l = 1$. But the spin-orbit interaction makes this description incomplete, so the value of j is added to the symbol as a subscript ($3p_{3/2}$ denotes $n = 3$, $l = 1, j = \frac{3}{2}$). Finally, as a reminder that the fine structure turns each level of a given l (except $l = 0$) into a doublet ($j = l \pm \frac{1}{2}$) the figure 2 is written as a superscript just in front of the l symbol, denoting what is called the *multiplicity* of a level of given l. Thus a level belonging to $n = 3, l = 1, j = \frac{3}{2}$ is written[6] as $3^2p_{3/2}$.

[6]It might seem that the multiplicity number 2 is superfluous. This is true if we are concerned only with states of a single electron, since all levels are then doublets unless $l = 0$. But if we have a possible combined state of two or more electrons, the multiplicity of a state of given total orbital angular momentum may assume other values.

In hydrogen, and in many other atomic systems, we think of a single electron as being responsible, by its quantum jumps, for the observed spectral lines, and the values of orbital and total angular momentum for the whole electron structure of the atom are identical with those of the single electron. But to give expression to the fact that a quantum jump is a change of state of the atom as a whole, it is written as a transition between states of the complete atom. This is done by using a *capital* letter to denote the orbital angular momentum. Thus, for example, a sodium atom with its single valency electron in a state with $n = 3$, $l = 1$, $j = \frac{1}{2}$ is designated by the notation $3^2 P_{1/2}$. (The electrons in the filled inner shells of sodium are paired off in such a way that their resultant total spin and total orbital momentum are zero and do not contribute to the total angular momentum of the atom.)

When we come to systems in which more than one electron contributes significantly to the total state of the atom (for example, calcium or mercury, each with two valency electrons) there may arise a real distinction between the quantum numbers of the atomic state as a whole (designated with capital letters for orbital angular momentum) and the quantum numbers of the individual electrons giving rise to it (designated with a lower case letter). For the present, however, we shall not need to consider this extra complication.

12-8 FINE STRUCTURE OF HYDROGEN ENERGY LEVELS

We have seen, using the simple Coulomb model, that the coarse structure of the energy levels of an electron in the hydrogen atom is given by the following set of energy eigenvalues:

$$\epsilon_n = -\frac{1}{n^2}$$

that is,

$$E_n = -\frac{M_e e^4}{2\hbar^2} \cdot \frac{1}{n^2} \quad \text{(cgs)} \quad (n = 1, 2, 3, \ldots) \quad (12\text{-}19)$$

Quantum jumps between these levels correspond closely to the Balmer and other series that we discussed on the basis of

Quantum states of three-dimensional systems

the original Bohr model in Chapter 1. However, we have remarked repeatedly that the Coulomb model does not tell the whole story. Since the hydrogen atom is such a basic and important system, it is worth looking more closely at some of the finer details.

The first point to consider is that an electron in one of the lower states (small n) in a hydrogen atom has a mean kinetic energy that is high enough to involve small but significant corrections due to relativistic increase of mass. The order of magnitude of this effect can be simply estimated using the Bohr model. From Section 1-7, the speed of the electron in the circular orbit having $n = 1$ is given by

$$v = \frac{e^2}{\hbar} = 2.19 \times 10^8 \text{ cm/sec}$$

This is nearly one percent of the speed of light. We have, in fact,

$$\frac{v}{c} = \frac{e^2}{\hbar c} \equiv \alpha \approx \frac{1}{137}$$

This pure number, $e^2/\hbar c$, is a very important quantity in atomic physics; it is called the *fine-structure constant* and is given the symbol α.

Now it is typical of relativistic effects that they involve fractional corrections to the energy of the order of v^2/c^2 with respect to the result of a nonrelativistic calculation, at least for speeds such that v/c is small. Thus, in the present case we may expect corrections of the order of $(1/137)^2$ or about 10^{-4} of the previously calculated value of E_n for $n = 1$, and smaller corrections for the higher levels. This is comparable to the size of the spin-orbit energy correction discussed in Chapter 11 (Section 11-5). The fact that the relativistic and spin-orbit energy effects are of comparable magnitude is no accident (see the exercises).

The relativistic corrections were first investigated by A. Sommerfeld in 1916, using an elaborate outgrowth of the Bohr model in which elliptic as well as circular orbits were considered. He employed, in addition to the principal quantum number n, a second quantum number that was, in effect, the forerunner of the quantum number l of the wave-mechanical analysis. For a given value of n this second quantum number

12-8   Fine structure of hydrogen energy levels

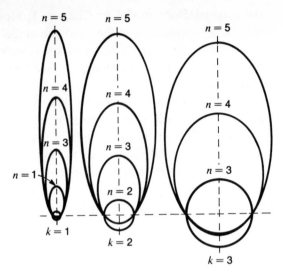

*Fig. 12-8 Som-
merfeld's elliptic
orbits for hydrogen.
Note that the orbits
for a given n have
the same major axis
and thus, clas-
sically, the same en-
ergy. Sommerfeld's
quantum number k
is related to the
modern quantum
number l: k = l + 1.*

could take on n different values, and it defined a set of ellipses of differing eccentricities but having major axes of equal length and therefore, according to classical mechanics, having the same energy. With the relativistic corrections included, the energy degeneracy was removed in such a way that the flattest ellipse came lowest in energy and the circular orbit highest, with an overall fractional splitting of the order of α^2/n. The corresponding wave-mechanical calculation (ignoring spin) leads to a very similar result, with the level for $l = 0$ coming lowest. See Figure 12-8.

We have mentioned that a thoroughgoing relativistic theory of the electron was developed by Dirac. When this was applied to the electron in a Coulomb field it yielded at one stroke the combined effect of the relativistic mass changes and the spin-orbit interaction. Dirac found that the energy levels did not depend explicitly on l, but were completely specified by the values of n and j, according to the following approximate equation:

$$E_{n,\,j} = E_n \left[1 + \frac{\alpha^2}{n} \cdot \left(\frac{1}{j + \frac{1}{2}} - \frac{3}{4n} \right) \right] \qquad (12\text{-}20)$$

where E_n is the energy as given by the Bohr theory and by nonrelativistic quantum mechanics. The way in which this compares with the results of Sommerfeld's relativistic calculation (without any consideration of spin) is shown in Figure 12-9. The Dirac result is a remarkable one because it brings levels of the same n, j but different l close together. Indeed, according to Dirac's theory they coincide exactly. But one of the most impressive pieces of experimental research in atomic physics in this century has shown that this is not quite correct. In 1947, W. E. Lamb and R. C. Retherford were able to prove experimentally that there is a small energy difference between the two levels for $n = 2$ having $j = \frac{1}{2}$ (but $l = 0$ and $l = 1$, respectively).

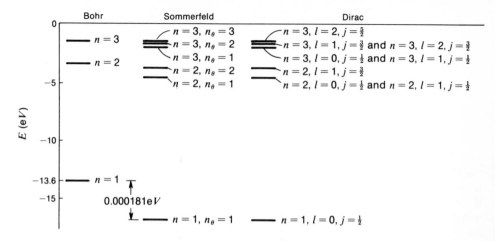

Fig. 12-9 Relation between the predictions of the Bohr theory, the Sommerfeld theory, and the Dirac theory of hydrogen. The displacements from the Bohr energy levels are exaggerated by a factor of (137)². [Note: The Sommerfeld constant n_θ in this figure is the same as the constant k in Figure 12-8.]

In the spectroscopic notation of the preceding section, these levels are designated $2S_{1/2}$ and $2P_{1/2}$. This splitting had been suspected on the basis of the best measurements possible through conventional spectroscopy. But Lamb and Retherford, using the new techniques of microwave spectroscopy, were able, in their very first experiments, to measure the energy difference to an accuracy of 10 percent, and ultimately they pushed the accuracy to better than one part in 10^4. Since the splitting itself corresponds to an energy difference of only about $4 \cdot 10^{-6}$ eV, its determination to 0.01 percent accuracy is impressive indeed.[7] The existence of this splitting (now known as the *Lamb shift*) means, in effect, that the electrostatic interaction between proton and electron is not described with complete accuracy by Coulomb's law—although there is no implication that Coulomb's law fails as a statement of the force between ideal point charges. The modified result can be understood in terms of the theory called quantum electrodynamics. From this same theory (as mentioned in footnote 11 in Section 11-5) came an explanation of the fact that the *g* factor for the electron spin is not exactly 2. That result is very relevant here, for if this *g* factor were indeed exactly 2, the splitting of the two levels of the same *j* but different *l* would not occur.

[7]For this work Lamb received the Nobel Prize for physics in 1953. His experiments with Retherford are described in a set of classic papers: Phys. Rev. **72**, 241 (1947); **75**, 1825 (1949); **79**, 549 (1950); **81**, 222 (1957). Lamb's Nobel lecture is reprinted in H. A. Boorse and L. Motz, *The World of the Atom*, Vol. II, Basic Books, New York, 1966, p. 1499.

12-8 Fine structure of hydrogen energy levels

Subsequent to the measurements of hydrogen fine structure by Lamb and Retherford, the techniques of optical spectroscopy were raised to a new order of accuracy by the use of lasers, which act as sources of spectral lines with unprecedentedly small widths. Figure 12-10a shows a profile of the H_α "line" clearly resolved into a number of separate components by these new techniques, and Figure 12-10b relates these components to transitions between the various fine-structure energy levels belonging to $n = 3$ and $n = 2$.

*Fig. 12-10   The H_α "line" of hydrogen resolved into separate components with the use of lasers. (a) The spectrum. (b) The corresponding transitions. [Adapted from T. W. Hänsch, I. S. Shahin, and A. L. Schawlow, Nature Phys. Sci. **235**, 63 (1972).]*

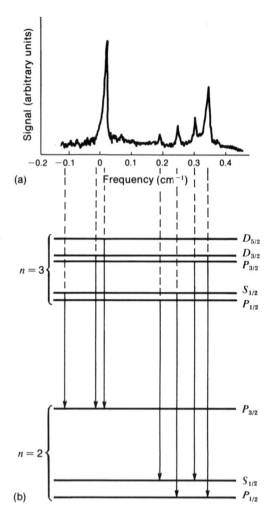

12-9 ISOTOPIC FINE STRUCTURE: HEAVY HYDROGEN

In the preceding section we were concerned with the fine structure associated with the details of the energy levels for

Quantum states of three-dimensional systems

hydrogen with a proton as nucleus. But there is another type of fine structure of a very simple kind, stemming directly from the fact that hydrogen, although it is a unique element chemically, exists in two different isotopic forms. Besides the common form, in which the nucleus of the atom is a single proton, there is a much rarer stable isotope in which the nucleus contains a proton and a neutron bound together in the combination known as a *deuteron*. This other isotope, called *heavy hydrogen* or *deuterium*, has an atomic mass close to 2 amu and is present in all naturally occurring hydrogen to the extent of about one atom in 7000.

Actually, every result obtained for hydrogen in this chapter must be corrected slightly to take account of the fact that the proton is not infinitely massive, as was implied by assuming a fixed center of force in setting up the equations. Just as in our discussion of vibrating molecules (Section 4-4) and rotating molecules (Section 11-3), we must replace the mass of the less massive particle by the *reduced mass* μ. In the case of hydrogen, the reduced mass is

$$\mu = \frac{M_e M_n}{M_e + M_n} = \frac{M_e}{(1 + M_e/M_n)} \tag{12-21}$$

where M_e is the electron mass and M_n is the mass of the nucleus, whether it is a proton or a deuteron. Mass-corrected values of all expressions then result if the mass of the electron is simply replaced by μ.

In particular, the energy levels (Eq. 12-19) are given by

$$E_n = -\frac{M_e e^4}{2\hbar^2 n^2} \cdot \frac{1}{1 + M_e/M_n} \tag{12-22}$$

Putting $M_n = 1$ atomic mass unit (amu) and 2 amu successively (with $M_e \approx 1/1840$ amu), one deduces that, for every spectral line due to ordinary hydrogen, there should be a corresponding line from deuterium with a wavelength shorter by about one part in 3680. This fractional difference implies, for example, a wavelength difference of 1.32 Å for the line at a wavelength 4861 Å representing the transition $n = 4$ to $n = 2$ (the so-called H_β line).

Historically, when the existence of deuterium began to be suspected as a result of the discovery of the neutron in 1932,

12-9   Isotopic fine structure: heavy hydrogen

an intensive search for it was begun. The strategy was to use thermal diffusion techniques to obtain samples of hydrogen enriched in the heavier isotope (if it was present at all in the ordinary hydrogen that provided the starting point) and then to test spectroscopically for the presence of faint, shifted Balmer lines at the expected wavelengths. H. C. Urey succeeded in this effort in 1932 and was awarded the Nobel Prize in 1934.[8] Figure 12-11 shows how the intensity of one of the Balmer lines of deuterium was observed to grow in progressively enriched samples of hydrogen gas—a beautiful example of how spectroscopy can be used as a delicate diagnostic tool, and a very nice verification of the atomic dynamics on which the calculated isotope effect was based.

Of course, all other atomic species having more than one isotope exhibit this isotopic effect in their spectra, but the fractional wavelength shift (for a mass difference of 1 amu between isotopes) falls off as $1/M^2$ (see the exercises) and thus for heavier atoms becomes a very small effect indeed.

12-10 OTHER HYDROGEN-LIKE SYSTEMS

As with the original Bohr theory (Chapter 1), it is an easy matter to adapt the wave-mechanical calculations of this chapter to the case of a single electron in the field of an arbitrary central charge, $+Ze$. Allowing for the effect of finite nuclear mass, but ignoring the effects of relativity and electron spin, the energy levels are given by

$$E_n = -\frac{M_e Z^2 e^4}{2\hbar^2(1 + M_e/M_A)} \cdot \frac{1}{n^2} \tag{12-23}$$

where M_A is the nuclear mass ($\approx A$ amu) for an atom of mass number A. Photon energies and wavelengths calculated from Eq. 12-23 represent quite well the observed spectra of atoms that are so highly ionized that only one electron is left—He^+, Li^{2+}, C^{5+}, and so on. It should be noted that the Balmer lines in these spectra (resulting from transitions to the $n = 2$ level from higher levels) move rapidly through the ultraviolet and toward the region of x-ray wavelengths as the atomic number Z

---

[8]H. C. Urey, F. G. Brickwedde, and G. M. Murphy, Phys. Rev. **40**, 1 (1932).

Quantum states of three-dimensional systems

Fig. 12-11 High-
resolution spectrum
of the H_β line
for (a) ordinary
hydrogen
and (b),(c) hy-
drogen evaporated
just above its triple
point, to concen-
trate deuterium.
Notice that no line
is found for tri-
tium—hydrogen
with a triton for a
nucleus (two neu-
trons and one pro-
ton). Wavelengths:
ordinary hydrogen
  H^1_β 4861.326 Å
deuterium
  H^2_β 4860.000 Å
Tritium (expected)
  H^3_β 4859.567 Å
[After H. C. Urey,
F. G. Brickwedde,
and G. M. Murphy,
Phys. Rev. **40**, 1
(1932).]

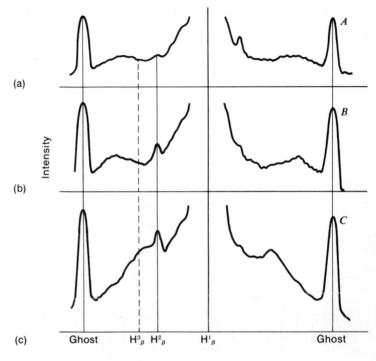

(a)

(b)

(c)

Intensity

A

B

C

Ghost H^3_β H^2_β H^1_β Ghost

increases. For a given transition (that is, given values of n_1 and n_2), the frequency of emitted radiation is proportional to Z^2.

A similar situation exists for the inner shells of many-elec-tron atoms. The electrons near the nucleus feel nearly the full effects of the nuclear charge and so can be described approxi-mately by a hydrogen-like energy scheme. This explains the pattern (discussed in Chapter 1) of characteristic K_α and K_β x-ray lines from atoms that have been ionized by the expulsion of an electron from the lowest energy level ($n = 1$)—that is, a K electron.

In the realm of ordinary optical (visible) spectra, some re-semblance to the hydrogen spectrum is exhibited also by cer-tain series of lines from the alkali metals. In the atom of an alkali metal, the single valency electron can be crudely pic-tured as moving in the field of a central charge of net magni-tude e, made up of the nuclear charge Ze reduced by the charge of the other $Z - 1$ electrons which effectively shield the valency electron from the nucleus. Clearly this picture can apply with accuracy only to those states of the valency elec-tron that do not involve any significant penetration into the

12-10   Other hydrogen-like systems

charge cloud of the other electrons, which exposes the valency electron to a larger effective charge. We shall consider this matter more fully in Chapter 13, but from a knowledge of the hydrogen wave functions one is led to expect the greatest anomalies with transitions involving states with $l = 0$. For $l = 0$ states, the probability density is relatively high near the nucleus, compared to states of the same n but nonzero l. At the other extreme, for states with large l (and hence also with large n, since $n \geq l + 1$) one can expect good agreement between the energy levels of hydrogen and the alkali metals. This is borne out in practice. The energy levels for various n for a valency electron with $l = 2$ or $l = 3$ in lithium, sodium, or potassium are found to lie quite close to the levels of the same n in hydrogen. This is shown in Figure 12-12 (based on evidence

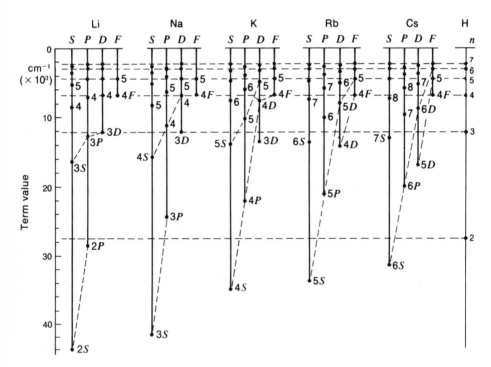

Fig. 12-12   Comparison of alkali atom energy levels with energy levels for hydrogen (right-hand column). Notice that for higher angular momentum states the energies approach those of the hydrogen atom for the same principal quantum number. (After   F. K. Richtmeyer, E. H. Kennard, and T. Lauritsen, Introduction to Modern Physics, McGraw-Hill, New York, 1955.)

Quantum states of three-dimensional systems

from the observed emission spectra of the alkali metals in vapor form).

One final point deserving special attention in the case of a hydrogen-like system with a high central charge Ze is the increased magnitude of relativistic effects. We have seen that in the hydrogen atom, as described by the Bohr theory, the value of v/c for the innermost orbit (for which v is greatest) is only $1/137$. However, in the general case, for the orbit $n = 1$, we have

$$\frac{v}{c} = \frac{Qe}{\hbar c} \approx \frac{Z}{137} \tag{12-24}$$

Thus, to take an extreme example, if we consider an electron in the $n = 1$ orbit of a uranium atom, with $Z = 92$ and $Q \approx 91e$ (since the inner orbit is shared with another electron) the value of v/c is about $\frac{2}{3}$, according to Eq. 12-24. In this case, relativistic effects are clearly very important; indeed one can recognize that a thoroughgoing relativistic analysis from the beginning is obligatory, rather than an analysis based on the nonrelativistic Schrödinger treatment with minor relativistic corrections added later. One can guess that in such circumstances the splitting of levels of the same n but different j will be relatively much greater than in hydrogen, more or less in proportion to Z^2. The energy levels according to the exact form of Dirac's relativistic theory (ignoring the nuclear mass correction) are in fact given by the following formula:

$$\frac{E_{n,j}}{M_e c^2} = \left\{ 1 + \frac{\alpha^2 Z^2}{(n - j - \frac{1}{2}) + [(j + \frac{1}{2})^2 - \alpha^2 Z^2]^{1/2}} \right\}^{-1/2} - 1 \tag{12-25}$$

For $Z \ll 1$, this reduces in the first approximation to the energy levels as given by the simple Bohr theory. In slightly better approximation it gives

$$E_{n,j} \approx E_n \left[1 + \frac{\alpha^2 Z^2}{n} \cdot \left(\frac{1}{j + \frac{1}{2}} - \frac{3}{4n} \right) \right] \tag{12-26}$$

which is just like the result we quoted earlier for the Dirac theory of hydrogen fine structure (Eq. 12-20), but with the relative fine-structure splittings increased by the expected factor equal to the square of the central charge (measured in units of e).

12-1 *The infinitely deep spherical well.* Consider a bound particle in an infinitely deep potential well which has spherical symmetry around a binding center: $V(r) = 0$ for $0 < r < R$; $V(r) \to \infty$ for $r > R$. A complete set of states can be chosen which are eigenstates of both energy and angular momentum.

(a) Find the spectrum of allowed energy values for $l = 0$. (Remember the requirement that ψ be finite everywhere.)

(b) Draw an effective potential energy curve for $l \neq 0$.

(c) Using the knowledge that the lowest possible energy state for $l \neq 0$ must lie above the bottom of the effective potential energy well, show that the lowest state with $l = 6$ must lie above the second energy level for $l = 0$.

(d) See if, *without* involved calculation, you can convince yourself that the result of (c) is too conservative—that is, the first state for some l value *less* than 6 stll lies above the second state for $l = 0$. (You may be able to make a good case for $l = 4$—and detailed analysis shows that even $l = 3$ satisfies this condition.)

12-2 *The spherical well of finite depth.* Consider a particle in a spherical potential well of finite depth:

$$V(r) = 0, \qquad 0 < r < R$$
$$V(r) = \frac{\alpha^2 h^2}{8mR^2}, \qquad r \geq R$$

[The parameter α^2 is really a dimensionless unit of energy. Recall that the energy levels of an infinite one-dimensional square well of width L are $n^2 h^2/(8mL^2)$.]

(a) Find the dimensionless radial equation for states with $L^2 = l(l + 1)\hbar^2$.

(b) How many bound energy values with $l = 0$ exist in this potential? (Similar considerations appeared in Exercise 5-7.) Evaluate your result for $\alpha = 1.6$.

(c) Carefully examine the expression for the effective potential. What is the smallest l value for which you can rule out the possibility of a bound state? Evaluate your result for $\alpha = 1.6$.

(d) For what values of α can you rule out the existence of any bound state with $l = 2$?

12-3 *Solutions of the radial wave equation for hydrogen.*

(a) Refer to Section 12-3 concerning radial wave functions for hydrogen, and verify that a solution of the form $(c_0 - c_1\rho)\, \rho^{l+1}\, e^{-b\rho}$ corresponds to the energy eigenvalue $-1/(l + 2)^2$ in dimensionless units.

Quantum states of three-dimensional systems

(b) Putting $l = 1$, sketch (i) the form of the wave function; and (ii) the effective potential-energy curve with the energy eigenvalue superimposed as a horizontal line.

12-4 *Radial probability density for a hydrogen state.* Consider the spherically symmetric hydrogen-atom state for $n = 2$. The (unnormalized) radial function is

$$r\psi_2 = u_2(r) = re^{-r/2a_0}\left(1 - \frac{r}{2a_0}\right)$$

Sketch the radial probability distribution $w(r) \sim r^2\psi^2$ for this state and find the value of r for which it has a maximum.

12-5 *Orthogonality of hydrogen-atom wave functions.* By direct integration show that the hydrogen state $|n = 1, l = m = 0\rangle$ is orthogonal to the state $|n = 2, l = m = 0\rangle$.
[*Note:* No need to use normalized functions.]

12-6 *Orthogonality of radial wave functions.* The hydrogenic eigenfunctions $\psi_{nlm}(\mathbf{r}) = R_{nl}(r)Y_{lm}(\theta, \phi)$ constitute an orthogonal set. That is, $\int \psi_{n'l'm'}\psi_{nlm} \, dV = 0$ unless all three of the quantum numbers match: $n' = n$, $l' = l$, $m' = m$. To satisfy this property, which of the following "radial integrals" must vanish? State a general criterion for your choices.

(a) $\displaystyle\int_0^\infty R_{31}^* R_{32} \, r^2 \, dr$

(b) $\displaystyle\int_0^\infty R_{31}^* R_{21} \, r^2 \, dr$

(c) $\displaystyle\int_0^\infty R_{53}^* R_{63} \, r^2 \, dr$

(d) $\displaystyle\int_0^\infty R_{52}^* R_{63} \, r^2 \, dr$

12-7 *Nonconservation of angular momentum in a rectangular box.* Consider a particle confined to a three-dimensional box: $V(x, y, z) = 0$ for $|x| \le a/2$, $|y| \le b/2$, and $|z| \le c/2$; and $B \to \infty$ otherwise. (We have centered the box on the coordinate origin.)

(a) Give a physical argument why a classical particle in a rectangular box *would not* have a constant angular momentum about the center, while the orbital angular momentum of a particle in a spherical well *would* be conserved.

(b) Give an argument based on the time-independent Schrödinger equation to show that there are no states that have both a definite energy and a definite magnitude of the orbital angular momentum about the origin.

12-8 *Expectation values for hydrogen: powers of the radius.*
Calculate the expectation values of r and of r^2 for the ground state of hydrogen. Note that

$$\int_0^\infty r^n e^{-\alpha r} \, dr = n!/\alpha^{n+1}$$

12-9 *Bohr orbits and the radial wave functions in hydrogen.*
(a) Verify Eq. 12-17 that, for hydrogen-atom states with the largest z components of angular momentum for a given l, the radial probability density is a maximum at the radius of the corresponding Bohr orbit.

(b) Calculate the expectation value $\langle r \rangle$ for the radial position of the electron in the above states. Discuss the difference between the results of (a) and (b). [Refer to integral formula in previous exercise.]

12-10 *Expectation values of potential and kinetic energy in hydrogen.*
(a) Calculate the expectation value of the potential energy of the electron in the ground state of hydrogen.

(b) From the conservation of energy, calculate the expectation value of the kinetic energy of the electron in the same state.

(c) Draw a diagram relating total energy, expectation value of potential energy, and expectation value of kinetic energy for this state.

12-11 *Electron inside the proton?* Calculate the probability that the electron is within the volume occupied by the proton in the ground state of the hydrogen atom. [Assume that the wave function for the electron can be adequately approximated by the solution for a point-charge nucleus. Verify that the proton radius ($\sim 10^{-13}$ cm) is so small that the value of the electron wave function everywhere inside the proton is very nearly its value at $r = 0$.]

12-12 *Bohr and Schrödinger models of the hydrogen atom.* Write a short paragraph or construct a table that compares and contrasts the descriptions of the lowest-energy state of the hydrogen atom according to the Bohr model and according to wave mechanics.

12-13 *Time-dependent state of a hydrogen atom.* Consider a hydrogen atom described by the following wave function (ignoring spin):

$$\Psi(\mathbf{r}, t) = \sqrt{\tfrac{2}{3}} \, \psi_{100}(\mathbf{r}) \, e^{-iE_1 t/\hbar} + \sqrt{\tfrac{1}{3}} \psi_{211}(\mathbf{r}) \, e^{-iE_2 t/\hbar}$$

Quantum states of three-dimensional systems

(a) What is the probability that measurements on this state will yield the results $E = E_2$, $L^2 = 2\hbar^2$, and $L_z = \hbar$?

(b) What is the expectation value of E? of L^2? of L_z?

(c) Does $|\Psi|^2$ depend on t? on r? on θ? on ϕ? If it depends on t, what is the repetition period?

12-14 *Hydrogen-atom states including electron spin.* A hydrogen atom is in the $4F$ state.

(a) What are the possible values for the total angular momentum of the electron (including spin)?

(b) What are the possible z components of angular momentum?

(c) List all the possible substates in spectroscopic notation.

12-15 *Spectroscopic notation and addition of angular momenta.* "Translate" the angular momentum states listed below (in spectroscopic notation) into three quantum numbers L, S, and J. Which of these "candidate" states correspond to possible values of L, S, and J for a *one-electron* atom? For each *rejected* set (L, S, J), give at least one reason why it cannot be realized.

Candidate list: $^2S_{1/2}$; $^2D_{1/2}$; $^2P_0$; $^3P_1$; $^2P_{1/2}$; $^2F_{7/2}$; $^1S_0$; $^2S_{3/2}$; $^2D_{5/2}$; $^2P_{3/2}$; $^2P_{5/2}$; $^2D_{3/2}$; $^2G_{5/2}$.

12-16 *Comparison of relativistic and spin-orbit effects.* This quote is from Section 12-8, "The fact that the relativistic and spin-orbit energy effects for low-energy states in hydrogen are of comparable magnitude is no accident." Review the discussion of spin-orbit coupling in Section 11-5 and express in terms of fundamental constants the fractional energy shift $|\Delta E/E|$ of the lowest Bohr orbit. Show that this shift is equal to $\alpha^2 = (e^2/\hbar c)^2$, which is the same as the order-of-magnitude estimate of the relativistic energy shift made in Section 12-8.

12-17 *Dirac theory of hydrogenic levels.* Which two of the following hydrogenic transitions have the same wavelength, ignoring the Lamb shift? Refer to Eq. 12-20 or to Eq. 12-25. (i) $3\,^2P_{1/2} \to 2\,^2S_{1/2}$; (ii) $3\,^2P_{3/2} \to 2\,^2S_{1/2}$; (iii) $3\,^2S_{1/2} \to 2\,^2P_{1/2}$; (iv) $3\,^2S_{1/2} \to 2\,^2P_{3/2}$.

12-18 *Relativistic theory of the hydrogen atom.* We pointed out in Section 12-8 that a remarkably successful theory of fine structure in hydrogen was developed by Sommerfeld, well before the advent of wave mechanics, by extending the Bohr theory to quantized elliptic orbits and by incorporating relativistic corrections to the energy (see Figures 12-8 and 12-9). The form of Eq. 12-20, although it is presented in the text as an approximation to Dirac's exact solution, in fact corresponds exactly to the form of Sommerfeld's results, although the notation and the physical significance attached to the quantum numbers is different in the two theories.

Show that Dirac's exact solution (Eq. 12-25) leads to the simpler form, Eq. 12-26, by suitable approximation ($\alpha \ll 1$) and hence to the Sommerfeld form of solution (Eq. 12-20) for the hydrogen fine structure.

12-19 Hyperfine structure. Many nuclei have magnetic moments. *Atomic hyperfine structure* refers to the effects on atomic spectra caused by the presence of this nuclear magnetic moment. Hyperfine structure in hydrogen is smaller than its fine structure by approximately the ratio m/M, where m is the mass of the electron and M the proton mass. Explain this result. The discussion in Section 10-4 may be useful.

12-20 Positronium. Positronium is a short-lived "atomic" system consisting of an electron bound electrically with a positron (same mass as the electron, positive charge e). Use the reduced-mass analysis of hydrogen (Section 12-9) to predict the ionization energy of positronium and the wavelength of light given off when positronium drops from state $n = 2$ to $n = 1$. Neglect electron and positron spin. Compare to the wavelength of the corresponding transition in hydrogen (Lyman alpha: $\lambda = 1216$ Å). Should any of the transitions of this atomic system lie in the visible region of the spectrum?

12-21 Isotopic spectral shifts for heavy atoms. Show that the fractional wavelength shift $\Delta\lambda/\lambda$ in the spectrum of an atom due to a mass difference ΔM_A between isotopes is proportional to $\Delta M_A/M_A^2$, where M_A is the nuclear mass. Assume that the permitted energies of the atom (when the nucleus is assumed to be fixed) are proportional to M_e, the mass of an electron, as is the case for hydrogen. The $1/M_A^2$ dependence means that the isotopic shift becomes very small for heavy elements.

12-22 Three-dimensional isotropic simple harmonic oscillator. Consider the isotropic three-dimensional simple harmonic oscillator whose potential energy function has the form

$$V = \tfrac{1}{2}m\omega_o^2(x^2 + y^2 + z^2) = \tfrac{1}{2}m\omega_o^2 r^2$$

and is therefore a central potential.

(a) Separate the Schrödinger equation in Cartesian coordinates (x, y, z). Show that the permitted energies are

$$E_n = (n + \tfrac{3}{2})\hbar\omega_o$$

Quantum states of three-dimensional systems

Write down the degeneracy of each energy level for $n \leq 4$ [to be used in part (d) below].

(b) Now consider the solution in spherical polar coordinates. Will the Schrödinger equation be separable in spherical polar coordinates? If so, name the angular solutions and write down the equation for the radial function.

(c) Consider the s states ($l = 0$). Show that for this case the radial equation for the function $u(r) = r R(r)$ is identical in form to that of the *one*-dimensional simple harmonic oscillator, so that the one-dimensional solutions should apply (Chapter 4, especially Table 4-1). However, show that the acceptable one-dimensional solutions are limited by the additional requirement that $R(r)$ be finite at the origin. Show that the resulting values of the energy for s states are limited to those for *even* values of n (including zero) in the equation given in (a).

(d) Without solving the radial equation, you can *guess* dependably and accurately the values of n, l, and m for each state of the three-dimensional oscillator. Strategy: For each energy, choose the lowest permissible l values [part (c)] that give the total number of states equal to the degeneracy for that energy [part (a)]. Carry out this choice for $n \leq 4$. [Partial answer: The $n = 3$ energy level is tenfold degenerate—part (a). The value $l = 0$ is not allowed, since n is odd —part (c). Inventory: the value $l = 1$ is threefold degenerate ($m = 1, 0, -1$). Similarly, the value $l = 2$ is fivefold degenerate and the value $l = 3$ is sevenfold degenerate. Choosing among these, only the pair $l = 1$ and $l = 3$ yield a total of ten states for $n = 3$, as required.]

(e) Construct an energy-level diagram for the isotropic oscillator for $n \leq 4$. Use a separate column in your diagram for each value of l, as is done in Figure 12-12. By inspection, extend your diagram to $n = 5$ and check that the degeneracy is right. Notice the systematic properties of the l values for the successive values of n. This reflects the parity of the wave functions for a given value of n. (see the subsection "Properties of the eigenfunctions" in Section 11-2).

A new phase of my scientific life began when I met Niels Bohr personally for the first time. This was in 1922, when he gave a series of guest lectures at Göttingen, in which he reported on his theoretical investigations on the Periodic System of Elements ... The question as to why all electrons for an atom in its ground state were not bound in the innermost shell had already been emphasized by Bohr as a fundamental problem ... It made a strong impression on me that Bohr at that time and in later discussions was looking for a general *explanation which should hold for the closing of* every *electron shell ...*

WOLFGANG PAULI, Nobel Lecture (1946)

13

Identical particles and atomic structure

13-1   INTRODUCTION

Extending the one-particle Schrödinger equation from one to three spatial dimensions enabled us to describe electron states in the hydrogen atom and in other hydrogen-like systems. In the present chapter we further generalize the Schrödinger equation from one to two and then to many particles bound as a single system. This will enable us to account in considerable detail for the states of the helium atom, and to understand the main features of the electronic structures of other atoms. Central to this generalization is a quantum effect that has no classical analog whatever. This effect, described by the *Pauli exclusion principle*, is that no two electrons in the same system can be in the same quantum state. The structure of any atom with more than one electron—and therefore the structure of the physical world—depends crucially on the exclusion effect.

The exclusion effect itself, as we shall see shortly, is a consequence of the fundamental property of electrons (and other particles) already used in describing one-particle quantum states (Section 5-7): Electrons are identical in the sense that they are *completely indistinguishable* from one another. When the wave function of a multielectron atom is written so as to embody this indistinguishability, the exclusion principle is automatically satisfied. In this chapter we shall trace the profound effects of these apparently simple features of the properties of electrons.

The two-electron system, such as helium, simple as it is, is too complicated to allow closed analytic solutions for the wave functions. Of course the same is true for all other multielectron atoms. The fundamental reason for this complication is that the electrons repel one another electrically. Roughly speaking, we cannot calculate the state of one electron in the atom until we know the states of all other electrons in the atom, but the states of these other electrons depend on the state of the electron we are considering. Added to this is the fundamental complication that the electrons are indistinguishable! The result of electron interaction and indistinguishability is that there is no natural way to break the multielectron problem down into smaller subproblems without introducing approximations. Even these approximations involve considerable numerical computation. Happily a good deal of intuitive insight can be provided by such approximations, even for atoms with very many electrons. Nevertheless, for atoms with two or more electrons, exact wave functions cannot be found.

Before taking up the problem of the helium atom itself, we consider in more general and basic terms the extension of Schrödinger's equation to two particles.

13-2   SCHRÖDINGER'S EQUATION FOR TWO NONINTERACTING PARTICLES

Suppose that two *noninteracting* particles are bound in the same environment, consisting, for example, of an external electric field. Let the masses of the particles be m_1 and m_2. For the moment we limit attention to a one-dimensional system in which the positions of the particles are x_1 and x_2. The potential energies of the particles due to their environment can then be written as $V_1(x_1)$ and $V_2(x_2)$ respectively. In general, even though the environment is the same for both, the values of V_1 and V_2 need not be the same, even when $x_1 = x_2$; the particles might, for example, be a proton (charge $+e$) and an alpha particle (charge $+2e$). The cases of most interest to us will in fact be those for which the particles are identical, so that $q_1 = q_2$, $m_1 = m_2$, and $V_1(x) = V_2(x)$, but we do not introduce these conditions just yet.

Classically, the total energy of the system described

Identical particles and atomic structure

above will be given by the following equation:

$$\frac{p_1^2}{2m_1} + \frac{p_2^2}{2m_2} + V_1(x_1) + V_2(x_2) = E \tag{13-1}$$

We now suppose that the state of the two particles together can be described by a certain function Ψ that depends on x_1, x_2, and t. Introducing the operators corresponding to p_1, p_2, and E, we deduce that Ψ must obey the following form of the time-dependent Schrödinger equation:

$$-\frac{\hbar^2}{2m_1}\frac{\partial^2\Psi}{\partial x_1^2} - \frac{\hbar^2}{2m_2}\frac{\partial^2\Psi}{\partial x_2^2} + V_1(x_1)\Psi + V_2(x_2)\Psi = i\hbar\frac{\partial\Psi}{\partial t} \tag{13-2}$$

Now each kinetic and potential energy operator is a function of one coordinate only, either x_1 or x_2 but not both. Moreover, the term on the right side of Eq. 13-2 involves an operator that is a function of time alone. When this type of condition occurred before, for example with the x, y, and z coordinates of a particle in a rectangular box (Section 5-4), the x and y coordinates for a two-dimensional harmonic oscillator (Section 10-8), or with the r, θ, ϕ, coordinates of the electron in the hydrogen atom (Section 12-2), it was possible to separate the variables by substituting a product solution, each factor of which is a function of only one variable. In the present case we look for solutions Ψ which can be expressed as products of three separate functions of the individual independent variables x_1, x_2, and t:

$$\Psi(x_1, x_2, t) = \psi_A(x_1) \cdot \psi_B(x_2) \cdot f(t) \tag{13-3}$$

Substituting this into Eq. 13-2, we then obtain the following equation:

$$\left[-\frac{\hbar^2}{2m_1}\frac{d^2\psi_A}{dx_1^2} + V_1(x_1)\psi_A\right]\psi_B f$$
$$+ \left[-\frac{\hbar^2}{2m_2}\frac{d^2\psi_B}{dx_2^2} + V_2(x_2)\psi_B\right]\psi_A f = i\hbar\psi_A\psi_B\frac{df}{dt}$$

We can then see that this factorization permits solutions in which ψ_A and ψ_B represent solutions of time-independent Schrödinger equations for the particles individually. (This is not surprising since the particles are assumed to be nonin-

13-2 Two noninteracting particles

teracting.) We can, in fact, put

$$-\frac{\hbar^2}{2m_1}\frac{d^2\psi_A}{dx_1^2} + V_1(x_1)\psi_A = E_A\psi_A$$

and

$$-\frac{\hbar^2}{2m_2}\frac{d^2\psi_B}{dx_2^2} + V_2(x_2)\psi_B = E_B\psi_B$$

Substituting these in the previous equation, we arrive at the following simple differential equation for the time-dependent factor f:

$$(E_A + E_B)f = i\hbar\frac{df}{dt}$$

leading to the solution

$$f(t) = e^{-i(E_A + E_B)t/\hbar}$$

This means that the Schrödinger wave function Ψ can be written in the form

$$\Psi(x_1, x_2, t) = \psi_A(x_1) \cdot \psi_B(x_2) e^{-iEt/\hbar} \tag{13-4}$$

where $E = E_A + E_B$. Thus we have an overall solution that is a product of single-particle space functions, but whose time dependence is defined by the total energy of the two particles.

The interpretation of Ψ as a probability amplitude involves the specification of the positions of *both* particles. The probability that particle number 1 is within a small range of positions dx_1 at x_1 *and* that particle number 2 is within a small range of positions dx_2 at x_2 is given by

$$dP = |\Psi|^2 dx_1 dx_2 = |\psi_A(x_1)|^2 dx_1 \cdot |\psi_B(x_2)|^2 dx_2$$

Note that this expression conforms to the rule for the joint occurrence of two independent events, that is, the probability of occurrence of both of two independent events is the product of the individual probabilities.

Identical particles and atomic structure

We now add the condition that the two particles are identical and therefore truly indistinguishable. It then follows that although the symbols x_1 and x_2 continue to denote the positions (in general different) of the two particles, we can no longer interpret them as meaning that particle number 1 is at x_1 and particle number 2 is at x_2; it could just as well be the other way around. And what this means is that the combined states hitherto described by $\psi_A(x_1) \cdot \psi_B(x_2)$ and $\psi_A(x_2) \cdot \psi_B(x_1)$ must be physically indistinguishable. However, the product functions as they stand are not consistent with this indistinguishability. We illustrate this with a specific example.[1]

Suppose that our two particles are confined to the one-dimensional box with infinite walls, and that ψ_A and ψ_B represent the first and second energy states in this potential. Then we shall have (see Chapter 3),

$$\psi_A(x_1)\psi_B(x_2) \sim \sin\left(\frac{\pi x_1}{L}\right) \cdot \sin\left(\frac{2\pi x_2}{L}\right)$$

and

$$\psi_A(x_2)\psi_B(x_1) \sim \sin\left(\frac{\pi x_2}{L}\right) \cdot \sin\left(\frac{2\pi x_1}{L}\right)$$

If we pick arbitrary values of x_1 and x_2, the values of these two products are different; for example, if $x_1 = L/2$ and $x_2 = L/4$ the first product of sines is unity and the second is zero. We can give a vivid picture of the overall situation by constructing a sort of contour map, as shown in Figure 13-1, in which the individual coordinates x_1 and x_2 are displayed on two perpendicular axes and the values of $\psi_A \cdot \psi_B$ are plotted as positive or negative magnitudes with respect to the plane of the paper. The contours are lines of constant $\psi_A \cdot \psi_B$. Each of the two possible products then gives a hill and a valley in this representation of $\psi_A \cdot \psi_B$, but the two diagrams imply different probability densities, in general, for an arbitrary choice of the pair of coordinates (x_1, x_2). This cannot be tolerated if the particles are indistinguishable. Any acceptable wave function $\psi(x_1, x_2)$ for the system must have $|\psi(x_1, x_2)|^2 = |\psi(x_2, x_1)|^2$.

The kind of contradiction shown in the above example can be removed if we construct a wave function that is either the

---

[1]Our treatment from here on in this section owes a large debt to a discussion of the same problem in C. W. Sherwin, *Introduction to Quantum Mechanics*, Holt-Dryden, New York, 1959.

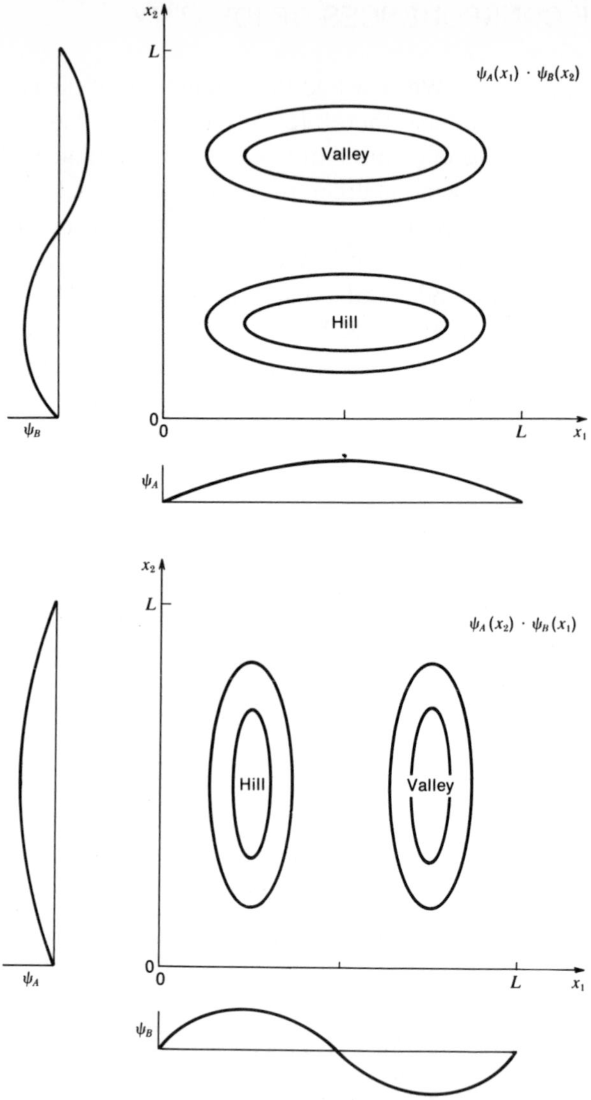

Fig. 13-1 Product
wave functions for
two noninteracting
particles in an infi-
nite square well.
One of the particles
is in the lowest-
energy state, the
other is in the sec-
ond energy state.
The product func-
tions do not yield
probability densities
that are indistin-
guishable with re-
spect to exchange of
the two particles
and therefore can-
not be valid wave
functions. (After
Chalmers W. Sher-
win, Introduction to
Quantum Me-
chanics, Holt-
Dryden, New York,
1959.)

sum or the difference of the product functions with x_1 and x_2 in-
terchanged. Each of these combinations gives a probability
distribution that is unaffected by an interchange of the two particles. Allowing for correct normalization, these combinations
are as follows:

Symmetric:

$$\psi_s(x_1, x_2) = \frac{1}{\sqrt{2}} [\psi_A(x_1) \cdot \psi_B(x_2) + \psi_A(x_2) \cdot \psi_B(x_1)] \qquad (13\text{-}5a)$$

Identical particles and atomic structure

Antisymmetric:

$$\psi_a(x_1, x_2) = \frac{1}{\sqrt{2}} [\psi_A(x_1) \cdot \psi_B(x_2) - \psi_A(x_2) \cdot \psi_B(x_1)] \qquad (13\text{-}5b)$$

The labels *symmetric* and *antisymmetric* arise from the fact that an interchange of the coordinates x_1 and x_2 leaves the first combination unchanged, but leads to a reversal of sign in the second combination. The values of the probability density $|\psi|^2$ are thus invariant with respect to the interchange in both cases.

The resulting contour maps of the probability amplitudes of the symmetric and antisymmetric functions are somewhat as shown in the two diagrams of Figure 13-2. We note some

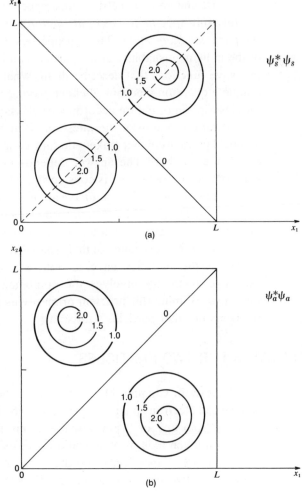

Fig. 13-2 (a) The joint probability distribution for two identical particles in a symmetric combination of the two lowest states in an infinite square well. (b) The joint probability distribution for the antisymmetric combination of the same two states.

13-3   The consequences of identity

very striking features. In the symmetric state the maximum probability occurs for two points at which $x_1 = x_2 = L/4$ or $3L/4$, and quite generally the probability is large only if the difference between x_1 and x_2 is small (particles close together). In the antisymmetric state, on the other hand, the probability is *zero* everywhere along the diagonal that corresponds to $x_1 = x_2$, and the peaks of the probability distribution occur for situations in which x_1 and x_2 are widely different (particles far apart). Thus if the particles are described by a symmetric function of the space coordinates they tend to huddle together, whereas if they are described by an antisymmetric function they act as if they were repelling one another.

It is important to recognize that the properties of "huddling together" or "avoiding one another" that characterize symmetric and antisymmetric states respectively arise *without our invoking specific forces of attraction or repulsion between the particles themselves.* These properties arise naturally out of the indistinguishability of the particles and the symmetry of the wave functions that describe them. What we see here is a uniquely quantum-mechanical effect, having no counterpart in the description of nature according to classical physics.

Naturally one asks the question: Which choice do two identical particles in fact make—do they congregate or do they avoid one another? The answer depends in a basic way on what type of particles one is considering. However, the question, as posed above, does not have an unequivocal answer; its scope is too limited. What is involved is a symmetry property that applies to a *complete* interchange of the roles of the two particles, not just exchange of their spatial coordinates. In particular, if the particles have an intrinsic spin, the overall exchange symmetry involves the combined spin state. We shall now examine this problem primarily as it applies to two electrons or other particles with spin $\frac{1}{2}$.

13-4 SPIN STATES FOR TWO PARTICLES

We saw in Chapter 11 (Section 11-4) how the property that we call electron spin is expressed in two possible configurations—"spin up" or "spin down"—of an electron with respect to a given axis of quantization. Let us label these spin states α and β, respectively, and then consider the spin states available to two electrons together. We shall label the elec-

Identical particles and atomic structure

trons as number 1 and number 2, so that, for example, the combination $\alpha(1) \cdot \beta(2)$ describes a situation in which number 1 has spin up and number 2 has spin down. The possible combinations of the individual spin states are then the following:

$$\alpha(1) \cdot \alpha(2) \qquad \alpha(1) \cdot \beta(2) \qquad \alpha(2) \cdot \beta(1) \qquad \beta(1) \cdot \beta(2)$$

The first and the last are automatically symmetric with respect to an interchange of the two electrons, but the second and third, like the products $\psi_A \psi_B$ of the space functions, are neither symmetric nor antisymmetric. However, from these second and third products we can, just as with the space functions, construct two other spin functions that have definite symmetry properties:

Symmetric:

$$\frac{1}{\sqrt{2}} [\alpha(1)\beta(2) + \alpha(2)\beta(1)]$$

Antisymmetric:

$$\frac{1}{\sqrt{2}} [\alpha(1)\beta(2) - \alpha(2)\beta(1)]$$

We then observe that we have a total of four combined spin states with definite exchange symmetry, three of them being symmetric and one antisymmetric:

Symmetric	*Antisymmetric*
$\alpha(1)\alpha(2)$	
$\frac{1}{\sqrt{2}} [\alpha(1)\beta(2) + \alpha(2)\beta(1)]$	$\frac{1}{\sqrt{2}} [\alpha(1)\beta(2) - \alpha(2)\beta(1)]$
$\beta(1)\beta(2)$	

Looking at the *symmetric* spin states, we notice that one of them describes both spins up, one describes both spins down, and one is a superposition of up-and-down spins. The values of the z component of the *combined* spin defined by these functions correspond to resultant z components of spin equal to $+1$, 0, and -1. In contrast, the single *antisymmetric* spin function with its equal superposition of up and down spins corresponds to a spin z component of zero only. These respective values of z angular momentum are those we would associate, by analogy with the properties of orbital angular momentum, with combined spin quantum numbers $S = 1$ for the symmetric states, and $S = 0$ for the antisymmetric. Thus for two electrons, or for a pair of any particles of half-integer spin, there are two basic kinds of spin states: a group of three symmetric states (a *triplet*) belonging to $S = 1$, and one antisymmetric state (a *singlet*) belonging to $S = 0$. We shall denote the complete set of spin functions of these two different kinds of states by χ_s and χ_a, respectively.

If we had considered two spinless particles, for example two alpha particles, there would of course be no spin states and the above complications would never arise. On the other hand, for two particles of intrinsic spin greater than $\frac{1}{2}$, we should have a classification into symmetric and antisymmetric states much as we have found for electrons, with singlets, triplets, and this time also with *multiplets*. For example, for two particles of intrinsic spin one (e.g., deuterons), their combination can have $S = 0$ (one state), $S = 1$ (three states), and $S = 2$ (five states).

The total wave function for two identical particles is then the product of a spatial function ψ and one of the possible spin functions χ. It is the *overall* symmetry of this *total* wave function that we shall now consider.

13-5 EXCHANGE SYMMETRY AND THE PAULI PRINCIPLE

Overall Symmetry of the Wave Function

Given the existence of symmetric and antisymmetric space functions (ψ_s, ψ_a) and symmetric and antisymmetric spin functions (χ_s, χ_a), it is possible to construct symmetric and antisymmetric total wave functions as follows:

Symmetric:

$\psi_s\chi_s$ *or* $\psi_a\chi_a$

Antisymmetric:

$\psi_s\chi_a$ *or* $\psi_a\chi_s$

All of these wave functions satisfy the property of indistin-

Identical particles and atomic structure

TABLE 13-1 Some Fermions and Bosons

(Intrinsic Spin Given in Parenthesis)

Fermions	Bosons
Antisymmetric with respect to complete exchange	Symmetric with respect to complete exchange
electron ($\frac{1}{2}$)	pi meson (0)
proton ($\frac{1}{2}$)	alpha particle (0)
neutron ($\frac{1}{2}$)	He^4 atom (0)
He^3 atom ($\frac{1}{2}$)	deuteron (1)

guishability between the two particles involved. Therefore, indistinguishability alone gives us no basis for deciding which of these functions is valid for any species of particle. However, these different functions predict different *properties* of the two-particle systems, in particular the quantized energies of the combined system. For two electrons, the only states found in nature are those for which the total wave function is *antisymmetric*. This fact was first recognized (in 1926) by Heisenberg,[2] and led to a detailed understanding of the energy-level system of the helium atom (which we shall discuss in Section 13-8). The condition of antisymmetry eliminates half of the combined states that would otherwise be possible for two electrons.

Not all wave functions for identical particles are antisymmetric, but the study of observed states shows that, for a given kind of particle, states of only one overall symmetry exist: the states of a given species are characteristically all symmetric *or* all antisymmetric. For all particles of half-integral spin, the total wave function, as for electrons, is always antisymmetric with respect to (complete) interchange, whereas with all particles of integral spin the total wave function is always symmetric. Table 13-1 lists a selection of particles of each main type. In this classification, particles of half-integral spin (such as electrons) are called *fermions* after Enrico Fermi; particles with the other kind of symmetry are called *bosons* after the Indian physicist S. N. Bose.

Overall Symmetry and the Exclusion Principle

As far as atomic structure is concerned, the most impor-

[2]W. Heisenberg, Z. Phys. **39**, 499 (1926).

13-5   Exchange symmetry and the Pauli principle

tant manifestation of these characteristic symmetry properties of identical particles is summarized in the *Pauli exclusion principle*. One can see the basis of this in a property of the antisymmetric space function ψ_a of Eq. 13-5b:

$$\psi_a(1, 2) = \frac{1}{\sqrt{2}}[\psi_A(1)\psi_B(2) - \psi_A(2)\psi_B(1)]$$

If in this expression we choose the states A and B to be the *same* single-particle state, then the total wave function vanishes identically. This is another way of saying that no such total state exists. If we now extend the meaning of the symbols ψ_A and ψ_B to refer not just to the spatial factors but to the complete space-and-spin state functions of individual particles, then ψ_a vanishes whenever the states A and B are taken to be the same in all respects (now including spin orientation). *The Pauli exclusion principle* spells this out by saying that *no two electrons in the same atom can have all quantum numbers the same*. This result was in fact discovered by W. Pauli in 1924[3] before an understanding of it in terms of the symmetry of wave functions had been developed. Pauli simply inferred, from the detailed structure of atomic energy levels as revealed by spectroscopy, that electrons must have a two-valued quantum number (later to be associated with spin) in addition to the three quantum numbers (n, l, m_l) previously known. He inferred, further, that if the three spatial quantum numbers for two electrons in an atom were the same, the fourth (spin) quantum number must be different, which we now describe by saying that their spins are *antiparallel*. Naturally, the same restriction, couched in somewhat different terms, applies if one writes the total wave function as the product of a total space function and a total spin function (see the exercises).

Exchange Energy

At the end of Section 13-3 we posed a question about the space properties of identical particles bound in a common potential: do they huddle together (symmetric space function) or

[3]W. Pauli, Z. Phys. **31**, 373 and 765 (1924). For an interesting account of the historical development of the ideas of the exclusion principle and spin, see the articles by R. Kronig and B. L. van der Waerden in *Theoretical Physics in the Twentieth Century* (a memorial volume to Wolfgang Pauli) ed. M. Fierz and V. F. Weisskopf, Interscience Publications, New York, 1960.

Identical particles and atomic structure

do they avoid one another (antisymmetric space function)? We are now able to answer this question in terms of the kind of particles involved. Any two electrons in a single system, for example, must have an *overall* antisymmetric wave function. There are four ways this can occur: $\psi_s\chi_a$ with a symmetric space function ψ_s for which there is only one antisymmetric spin state χ_a (a *singlet state*); or $\psi_a\chi_s$ with an antisymmetric space function ψ_a for which there are three symmetric spin states (a *triplet state*). Two electrons in the singlet state will tend to huddle together because of their symmetric space function. In contrast two electrons in a triplet state will tend to avoid one another because of their antisymmetric space function. In developing the model that led to this result, we assumed that there was no electron-electron Coulomb interaction; any "huddling" or "avoidance" of electrons in this model is solely a quantum effect. Of course, in a more complete theory, the Coulomb repulsion between electrons must be taken into account. When we include the Coulomb energy in the model of the helium atom (Section 13-8), it will lead to a slightly higher energy for the system in the singlet state (in which electrons huddle) than for the triplet state (in which electrons avoid one another). This difference of energy in the two cases is called the *exchange energy* because it arises from the symmetry properties of electron wave functions under an assumed exchange of space and spin coordinates.

13-6 WHEN DOES SYMMETRY OR ANTISYMMETRY MATTER?

In principle the symmetry or antisymmetry of the wave function for two or more identical particles must be involved for every physical system. In practice, however, it will often be unnecessary (as well as impracticable) to apply the symmetry condition explicitly. Consider, for example, two separate hydrogen atoms, both in their ground state. Then, strictly speaking, we ought to construct for this system a total wave function that is antisymmetric with respect to exchange of the two electrons (and of the two protons also). However, it is intuitively more or less obvious that if the atoms are widely separated (by more than a few angstroms), they act as entirely independent systems. The fact that each electron is in the same quantum state with respect to its own proton is in no sense a violation of the Pauli exclusion principle. On the other hand, if the two atoms are brought so close that they begin to interact (the ultimate result perhaps being the formation of a hydrogen molecule) then the construction of an antisymmetrized wave function is essential to a correct description of the complete system.

There is a simple criterion for whether or not the symmetry must be considered. If the individual-particle wave functions *overlap* significantly, in the sense that there are regions in which both have appreciable magnitude, then the symmetry or antisymmetry is important. If the amount of overlap is negligible, then one can use a simple product wave function that is neither symmetric nor antisymmetric. The formal basis of this criterion can be made apparent if one writes the total space wave function for a system of two particles:

$$\psi(x_1, x_2) = \frac{1}{\sqrt{2}} \left[\psi_A(x_1)\psi_B(x_2) \pm \psi_B(x_1)\psi_A(x_2) \right]$$

Forming from this the probability density, we have

$$|\psi(x_1, x_2)|^2 = \tfrac{1}{2}|\psi_A(x_1)\psi_B(x_2)|^2 + \tfrac{1}{2}|\psi_B(x_1)\psi_A(x_2)|^2$$
$$\pm \tfrac{1}{2}\{\psi_A^*(x_1)\psi_B(x_1)\psi_B^*(x_2)\psi_A(x_2)$$
$$+ \psi_B^*(x_1)\psi_A(x_1)\psi_A^*(x_2)\psi_B(x_2)\}$$

The term within braces embodies the essential consequence of identity, that each particle must be associated with both of the component wave functions. However, if ψ_A is negligible at values of x where ψ_B is large, and vice versa, the products in this term—$\psi_A^*(x_1)\psi_B(x_1)$, for example—are negligible because the *same* value of x is involved in both factors in the product. In this case, only the first two terms in the above equation are significant. Then we are left with a probability function $|\psi|^2$ given by a sum of squared magnitudes, both of which assign each particle separately to a particular state. In this case the value of the probability density, for given values of x_1 and x_2, is indifferent to whether the total wave function is symmetric, antisymmetric, or merely a product function with no particular symmetry.

At the end of Chapter 12, we called alkali metals "hydrogen-like." In a neutral alkali atom the nuclear charge $+Ze$ is shielded from the outer "valence" electron by $Z-1$ electrons in completed shells. This outer electron then acts somewhat like an electron in a hydrogen state. This will be discussed somewhat more rigorously in Section 13-9 below. The important point here is that talking about "the valence electron" as if it had an identity separate from the other electrons in the atom is justified only to the extent that the probability overlap between the valence state and other electron states is small. We saw at the end of Chapter 12 that the electrostatic energy associated with this overlap—and thus presumably the overlap itself—is smaller for larger values of orbital angular momentum for the valence state. Electron indistinguishability and wave function symmetry are less important considerations for these higher-l states.

We can see, then, that although the Pauli exclusion principle is a direct expression of the requirement of antisymmetry for a state of two or more electrons, it acquires physical importance only to the degree that the electrons in question have overlapping wave functions, whether or not they are bound in the same atom.

13-7 MEASURABILITY OF THE SYMMETRY CHARACTER

The Periodic Table

For electrons, the facts of atomic structure supply the main evidence that only antisymmetric states occur. It is only on this basis that we can understand the building up (German: *Aufbau*) of the total electron configuration of a many-electron atom. Only $2(2l + 1)$ electrons can be accommodated with different values of m_l and m_s in a shell associated with particular values of n and l—one electron for each of the states discussed in Chapter 12 (Section 12-6). We shall, however, defer our detailed discussion of this until later in the chapter (Section 13-9).

Electron Gas in a Metal

For a confined system of very many identical particles, the properties of the system can be profoundly influenced according to whether the overall wave function is symmetric or antisymmetric. If the basic character of the particles is such that the wave function is symmetric, this means that an unlimited number of particles can be accommodated in the lowest possible energy level. If, on the other hand, the particles are such that the overall wave function is antisymmetric, the number of particles in each energy level is strictly limited (to a total of $2s + 1$, where s is the spin quantum number). In highly condensed systems (many particles per unit volume) this has the consequence that fermions "fill up" the lower energy states so that some are driven to occupy eigenstates of high energy, whereas bosons can congregate in states of low energy (see the exercises).

The behavior of the "gas" of conduction electrons in a metal is in accord with the proposition that electrons are fermions obeying the Pauli principle. In a metal at room temperature, electrons are forced to occupy states up to energies corresponding to tens of thousands of degrees Kelvin. One consequence of this is that, if the temperature of the metal is raised through, say, 100 K, the energy distribution of the electrons is scarcely affected. This means that the electron gas does not give rise to any significant contribution to the specific heat of a metal, even though there is of the order of one conduction electron for every atom in the metal (hence about a mole of free electrons in each gram-atom of metal). The absence of any such contribution is inexplicable by classical physics, but can be viewed as further evidence that many-electron wave functions are antisymmetric.

Superfluidity

Limitations imposed by exchange symmetry apply not only to electrons in a single atom but also to whole atoms that are part of a larger system. For example, a system of identical *spinless* atoms (which are therefore bosons) has total wave functions that are symmetric with respect to space exchange alone. The outstanding example of this is helium (He^4) in bulk. The exchange symmetry encourages all of the atoms to be in the same overall quantum state, including both the internal state and the translational state of the atom. One consequence of this is the *superfluidity* of liquid helium, which at temperatures near absolute zero encounters no resistance to its flow through microscopic holes.

A system containing identical photons, too, exhibits this overall symmetric property in its total wave functions. The unique energy spectrum of radiation within an enclosure at a given temperature (*black-body radiation*) is a result of this symmetry.

Identical particles and atomic structure

Scattering

Perhaps the nearest to a direct measurement of the symmetry character occurs in the analysis of elastic collisions between pairs of identical particles. Suppose that two such particles approach one another and mutually scatter. In the center-of-mass frame the results of the elastic collision are that the velocity of each particle is changed in direction without being changed in magnitude. However, if the particles are identical, a collision involving a change of direction by θ is indistinguishable from one involving a change of $\pi - \theta$ (Figure 13-3). Classically, the deflections through θ and $\pi - \theta$ would represent physically separate events, and the observed intensity at θ would be the sum of these independent contributions. If $p(\theta)$ and $p(\pi - \theta)$ are the probabilities of these processes, then the total probability of a particle emerging from the collision in the direction θ with respect, say, to the particles incident from the left would be given by

$$P_{\text{classical}}(\theta) = p(\theta) + p(\pi - \theta)$$

Quantum-mechanically, however, the scattering probability is the square of the modulus of a net probability *amplitude* which is a coherent superposition of the amplitudes belonging to θ and $\pi - \theta$ separately. This is yet another example of superposing amplitudes for alternative ways to the same outcome, as with double-slit and polarized-photon interference experiments. Thus we have

$$P_{\text{quantum}}(\theta) = |f(\theta) \pm f(\pi - \theta)|^2 \tag{13-6}$$

where the $+$ sign corresponds to a symmetric space wave function for the two particles, and the $-$ sign to an antisymmetric

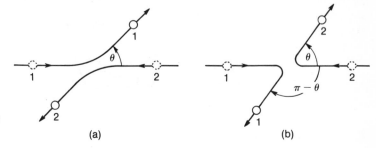

Fig. 13-3 Collision processes which are indistinguishable if particles 1 and 2 are identical.

(a)   (b)

     13-7   Measurability of the symmetry character

space function. The classical result, if expressed in terms of the squares of probability amplitudes, would be

$$P_{\text{classical}}(\theta) = |f(\theta)|^2 + |f(\pi - \theta)|^2$$

In the particular case of spinless particles, for which the combined spin function is necessarily symmetric, the results of the scattering will then reveal whether the space function is symmetric or antisymmetric, corresponding to the plus or minus sign in Eq. 13-6. Notice in particular that if we put $\theta = \pi/2$, then two very clear alternatives develop:

For symmetric states:

$$P\left(\frac{\pi}{2}\right) = 4\left|f\left(\frac{\pi}{2}\right)\right|^2$$

For antisymmetric states:

$$P\left(\frac{\pi}{2}\right) = 0$$

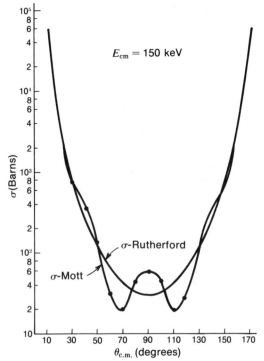

Fig. 13-4 Elastic-scattering cross section for alpha particles incident on He^4. The Rutherford theory is classical; the Mott theory is quantum-mechanical (1 barn $= 10^{-24}$ cm^2). [After N. P. Heydenburg and G. M. Temmer, Phys. Rev. 104, 123 (1956).]

Identical particles and atomic structure

The corresponding classical result for $\theta = \pi/2$ would be $2|f(\pi/2)|^2$. Thus for exchange-symmetric particles the scattering at 90° would be just twice as intense as a classical calculation would predict. This is precisely what experiment shows with alpha particles; see Figure 13-4, which shows an angular distribution of alpha particles scattered by helium.

13-8 STATES OF THE HELIUM ATOM

Energy Levels of the Helium Atom

The helium atom provides us with the clearest example of a system of two identical particles in a common potential. The components of the neutral helium atom are two electrons bound in the field of a nucleus consisting of two protons plus either two neutrons (the principal isotope, helium-4) or one neutron (helium-3: relative natural abundance 0.00013 percent). The analysis of the optical spectrum of helium shows that the spectral lines can be classified with the help of two virtually separate energy-level diagrams (Figure 13-5), with almost no transitions taking place from an energy level in one set to levels in the other set.

At one time it was believed that these separate sets of levels belonged to two different and distinct elements, which were given the names ortho-helium and para-helium. But with the discovery of electron spin, it came to be realized that ortho-helium and para-helium were not two distinct substances, but were simply the manifestation of two different classes of energy levels into which the states of any helium atom could be divided, according to whether the electron spins were parallel ($S = 1$) or antiparallel ($S = 0$). The total angular momentum **J** of any state is then the vector combination of the resultant spin **S** with the resultant orbital angular momentum **L** of the two electrons.[4] For $S = 0$ this gives only one possibility ($J = L$) but for $S = 1$ there are (except for $L = 0$) three possibilities ($J = L - 1$, L, $L + 1$) with slightly different energies. Combined electron states having $S = 0$ are thus singlets, whereas states having $S = 1$ are in general *triplets*.

This hypothesis concerning the energy-level structure of

[4]Here we are using the italic capital letters L, S, and J to denote the quantum *numbers* associated with the orbital, spin, and total angular momentum of a two-electron system. The *vectors* **L**, **S**, **J** are the actual angular momenta so that, for example, $|\mathbf{L}| = \sqrt{L(L + 1)}\hbar$.

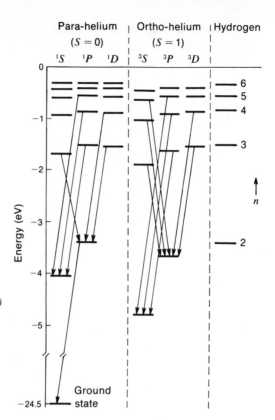

Fig. 13-5 Energy levels of helium, showing the transitions observed in its optical spectrum. The zero of the energy scale corresponds to singly-ionized helium (He⁺) in its ground state, plus a free electron with zero kinetic energy. Hydrogen energy levels are shown for comparison.

the helium atom was further justified when fine structure of the "triplet lines" was detected. The effectively complete separation between the para-helium ($S = 0$) and ortho-helium ($S = 1$) systems of energy levels is the result of the lack of any efficient radiative mechanism by which electrons of a helium atom, after once being excited to a state having $S = 1$, can find their way back to a state with $S = 0$ (or vice versa). Any such transition must involve a "spin flip" which (as we shall see in Chapter 14) is a very improbable process compared to the electric dipole transitions through which almost all optical spectra are produced.

The problem of accounting in detail for the energy level structure of helium is a complicated one, but is worth discussing because it illustrates so clearly the quantum-mechanical consequences of identity. However, because of the complexity of the problem, we shall approach it via some simpler models which, although less accurate, provide a useful background to, and are consistent with, the more rigorous analysis.

Identical particles and atomic structure

Our discussion begins with simple semiclassical models, first for the ground state and then for the excited states. We will then present the formulation and the main results of the quantum-mechanical theory of the helium atom.

Helium Ground State: Simplest Models

In the ground state of helium, both electrons are in a lowest possible energy state, corresponding in first approximation to the state described by $n = 1$, $l = 0$, and $m_l = 0$ for a single-electron system. This situation requires a *symmetric* space function for the two electrons together, since an antisymmetric space function with equal quantum numbers n, l, m_l would be identically zero (Eq. 13-5b). A symmetric space function is possible provided that the combined spin function is antisymmetric. This means that the ground state of helium belongs to the para-helium ($S = 0$) energy-level structure, as is shown in Figure 13-5. The ionization energy of helium—the energy required to remove one electron from an atom initially in the ground state—is observed to be 24.5 eV. This is the first quantitative result that any theoretical model of the helium ground state must produce.

If we could regard the helium atom as consisting of two electrons each interacting with the central charge $2e$, but having no interaction with one another, then the energy levels would correspond simply to the combination of two hydrogen-like states. The energy of an individual electron would be that of the hydrogen-like ground state[5] with $Z = 2$:

$$E_n = -\frac{Z^2 m e^4}{2\hbar^2} \cdot \frac{1}{n^2} = -13.6 \frac{Z^2}{n^2} \text{ eV}$$

Thus for the ground state, with $n = 1$ for both electrons, we should have

$$E = -54.4 \text{ eV} \qquad \text{(for either electron)}$$

The energy needed to ionize the atom by removing one electron would then be 54.4 eV. The fact that this is in profound disagreement with the actual ionization energy (24.5 eV) is not surprising since the Coulomb interaction between the two

---

[5]In this discussion we shall, for simplicity, be using cgs units only.

13-8   States of the helium atom

electrons is clearly a very important contribution to the total energy of the system.

A first simple step toward a better model of the ground state is to include the Coulomb repulsion between the electrons by assuming that its existence causes the electrons to stay as far apart as possible. The most primitive way of doing this is to think in terms of Bohr orbits with the electrons at opposite ends of a line passing through the nucleus (Figure 13-6), and to assume that the actual orbit radius of each electron is the same as it would be if the other electron were absent. For the lowest state this radius is $a_0/2$, where a_0 is the Bohr radius for hydrogen. The positive energy of the electron-electron interaction in this situation would therefore be e^2/a_0, which is equal to 27.2 eV. On this model, however, the net energy of

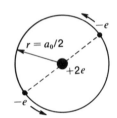

Fig. 13-6 *Simplest Bohr model of the helium-atom ground state.*

each electron orbiting in the field of the nucleus is -54.4 eV (that is, -108.8 eV for both together). The *total* energy of the system would thus be -81.6 eV. The ionization energy of helium would be the difference between this ground-state energy and the combined energy of a free electron and a He^+ ion containing a single electron in its lowest state at -54.4 eV. This difference is 27.2 eV, as against the observed value of 24.5 eV—better agreement than we might have expected.

Bohr himself made a somewhat more careful calculation along these lines, incorporating the repulsion between the electrons as an essential part of the dynamics of the quantized system (see the exercises). Unfortunately, the result of the more careful calculation is slightly *worse* than our result above. The fact is that the semiclassical approach simply does not work, even for this minimally complex system.

Helium Excited States: Simple Model

It is the ground state of helium that is least amenable to semiclassical calculation because the strong interaction between the electrons (or the overlap of their wave functions)

Identical particles and atomic structure

makes the consequences of identity and antisymmetry most crucial to the precision of the result. When we come to excited states, a Bohr-type model achieves fair success, although it is still not nearly good enough to be regarded as fully acceptable.

The basis of a simple calculation of the excited states is the assumption that these excitations involve only *one* electron, while the other remains in its lowest possible state, with $n = 1, l = 0$. A strong clue to the correctness of this (at least as an approximation) is that the correct *higher* levels of helium, whether in the para- or the ortho-system, agree quite closely in energy with the levels of the hydrogen atom (See Figure 13-5). That is just what one would expect if one electron remained close to the helium nucleus in an $n = 1$ state. Under these conditions the second electron, if in an orbit of larger n and correspondingly larger radius, sees a net central charge of $+e$ only—just like the electron in a hydrogen atom—because the other electron screens the nuclear charge $+2e$ with its own charge $-e$. This picture will become more and more nearly valid as the excited electron goes to states of higher n. Even for its lowest possible excitation ($n = 2$), however, this analysis is surprisingly good. The reason for this is indicated pictorially in Figure 13-7. On a Bohr model of the situation, the excited electron is in an orbit of radius $4a_0$ (since the Bohr orbit radii for a net central charge $+e$, as in hydrogen, are given by $n^2 a_0$). But the inner, unexcited electron is in the orbit corresponding to $n = 1$ for a central charge equal to the full nuclear charge $2e$, and the radius of this orbit is $a_0/2$—only one-eighth of the other. As Figure 13-7 indicates, the net central charge, from

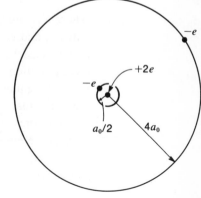

Fig. 13-7 *Simplest Bohr model of the helium atom for the first excited state. This figure is drawn to a different scale than Figure 13-6. The orbit radii for the two electrons are drawn in the correct proportion according to the Bohr theory for this system.*

13-8 States of the helium atom

the point of view of the excited electron, is highly concentrated.

A better analysis, in terms of electron wave functions rather than Bohr orbits, will recognize the fact that the energy levels will depend to some extent on the l value, as well as the n value, of the excited electron. In particular, for $l = 0$, the wave function (as we saw in discussing the Coulomb model in Chapter 12) gives a significant electron probability density at the position of the nucleus; this corresponds, in physical terms, to some penetration through the screening provided by the inner electron. In consequence the electron having $n > 1$ but $l = 0$ will be more fully exposed to the attractive potential of the full nuclear charge, and its energy will be lowered accordingly. This effect is apparent in Figure 13-5; the levels for which both electrons have $l = 0$ are pulled down significantly compared to the corresponding levels in hydrogen.

After this limited success in describing the energy-level structure of helium in very simple terms, we shall now turn to the more rigorous quantum-mechanical description. Here again we begin with the simplest possible model by initially ignoring electron spin and exchange effects. Later we shall add these effects to the analysis.

Quantum Model without Exchange Energy

Any acceptable solution for the electronic structure of helium must be based on the Schrödinger equation as it applies to this system. We have

$$\left[\left(-\frac{\hbar^2}{2m_1}\nabla_1^2 - \frac{2e^2}{r_1}\right) + \left(-\frac{\hbar^2}{2m_2}\nabla_2^2 - \frac{2e^2}{r_2}\right) + \frac{e^2}{r_{12}}\right]\psi = E\psi \quad (13\text{-}7)$$

where ∇_1^2 and ∇_2^2 are the Laplacian operators for the coordinates $\mathbf{r}_1$ and $\mathbf{r}_2$ of the separate electrons, respectively, and r_{12} is the (scalar) distance between the electrons. The presence of the interaction term between electrons, e^2/r_{12}, immediately prevents the equation from being separable in the coordinates of the individual electrons; therefore there is no closed analytic solution, only a process of numerical approximation. For the present we ignore electron spin and exchange effects (that is, the consequences of identity and antisymmetrization, to be examined later) and concentrate on approximation methods that

Identical particles and atomic structure

will allow a crude quantum estimate of the energies of lower levels in helium.

We apply first a technique known as *perturbation theory* that parallels closely the semiclassical Bohr analysis of helium given above. Starting with the initial assumption that the electrons do not interact allows us to use exact wave functions $\psi_i(\mathbf{r}_1)$ and $\psi_k(\mathbf{r}_2)$ for the separate electrons that are identical to the hydrogen atom solutions modified for the nuclear charge $2e$. Here i and k each specify values of n, l, and m_l for the separate states. By ignoring electron-electron interaction and the effects of spin, one can separate Eq. 13-7 if the wave function is written as a simple product $\psi_i(\mathbf{r}_1)\psi_k(\mathbf{r}_2)$ of the hydrogen-like solutions of the separate electrons. Then it is plausible, and more detailed analysis confirms,[6] that the displacement of energy ΔE_1 caused by the electron-electron interaction is given, to first approximation, by multiplying the interaction energy $V(r_{12})$ by the probability that the electrons have the particular separation r_{12}, as given by $|\psi_i(\mathbf{r}_1) \cdot \psi_k(\mathbf{r}_2)|^2 \, d\tau_1 \, d\tau_2$, and then integrating over all the possible positions r_1 and r_2. That is,[7]

$$\Delta E_1 \approx \int\int V(r_{12})|\psi_i(\mathbf{r}_1) \cdot \psi_k(\mathbf{r}_2)|^2 \, d\tau_1 \, d\tau_2 \tag{13-8}$$

This perturbation technique is based on the principle that the effect of a small disturbing interaction can be calculated quite well by applying it to the undisturbed state of the system, ignoring the changes that the perturbing interaction causes in the state itself. (The method was extensively used, long before the advent of quantum mechanics, to deal with comparable problems in classical mechanics, for example the effect of the gravitational interaction between two planets, assuming in the first instance that each orbit is determined by the sun's attraction alone.) Since the functions ψ_i and ψ_k in Eq. 13-8 are known exactly, the integration can be carried out numerically to as high an accuracy as one is willing to pay for. The results are shown in column 3 of Figure 13-8, where ΔE_1 for the pair of states i and k is added to each energy in column 1. Notice

[6]See, for example, R. M. Eisberg, *Fundamentals of Modern Physics*, John Wiley, New York, 1961, Chaps. 9 and 12.

[7]We use $d\tau$ here for volume element (instead of dV) to avoid confusion with the use of V for potential energy.

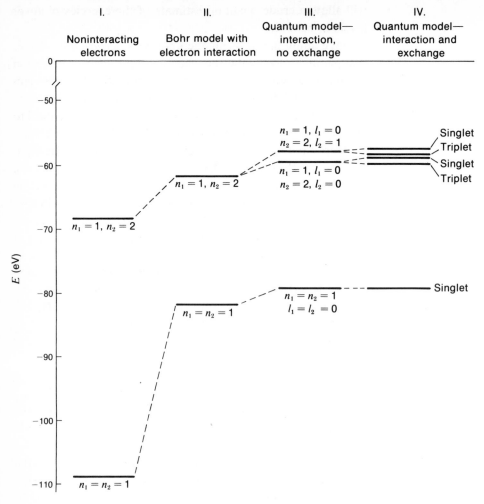

Fig. 13-8   Comparison of theoretical predictions for
energy levels of the helium atom according to different
models. The zero of energy in this diagram is not the same
as in Fig. 13-5; it corresponds to complete separation of
the atom into the helium nucleus (He^{++}) and two free
electrons with zero kinetic energy. (Note also that the vertical energy scale is contracted relative to that in Figure 13-5.)

that the second level is split into two by this interaction. This is
precisely the effect that we have already described qualitatively, namely the removal of the energy degeneracy between
states of $l = 0$ and $l = 1$ (for the outer electron in its $n = 2$
states) by virtue of the reduced effectiveness of the electrostatic screening for the $l = 0$ state.

Identical particles and atomic structure

Perturbation theory, which we have invoked in this analysis, assumes that for *small* interactive energies one can use wave functions that are exact in the absence of the interaction. Since the changes in energy between column 1 and column 2 of Figure 13-8 show that the electron-electron interaction energies are *not* small, we see that perturbation theory cannot yield accurate values of energy levels. More sophisticated perturbation methods can be employed, but the one we have described shows all essential features of this particular model and its consequences.

Quantum Model with Exchange Energy

Our discussion of the helium atom up to this point has, for the sake of simplicity, ignored the indistinguishability of the two electrons in the system. We have talked as if each electron could be assigned to a separate orbit or state. However, we cannot escape the fundamental fact that identity must be incorporated in the basic form of the wave function. We were not in fact justified in using descriptions of the kind "the first electron remains in its ground state and the second electron is raised to an excited state." All we can ever do is to speak of two identical electrons sharing a total state that may be based on two single-particle states. When we include this feature formally, there appears in the calculation of the total energy an *exchange term* that results in a difference of energies between symmetric and antisymmetric space states constructed from the same sets of space-state quantum numbers. It is the purpose of the present discussion to show how this correction comes about.

The Schrödinger equation for helium (Eq. 13-7) contains no explicit mention of the spin states of the electrons. However, the spin states play an essential role through the requirement that the space function ψ be either symmetric (for singlet states, $S = 0$) or antisymmetric (for triplet states, $S = 1$). These symmetry considerations place certain limitations (as expressed in the Pauli principle) on the values of n, l, and m_l that the two electrons may have. More importantly, however, they lead to the above-mentioned exchange term in the expression for the total energy of the system. The essential character of this energy difference can be understood along the lines of the earlier discussion of one-dimensional systems (Sections 13-3 through 13-5). If the electrons are in the spin state $S = 0$, which makes the space wave function symmetric, they tend to

13-8   States of the helium atom

huddle close together, whereas in the $S = 1$ state involving the same two sets of quantum numbers n, l, m (but in this case an antisymmetric space state) they tend to stay apart. Since the Coulomb energy of interaction between electrons is positive and increases with decreasing separation, the $S = 0$ state will be pushed somewhat higher in energy than the corresponding $S = 1$ state. (In addition, *both* states will be raised in energy with respect to what one would have in the absence of the electron-electron repulsion.)

To obtain a more formal description of these effects, suppose that the electrons are in space states labeled i and k, where these indices identify the complete sets of quantum numbers n, l, m_l. The normalized symmetric and antisymmetric space functions constructed from these are the following:

$$\psi_s = \frac{1}{\sqrt{2}}[\psi_i(\mathbf{r}_1)\psi_k(\mathbf{r}_2) + \psi_i(\mathbf{r}_2)\psi_k(\mathbf{r}_1)] \qquad \text{for } S = 0 \text{ (singlet)}$$

$$\psi_a = \frac{1}{\sqrt{2}}[\psi_i(\mathbf{r}_1)\psi_k(\mathbf{r}_2) - \psi_i(\mathbf{r}_2)\psi_k(\mathbf{r}_1)] \qquad \text{for } S = 1 \text{ (triplet)}$$

$$(13\text{-}9)$$

As before, the probability that one electron is in a volume element $d\tau_1$ at $\mathbf{r}_1$ and the other in a volume element $d\tau_2$ at $\mathbf{r}_2$ is given by $\psi^*\psi d\tau_1 d\tau_2$ where ψ is ψ_s or ψ_a as the case may be. We can substitute from Eq. 13-9 into Eq. 13-8 and expand to obtain

$$\Delta E = \frac{1}{2}\int\int |\psi_i(\mathbf{r}_1)|^2 |\psi_k(\mathbf{r}_2)|^2 V(r_{12}) d\tau_1 d\tau_2$$

$$+ \frac{1}{2}\int\int |\psi_i(\mathbf{r}_2)|^2 |\psi_k(\mathbf{r}_1)|^2 V(r_{12}) d\tau_1 d\tau_2$$

$$\pm \frac{1}{2}\int\int \psi_i^*(\mathbf{r}_1)\psi_k(\mathbf{r}_1)\psi_k^*(\mathbf{r}_2)\psi_i(\mathbf{r}_2) V(r_{12}) d\tau_1 d\tau_2$$

$$\pm \frac{1}{2}\int\int \psi_k^*(\mathbf{r}_1)\psi_i(\mathbf{r}_1)\psi_i^*(\mathbf{r}_2)\psi_k(\mathbf{r}_2) V(r_{12}) d\tau_1 d\tau_2$$

where the $\pm$ refer to the symmetric and antisymmetric space states, respectively. This looks like a very formidable expression, but the first two integrals are exactly equal to one another and, moreover, when added together equal the quantity ΔE_1 which we discussed (Eq. 13-8) in the case in which spin was ig-

Identical particles and atomic structure

nored. The last two terms are equal to one another also, since there is a complete symmetry between the roles of r_1 and r_2. Thus we can put

$$\Delta E = \Delta E_1 \pm \Delta E_2 \qquad (13\text{-}10)$$

where

$$\Delta E_1 = \int\int |\psi_i(\mathbf{r}_1)|^2 |\psi_k(\mathbf{r}_2)|^2 V(r_{12}) d\tau_1 d\tau_2 \qquad (13\text{-}11a)$$

and

$$\Delta E_2 = \int\int \psi_i^*(\mathbf{r}_1)\psi_k(\mathbf{r}_1)\psi_k^*(\mathbf{r}_2)\psi_i(\mathbf{r}_2) V(r_{12}) d\tau_1 d\tau_2 \qquad (13\text{-}11b)$$

The term ΔE_1 then represents a general upward shift of the energy levels as a result of the electron-electron repulsion, while the further correction, $\pm \Delta E_2$, is the exchange term, involving (as the explicit expression for it shows) the association of each electron with both of the component one-electron states. The resultant modification of energy levels is indicated in column 4 of Figure 13-8. The agreement with observation is good.

Using perturbation theory for the lower-energy levels and a hydrogen-like approximation for the higher-energy levels, we have obtained a respectable description of the electronic structure of helium. More accurate approximation methods have been developed that yield results very much closer to the observed energies. As important as the energy values, however, are the physical arguments in our analysis that accounted qualitatively for the various energy splittings. These arguments will have further application in the analysis of multielectron atoms.

Classification of Helium Levels

We end this discussion of the helium atom with a brief description of the spectroscopic classification of the energy levels. The basic feature is that the classification is in terms of the resultant angular momenta of the two electrons, not their individual angular momenta. The vector model for the combination of *orbital* angular momenta, as discussed in Chapter 11 (Section 11-6) applies equally well here. If the states of the individual electrons correspond to orbital angular momentum quantum numbers l_1, l_2, then the resultant orbital angular momentum is characterized by the quantum number L, where

$$|l_1 - l_2| \leq L \leq l_1 + l_2$$

The *total* angular momentum J of the two electrons together is then obtained by the vector combination of the orbital angular momentum, $L(|L| = \sqrt{L(L+1)}\hbar)$ with the total spin angular momentum. For the singlet spin states $(S = 0)$ we of course have $J = L$ simply, but for the triplet spin states $(S = 1)$ there are in general three different values of J for each value of L (other than $L = 0$). To take a specific example, consider the various states that might be based on the single-electron states $n_1 = n_2 = 2$, $l_1 = l_2 = 1$. Then we can have

$$L = 0, 1, 2$$

For $S = 0$ we thus have

$$J = 0, 1, 2$$

Using the spectroscopic notation, we designate the possible states as $^1S_0$, $^1P_1$, $^1D_2$, where (remember) S, P, D refer to the total orbital angular momentum numbers $L = 0$, 1, and 2, respectively, the superscript (equal to $2S + 1$) indicates that the level is a spin singlet, and the subscript gives the value of J. For $S = 1$ the possibilities are considerably more numerous:

$$L = 0: \quad J = 1 \qquad (^3S_1)$$
$$L = 1: \quad J = 0, 1, 2 \quad (^3P_0, \, ^3P_1, \, ^3P_2)$$
$$L = 2: \quad J = 1, 2, 3 \quad (^3D_1, \, ^3D_2, \, ^3D_3)$$

The energy levels of helium are identified in this notation on Figure 13-9. The energy differences between states of the same L and S but different J are quite small, and they are not shown as separate in the level scheme of the figure.

To complete the classification of these two-electron states in helium, it is convenient to introduce an *effective principal quantum number* n such that $n = 1$ for the lowest state of the atom (a $^1S_0$ state). We can do this by defining

$$n = n_1 + n_2 - 1$$

In crudest approximation, the levels for all the different values

Identical particles and atomic structure

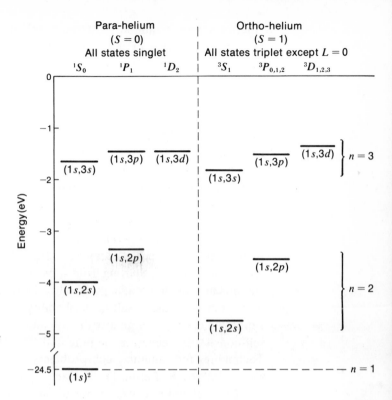

Fig. 13-9 Clas-
sification of the
lowest-energy states
of the helium atom.
(Same energy scale
as Figure 13-5.)

of L and S can then be grouped as belonging to $n = 1$, $n = 2$,
$n = 3$, and so on. This, too, is indicated on Figure 13-9. The effectiveness of this classification in terms of n is due, as we
mentioned earlier, to the fact that the states shown in Figure
13-9 all derive from the excitation of one electron only, leaving
the other one in the lowest state ($n_1 = 1$, $l_1 = 0$) to screen the
nuclear charge, and thus yielding a hydrogen-like configuration
that is increasingly accurate for higher levels of excitation of
the outer electron.

13-9 MANY-ELECTRON ATOMS

The Self-Consistent Field Method

The analysis of the two-electron system in helium is already complicated. The detailed consideration of electron
states in atoms with still more electrons is prohibitively difficult if one tries to proceed along similar lines. There are, however, other approaches that thrive on the old principle that

"there's safety in numbers." These methods all make use in one way or another of the assumption that, in a many-electron atom, any one electron finds itself in a spherically symmetric field due to the combined effect of the nuclear charge and all the other electrons.

All such methods must contain a considerable element of trial and error, because a knowledge of the potential $V(r)$ in which any one electron finds itself requires knowledge (or at least an assumption) about the radial distribution of electric charge density that corresponds to the position probability distributions for all the other electrons. But these latter, in turn, must represent acceptable solutions of the Schrödinger equation for the potential $V(r)$. Thus the analysis involves the search, through successive approximations, for what is called a *self-consistent field*. Starting from some assumed form of $V(r)$, we can calculate the spatial probability distribution of the electrons, but then the resulting probability distribution, which defines an electric charge distribution, must—if the analysis is self-consistent—generate the potential $V(r)$ that we assumed. Techniques for handling such problems were first developed about 1928 by E. Fermi, D. R. Hartree, and L. H. Thomas.

In what follows, we do not discuss self-consistent field methods in any depth or detail. Rather, we shall simply indicate how the picture of a many-electron atom can be approximated in terms of the solutions to the one-electron problem, in conjunction with the Pauli exclusion principle. There will be no attempt to construct antisymmetrized wave functions as such. Instead, the Pauli principle is used simply as a basis for determining how many electrons can be accommodated in a "shell" made up of electrons all having the same principal quantum number. (The reason why it is appropriate to speak of electron shells will become apparent in the course of the discussion.) Our concern will be with the ground state of a neutral atom of atomic number Z. The energy of this state will be the lowest possible energy of a system made up of the central charge Ze together with the Z electrons—subject, of course, to the limitations imposed by the Pauli principle. For the purposes of finding this lowest level, we can imagine building up the atom from scratch, as it were, starting with the nucleus alone and adding the electrons, each of which falls into the lowest available level.

Identical particles and atomic structure

The K Shell

From our discussion of the helium atom, it is clear that the first two electrons will fall into the $1s$ level (each with $n = 1$, $l = 0$), with their spins antiparallel. This completed inner level is called the K *shell*. Addition of a third electron to the $1s$ state is impossible because it would be obliged to have the same set of quantum numbers (including spin) as one of the two electrons already present. Thus the third electron must enter a state with $n = 2$ and $l = 0$ or 1. Before considering the energy of this third electron, we note that the energy of the first two electrons is not quite as negative as it would be if each moved alone in the field of the central charge. A single $1s$ electron would have a characteristic distance $a_1 = a_0/Z$ from the nucleus, a potential energy $-Ze^2/a_1 = -Z^2e^2/a_0$ and a radial wave function of the form e^{-r/a_1}. The presence of the second electron will raise the energy of each, and will also keep each electron farther from the nucleus, on the average, than it would otherwise be. Both of these effects can be approximately described, so far as either electron is concerned, in terms of a partial screening of the nuclear charge by the other electron. If the screening were complete, the effective value of the central charge would be reduced to $Z - 1$, and the parameter a_1 increased to $a_0/(Z - 1)$. We can anticipate that a more careful self-consistent field calculation will yield results somewhere between the no-shielding and the complete shielding limits discussed here.

The L Shell

When we consider, now, the energy of the third electron, it is not far from the truth to say that it "sees" a central charge $(Z - 2)e$ unless it penetrates within the tiny charge cloud represented by the first two electrons in the K shell. However, to the degree that it does penetrate, it will be exposed more nearly to the full positive charge Ze of the nucleus and its energy will be lowered. If the third electron is in a $2s$ state, for which ψ remains nonzero at $r = 0$, we have such penetration (see Figure 13-10). We can thus infer that the third electron will fall into the $2s$ state ($l=0$) rather than the $2p$ state ($l=1$), which has zero probability at the nucleus. The fourth electron will also fall into the $2s$ state, with spin antiparallel to the third

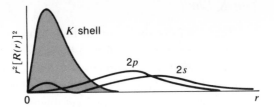

Fig. 13-10 Comparison of K-shell penetration by 2s and 2p radial probability densities.

electron. After this point the Pauli principle again intervenes, and the fifth electron will go into the slightly higher $2p$ state ($n = 2$, $l = 1$).

The energy difference between $2s$ and $2p$ levels can be further illuminated if we backtrack for a moment and consider the case $Z = 3$ (lithium). For lithium the third electron completes a neutral atom and the ground state can be described by the notation $(1s)^2 (2s)$, where the superscript tells the number of electrons in each state and the absence of superscript implies only one electron in that state. The first excitation of the lithium atom will be obtained by raising the third electron to the relatively nonpenetrating state $2p$, and the energy of this state will correspond quite closely to that of the $n = 2$ states in a hydrogen atom, since in both cases the electron moves in the field of an effective central charge $+e$. Figure 13-11 illustrates this point. (The situation is like the one we discussed in connection with excited states of the helium atom.)

Returning now to the case where Z is large enough to accommodate many electrons, we can see that it is at least plausible that the filling of the two $2s$ states will be followed by the

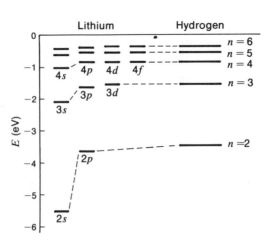

Fig. 13-11 Energy levels of lithium compared with hydrogen. Note that for higher orbital angular momentum of the valence electron in lithium its energy more nearly equals the corresponding energy in hydrogen.

Identical particles and atomic structure

filling of the 2p levels. There will be six of these latter (two states of opposite spin for each of the three substates for $l = 1$: namely, $m_l = +1$, 0, and -1). Figure 13-12 indicates how the mean radii of the position probability distributions depend much more on n than on l. Thus, for $n = 2$, the two 2s electrons and the six 2p electrons form, in effect, a "shell" of electric charge, significantly further out from the nucleus than the 1s electrons, and characterized roughly by a Bohr-atom orbit radius calculated for $n = 2$, with a central charge $(Z - 2)e$. This completed second set of levels is called the L shell.

Sodium

The screening of the nuclear charge by the L shell, when this is completely filled, has an important effect on the energies

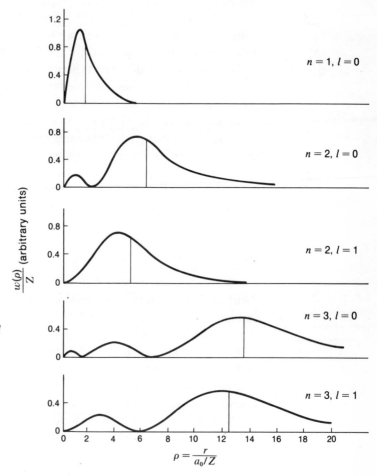

Fig. 13-12 Radial probability densities w for various n and l for hydrogen-like systems, showing that l has less effect than n on mean radius. (Mean radius in each case shown by vertical line.) The horizontal axis is in units a_0/Z.

13-9 **Many-electron atoms**

of electrons added still later. Together with the charge $-2e$ represented by the K shell, it reduces the effective central charge to $(Z - 10)e$ with respect to any electron that does not penetrate the K and L shells. However, the first electron to be added after the $n = 2$ states are all filled will be a $3s$ electron which penetrates the K and L shells, yielding a more negative interaction energy with the nucleus than a nonpenetrating state would do. In the case of sodium ($Z = 11$) the addition of this electron yields a neutral atom, and the $3s$ electron plays the role of valency electron in the atom, being loosely bound compared to any of the others. (The energy needed to remove the outer $3s$ electron from sodium is 5.14 eV; the energy required to remove a second electron is nearly 10 times greater—47.3 eV.)

In the above description of sodium, we talk about the "outer electron in the $3s$ state" as if it were distinguishable from other electrons in the atom. In reality, we know, electrons are indistinguishable, so that the correct wave function must combine the one-electron wave functions in such a way as to yield an overall wave function antisymmetric with respect to exchange of any two electrons. As mentioned at the end of Section 13-6, we are justified in ignoring this necessity to the extent that the electron wave functions for the respective states do not overlap. When this overlap is small, the practical consequences of explicitly recognizing the identity of electrons and setting up an antisymmetric total wave function are then quite slight because the overlap integrals that express the exchange energies—similar to energy ΔE_2 in Eq. 13-10—become very small.

Higher Shells

By this point in the argument the main features of the "Aufbau" of the electron systems in atoms are fairly clear. By and large the electron states will be filled in order of increasing n, and in order of increasing l for each n. However, the pulling down of energy for the penetrating s states becomes more and more important as the central charge Ze increases, and ultimately this has the effect that an s state for a certain n has a lower energy than that of states of larger l belonging to $n - 1$. The first example of this effect occurs with potassium ($Z = 19$) after completion of the "subshell" of six electrons in the $3p$ states, giving 18 electrons up to this point. The next

Identical particles and atomic structure

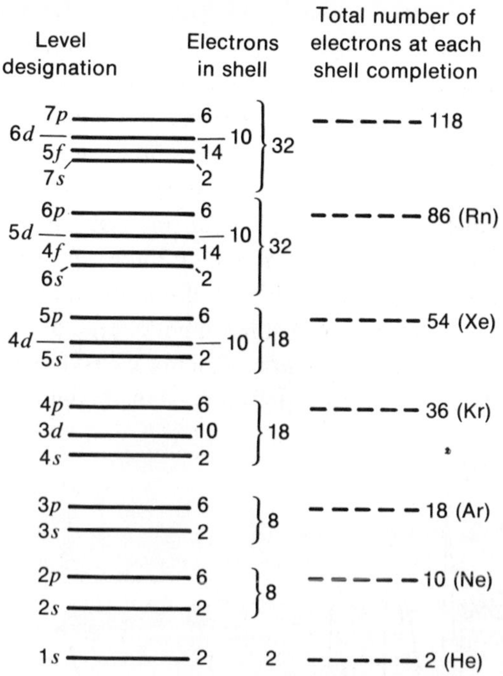

Total number of
| Level designation | Electrons in shell | electrons at each shell completion |

7p ———— 6
6d — ———— 10 } 32
5f ———— 14
7s ———— 2   ------ 118

6p ———— 6
5d — ———— 10 } 32
4f ———— 14
6s ———— 2   ------ 86 (Rn)

5p ———— 6
4d — ———— 10 } 18
5s ———— 2   ------ 54 (Xe)

4p ———— 6
3d ———— 10 } 18
4s ———— 2   ------ 36 (Kr)

3p ———— 6
3s ———— 2 } 8   ------ 18 (Ar)

2p ———— 6
2s ———— 2 } 8   ------ 10 (Ne)

1s ———— 2   2   ------ 2 (He)

Fig. 13-13 Electronic shell structure.

electron to be added (the nineteenth) enters a $4s$ state instead of $3d$, and is the valency electron of the potassium atom. Thereafter as Z increases beyond potassium a certain amount of competition goes on between several possible subshells for each new electron added. However, the configurations at completion of particularly stable structures, as represented by the noble gases, are well defined and correspond to the ordering of levels shown in Figure 13-13. It is in these completed structures (Table 13-2) that we see the groupings that characterize the periodic table of the elements, marked by the noble gases themselves at $Z = 2$ (He), 10 (Ne), 18 (Ar), 36(Kr), 54 (Xe), and 86 (Rn).

TABLE 13-2 Electron Shell Structure

Values of (n, l)	Shell Capacity	Cumulative Total
(1, 0)	2	2
(2, 0) + (2, 1)	8	10
(3, 0) + (3, 1)	8	18
(3, 2) + (4, 0) + (4, 1)	18	36
(4, 2) + (5, 0) + (5, 1)	18	54
(4, 3) + (5, 2) + (6, 0) + (6, 1)	32	86

In terms of observable physical properties, the approach to a major shell closure is marked by an increase in the first ionization potential of the atom [(the energy required to remove one electron from the neutral atom—see Figure 13-14a]. The ionization potential then drops to an especially low value for each element for which Z corresponds to a closed shell plus one electron. These elements are the alkali metals (Li, Na, K, Rb, Cs); they represent systems in which the last electron finds itself mainly outside a relatively tightly bound structure that has a net charge of $+e$. The valency electron, in consequence of being weakly bound, also has a tendency to be especially far from the center of the atom. As a

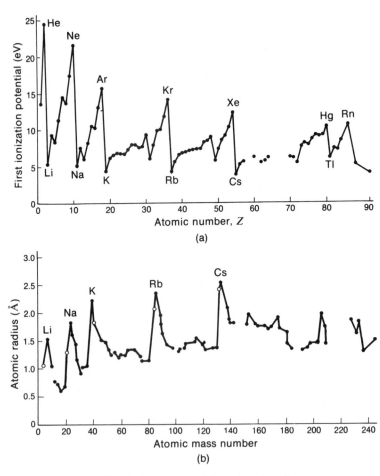

Fig. 13-14   (a) First ionization potential as a function of Z.   (b) Atomic radius as a function of atomic mass number.

Identical particles and atomic structure

result the alkalis are the largest of the atoms in the periodic table (Figure 13-14b).

It is worth noting that, although there are clear periodicities in atomic radii, the atoms—all the way from hydrogen to uranium—show remarkably little variation in size, and in particular no important tendency to be bigger if they are more massive. In a very qualitative way, this can be understood as a consequence of the large central charge pulling the electrons into orbits of much smaller radius than the corresponding ones in hydrogen, so that even though shells belonging to $n > 1$ are involved, the outer radius of the overall charge distribution is not much affected.

The picture of groups of electrons forming shells or subshells at fairly well-defined radii is well grounded theoretically and has also been verified experimentally. Figure 13-15a shows the results of some calculations of the radial probability distribution (and hence the charge distribution) for the complete set of 18 electrons in the argon atom. One curve in the figure shows the result of a crude calculation in which the electrons in any one shell are assumed to have hydrogen-like wave functions defined for a central charge Ze diminished by the charge of the electrons in completed shells nearer the nucleus. A second curve in the figure shows the result of a better calculation embodying the requirements of the self-consistent field. The difference between the two curves is not great, but it is significant. Figure 13-15b shows the result of an experiment in which the radial charge distribution in argon was derived from a study of the scattering of electrons from the atoms in argon gas (similar to an x-ray diffraction analysis). The agreement between experiment and the better theoretical distribution is very good. Thus in various ways one finds reason to believe that the theory of many-electron atoms, using quantum mechanics, is a reflection of reality. In fact our brief account here does not begin to do justice to the extent to which the detailed physical properties of atoms (and especially their spectra) have been successfully interpreted with the help of quantum theory.

13-10   GENERAL STRUCTURE OF A MASSIVE ATOM

To bring together the results discussed above, it may be helpful to consider the complete picture of a particular mas-

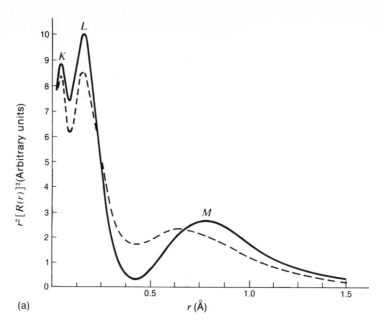

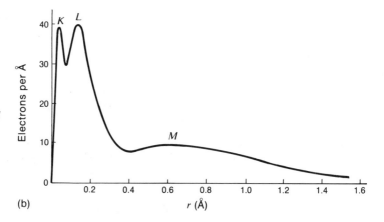

Fig. 13-15 Total radial distribution of electrons in the argon atom. (a) Theoretical. Solid curve: simple shell model. Broken curve: self-consistent field approximation. (b) Experimental, deduced from electron scattering. [After L. S. Bartell and L. O. Brockway, Phys. Rev. **90**, 833 (1953)]

(a)

(b)

sive atom with many electrons. Let us take the most abundant isotope of tin, with $Z = 50$, $A = 120$.

The nucleus of this atom has a charge of $+ 50e$; it contains 50 protons and a number of neutrons equal to $A - Z$, i.e., 70, but the nuclear structure is not our concern here. Closest to the nucleus (on the average) are two electrons in states with $n = 1$; these electrons are exposed to almost the full force of the nuclear electric field. According to the simple theory of hydrogen-like atoms, the energy of an electron in the field of an effective central charge Qe is given by

Identical particles and atomic structure

$$E(Q,n) = -\frac{2\pi^2 mQ^2 e^4}{n^2 h^2} = -13.6 \frac{Q^2}{n^2} \text{ eV} \qquad (13\text{-}12)$$

Thus, for $n = 1$, $Q \approx 50$, each electron would be bound with a negative energy equal to almost 2500 times the binding energy of the electron in the hydrogen atom (13.6 eV). This would be on the order of 30 keV. Direct measurements show that it takes x rays with a quantum energy of 29.2 keV to dislodge an electron from this innermost shell. The average distance of these electrons from the nucleus is comparable to the orbit radius calculated from the Bohr theory for $n = 1$, $Q = 50$; this is about 0.01 Å, or 10^{-12} m—about 150 times the nuclear radius of tin.

Going to the next electron shell (cf. Table 13-2), we have eight electrons for which $n = 2$ and for which the nuclear charge is significantly shielded, first by the two inner-most electrons, secondly by the other electrons in this same shell, and thirdly by penetrating electrons from shells still farther out. The effective central charge is not easy to estimate in this case, but experiment shows that x rays of about 4 keV can eject electrons from this shell. (For experimental evidence of this, see the graph of x-ray absorption edges in Chapter 1, Figure 1-18.) Actually, this and subsequent shells have a sub-structure, based on the involvement of two or more values of the quantum number l, that we shall not go into here. Putting $n = 2$ in Eq. 13-12, and using the observed electron binding energy of about 4 keV, one would infer an effective central charge of about $35e$ and a mean shell radius of about 0.06 Å.

Proceeding in the same way, one finds that the third shell (8 electrons, $n = 3$) has a critical x-ray absorption energy of about 0.9 keV, corresponding to $Q \approx 24$, $r \approx 0.2$ Å; and the fourth shell (18 electrons) has a critical x-ray absorption energy of about 120 eV, corresponding to $Q \approx 12$, $r \approx 0.7$ Å.

This accounts for 36 out of the 50 electrons in the atom. The remaining 14 belong to the fifth electron shell, which can accommodate up to 18 electrons. The situation in this region of the atom is complicated, but we know that these electrons are the main contributors to the outer parts of the atomic charge cloud, which extends to a radius of about 1.5 Å (see Figure 13-14b). To remove one of these electrons requires an energy of 7.3 eV (the first ionization potential of tin). Thus we see that the energy-level structure within the Sn atom

13-10   General structure of a massive atom

ranges all the way from weakly bound electrons (less than 10 eV) to very tightly bound electrons (tens of keV).

The chemical and spectroscopic characteristics of an element depend on the details of the quantum state of its outermost electrons. In particular, the chemical valence depends upon the extent to which the total number of electrons in an atom represents an excess or a deficit with respect to the more stable configuration of a completed shell. In the case of tin, for example, the total of 50 electrons is four short of completing the fifth shell of Table 13-2. In these terms one can understand why tin is quadrivalent (although it takes a study of finer details to understand in physical terms why it also exhibits divalency). In such matters as these, however, although the properties certainly have their complete basis in electric forces and quantum theory, the theoretical analysis is at best semi-empirical.

EXERCISES

13-1 *Unlike particles in a box.* Consider two dissimilar particles, of masses m_0 and $4m_0$ (for example, a neutron and an alpha particle) confined in the same one-dimensional box of length L. Ignoring their mutual interactions, write down and solve the Schrödinger equation for this system and obtain an expression for the total energy. Consider the conditions for energy degeneracy of the system (that is, different combinations of quantum numbers resulting in the same total energy).

13-2 *Like particles in a box: I.*
(a) Consider two particles of the same mass in a one-dimensional box of length L. Sketch the "contour map," analogous to Figure 13-1, for the two possible wave functions associated with having one particle in the state $n = 1$ and the other particle in the state $n = 3$.
(b) Now impose the condition that the particles are identical and construct the contour maps of probability density, analogous to Figure 13-2, for the symmetric and antisymmetric space states belonging to the same pair of quantum numbers.

13-3 *Like particles in a box: II.* Two particles in a one-dimensional box are in a combined quantum state belonging to the quantum numbers $n = 1$ and $n = 2$.
(a) Ignoring the need to make the total wave function either symmetric or antisymmetric, calculate the probability that *both* particles are within a distance $\pm L/20$ of the point $x = L/4$.

Identical particles and atomic structure

(b) Assume now that the total space wave function must be symmetric, and calculate the probability in this case.

(c) Compare the results of (a) and (b) with the corresponding probability for two classical Newtonian particles simply bouncing back and forth between the ends of the box with incommensurable constant speeds.

13-4 *Combined spin states.*

(a) Consider the combined spin states for a particle of spin 1 and a particle of spin $\frac{1}{2}$. Tabulate all the possible z components of the resultant spin and show that they correspond to the array of spin states arising from total spins of $\frac{3}{2}$ and $\frac{1}{2}$.

(b) Suppose that the particles in question are a deuteron and a neutron. The deuteron is the combination of a proton with a neutron, and with the addition of another neutron it forms the nuclide hydrogen-3 (tritium). Taking account of the Pauli principle, what possible spin(s) would you predict for the tritium nucleus in its ground state and in a state in which one of the three nucleons is raised to an excited energy level?

13-5 *Antisymmetric wave functions and the Pauli principle.* Consider a wave function for two electrons written explicitly as the product of a symmetric space function and an antisymmetric spin function, or vice versa:

$$\psi(\mathbf{r}_1, \sigma_1, \mathbf{r}_2, \sigma_2) \sim [\psi_A(\mathbf{r}_1)\psi_B(\mathbf{r}_2) \pm \psi_B(\mathbf{r}_1)\psi_A(\mathbf{r}_2)]$$
$$\times [\alpha(1)\beta(2) \mp \beta(1)\alpha(2)]$$

The quantities σ_1, σ_2 denote "spin coordinates" corresponding to the single-particle spin functions α and β. Multiply out this expression and show that it can be rewritten as the sum of two expressions each of the form

$$\psi_A(1)\psi_B(2) - \psi_B(1)\psi_A(2)$$

where the subscripts A and B now refer to a complete set of quantum numbers (space + spin) for one particle. This gives an alternative way of verifying that the total wave function vanishes identically if A and B represent identical sets of space and spin quantum numbers.

13-6 *Pauli's principle and the energy of a many-particle system.* The Pauli principle allows two electrons (or other spin-$\frac{1}{2}$ particles) with opposite spins to occupy the same space state and hence have the same energy.

(a) Compare the minimum possible *total* energy for 10 electrons confined within a single one-dimensional box to the total energy for 10 electrons distributed in pairs in five similar boxes isolated from one another. (Ignore interactions between the electrons.)

(b) Now consider a total of $2N$ electrons in the same one-dimensional box. Assuming $N \gg 1$, show that the minimum possible total energy exceeds by a factor of approximately $N^2/3$ the minimum energy that the system could have if the restrictions expressed by the Pauli principle did not apply. (Convert sums over discrete integers into integrals.)

13-7 *One-dimensional model of electrons in metal.* Another way of expressing the result of the previous problem is to say that the minimum mean energy $\bar{E}$ per particle for $2N$ fermions in a one-dimensional box of length L is given by

$$\bar{E} \approx \frac{1}{3} N^2 \frac{h^2}{8mL^2}$$

This result implies that $\bar{E}$ is controlled by the ratio N/L, or in other words the *linear density* of particles, rather than by N and L separately.

The conduction electrons in a metal are an example of such a system (except of course that the problem is really three-dimensional). Assuming one conduction electron per atom and an interatomic spacing of about 2 Å, show that $\bar{E}$ according to the above formula is of the order of 10 eV.

[This shows that the highly condensed state of electrons in a metal makes them very hot. An average energy per particle of 10 eV corresponds to a temperature of about 100,000 K—verify this.]

13-8 *The Pauli principle and the nucleus.* To some approximation, a medium-weight nucleus can be regarded as a flat-bottomed potential well with rigid walls. To simplify the picture still further, model a nucleus as a cubical box of edge-length equal to the nuclear diameter. Each space state can, by the Pauli principle, accommodate two protons and two neutrons. Consider, in particular, the nucleus of iron-56 which has 28 protons and 28 neutrons.

(a) Taking account of the energy degeneracies (Section 5-4 and Exercise 5-4) estimate the kinetic energy of the highest-energy nucleon in the nucleus. Assume a nuclear radius of 5×10^{-13} cm. [Tabulate the individual particle states in terms of the quantum numbers (n_1, n_2, n_3), beginning with $(1, 1, 1)$.]

(b) Calculate the *average* kinetic energy per nucleon for the nucleus as a whole.

13-9 *Refinement of Bohr model for helium ground state.* Imagine that, in the helium ground state, the electrons travel in a circle at opposite ends of a diameter passing through the helium nucleus. Assume that each electron has one unit of angular momentum $\hbar$ (which corresponds to the circumference of the orbit being equal to one de Broglie wavelength).

(a) Solve for the orbit radius, taking into account the Coulomb repulsion between the electrons as well as the attraction of each toward the nucleus.

(b) Deduce what would be the total energy of the ground state and the energy needed to remove one electron, according to this model.

13-10 *Energy perturbations of a two-particle state.* In Section 13-8 we indicate how one can calculate (approximately) the energy shifts for levels of the helium atom due to the mutual interactions of the electrons and exchange effects. As a simpler, explicit example of such calculations, consider two identical particles sharing the two lowest states ($n = 1$ and $n = 2$) in a one-dimensional box of length L. Suppose that the interaction between them is represented by a small constant potential energy V_0 if they are separated by a distance $\pm b/2$ ($b \ll L$) and zero for all larger separations.

(a) Referring to Eq. 13-8, satisfy yourself that the general upward shift of energy resulting from the interaction is given in this case by

$$\Delta E_1 = \frac{4V_0}{L^2} \int_{x_1=0}^{L} \sin^2\left(\frac{2\pi x_1}{L}\right) dx_1 \int_{x_2=x_1-b/2}^{x_1+b/2} \sin^2\left(\frac{\pi x_2}{L}\right) dx_2$$

(b) Carry out the integrations and show that $\Delta E_1 = bV_0/L$.

(c) Referring to Eq. 13-11b, satisfy yourself that the further shift of energy due to the exchange effects for identical particles is given by

$$\Delta E_2 = \pm \frac{4V_0}{L^2} \int_{x_1=0}^{L} \sin\left(\frac{2\pi x_1}{L}\right) \sin\left(\frac{\pi x_1}{L}\right) dx_1$$
$$\times \int_{x_2=x_1-b/2}^{x_1+b/2} \sin\left(\frac{2\pi x_2}{L}\right) \sin\left(\frac{\pi x_2}{L}\right) dx_2$$

(d) Carry out the integrations and show that

$$\Delta E_2 = \pm(4V_0/3\pi) \sin(\pi b/L)$$

13-11 *Electron configurations of many-electron atoms.* The electron configuration of a many-electron atom is written in a way that identifies all the one-electron states involved. For example, the ground state of boron ($Z = 5$) is written $(1s)^2 (2s)^2 (2p)$, indicating that it has two electrons in the $1s$ state (K shell) with $n = 1, l = 0$, two electrons in the $2s$ state (L shell) with $n = 2, l = 0$, and one electron in the $2p$ state (L shell) with $n = 2, l = 1$.

(a) Using this notation, and with the help of Figure 13-13, write out the complete electron configurations for the ground states of the following atoms: nitrogen ($Z = 7$); silicon ($Z = 14$); arsenic ($Z = 33$); silver ($Z = 47$).

(b) Propose for each of these atoms a plausible overall spectral designation (see Section 13-8) showing the resultant L value, the multiplicity, and the value of J.

13-12 *Electron configurations of excited atoms.* In astronomy and astrophysics, the observed spectra from many stars contain lines from highly excited or ionized atoms that reflect the violent conditions at the surface of the star. It is therefore important to be able to identify such spectra with the help of our understanding of electron configurations in general.

(a) With the help of Figure 13-13, make your best guesses as to the electronic configurations of the ground state and the first excited state of singly ionized neon ($Z = 10$, with nine electrons in all).

(b) Try to make a reasoned estimate of the energy of a radiative transition between these states. Which region of the spectrum would you expect it to be in—infrared, visible, or ultraviolet?

(c) Which normal, neutral atom would you expect to have a spectrum most nearly resembling that of Ne$^+$? Can you venture any quantitative estimate of the relation between the wavelengths of corresponding transitions in the two atoms?

13-13 *Screening in a many-electron atom.*

(a) We know that the valence electron of sodium ($Z = 11$) is in a $3s$ state, and that the ionization potential of sodium is 5.14 eV. We also know that the D lines of sodium arise from transitions between $3P$ and $3S$ states of the atom, and have a wavelength of about 5900 Å. What do these facts tell you about the effective nuclear charge as seen by the valence electron in sodium?

(b) It is noted in the text (Section 13-9) that the *second* ionization potential of sodium (that is, to remove a further electron from singly ionized sodium in its ground state) is 47.3 eV. Consider the implications of this result.

13-14 *Energy of a K electron in a many-electron atom.* Consider one of the two K electrons in an atom of fairly large Z (say $Z \approx 50$).

(a) Make a first estimate of the energy of this electron by using the Bohr model for a single electron in the field of a central charge equal to $(Z - 1)e$.

(b) Develop qualitative arguments to show whether the result of (a) must be corrected upward (less negative) or downward (more negative) by making the necessary allowances for (i) relativistic effects (ii) the finite size of the nucleus, and (iii) partial screening of the nuclear charge by electrons outside the K shell.

Identical particles and atomic structure

13-15   *The nuclear shell model and "magic numbers."* The neutrons and protons in a nucleus are, like electrons, fermions of spin $\frac{1}{2}$ that obey the Pauli principle and exhibit spin-orbit coupling. A remarkably successful semi-empirical model of nucleon configurations in nuclei was developed from these facts,[8] and led (amongst other things) to an understanding of why nuclei with certain "magic" numbers of protons or neutrons (2, 8, 20, 28, 50, 82, 126) are especially stable. The figure indicates part of the sequence of energy levels calculated from the Schrödinger equation for a single particle in a spherical potential well, ignoring the spin-orbit effects. Add to this picture a spin-orbit splitting with the properties that (i) the states with $j = l + \frac{1}{2}$ become lower in energy than those with $j = l - \frac{1}{2}$, and (ii) the magnitude of the spin-orbit splitting increases with l. See how many of the "magic numbers" you can account for on these grounds.

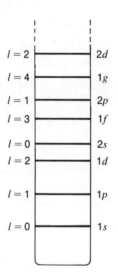

13-16   *The electron cloud of a many-electron atom.* Figure 13-15 (a) shows the results of some calculations of the overall radial charge distribution for the argon atom.

(a) Try carrying out a similar calculation for the simpler case of neon by just superposing the single-electron probability distributions for the different shells. (You can use the probability distributions shown in Figure 12-5.)

(b) Add to this picture the probability distribution for a single extra electron in a $3s$ state, such as one would have in the next atom (sodium). This will help to give a feeling for the physical basis for the abrupt increase in atomic radius in going from the closed-shell configuration of neon to the closed-shell-plus-one configuration of sodium.

[8]O. Haxel, J. H. Jensen, and H. E. Suess, Phys. Rev. **75**, 1766, (1949).

There appears to me one grave difficulty in your hypothesis, which I have no doubt you fully realise, namely, how does an electron decide what frequency it is going to vibrate at when it passes from one stationary state to the other? It seems to me that you would have to assume that the electron knows beforehand where it is going to stop.

ERNEST RUTHERFORD, *letter to Niels Bohr* (1913)

If we are going to stick to this damned quantum-jumping, then I regret that I ever had anything to do with quantum theory.

E. SCHRÖDINGER *(quoted by W. Heisenberg), Niels Bohr and the Development of Physics (ed. W. Pauli) (1955)*

14

Radiation by atoms

14-1 INTRODUCTION

Almost everything we know about atomic energy levels comes from the spectra resulting from radiative transitions between these levels. Although a rigorous analysis of the process by which excited atoms emit photons is beyond the scope of this book, the essential role that spectra play in the investigation of atomic structure makes it desirable to provide a meaningful although elementary treatment. One of our main goals will be to develop criteria, known as *selection rules*, that tell us whether or not, in principle, a radiative transition between two given energy levels is possible.

In our treatment of radiation we shall rely heavily on various aspects of the classical theory of radiation by oscillating systems of charges. To begin our discussion, therefore, we review some of the main features of the classical results.

14-2 THE CLASSICAL HERTZIAN DIPOLE

Classically the radiating dipole is a system made up of two equal and opposite charges that move back and forth in opposite directions, each one performing simple harmonic motion along a straight line. (Historically, this was the first radiating system to which Maxwell's electromagnetic theory was applied—by Heinrich Hertz in 1886. Hence the name the *Hertzian dipole*.) If the charged particles constitute an isolated system, we can find an inertial frame in which their center of mass remains at rest as they oscillate. If the charged particles have equal mass, their displacements will also be equal and opposite relative to this center of mass. However, if one of the particles is very much more massive than the other, we can treat the more massive particle as fixed; the oscillating dipole is then the result of the motion of the lighter particle alone. This latter situation is the relevant one in the case of atomic radiation, since to a good approximation the nucleus can be regarded as fixed, and it is (classically speaking) an oscillating electron that generates the radiation field. We shall find that, for most atomic spectra, the principal mechanism of emission corresponds to classical electric dipole radiation of this type.

Consider now a charge q (in most cases this will be an electron for which $q = -e$) that is oscillating up and down the z axis with maximum displacement z_o and angular frequency ω (Figure 14-1a). (As a model of the atom, this classical picture has the awkward property that the electron oscillates right through the position of the nucleus; the quantum-mechanical

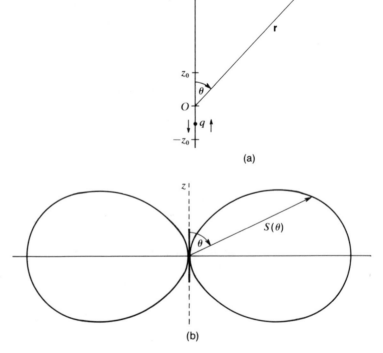

Fig. 14-1 (a) An oscillating electric dipole, showing distant transverse electric field. (b) A polar diagram showing the angular distribution of radiated intensity at distances $r \gg \lambda$ for the dipole of (a).

(a)

(b)

Radiation by atoms

model in Section 14-4 will not have this defect.) The results of classical electromagnetic theory for this system can be summarized as follows:[1] At the time t the transverse electric field at some distant point P with coordinates r, θ is proportional to the acceleration $a_{(t-r/c)}$ of the charge at the earlier time $t - r/c$ (called the *retarded time*), and is given explicitly by the equation[2]

$$\mathcal{E}(r, \theta, t) = \frac{q a_{(t-r/c)} \sin \theta}{c^2 r} \quad \text{(cgs)} \tag{14-1}$$

For a harmonically oscillating charge, we put

$$z = z_0 \cos \omega t$$

so that the acceleration is

$$a = -\omega^2 z_0 \cos \omega t$$

Hence

$$\mathcal{E}(r, \theta, t) = -\frac{\omega^2 q z_0 \sin \theta}{c^2 r} \cos \omega \left(t - \frac{r}{c} \right)$$

The product $q z_0$ is the maximum value D_E of the electric dipole moment of the system (we assume that a charge $-q$ remains at rest at $z = 0$). Thus, finally, the transverse electric field at a distant point P is given by the equation

$$\mathcal{E}(r, \theta, t) = -\frac{\omega^2 D_E \sin \theta}{c^2 r} \cos \omega \left(t - \frac{r}{c} \right)$$

The instantaneous rate of radial energy flow per unit area in the direction θ is equal to the magnitude of the Poynting vector **S**:

$$S = \frac{c}{4\pi} \mathcal{E}^2 = \frac{\omega^4 D_E^2 \sin^2 \theta}{4\pi c^3 r^2} \cos^2 \omega \left(t - \frac{r}{c} \right)$$

[1]See, for example, B. Rossi, *Optics*, Addison-Wesley, Reading, Mass., 1957.

[2]In this chapter, to reduce complication, all equations and formulas will be given in cgs units only.

14-2   The classical Hertzian dipole

The time average of this rate of energy flow per unit area is thus given by

$$S_{av} = \frac{c}{4\pi} (\mathscr{E}^2)_{av} = \frac{\omega^4 D_E{}^2 \sin^2 \theta}{8\pi c^3 r^2}$$

This radiation varies as a function of direction (Figure 14-1b). The total rate of radiation, W, is obtained by integrating S_{av} over the entire surface of the sphere of radius r:

$$W = \int_{\theta=0}^{\pi} S_{av} \, 2\pi r^2 \sin \theta \, d\theta = \frac{\omega^4 D_E{}^2}{4c^3} \int_0^{\pi} \sin^2 \theta \sin \theta \, d\theta$$

Carrying out the integration gives

$$W = \frac{\omega^4 D_E{}^2}{3c^3} \tag{14-2}$$

in units of erg/sec.

This is the classical result for radiation from a Hertzian dipole that we will adapt below to the quantum-mechanical analysis.

14-3 RADIATION FROM AN ARBITRARY CHARGE DISTRIBUTION

In the case of an arbitrary charge distribution, described by a distributed charge density ρ, the electric dipole moment can be defined in terms of an integral over the distribution:

$$\mathbf{D} = \int \mathbf{r}\rho(r) \, dV \tag{14-3}$$

This vector dipole moment can be analyzed into components:

$$D_x = \int x\rho(r) \, dV \tag{14-4}$$

with similar expressions for the components D_y and D_z. If the charge distribution is undergoing harmonic oscillations, then we can adapt Eq. 14-2 to calculate the rate of radiation of energy due to the oscillating electric dipole moment of the distribution. However, the distant field of an arbitrary charge distribution is not due simply to the electric dipole moment; other contributions can be analyzed in terms of the other so-called

Radiation by atoms

multipole moments of the distribution. These higher moments are classified as electric quadrupole, electric octopole, and so forth, along with magnetic dipole, magnetic quadrupole, and so forth. (If the system has a net charge, there is also an "electric monopole moment" that causes a field at distant points. But since the net charge is independent of time, this monopole moment does not contribute to the *radiation* field.) It is possible that the electric dipole moment of a charge distribution may vanish, in which case the radiation depends on the higher moments. But we shall now show, at least on the basis of a classical analysis, that these higher multipoles represent a relatively feeble contribution to the radiation of visible light from atomic systems.

Electric Quadrupole Radiation

We shall compare the field strengths at a distant point due to (a) a harmonically oscillating electric dipole contained within the limits of a certain charge distribution, and (b) an electric quadrupole made up of two such dipoles parallel to each other and a distance d apart. We assume that these two paired dipoles are oscillating in antiphase, so that their *net dipole moment* is zero. The configurations (a) and (b) are shown in Figure 14-2. Let the field at a distant point P due to the single dipole of Figure 14-2a be $\mathscr{E}_{10} \cos \omega t$. Then the field $\mathscr{E}_2(t)$ at a distant point along the axis of separation of the two dipoles (Figure 14-2b) is given by

$$\mathscr{E}_2(t) = \mathscr{E}_{10} \cos \left(\omega t - \frac{2\pi}{\lambda} \frac{d}{2} \right) + \mathscr{E}_{10} \cos \left(\omega t + \pi + \frac{2\pi}{\lambda} \frac{d}{2} \right)$$

(account being taken of the 180° phase difference between the two dipoles). Combining the two terms on the right, we have

$$\mathscr{E}_2(t) = 2\mathscr{E}_{10} \sin \left(\frac{\pi d}{\lambda} \right) \sin \omega t$$

This defines a quadrupole radiation field strength given by

$$\mathscr{E}_{20} = 2\mathscr{E}_{10} \sin \left(\frac{\pi d}{\lambda} \right) \tag{14-5}$$

Now if for d we put the characteristic linear dimension of an

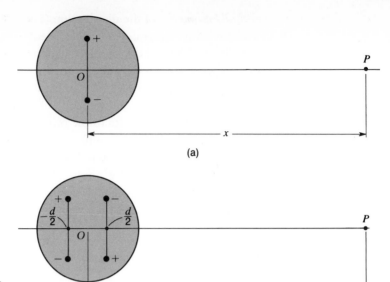

Fig. 14-2 (a) A single electric dipole oscillator. (b) The combination of two dipoles to form a quadrupole.

atom (≈ 1 Å) and for λ we put the wavelength of visible light (≈ 6000 Å), we have $\pi d/\lambda \approx \frac{1}{2} \times 10^{-3}$. Since, again in classical terms, the rate of radiation is proportional to the square of the field amplitude, the quadrupole system would be radiatively weaker than the dipole by a factor of the order of $4\sin^2(\frac{1}{2} \times 10^{-3}) \approx 4(\frac{1}{2} \times 10^{-3})^2 = 10^{-6}$. Assuming that this feature applies also to the quantum description of atomic systems, we can conclude that if electric dipole radiation is possible at all we can ignore the contribution from the quadrupole and higher electric multipole components of the charge distribution.[3] And for cases in which the electric dipole is zero, then, to some fairly close approximation, the emission of radiation can be regarded as impossible, or "forbidden."

Magnetic Dipole Radiation

To complete this analysis we should also consider those components of the oscillating motion of the charge distribution that correspond to *magnetic* dipole, quadrupole, and so on. Classically, the source of magnetic dipole radiation is a circulating current that reverses direction harmonically, as in a

---

[3]Notice, however, that if we are considering x radiation, with $\lambda \approx 1$Å $\approx d$, the higher multipoles will be important.

Radiation by atoms

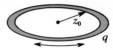

Fig. 14-3 A uniformly charged ring of total charge q oscillating harmonically about its axis to create an oscillating magnetic dipole.

loop antenna (Figure 14-3). Since a steady-current loop can be regarded as equivalent to a static magnetic dipole, the oscillating current is equivalent to an oscillating dipole.

In order to compare the intensities of electric and magnetic dipole emission we relate the electric dipole emission of a point charge q oscillating along a line with amplitude z_0 (Figure 14-1) to the magnetic dipole emission of a uniformly charged ring of the same radius z_0 and total charge q rotating back and forth harmonically about its axis (Figure 14-3) like the balance wheel of a mechanical watch. The magnetic dipole moment is given by

$$D_M = \frac{\text{(current)} \times \text{(area)}}{c} = \frac{q}{c} \frac{v}{2\pi z_0} \cdot \pi z_0{}^2 = \frac{qvz_0}{2c} \qquad (14\text{-}6)$$

where v is the maximum speed of rotation. The *electric* dipole moment for the linearly oscillating point charge is

$$D_E = qz_0$$

These magnetic and electric dipole moments are dimensionally similar in cgs units and their ratio is given by[4]

$$\frac{D_M}{D_E} \approx \frac{v}{2c} \qquad (14\text{-}7)$$

If the frequency of oscillation is the same for both dipoles, the strengths of the radiation fields are in the same ratio as these two dipole moments, and the radiated intensities are as the square of this ratio—that is, v^2/c^2 in order of magnitude. (These results will be discussed more fully later in the chapter—see Section 14-9.) Since, as we have seen, the value of v/c for an electron in a hydrogen atom is at most $1/137 \ (= e^2/\hbar c)$, the magnetic dipole radiation would be less than 10^{-4} of the intensity of electric dipole radiation and could be ignored if both types of radiation were possible. If the electric dipole radiation

[4]In mks units the ratio D_M/D_E has units of velocity. The factor v^2/c^2 in radiated power results from an additional $1/c^2$ in the mks source equation for **B**.

14-3 Radiation from arbitrary charge distribution

is impossible, but magnetic dipole radiation can occur, the radiative process would be classed as "forbidden" because of its low intrinsic intensity.

14-4   RADIATING DIPOLES ACCORDING TO WAVE MECHANICS

Electric Dipole Moment of an Atom

Accepting the foregoing indications from classical physics, we fix attention on the electric dipole moment of a radiating atom according to quantum mechanics. The description of an electron in an atom by a complete wave function ψ implies, as we have seen, a spatial distribution of probability density $\psi^*\psi$. Given this picture, we can regard the value of $-e\psi^*\psi$ at any point as being equivalent to a local *charge* density ρ. Now, in classical physics, the electric dipole moment of a charge distribution ρ is given by Eq. 14-3,

$$\mathbf{D} = \int \mathbf{r}\rho(r)\,dV$$

It is reasonable, then, to define a dipole moment in wave mechanics as

$$\mathbf{D} = -e \int \mathbf{r}\psi^*\psi\,dV \qquad (14\text{-}8)$$

Electrons in Energy Eigenstates Do Not Radiate

In the particular case of an energy eigenstate of an electron in an atom we have

$$\psi(\mathbf{r},\,t) = \psi(\mathbf{r})e^{-iEt/\hbar}$$

For such a stationary state, the exponential time factors in the product $\psi^*\psi$ cancel and the dipole moment $\mathbf{D}$ is independent of time. Thus we automatically satisfy the requirement (which is arbitrary and impossible to justify in the Bohr model) that an electron does not radiate so long as it remains in a single stationary state.

As a matter of fact, the dipole moment of an atom in a definite energy state is not merely *constant*, it is *zero*. A look at the probability density plots for hydrogen given in Figure 12-7 will verify this result for that atom: the probability density $\psi^*\psi$ at any point is matched by an equal probability density an equal distance from the nucleus at a point diametrically opposite to it. Therefore the contributions of these two regions cancel in Eq. 14-8, as do every other similar pair of regions, so the dipole moment is zero. A more analytic proof of the same result can be constructed using the symmetry of the energy eigenfunctions.

The Wave Functions "During" a Transition

In order to construct a time-dependent dipole moment we must consider wave functions other than pure stationary state functions. Until now we have implicitly assumed that the wave function describing a radiating atom changes instantaneously from the initial to the final state with no time spent between states. If this were literally true, then the wave function would always describe a pure stationary state and there would never be any radiation at all. But if we relax our assumption and permit the wave function to spend some time with, so to speak, one foot in each of the states (initial and final) then during the transition the wave function can be thought of as a superposition of the initial state Ψ_i and the final state Ψ_f.

$$\Psi(\mathbf{r}, t) = a_i(t)\psi_i(\mathbf{r})e^{-iE_i t/\hbar} + a_f(t)\psi_f(\mathbf{r})e^{-iE_f t/\hbar} \qquad (14\text{-}9)$$

(The relation between this wave function and the behavior of individual atoms is discussed more fully below.) *Before* the transition the coefficient a_i is unity and a_f is zero. During the transition these coefficients change with time while maintaining the sum of their squared moduli equal to unity. [We assume, however, that the time scale for the changes of a_i and a_f is very long compared to the period of oscillation of the emitted radiation, so that the time dependence of Ψ is mostly in the exponential factors of Eq. 14-9.] *After* the transition, a_i equals zero, a_f equals unity, and the atom is once again in a stationary state in which no dipole radiation is possible.

During the transition, when both terms in Eq. 14-9 are present, the dipole moment (Eq. 14-8) has the form

$$\mathbf{D}_{if} = a_i{}^* a_i(-e) \int \mathbf{r} \, |\psi_i|^2 \, dV + a_f{}^* a_f(-e) \int \mathbf{r} \, |\psi_f|^2 \, dV$$

$$+ a_i{}^* a_f(-e) \int \mathbf{r}\psi_i{}^* \psi_f dV e^{+i(E_i - E_f)t/\hbar}$$

$$+ a_i a_f{}^*(-e) \int \mathbf{r}\psi_i \psi_f{}^* \, dV e^{-i(E_i - E_f)t/\hbar}$$

14-4   Radiating dipoles according to wave mechanics

The first two terms vanish (stationary states of atoms have zero electric dipole moment). The last term is just the complex conjugate of the third term, and $\mathbf{D}_{if}$ can be written in the following form:

$$\mathbf{D}_{if} = a_i{}^* a_f \mathbf{D}_o e^{i\omega t} + (a_i{}^* a_f \mathbf{D}_o e^{i\omega t})^*$$

where

$$\mathbf{D}_o = (-e) \int \mathbf{r}\psi_i{}^* \psi_f \, dV \qquad (14\text{-}10)$$

and

$$\omega = \frac{E_i - E_f}{\hbar}$$

Since the sum of a number and its complex conjugate is just twice the real part of the number, we can put

$$\mathbf{D}_{if} = 2 Re\left[(a_i{}^* a_f) \cdot \mathbf{D}_o \cdot e^{i\omega t}\right] \qquad (14\text{-}11)$$

Equation 14-11 embodies the quantum-mechanical counterpart of the radiating classical electric dipole. The expression between square brackets in this equation has three factors, separated by dots. Each factor is in general a complex function. The real part of the last factor, namely cos ωt, oscillates harmonically with a frequency corresponding to the emitted photon energy $E_i - E_f$. The first factor $a_i{}^* a_f$ also changes with time, but very much more slowly; it is nonzero only during the transition when both a_i and a_f are different from zero.[5] The middle factor $\mathbf{D}_o$ is a measure of the peak magnitude of the dipole oscillations; it can be thought of as measuring the effectiveness of overlap of the initial and final state wave functions, or of the corresponding charge distributions, with respect to generation of an electric dipole moment. Hence the expression $\mathbf{D}_o$ given in Eq. 14-10 is conventionally called the *overlap integral*.

From the picture that arises from Eq. 14-11 we can es-

[5]Actually, the *phase* of the harmonic oscillation of $\mathbf{D}_{if}$ depends also on the phases of the first two complex factors in the parenthesis.

Radiation by atoms

timate values for lifetimes of excited atomic states and derive some of the selection rules that determine which of the conceivable transitions will spontaneously take place in atoms.

14-5 RADIATION RATES AND ATOMIC LIFETIMES

Estimate of Radiative Lifetimes

How long, on the average, does it take a sample of atoms initially in the excited state i to drop directly to a lower state f? According to Eq. 14-11 the atomic dipole is different from zero—and the atom can radiate—only during the time that it is in a superposition of initial and final states. This time is described by the coefficients $a_i(t)$ and $a_f(t)$ that govern the superposition Eq. 14-9. The squared moduli of these coefficients add up to unity at all times. but it is their *product* that determines the radiation time in Eq. 14-11. In this treatment we have no way to calculate these coefficients directly, but we can use a combination of classical and quantum results to calculate transition times. To do this we assume that the quantity $|2a_i^* a_f|$ in Eq. 14-11 has the value unity for the transition time (call it τ) and the value zero at all other times. Then the electric dipole has the magnitude of the overlap integral D_o during the transition time. For the rate of radiation of the atom during this time we take the classical expression Eq. 14-2:

$$\text{Rate of energy radiation} = \frac{\omega^4 D_o{}^2}{3c^3}$$

However, a quantum jump involves the radiation of an amount of energy equal to $\hbar\omega$. If we assume that this is describable in terms of the classical radiation rate continuing for the radiation time τ, we have

$$\frac{\omega^4 D_o{}^2}{3c^3} \tau = \hbar\omega$$

This then gives

$$\text{Radiative lifetime} = \tau = \frac{3\hbar c^3}{\omega^3 D_o{}^2} \qquad (14\text{-}12a)$$

In terms of the wavelength λ of the emitted radiation, we have

$$\omega = 2\pi\nu = \frac{2\pi c}{\lambda}$$

Thus Eq. 14-12a can be rewritten

$$\tau = \frac{3\hbar}{8\pi^3}\frac{\lambda^3}{D_o{}^2} \tag{14-12b}$$

Now although an accurate calculation of the radiative life-time τ requires the detailed evaluation of the overlap integral of Eq. 14-10, an order-of-magnitude estimate, for cases in which the overlap is favorable to a large value of the dipole moment, is provided by putting

$$D_o \approx eR$$

where R is the atomic radius. Substituting $e = 4.8 \times 10^{-10}$ esu and $R \approx 1$ Å $= 10^{-8}$ cm, we thus have

$$D_o \approx 5 \times 10^{-18} \quad \text{esu-cm}$$

If we substitute this value of D_o in Eq. 14-12b together with a typical value of λ for an optical transition (say $\lambda \approx 5 \times 10^{-5}$ cm) we obtain

$$\tau = \frac{3 \times 10^{-27}}{250} \cdot \frac{1.25 \times 10^{-13}}{2.5 \times 10^{-35}} \approx 10^{-7} \text{ sec}$$

This time is very much greater than the period $\lambda/c \approx 10^{-15}$ sec of the emitted radiation, which justifies our assumption that the time scale for changes in a_i and a_f is very long compared to the oscillation period of the emitted radiation.

The Meaning of Radiative Lifetimes

The value 10^{-7} sec is comparable to the directly measured mean lifetimes of excited states of atoms—see Table 14-1. Indeed, the mean lifetime of an atom in an excited state is just equal to the "radiation time" required for a transition whose value we have been estimating. We have already discussed such matters to some extent in Chapter 8 (Section 8-10). We

Radiation by atoms

TABLE 14-1 Lifetimes of Some Excited Atomic States

Atom	State	Measured Lifetime (in units of 10^{-7} sec)
Na	$3P_{3/2}$	0.16[a]
K	$5P_{3/2}$	1.41[b]
Rb	$5P_{1/2}$	0.28[c]
Rb	$6P_{3/2}$	1.15[d]

[a]J. K. Link, J. Opt. Soc. Am. **56**, 1195, 1966.
[b]R. W. Schmieder et al., Phys. Rev. **173**, 76, 1968.
[c]A. Gallagher, Phys. Rev. **157**, 68, 1967.
[d]H. Bucka et al., Z. Phys. **194**, 193, 1966.

saw there that the mean lifetime τ is equal to the reciprocal of the probability per unit time for the atom to emit a photon. In the context of the present discussion we shall denote this probability per unit time by p_{if} (replacing the decay constant γ of Eq. 8-30). The change in the number of excited atoms during time dt is thus given by

$$dN = -p_{if}N\,dt$$

which when integrated gives

$$N(t) = N_o e^{-p_{if}t} \equiv N_o e^{-t/\tau}$$

Thus for electric dipole radiation, using Eq. 14-12a, we have

$$p_{if} = \frac{1}{\tau} \approx \frac{\omega^3 D_o{}^2}{3\hbar c^3} \tag{14-13}$$

Some of the best measurements of lifetime of atomic excited states come from direct observation of the exponential decay of emitted intensity; refer back to Figure 8-10 for an example.

14-6 SELECTION RULES AND RADIATION PATTERNS

Under certain conditions the overlap integral, Eq. 14-10, between a particular pair of initial and final states is equal to zero. In this case there is no possibility of a transition between these initial and final states via electric dipole radiation. Even in some cases where the overlap integral does not vanish, the details of the radiative process can be highly dependent on the quantum numbers of the initial and final states. By introducing specific forms of the wave functions ψ_i and ψ_f we can obtain the quantum-mechanical selection rules that govern the possibility of electric dipole transitions.

We shall illustrate these selection rules in their simplest form by taking as our radiating system the hydrogen atom, using the Coulomb model. In this section we shall deal only with transitions between major energy levels and will therefore ignore the effects of electron spin. In this case the possible wave functions are symbolized by the expression (recall Eq. 12-13 of Chapter 12):

$$\psi_{n,l,m}(r, \theta, \phi) = R_{n,l}(r) P_{l,m}(\theta) e^{im\phi}$$

In what follows the subscripts i and f for initial and final states will be shorthand for the complete set (or the relevant subset) of the quantum numbers n_i, l_i, m_i and n_f, l_f, m_f, respectively.

Selection Rules on m

To begin, we consider explicitly the selection rules depending on the quantum number m, since the results for this are especially clearcut and easy to derive. In order to do this we resolve the overlap integral $\mathbf{D}_o$ into its Cartesian components D_x, D_y, and D_z. In terms of the spherical polar coordinate system we have

$$x = r \sin \theta \cos \phi, \qquad y = r \sin \theta \sin \phi, \qquad z = r \cos \theta$$

Then the x component of the overlap integral in Eq. 14-10 takes on the form

$$D_x = (-e) \int x \psi_i^* \psi_f \, dV$$

or

$$D_x = (-e) \int r \sin \theta \cos \phi \, R_i(r) P_i(\theta) e^{-im_i \phi}$$

$$\times R_f(r) P_f(\theta) e^{im_f \phi} \, dV \tag{14-14a}$$

where the volume element dV is given by

$$dV = r^2 dr \sin \theta \, d\theta \, d\phi$$

Radiation by atoms

Similarly, for the y and z components of $\mathbf{D}_o$ we have

$$D_y = (-e) \int r \sin \theta \sin \phi \, R_i(r) P_i(\theta) e^{-im_i \phi}$$

$$\times R_f(r) P_f(\theta) e^{im_f \phi} \, dV \tag{14-14b}$$

$$D_z = (-e) \int r \cos \theta \, R_i(r) P_i(\theta) e^{-im_i \phi}$$

$$\times R_f(r) P_f(\theta) e^{im_f \phi} \, dV \tag{14-14c}$$

These expressions for the components of $\mathbf{D}_o$ look rather complicated, and are; but each of them can be factored into three integrals, one over each spherical coordinate. Then if we want to study the selection rules for the quantum number m, we are concerned only with the ϕ-dependent parts of Eqs. 14-14a, 14-14b, and 14-14c. We can, in fact, abbreviate the expressions for the components of $\mathbf{D}_o$ as follows:

$$D_x = A \int_0^{2\pi} \cos \phi \, e^{i(m_f - m_i)\phi} \, d\phi$$

$$= \frac{1}{2} A \int_0^{2\pi} \left[e^{i(m_f - m_i + 1)\phi} + e^{i(m_f - m_i - 1)\phi} \right] d\phi$$

$$D_y = A \int_0^{2\pi} \sin \phi \, e^{i(m_f - m_i)\phi} \, d\phi \tag{14-15}$$

$$= -\frac{i}{2} A \int_0^{2\pi} \left[e^{i(m_f - m_i + 1)\phi} - e^{i(m_f - m_i - 1)\phi} \right] d\phi$$

$$D_z = B \int_0^{2\pi} e^{i(m_f - m_i)\phi} \, d\phi$$

The above expressions show identical numerical coefficients A for components D_x and D_y because they involve identical integrals over r and θ. (Satisfy yourself that this is so.) The component D_z, however, involves a different integral and therefore a different coefficient B.

Now any integral of the form

$$\int_0^{2\pi} e^{ib\phi} \, d\phi$$

where b is an integer, is zero unless b itself is zero. And for $b = 0$ the integral is equal to 2π. Thus we see, by reference to the equations for D_x, D_y, and D_z, that there are only *three* con-

14-6 Selection rules and radiation patterns

ditions under which $\mathbf{D}_o$ can be different from zero, namely $m_i - m_f = \pm 1$ or 0. In detail they lead to dipole moment components as shown in Table 14-2.

As we shall now show, these selection rules not only define the conditions under which an electric dipole transition can occur spontaneously, they also predict a particular polarization for radiation emitted in a transition involving a given change of m. For we can see that, if $m_f = m_i \pm 1$, the components D_x and D_y are equal in magnitude, but D_y has an additional factor $\pm i$. If we consider what this means in terms of simple harmonic oscillations of the type $e^{i\omega t}$, we see that $\pm i \ (= e^{\pm i\pi/2})$ signifies that the oscillations of D_y either lead or lag those of D_x by 90° phase angle ($\pi/2$ rad). The combination of two such perpendicular vibrations of equal amplitude, differing in phase by 90°, yields a dipole moment rotating in the xy plane. For $m_f = m_i$, on the other hand, we have only a single linear vibration along the z axis.

In the light of these results, consider what one might expect to observe if the radiating atoms are in a region in which there is a uniform magnetic field that physically defines the z direction. Because of the association of magnetic moment with angular momentum, the states of different m now have slightly different energies of orientation in the magnetic field. That is, the various m states for given n, l are no longer degenerate. Thus, for a given value of m_i, photons of three different energies are emitted, corresponding to the different values of m_f. What would have been a single spectral line in the absence of the magnetic field is now split into three lines of slightly different wavelengths. Moreover, each line has its own characteristic polarization, which we can predict on the basis of Table 14-2. Consideration of the energy levels in the magnetic field shows that the transition $m_f = m_i - 1$ gives us the highest frequency, and the observed spectrum would have the following characteristics:

Frequency	*Emitted light is*
Highest frequency $m_f = m_i - 1$	Circularly polarized in plane perpendicular to magnetic field
Lowest frequency $m_f = m_i + 1$	Circularly polarized in plane perpendicular to magnetic field but in opposite sense to that of previous case
Intermediate frequency $m_f = m_i$	Linearly polarized parallel to magnetic field

These results correspond precisely to what is actually observed in some atomic transitions, and is known as the *Zeeman effect* (after P. Zeeman, who discovered it in 1890). Figure 14-4 indicates what is observed if the emitted radiation is viewed along

Radiation by atoms

TABLE 14-2 Components of the Overlap Integral
D_o **for States of Different** m **in Hydrogen**

m_f	D_x	D_y	D_z
$m_i + 1$	πA	$i\pi A$	0
$m_i - 1$	πA	$-i\pi A$	0
m_i	0	0	$2\pi B$

the direction of the magnetic field and perpendicular to it. An oscillating linear dipole emits no radiation along the direction of its axis of vibration. Thus, looking *along* the field direction, one sees only the two oppositely circular-polarized components. Looking perpendicular to the field, however, one sees the radiation due to D_z, linearly polarized parallel to the field, and also a pair of lines polarized *perpendicular* to the field, resulting from viewing the circularly polarized components in their own plane.

Selection Rules on l

Even when an electric dipole transition is permitted by selection rules on the changes in quantum number m, the transition may still be forbidden by selection rules relating to the changes in other quantum numbers. These selection rules are embodied in the values of the integrals included in the constants A and B in Eqs. 14-15. We now consider the selection rules on the quantum number l. The general result is easily stated, although to prove it involves manipulation of the mathematical properties of the associated Legendre functions. The result is simply that allowed transitions must satisfy the relation

$$l_f = l_i \pm 1 \quad \text{or} \quad \Delta l = \pm 1 \tag{14-16}$$

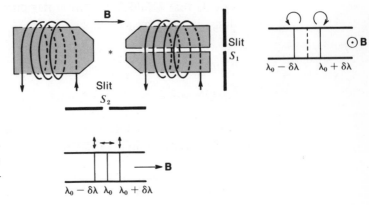

Fig. 14-4 The Zeeman effect, with spectral shifts and polarization effects observed in directions parallel and perpendicular to the magnetic field direction.

14-6 Selection rules and radiation patterns

This selection rule embodies a very basic property of electric dipole transitions: Such transitions can take place only between quantum states of opposite parity. We saw, in discussing the properties of the eigenfunctions in a central potential (Section 11-2) that the complete eigenfunctions $\psi_{nlm}(\mathbf{r})$ have either even or odd parity according to whether the orbital angular momentum quantum number l is even or odd. Now the integral defining the magnitude of the electric dipole moment for a transition is

$$\mathbf{D}_{if} = -e \int \psi_{nlm}^* \, \mathbf{r} \, \psi_{n'l'm'} \, dV$$

In the integrand of this expression, the factor $\mathbf{r}$ itself is clearly an *odd* function. Thus, if the integral is to be nonzero, the *product* $\psi_{nlm}^* \psi_{n'l'm'}$ must also be odd, so that the complete integrand is even. (If the complete integrand were odd, the contribution to the integral from one hemisphere would be exactly canceled by that from the opposite hemisphere.) Hence, if l is even, l' must be odd, and vice versa. This condition is not as restrictive as that of Eq. 14-16, but it does take us part way toward appreciating its physical basis. We shall be able to come even closer to seeing the basis of the l selection rule when we have discussed the angular momentum of individual photons (Section 14-8).

The selection rule $\Delta l = \pm 1$ for electric dipole transitions is general[6], but by considering a specific case we can bring out some additional interesting features. To make it as simple as possible, we suppose that the final state has $l_f = 0$, and hence also the unique value $m_f = 0$. This allows for the initial state the values $m_i = \pm 1$ or 0, which are actually the values associated with $l = 1$. Thus we put $l_i = 1$, and look at the components of $\mathbf{D}_0$ that would follow from this choice. The relevant functions, $P_{l,m}(\theta)$, are as follows:

$$P_{1,1}(\theta) = P_{1,-1}(\theta) = \frac{1}{\sqrt{2}} \sin \theta$$
$$P_{1,0}(\theta) = \cos \theta$$

The way in which these enter into the calculations of D_x, D_y, and D_z is already limited by the selection rules on m. Taking

[6]For a detailed discussion, see R. M. Eisberg, *Fundamentals of Modern Physics*, 1st ed., p 464, Wiley, 1961.

Radiation by atoms

account of this and referring back to Eqs. 14-14, we find the following possibilities:

$$\underline{m_i = \pm 1}: \quad D_x = C \int_0^\pi \sin\theta \cdot \frac{1}{\sqrt{2}} \sin\theta \cdot \sin\theta \, d\theta = \frac{2\sqrt{2}}{3} C$$

$$D_y = \pm iC \int_0^\pi \sin\theta \frac{1}{\sqrt{2}} \sin\theta \cdot \sin\theta \, d\theta = \pm i \frac{2\sqrt{2}}{3} C$$

$$D_z = 0$$

where

$$C \sim e \int_0^\infty r^3 R_f(r) \, R_i(r) \, dr \tag{14-17}$$

$$\underline{m_i = 0}: \quad D_x = 0$$
$$D_y = 0$$
$$D_z = 2C \int_0^\pi \cos\theta \cos\theta \sin\theta \, d\theta = \frac{4C}{3}$$

where the constant C has exactly the same significance as for $m_i = \pm 1$. If we denote the value of D_z for $m_i = 0$ by D, we then have the set of dipole moment components shown in Table 14-3.

Consider now what this means in terms of the angular distribution and relative intensity of the emitted radiation if the transitions for different m_i are separated into the individual Zeeman components by a magnetic field along z. We appeal to the fact that the radiation field of a linear dipole in a given direction is proportional to the sine of the angle between the direction and the dipole axis (Eq. 14-1), and that the radiated intensity is proportional to the square of the field. For $\Delta m = 0$ the radiation corresponds to that of a classical linear dipole D aligned along the z direction (Table 14-3). Thus the transverse radiation

TABLE 14-3 Components of the Overlap Integral for Transitions $l_i = 1 \rightarrow l_f = 0$ in Hydrogen

m_i	D_x	D_y	D_z
1	$\dfrac{1}{\sqrt{2}} D$	$\dfrac{i}{\sqrt{2}} D$	0
-1	$\dfrac{1}{\sqrt{2}} D$	$\dfrac{-i}{\sqrt{2}} D$	0
0	0	0	D

14-6 Selection rules and radiation patterns

field $\mathscr{E}$ in the direction θ is proportional to $D \sin \theta$ (Eq. 14-1), and is linearly polarized in the plane containing the z axis and the direction of observation, $\mathbf{r}$ (see Figure 14-5a). For $\Delta m = \pm 1$, the radiation in the direction θ is elliptically polarized because it comes from what corresponds classically to a radiating charge moving in a circle in the xy plane. This circular motion can be pictured in terms of two orthogonal oscillating linear dipoles, each of magnitude $D/\sqrt{2}$ (Table 14-3); one dipole lies in the plane defined by the z axis and $\mathbf{r}$, and the other is perpendicular to this plane (Figure 14-5b). These two dipoles are $90°$ out of phase. In the direction θ, the component perpendicular to the z-r plane contributes its full value, $D/\sqrt{2}$; the other one contributes only its component $(D \cos \theta)/\sqrt{2}$ transverse to $\mathbf{r}$. Taking these various factors into account, one arrives at the following relative measures of the radiated intensity in the three Zeeman components:

$$m_i = 0: \quad I(\theta) \sim \sin^2 \theta$$
$$m_i = \pm 1: \quad I(\theta) \sim \tfrac{1}{2}(1 + \cos^2 \theta)$$
(14-18)

Figure 14-6 shows the results of some actual measurements of the angular distribution of the atomic dipole radiation emitted in a particular case, exhibiting a rough correspondence with the theoretical pattern. (The discrepancies are instrumental, not due to the theory.) If no magnetic field were present, so that all the radiation occurred at the same wavelength, the intensities for the three different values of m_i would combine to give an isotropic radiated intensity, as one can see more or less by inspection from Eqs. 14-18.

Selection Rules Involving n and j

To round out this discussion of selection rules, we need to consider two other points: The selection rules for the principal quantum number n, and the consequences of having a well-

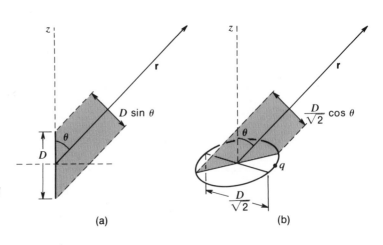

Fig. 14-5
(a) Classical model for $\Delta m = 0$ radiation.
(b) Classical model for $\Delta m = \pm 1$ radiation: circular motion decomposed into two linear dipoles.

(a)

(b)

Radiation by atoms

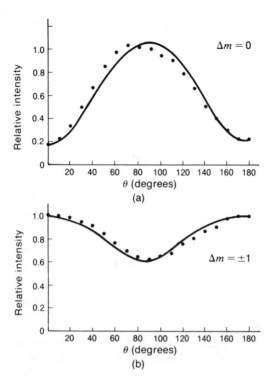

Fig. 14-6 Experimental results on angular distribution of intensity of separate components in the normal Zeeman effect for one of the lines in the spectrum of zinc ($\lambda = 4680$ Å). (a) Linearly polarized component ($\Delta m = 0$). (b) Circularly polarized components ($\Delta m = \pm 1$). (Education Research Center, MIT.)

defined value of the combined orbital and spin angular momentum, **J**.

The selection rule on n is easily stated. There is *no* general restriction on the relationship between the principal quantum numbers n_i and n_f in the production of electric dipole transitions. This might have been guessed from the existence of complete spectral series, such as the Balmer spectrum, involving arbitrary changes of n. Referring to Eqs. 14-14, this result can be expressed by saying that the radial integral $\int r^3 R_i R_f dr$ in the expression for the overlap integral does not automatically become zero for particular values of n_i, n_f and l_i, l_f.

Thus far in the discussion we have ignored electron spin. However, the existence of spin, although it scarcely affects the energies of the levels, does make a difference in the designation of those transitions that are permitted. The total angular momentum and its z component are properly characterized by j and m_j, and it is m_j, rather than m_l, that now defines the number of different substates and their energies in a magnetic field.

The selection rule on l still applies, but the derivations of

14-6   Selection rules and radiation patterns

selection rules on j and m_j are too complex to reproduce here. The results (for transitions involving a single electron) are

$$\Delta j = 0, \pm 1$$
$$\Delta l = \pm 1 \qquad\qquad (14\text{-}19)$$
$$\Delta m_j = 0, \pm 1$$

The details of the transition in the presence of a magnetic field now become more complicated and express themselves in what is called the *anomalous Zeeman effect*. The complexity, however, is mainly in increased numbers of different-wavelength components; the possible polarization states and angular distributions of intensity still conform to the scheme that we developed, ignoring spin, in terms of l and m_l alone. (For this effect with the D lines of sodium, see Figure 10-1d.)

14-7 SYSTEMATICS OF LINE SPECTRA

After the discussion of selection rules, we are better equipped to understand the existence of well-defined families of spectral lines from individual elements. We can illustrate this nicely in terms of the alkali metals.

In the absence of an external magnetic field, the only quantum numbers needed for labeling spectral lines are n, l, and j. In particular, the selection rule $\Delta l = \pm 1$ sharply limits the scheme of possible transitions. This is indicated in Figure 14-7, which shows a simplified version of the energy-level diagram for sodium. The diagram does not exhibit the doubling of the levels due to electron spin for all n and l (except $l = 0$) associated with the two possible values of j—this fine structure would not even be visible on the scale to which the diagram is drawn. A comparison with the corresponding levels in hydrogen (which has no significant splitting between states of different l with the same n) shows how, for sodium, the states of small l are pulled down in energy because of the extra negative energy that derives from closer approach to the strong attractive field of the nucleus, incompletely screened by other electrons. In general, the screening effect is a far more important source of energy splitting between states of different l than is the relativistic correction discussed in Chapter 12 (Section 12-8).

Applying the selection rule $\Delta l = \pm 1$, we then have the fol-

Radiation by atoms

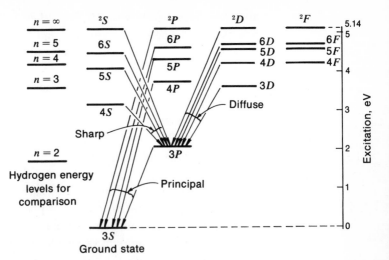

Fig. 14-7 The highest-energy permitted transitions of the sodium atom, showing the sharp, principal, and diffuse series.

lowing sets of allowed transitions that end on the levels $3P$ and $3S$ (the latter representing the normal ground state of the sodium atom):

(a) $nP \rightarrow 3S$ $(n = 3, 4, 5\text{---})$
(b) $nS \rightarrow 3P$ $(n = 4, 5, 6\text{---})$
(c) $nD \rightarrow 3P$ $(n = 3, 4, 5\text{---})$

The energy-level scheme of sodium has certain features that are typical of all the alkali atoms. The transitions $nP \rightarrow 3S$ provide the most energetic quanta permitted. The spectral lines of this type constitute the *principal series* that historically led to the use of the letter P for the levels $(l = 1)$ from which these transitions originate. The shortest wavelength possible corresponds to the full value of the ionization potential (5.13 eV) of the atom. The wavelength of the observed series limit for the principal series in sodium (that is, the wavelength of light given off when an unbound electron falls into the ground state) has the value $\lambda_\infty = 2412.6$ Å which is equivalent to 5.14 eV; the agreement is good. The well-known sodium D lines correspond to the transition $3P \rightarrow 3S$ in this series (the doublet composing two lines instead of one because of the spin-orbit splitting in the $3P$ state).

The other two series of lines that we have specially mentioned both terminate on the same $3P$ level. This means that they have the same series limit, although they do not match anywhere else. The transitions $nS \rightarrow 3P$ form the *sharp series*,

    14-7   Systematics of line spectra

and the transitions $nD \rightarrow 3P$ form the *diffuse series*. These names serve to define the general appearance of the lines and have led to the use of the letters $S(l = 0)$ and $D(l = 2)$ to define their parent levels. Finally there is a series $nF \rightarrow 3D$ which came to be called *fundamental* (because of its long wavelengths and low frequencies) and which led to the use of the symbol F to label $l = 3$.

14-8 ANGULAR MOMENTUM OF PHOTONS

The selection rules on l and m carry important implications for the dynamical properties of photons. Our analysis of the central-field problem demonstrated in clear terms the association of a definite amount of angular momentum with a particular value of l. Thus the fact that l necessarily changes by one unit in an electric dipole transition is tantamount to a requirement that the emitted photon must carry angular momentum with it. If, in particular, we consider a transition between states with $l = 1$ and $l = 0$, it becomes plain that in such a transition the photon must carry away one unit of quantized angular momentum if total angular momentum is to be conserved. More complete analysis shows that this same transfer of angular momentum occurs in all electric dipole transitions and corresponds to an intrinsic angular momentum of individual photons.

To discuss this more concretely, consider once again photons emitted along the direction of the magnetic field (the z axis) in a Zeeman experiment. As we have seen, all of these photons are circularly polarized. Moreover, not only is $\Delta l = \pm 1$ in the transitions in which these photons are emitted, but also $\Delta m = \pm 1$. In other words the z component of atomic angular momentum changes by one unit. In order to conserve angular momentum, the emitted circularly polarized photons must carry away a z component of angular momentum equal to $\pm \hbar$. But these photons are also *traveling* in the z direction. We thus arrive at a picture of circularly polarized photons carrying an angular momentum $\hbar$ along their direction of motion (or opposite to this direction, depending on which state of circular polarization they are in). Photons in either of these two states of circular polarization can be separated according to their energy, since, in a magnetic field, the $\Delta m = +1$ transition has a different energy than the $\Delta m = -1$ transition. More conveniently, photons of different circular polarization states can be sorted using an RL analyzer (Section 6-5). Every photon entering the analyzer will emerge in either the R channel or the L channel (Figure 6-7). In terms of the polarization convention stated and used in Chapter 6, L-polarized photons have angular momentum $+\hbar$ (in the *same* direction as their motion) and R-polarized photons have angular momentum $-\hbar$ (*opposite* to their direction of motion).

In the discussion of photons in Chapter 6 we emphasized that the states R and L form a *complete set* of polarization states: *all* incident photons emerge in one or the other channel of a RL analyzer and every possible photon polarization state can be expressed as a superposition of these two states alone. This leads to the perplexing conclusion that photons which carry one unit of angular momentum can have only the angular momentum components $\pm\hbar$ and not the component zero. This contrasts sharply with orbital angular momentum of atoms and molecules, whose states of unit angular momentum can have not only $L_z = \pm\hbar$ but also $L_z = 0$. The difference can be traced to the fact that photons are *relativistic* particles that move (in a vacuum) with the speed c and no other. [The emission of linearly polarized photons in a transition for which $\Delta m = 0$ (Section 14-6) does not conflict with this picture. In such transitions, no photons are emitted in the z direction along which the quantization of m is measured. And in any other direction, the linear polarization is describable as a superposition of states having angular momentum components $\pm\hbar$ with respect to the direction of travel of the photon.]

In view of these results, it is plausible that the states of circular polarization can be regarded as the truly natural choice of basis states for photon polarization, although mathematically any pair of mutually orthogonal states would fill this role. Then, for a photon traveling in an arbitrary direction with respect to the z (magnetic field) axis, we can express its polarization state as a linear superposition of the basic circular polarization states, in the way that we discussed in detail in Chapter 6. In the normal Zeeman effect, as we saw earlier (Section 14-6), the three components for different photon energies, viewed at an arbitrary angle θ to the beam, are elliptically polarized (highest and lowest energies) or linearly polarized (central component). Each of these is describable as an appropriate linear combination of the complete set of two basic circular polarization states; the coefficients in the combination depend on the direction, specified by θ, along which the photons are traveling.

The fact that photons possess one unit of intrinsic angular momentum has a clear bearing on the selection rule $\Delta l = \pm 1$ for electric dipole radiation (Section 14-6). We can see that this selection rule (coupled with the requirement of a parity change) finds a simple interpretation in terms of conservation of angular momentum, so that the initial angular momentum of the excited atom must be equal to the vector sum of the angular momentum in the de-excited state and the single unit of angular momentum carried away by the photon.

A further important consequence of this picture of the angular momentum states of photons is that, if we produce a beam of radiation that is circularly polarized in one sense only, we know that it carries with it an amount of angular momentum equal to $\hbar$ per photon projected along the direction of motion. Such a beam falling upon an absorber would thus be expected to exert a torque upon it. If the beam represents a rate of energy flow equal to W watts, and if the wavelength of the radiation is λ, the number of photons per second is equal to W divided by the photon energy $h\nu$ ($= hc/\lambda$). Thus we have

$$\text{Photons per second} = \frac{\lambda W}{hc} \tag{14-20}$$

If each photon carries angular momentum $\hbar$, clockwise or counterclockwise, about its direction of travel, the torque exerted on an absorber is thus given by

$$\text{Torque} = (\text{number of photons/sec})$$
$$\times (\text{angular momentum/photon}) = \frac{\lambda W}{2\pi c} \tag{14-21}$$

The absence of h from this expression implies, correctly, that a torque exerted by a circularly polarized beam is also to be expected from classical radiation theory.

The magnitude of the torque for a reasonable intensity of visible light is exceedingly small. For example, with $W = 1$ watt (about what one can obtain in a focused beam from a

Radiation by atoms

home movie projector lamp) and $\lambda = 6000$ Å, Eq. 14-21 gives

$$\text{Torque} = \frac{6 \cdot 10^{-5} \times 10^7}{2\pi \times 3 \times 10^{10}} \approx 3 \times 10^{-9} \quad \text{dyne-cm}$$

Despite the extreme smallness of the effect, its existence was quantitatively verified by R. A. Beth in 1936.[7] Since for a given beam power W the torque is proportional to λ, the effect can be far more easily detected if one uses microwaves (with a wavelength of several centimeters) instead of light. Such an experiment has more recently been performed and recorded on film.[8]

14-9 MAGNETIC DIPOLE RADIATION AND GALACTIC HYDROGEN

Our discussion of radiative transitions has focused almost entirely on electric dipole transitions because this accounts for almost all atomic radiation of any importance. However, some magnetic dipole transitions are important and we shall conclude this chapter by discussing a famous and fascinating case that has proved to be a prime source of information about the universe and in particular the structure of our own galaxy. This is the so-called *21-cm line of atomic hydrogen,* which is the result of a magnetic dipole transition produced when the electron in a hydrogen atom undergoes a "spin flip," reversing its spin direction with respect to the spin of the proton. (We have mentioned several times—such as in Section 13-4—that the proton is a spin-$\frac{1}{2}$ particle.)

Classical Background

Before considering the spin-flip transition itself, we shall extend our earlier discussion (Section 14-2) of magnetic dipole radiation in general. Although the physical origin of magnetic moments is circulating currents, the effect of an oscillating circular current (Figure 14-8a) can be calculated by considering an equivalent linear magnetic dipole (Figure 14-8b). For

[7]R. A. Beth, Phys. Rev. **50**, 115 (1936). Reprinted in *Quantum and Statistical Aspects of Light,* ed. P. Carruthers, American Institute of Physics, New York, 1950.

[8]Richard B. Anderson and Joseph S. Ladish, *The Angular Momentum of Circularly Polarized Radiation,* Education Development Center Inc., Newton, Mass.

14-9 Magnetic dipole radiation: galactic hydrogen

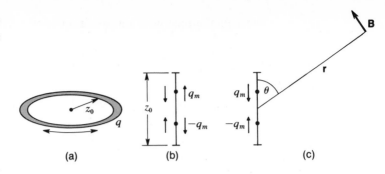

(a)   (b)   (c)

definiteness we assume that the magnetic dipole consists of "magnetic charges" $\pm q_m$ separated by a distance z_0 equal to the radius of the circle in which the electric charge q of Figure 14-8a actually moves. We then equate two alternative expressions for the magnetic moment:

$$D_M = \frac{(\text{maximum current}) \times (\text{area})}{c} = \left(\frac{qv}{2\pi z_0}\right)\frac{\pi z_0^2}{c} = \frac{v}{2c} q z_0$$

and

$$D_M = q_m z_0$$

This defines the magnitude that must be assigned to q_m:

$$q_m = \frac{v}{2c} q$$

The reason for introducing and evaluating the effective magnetic charge in this way (even though individual magnetic poles or charges do not exist) is that the radiation field of the magnetic dipole can then be calculated in complete analogy to our treatment of the electric dipole at the beginning of this chapter; the final results agree completely with a more physically correct analysis in terms of oscillating currents. At a distant point (see Figure 14-8c) the transverse *magnetic* field due to a magnetic charge q_m with acceleration a is given by

$$B(r, \theta) = \frac{q_m a \sin \theta}{c^2 r}$$

Assume that the magnetic charges are oscillating harmonically with amplitude z_0 and angular frequency ω. Then we can substitute the explicit expression for the acceleration a including the retarded time:

$$B(r, \theta, t) = -\frac{\omega^2 q_m z_0}{c^2 r} \sin \theta \cos \omega \left(t - \frac{r}{c} \right)$$

This magnetic radiation field is less, by the factor $q_m/q \simeq v/2c$, than the radiation field due to an *electric* dipole of magnitude $q z_0$ oscillating at the same frequency. We already established this result in Section 14-3. (Remember that in the Gaussian (cgs) system $\mathscr{E}$ and B are quantities of the same physical dimensions—force per charge.) This factor of order v/c is not surprising because it also defines the ratio of the magnitudes of magnetic and electric fields due to any moving charge traveling at speed v.

Now (again in terms of the Gaussian system) we know that the electric and magnetic fields at any point in an electromagnetic wave are *equal* in magnitude, and the energy flow, as given by the Poynting vector, can be written equally well in terms of $\mathscr{E}$ or B:

$$S = \frac{c}{4\pi} \mathscr{E}^2 = \frac{c}{4\pi} B^2$$

Thus we confirm the result, quoted and used earlier, that the intensities of magnetic and electric dipole radiation from a given system of charges are (if both are allowed)[9] approximately in the ratio v^2/c^2.

Electric and Magnetic Moments of Atoms

We now introduce the characteristic magnitudes of electric and magnetic dipole moments in atomic radiating systems. In this way we find some support for the applicability of the classical results obtained above.

For the electric dipole moment, an order of magnitude is defined by the electron charge e oscillating with an amplitude

---

[9]Actually, when we come to atomic systems, the quantum-mechanical selection rules have the consequence that magnetic dipole and electric dipole transitions are mutually exclusive.

14-9 Magnetic dipole radiation: galactic hydrogen

equal to the Bohr radius a_o $(= \hbar^2/me^2)$. Thus we put

$$D_E \approx ea_o = \frac{\hbar^2}{m_e e}$$

The order of magnitude of the magnetic dipole moment is given by the Bohr magneton

$$D_M \approx \mu_B = \frac{e\hbar}{2m_e c}$$

Hence

$$\frac{D_M}{D_E} \approx \frac{e\hbar}{2m_e c} \cdot \frac{m_e e}{\hbar^2} = \frac{1}{2} \frac{e^2}{\hbar c} = \frac{1}{2} \frac{v}{c} = \frac{1}{2} \alpha$$

It is hardly surprising, if one checks back to see how the Bohr magneton was calculated (Section 10-5), that the above ratio is, like the classical ratio, expressible as $v/2c$, where v is the speed of the electron in the lowest Bohr orbit and v/c $(= e^2/\hbar c)$ is numerically equal to the fine structure constant α (about 1/137). But essentially the same value for D_M/D_E will apply even if the magnetic dipole moment is due to intrinsic electron spin rather than to orbital motion, since we have seen that the spin magnetic moment is almost exactly equal to a Bohr magneton. Thus, for atomic radiation, we can conclude that the radiation rate for magnetic dipole radiation of a certain frequency should be smaller, by the factor $(1/274)^2$, than for electric dipole radiation of the same frequency. This is a reduction factor of about 10^{-5}.

In what follows we examine a magnetic dipole transition in hydrogen that involves the spin magnetic moments of both the electron and the proton. As we remarked in Chapter 10 (Section 10-4), the order of magnitude of the proton magnetic moment can be estimated by substituting the proton mass and charge into the expression above for the Bohr magneton. The larger proton mass makes the proton magnetic moment smaller by a factor 1/1840 than that of the electron. The resultant value, the nuclear magneton, is given the symbol μ_N:

$$\mu_N \approx \frac{\mu_B}{1840} \approx 5 \times 10^{-24} \quad \text{(cgs)}$$

Radiation by atoms

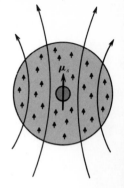

Fig. 14-9 *Schematic drawing of the hydrogen atom's ground state. The proton's magnetic moment is parallel to the **B** field produced by the electron's moment (this electron moment shown spread out over the "electron cloud").*

By observation, the proton's magnetic moment is found to be 2.8 nuclear magnetons.

The 21-cm Hydrogen Line

What we usually call the ground state of the hydrogen atom is actually double, according to whether the spins of the proton and the electron are parallel or antiparallel. The true lowest state is the antiparallel state, in which the proton magnetic moment points in the same direction as the magnetic field due to the electron (see Figure 14-9). The energy difference between these two states is an example of what is called *hyperfine splitting*. Hyperfine splitting can be calculated in a way analogous to the corresponding fine-structure splitting discussed at the end of Chapter 11[10]:

$$\Delta E = 2\mu_e B_p (= 2\mu_p B_e)$$

where μ_e = electron magnetic moment ($\approx$ 1 Bohr magneton, μ_B) and B_p is the magnetic field at the electron due to the proton magnetic moment. (As indicated by the alternative parenthetical expression, we can also calculate ΔE in terms of the energy of the proton magnetic moment in the magnetic field provided by the electron magnetic moment.) The order of magnitude of ΔE can be obtained by putting

$$B_p \approx \frac{2\mu_p}{a_0^3} \approx \frac{2 \times 2.8 \, \mu_N}{a_0^3}$$

[10]Note that there is no energy splitting due to spin-orbit coupling in the present case because $l = 0$ for the system.

14-9   Magnetic dipole radiation: galactic hydrogen

since the proton magnetic moment is known to be about 2.8 nuclear magnetons. Setting $\mu_N = \mu_B/1840 \approx 5 \times 10^{-24}$ cgs and $a_o \approx 5 \times 10^{-9}$ cm we have

$$B_p \approx \frac{5.6 \times 5 \times 10^{-24}}{1.25 \times 10^{-25}} \approx 250 \text{ gauss}$$

Hence

$$\Delta E \approx 2 \times 10^{-20} \times 250 \approx 5 \times 10^{-18} \text{ erg} \approx 3 \times 10^{-6} \text{ eV}$$

This is about 10^{-2} of the fine-structure splitting in the $n = 2$ state of hydrogen. In a transition between these two states the electron *spin* reverses direction but the *space* part of the electron wave function does not change: the electron remains in the state $n = 1$, $l = m = 0$. Therefore no *electric* dipole radiation can result from this (stationary) part of the wave function. The transition is a clear case of magnetic dipole radiation, with the electron spin and magnetic moment reversing direction. The calculated frequency and wavelength for such a transition are given, according to the above rough calculation, by

$$\nu = \frac{\Delta E}{h} \approx 10^9 \text{ Hz} = 1000 \text{ MHz}$$

and

$$\lambda = \frac{c}{\nu} \approx 30 \text{ cm}$$

More accurate values are $\nu = 1420$ MHz and $\lambda = 21$ cm.

The calculated *mean lifetime* τ for this transition is extraordinarily long. We could calculate it in a direct way by equating the energy radiated in time τ to the emitted photon energy:

$$\frac{\omega^4 D_M{}^2}{3c^3} \tau = \hbar\omega = \Delta E = 2\mu_e B_p$$

But it is perhaps simpler to estimate it by comparison with our previous estimate of 10^{-7} sec for an electric dipole transition in the visible range. For either transition, we can put

Radiation by atoms

$$\tau \sim \frac{1}{\omega^3 D^2} \sim \frac{\lambda^3}{D^2}$$

Hence

$$\frac{\tau_M(21 \text{ cm})}{\tau_E(\text{visible})} \approx \left[\frac{\lambda(21 \text{ cm})}{\lambda(\text{visible})}\right]^3 \cdot \left(\frac{D_E}{D_M}\right)^2$$

$$\approx \left(\frac{21}{5 \times 10^{-5}}\right)^3 \cdot (274)^2$$

$$\approx 5 \times 10^{21}$$

With $\tau_E \approx 10^{-7}$ sec, we thus have a lifetime

$$\tau_M(21 \text{ cm}) \approx 5 \times 10^{14} \text{ sec} \approx 10^7 \text{ years!}$$

Radiation from Galactic Hydrogen

With such a long lifetime, one might expect this 21-cm radiation to be far too improbable and feeble in intensity ever to be observed. However, even at the low hydrogen densities in interstellar space ($\approx$ one hydrogen atom per cm^3), there are immense numbers of hydrogen atoms available to emit this radiation (something like 10^{68} in our galaxy). Moreover, each atom experiences a collision with a neighbor every ten thousand years or so. Collisions can facilitate downward transitions and also help replenish the supply of atoms in the higher state, so that the radiation keeps coming. The result is that the 21-cm radiation is detectable even from distant galaxies. For studying the structure of our own galaxy the 21-cm radiation has the advantage that its long wavelength permits it to travel through dust clouds that are opaque to visible radiation. From exhaustive studies of the intensity of the radiation, coupled with observations of its Doppler shift (due to the radiating atoms having systematic motions toward or away from us) it has been possible to build up a rather detailed picture (Figure 14-10) of our galaxy as a rotating structure with spiral arms. Detection involves the use of radiotelescopes of great sensitivity with sharply tuned receivers.

14-10 CONCLUDING REMARKS

In this chapter we have concentrated exclusively on the *spontaneous emission* of radiation by atoms. We have not at-

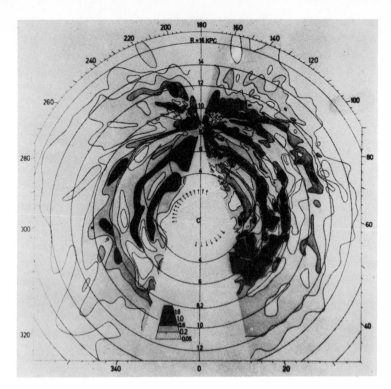

Fig. 14-10 Distribution of the atomic hydrogen in the Galaxy; scale in atoms/cm³. The map is based on the Doppler shifting of the 21-cm line, assuming a model for atomic velocity distribution as a function of position, including both rotation and expansion. (From F. J. Kerr and G. Westerhout, Distribution of Interstellar Hydrogen in the Stars and Stellar System, Vol. 5, University of Chicago Press, Chicago, 1965.)

tempted to discuss two other processes of great importance: the *absorption* of incident radiation by an atom that is thereby raised from a lower to a higher state, and the *stimulated emission* of radiation by an excited atom when exposed to radiation of the precise frequency corresponding to a possible spontaneous transition. Under special physical conditions absorption and stimulated emission can be combined to yield amplification and production of coherent signals across a wide range of frequencies in the devices called masers and lasers. Rather than give a sketchy and inadequate account here, we prefer to refer the reader to the now extensive literature on this subject.[11]

EXERCISES

14-1 *Catastrophe for the planetary atom (Rutherford without Bohr).* According to classical physics, the planetary (Rutherford)

[11]See, for example, *Lasers and Light* (Scientific American Readings) ed. A. L. Schawlow, W. H. Freeman, San Francisco, 1969; A. E. Siegman, *An Introduction to Lasers and Masers*, McGraw-Hill, New York, 1971; A. Yariv, *Introduction to Optical Electronics*, Holt, Rinehart and Winston, New York, 1971.

Radiation by atoms

atom will collapse spontaneously—and in a very short time! The logic is (classically) inescapable: an accelerated charge radiates electromagnetic energy; an orbiting electron in an atom must radiate, since it experiences centripetal acceleration; the radiated energy must come from the electron-nucleus system; a steadily decreasing system energy corresponds to orbits of smaller and smaller radius; therefore the electron must finally spiral into the nucleus.

Using the following outline or some other method, estimate an upper limit of the time for a hydrogen atom to collapse. Recall (Eq. 1-19) that the total energy of the electron in a circular orbit of hydrogen is

$$E = -e^2/(2r) \quad \text{(cgs)}$$

The rate at which energy is radiated by a charge experiencing acceleration a is given by the classical formula

$$-dE/dt = 2e^2a^2/(3c^3)$$

(which you should be able to relate to Eq. 14-2). In this analysis, assume that the orbit is initially circular and remains circular during collapse [see partial justification in part (e)].

(a) Calculate the numerical value of the rate at which energy is radiated by an electron in an orbit of radius 0.5 Å in hydrogen.

(b) Show that the rate of radiation for all circular orbits of *smaller* radius will be *greater* than that calculated in (a).

(c) Calculate the total energy of the orbiting electron in the instant before it crashes into the proton (radius about 0.8×10^{-13} cm).

(d) Assuming a constant radiation rate equal to that calculated in (a) and taking account of the result in (b), calculate an upper bound for the time-to-collapse for hydrogen as predicted by classical physics.

(e) More careful treatment (optional). You may want to find a more accurate prediction of the time-to-collapse by taking some account of the increasing radiation rate as the orbital radius shrinks. First satisfy yourself that the orbit is almost circular at all stages (for example, by estimating the number of orbits for the electron to lose 1 percent of its initial energy). Then you can simply express dE/dt as a function of r and integrate.

14-2 *Radiative transitions in highly excited atoms.*

(a) For large quantum numbers, we may expect that the classical expression for the radiation rate of the orbiting electron in a hydrogen atom (as given in Exercise 14-1) will be correct (in an "average" sense) for a quantum analysis. Assuming this to be the case, obtain an expression for the "radiation time" t_r required for the atom to radiate an amount of energy equal to the difference between adjacent levels,

$E_n - E_{n-1} = (2/n^3) \cdot (e^2/2a_o)$. Evaluate your result for $n = 50$, 100, and 500.

(b) Such transitions can produce a sharp spectral line only if (i) the highly excited hydrogen atoms are isolated in the first place and (ii) the atoms are not disturbed by collisions during the time t_r required to emit the transition radiation. (In terms of a classical model you can appreciate that a collision during radiation would distort the emitted wave, thus smearing the spectral line: this is called *collision broadening*.) The first restriction requires that $N r_n{}^3 < 1$, where N is the number density of gas atoms. Obtain a statement of the second restriction as follows: The typical interval t_c between collisions is $1/(N \sigma v_{th})$, where σ is the "collision cross section," which you may take simply as the geometrical cross section $\pi r_n{}^2$ and where v_{th} is a typical relative velocity due to thermal motions, which is roughly $\sqrt{2kT/m_p}$. Express the restriction $t_r < t_c$ in terms of N, T, and physical and atomic constants. Compare this restriction with the first one. One of these is always more severe than the other. Identify the more severe restriction. [*Note:* All physically relevant temperatures lie in the range 3–1000 K.]

(c) The number density of hydrogen atoms in one of the sources which emits lines corresponding to $n \approx 100$ is estimated to be roughly 10^3 atoms/cm³. Using the criterion found in (b), would you expect the emission line $n = 101$ to $n = 100$ to show collisional broadening?

14-3 *Radiation from various types of sources.* Estimate, on classical grounds, the *relative* intensities of electric dipole, magnetic dipole, and electric quadrupole radiation that one might typically expect in the following cases:

(a) Radiation corresponding to transitions between states of $n = 5$ and $n = 4$ for a single electron moving in the field of a carbon nucleus ($Z = 6$). (Use the Bohr model of circular orbits.)

(b) Radiation corresponding to transitions between states of $n = 2$ and $n = 1$ for a single proton inside a spherical well of radius 3×10^{-13} cm (that is, a light nucleus emitting gamma radiation). Your calculations must, of course, include estimates of the energy of the radiation and the speed of the radiating charge.

14-4 *Wave-mechanical calculation of an electric dipole moment.*

(a) Use the approach described in Section 14-4 (and culminating in Eqs. 14-10 and 14-11) to calculate the time-dependent electric dipole moment for a particle of charge q making a radiative transition between the states $n = 3$ and $n = 2$ of a one-dimensional box of length L.

(b) Evaluate this moment, the wavelength, and the mean lifetime for a particle of charge $q = e$ (for example, a proton) in a box of nuclear dimensions ($L = 10^{-12}$ cm).

Radiation by atoms

14-5 *Vibrational transitions.*

(a) Obtain an expression for the electric dipole moment associated with a transition between the states $n = 1$ and $n = 0$ of a simple harmonic oscillator. (Refer to the wave functions listed in Table 4-1.)

(b) Refer to the energy-level diagram for the two lowest vibrational levels of the CO molecule (Figure 4-5), and deduce the value of the length parameter a in the harmonic-oscillator wave functions. Compare this to the equilibrium separation of 1.13 Å between the nuclei in this molecule (Table 11-4).

(c) Combining the results of (a) and (b), make a theoretical estimate of the mean lifetime for the $n = 1 \rightarrow n = 0$ vibrational transition in the CO molecule.

14-6 *Gamma rays from a long-lived nuclear state.* The nucleus iron-57 has a first excited state only 14.4 keV above its ground state; the mean lifetime for decay of the excited state by gamma radiation is about 10^{-7} sec (see Exercise 8-22)

Assume that the radiation is due to a single proton, of kinetic energy about 10 MeV, making a transition between closely spaced nuclear energy levels 14.4 keV apart. Taking the nuclear diameter to be about 10^{-12} cm, make reasoned order-of-magnitude estimates of the mean lifetime that you might expect for this transition according to whether it is assumed to be electric dipole, magnetic dipole, or electric quadrupole. (The transition is in fact magnetic dipole.)

14-7 *Lifetime of a muonic atom.* In Exercise 1-26 (Chapter 1) it was stated that a muonic atom is relatively stable, in that the spontaneous decay of the muon, with a mean lifetime of 2.2×10^{-6} sec, is a very slow process compared to the orbital period of the muon around a nucleus. Examine this assertion in the case of a negative muon making an electric dipole transition between the orbits for $n = 2$ and $n = 1$ around a proton. What is the probability that the muon decays before it makes the quantum jump? (This probability is the ratio of the decay probability per unit time for decay to the radiative transition probability per unit time.)

14-8 *Radiative selection rules for a simple oscillator.* The general expression for the magnitude of the radiative electric dipole moment for a particle of charge q in a one-dimensional potential is

$$D_E = q \int \psi_i{}^* \psi_f \, dx$$

We saw in Chapter 3 that for a symmetrical potential $[V(-x) = V(x)]$ the functions ψ_i and ψ_f (if they are energy eigenfunctions) are either even or odd functions with respect to the central point of the system.

(a) For the simple harmonic oscillator, deduce a general selection rule governing electric dipole transitions between states of different n.

(b) Consider the similar selection rule for states of an infinite square well. (It may be useful, for this purpose, to place the origin, $x = 0$, at the center of the well, rather than at one side as we have done in previous discussions.)

14-9 *Selection rules for a central potential.* Consider a hydrogen atom whose electron is initially in a state with $n = 4$, $l = 2$, $m = 0$.

(a) Enumerate the lower states to which the atom may pass by means of an "allowed" (electric dipole) transition.

(b) In the absence of an external magnetic field, and ignoring spin-orbit splitting effects, how many different wavelengths would appear in the spectrum from a large population of hydrogen atoms all raised to this same initial state?

(c) How would the answer to (b) be changed if the system were not a proton and an electron, but a boron nucleus ($Z = 5$) surrounded by five electrons, of which the first four remain tightly bound near the nucleus and the fifth is the "optical" electron whose changes of quantum state lead to the radiative transitions?

14-10 *Zeeman splitting in hydrogen.* [*Note:* Electron spin is ignored in this problem; a correct treatment would have to include it.] The Coulomb model of hydrogen leads, as we have seen, to energy values that depend only on the quantum number n and not on l or m. The energies are *degenerate* with respect to l and m. When hydrogen is placed in an external magnetic field, however, the energy degeneracy is removed with respect to the m values (although it remains for l). The difference in energy between different m values is due to the interaction of the orbital magnetic moment μ_l with the external magnetic field. This energy is given by

$$E = -\mu_l \cdot \mathbf{B}$$

(a) For $l = 1$ states of hydrogen, estimate the numerical value of the energy differences between the three m states in an external magnetic field of 10,000 gauss (1 tesla).

(b) Taking account of the selection rules on l and m, show that the spectrum of transitions between states of different n will consist of *three* lines.

14-11 *The Zeeman pattern for the D lines of sodium.* The D lines of sodium arise from transitions between a pair of P levels ($P_{3/2}$ and $P_{1/2}$) and the ground state ($S_{1/2}$).

(a) Construct a qualitative energy-level diagram for these levels in an external magnetic field that removes the degeneracy between states of different m_j for the same j. Consider carefully which states should be higher in energy than others, and use the results of Exercise 11-17.

(b) Taking account of the selection rules for electric dipole transitions, deduce how many separate lines you would expect to see in the emission spectrum and indicate qualitatively the appearance of the spectrum, assuming that the splitting due to the external magnetic field is considerably less than the fine-structure splitting due to spin-orbit interaction. Compare your results with Figure 10-1d.

14-12 *Natural linewidth of the 21-cm line of neutral hydrogen.* As we saw in Section 14-9, the natural lifetime of the higher of the two hyperfine states in hydrogen is approximately 5×10^{14} sec (or nearly 20 million years)!

(a) If the only source of line broadening were the finite lifetime against radiative decay, what would be the fractional width $\Delta \nu / \nu_0$ of the hyperfine emission line of hydrogen?

(b) The hyperfine transitions from interstellar hydrogen gas (in our own Galaxy as well as other galaxies) are observed by radio astronomers. The gas is typically at a temperature of about 5°K. What is the fractional spreading $\Delta \nu / \nu_0$ of the observed line radiation due to the thermal Doppler motions in the interstellar gas? (This fractional spreading is of the order of $\bar{v}/c$, where $\bar{v}$ is the mean thermal speed of the atoms.) Compare your answer with the fractional width found in (a).

Answers to exercises

CHAPTER 1

1-1	Of the order of 10 Å
1-2	(a) 10^{20} (b) 10^{28} (c) 10^{50}
1-3	About 2 Å
1-4	2×10^{17} esu/g
1-5	2.2 Å
1-7	Lines at integral multiples of $\hbar\omega$
1-8	(a) 300 m (b) 4×10^{-9} eV (c) 7.5×10^{30} (d) 7.5×10^{24}
1-9	(a) 50 yr (b) 1 hr (c) 10^{-6} sec
1-10	(a) 3.3 eV (b) 2.9 eV
1-12	0.31 Å
1-13	(a) 7.1 fermi (10^{-13} cm) (b) 7000 (cf. 200) (c) 7 m
1-14	$n \geqslant 4$
1-15	(a) Be $(Z = 4)$ (b) 2540 Å
1-16	5×10^{-11} cm (b) 100 keV (approx.) (c) 0.67
1-17	(a) 1300 Å; 5300 Å; 0.013 mm
1-19	546 keV
1-20	Calcium $(Z = 20)$
1-21	(a) 0.22 Å (b) 0.19 Å
1-22	(b) About 30′ (c) 2 m
1-23	(b) 0.7 Å
1-24	(b) 2.7×10^{-4}
1-26	(a) $(n^2/207Z)a_o$ (b) $(207Z^2/n^2)E_R$ (c) 357 keV; 0.035 Å (d) Radius not small compared to 10^{-12} cm (Actual value 5.6×10^{-13} cm)

CHAPTER 2

2-1 $(v - c)/c \approx -\frac{1}{2}(m_o c^2/h\nu)^2$

2-2 (b) 400 (c) 1 cm/sec

2-3 (b) Yes (c) No

2-4 (a) Electrons: 120 Å, 12 Å, 1.2 Å, 0.12 Å, 870 F, 12 F, 0.12 F (b) Protons: 2.9 Å, 0.29 Å, 0.029 Å, 290 F, 29 F, 2.8 F, 0.11 F

2-5 About 4 kV

2-6 (a) Need QM (b) Need QM (c) Don't need QM

2-8 1.6 Å; 60° to normal

2-9 (a) 0.062 Å (b) 2.19 Å

2-10 (a) hD/pd (b) 0.06 Å, 10^{-4} cm (c) 18 cm, 32 km, 8.7 cm

2-11 (a) $d \gg \sqrt{D\lambda}$ (b) 0.7%

2-12 (a) $78/v$ (Å)

2-13 (a) 1.5×10^{-40} cm (b) 6.6×10^{-29} cm (c) 6×10^{-13} cm

2-14 (a) 5.9×10^{-13} cm (= 5.9 F) (b) Closest approach = 3.8×10^{-12} cm (c) Safety factor for classical calculation is certainly not large.

CHAPTER 3

3-3 All energies raised by V_o

3-4 (a) Displacement and slope continuous across boundary (b) Node separation and amplitude greater in Region B

3-5 3.3 Å (three times too large)

3-6 (a) 2 MeV (b) (i) 60 MeV (relativistic formulas necessary) (ii) 14 MeV

3-7 $\Delta x/\pi\sqrt{A^2 - x^2}$

3-8 (b) 3 Å (c) Irregularities $\approx$ 20 Å

3-10 (a) $\sqrt{2/L}$ (b) $(1/4) - (1/6\pi) \approx 0.2$

3-12 (a) Yes

3-13 (b) $\exp\left[-(4\sqrt{2mV_0}/\hbar)L\right]$ (c) $h^2/2V_oL^2$ (d) $h^2/8V_oL^2$ (e) Yes; No

3-14 No bound state for potential (b); at least one bound state for the others

3-18 (a) T (b) T (c) F (d) F (e) F (f) F (g) F (h) T

CHAPTER 4

4-1 (a) $2N$ (b) $\sqrt{2}N$ (c) $2\sqrt{2}N$ (d) N (both cases)

4-2 (b) $\tan(\sqrt{2mE/\hbar})L = -\sqrt{E/(V_o - E)}$ (f) Two

4-3 (a) $0.78V_o$ (b) $A \approx 5$

4-4 (f) N_o

4-6 (b) $b = -2/a^2$

4-7 (b) Same as for odd states of normal SHO

4-9 About 16%

4-10 (a) 3×10^{12} (b) 2×10^{-12} eV (c) 4×10^{-10} cm (d) No!

4-13 (a) $(nA/e^2)^{1/(n-1)}$ (c) $(n-1)e^2/r_o^3$ (d) $\Delta E = \hbar\sqrt{(n-1)e^2/\mu r_o^3}$; $E_o = \frac{1}{2}\Delta E$ (e) $n = 9.2$; $A = 1.7 \times 10^{-82}$ cgs (f) 5.1 eV (g) 3.7 eV

Answers to exercises

CHAPTER 5

5-1 (a) $\omega_o = \sqrt{C/m}$ (b) $\gamma = 2b/m$; $\omega = (b^2 + mC)^{1/2}/m$

5-3 (c) $\sqrt{5/2}$; 2

5-4 (a) $3h^2/8ma^2$ (b) $3h^2/4ma^2$ (c) Degeneracies: 1, 3, 3, 3, 1; $n^2 = 14$

5-6 (a) $Z/2a_o$; $-Z/2a_o$ (b) $Z/3a_o$; $-2Z/3a_o$; $2Z^2/27a_o^2$

5-7 (c) $E_n = n^2h^2/8mR^2$

5-8 About 35 MeV

5-9 (a) About 10^{-14} (b) 0.24 (c) $\frac{3}{2}a_o$ (d) Straight line through proton

5-10 $(1/2\sqrt{2\pi})(Z/a_o)^{3/2}$

5-12 (c) About 2×10^{-9} eV

5-13 (a) 0   (b) 0   (d) $\langle x^2 \rangle = a^2/2 = \hbar/2\sqrt{mC}$; $\langle V \rangle = \frac{1}{2}E_o$ (e) $\langle V \rangle = -mZ^2e^4/\hbar^2 = 2E_1$

CHAPTER 6

6-2 (a) $I_x = I_o \sin^2\theta$; $I_y = I_o \cos^2\theta$ (b) circularly polarized or unpolarized (c) elliptically polarized, or various mixtures

6-3 (a) 4×10^{-7} (b) 7×10^{10} photons/cm²/sec (c) Approx. 10^8, 10^6, 10^4; 2000; 100 (d) Of the order of 100/sec; 10^5

6-5 (a) 1; 0 (b) $\cos^2\theta_2$; $\cos^2\theta_1$; $\cos^4\theta_1$ (c) $\cos^2\theta_1 \cos^2(\theta_2 - \theta_1)$; $\theta_1 = \theta_2/2$

6-6 (a) Linearly polarized in θ direction (b) $\cos^{2N}(\theta/N)$ (c) 1 (d) No (e) $f^N\cos^{2N}(\theta/N)$ (f) $N = 3$

6-7 About 14 times as long

6-10 (a) Transmits 50% (b) No (c) Lucy is wrong

6-11 (a) Linearly polarized at $+45°$ to x axis (b) Unpolarized or circularly polarized (c) Partially polarized or elliptically polarized (d) Linearly polarized (e) Circularly polarized

CHAPTER 7

7-1 (a) 100% (b) L-polarized (c) Yes, the R channel in the R-L analyzer

7-2 (a) 1/4 (b) 1/4 (c) 0 (d) Same answers

7-4 (a) $I_o[\cos^2\alpha \cos^2(\theta - \alpha) + \sin^2\alpha \sin^2(\theta - \alpha)]$ (c) $I_o\cos^2\alpha \cos^2(\theta - \alpha)$ or $I_o\sin^2\alpha \sin^2(\theta - \alpha)$; $\alpha = \theta/2$; ratio $= \cos^4(\theta/2)$

7-6 (a) 1/20 (b) Same as (a) (c) Both 3/20

7-7 (a) No (b) Yes, at $-45°$ to x (c) Linear, at $-45°$ to x; elliptically polarized

7-8 (a) All (b) Some (c) All (d) Some (e) Some (f) All (g) None (h) All

7-9 (a) 16/25 (b) $(9 + 7\sin^2\theta)/25$ (c) $-24N\hbar/25$

7-10 (e) $|x\rangle(1/2) + |y\rangle(i\sqrt{3}/2)$

7-11 (a) Necessary (b) Necessary (c) Sufficient (d) Impossible (e) Impossible (f) Necessary (g) Irrelevant

7-12 (a) C only; A and B only; All three (c) Only condition (iv) is satisfied

CHAPTER 8

8-1 (d) $\frac{1}{2} + (4/3\pi) \cos[(E_2 - E_1)t/\hbar]$
(e) $\frac{1}{2}L - (16L/9\pi^2) \cos[(E_2 - E_1)t/\hbar]$

8-3 (a) $\sqrt{2/5L}$ (b) E_1 (probability 4/5) or E_2 (probability 1/5)
(c) $E_{av} = 8E_1/5$

8-5 (a) $B_n = (-1)^{(n-1)/2}(2/\sqrt{3L})[\sin(n\pi/6)]/(n\pi/6)$ for n odd;
zero for n even (b) B_1, B_3, B_5 are major terms
(c) $B_1 = 0.95/\sqrt{L}$, $B_3 = -0.74/\sqrt{L}$, $B_5 = 0.22/\sqrt{L}$

8-6 (a) 1/2 (b) $32/9\pi^2 \approx 1/3$ (c) Yes

8-7 (a) $1/\sqrt{2}$ (c) $2\pi/\omega$ (d) $\frac{3}{2}\hbar\omega$

8-10 (a) About 10^{-12} eV (b) About 10^{-3} cm (c) About 10^{-8} K

8-11 (c) $\Delta p \cdot \Delta x \approx 10^6 h$ (classical) (d) $m(\Delta x)^2 v/L$

8-12 (a) T (b) T (c) F (d) T (e) F (f) F (g) F

8-13 (b) Zero (c) Multiply by e^{-ikx_o}

8-14 (a) $(x_1 + x_2)/2$ (b) $\hbar k_o$ (c) $(\hbar k_o)^2/2m$

8-15 (b) $\sqrt{\Delta k}\, e^{ik(x+\beta)}[\sin(x+\beta)\Delta k]/[(x+\beta)\Delta k]$, where
$\Delta k = (k_2 - k_1)/2$ and $k = (k_2 + k_1)/2$ (d) $\hbar(k_1 + k_2)/2$
(e) $(\hbar^2/2m)(k_1^2 + k_1 k_2 + k_2^2)/3$

8-16 (b) $\sqrt{2/\pi}a^{3/2}/(\alpha^2 + k^2)$ (d) Zero (e) $\hbar^2\alpha^2/2m$

8-18 (a) 5.1×10^9 cm^{-1}($\lambda_o = 0.12$ Å) (b) 2.6×10^4 cm^{-1}; 3900 Å
(c) 3.3 μsec; 200 m

8-20 8.0×10^{-5} Å

8-21 $\tau \approx 6.5 \times 10^{-8}$ sec; $\Delta E \approx 1.0 \times 10^{-8}$ eV; $\Delta\lambda = 1.2 \times 10^{-5}$ Å

8-22 (a) $\Delta E/E_o = 2.1 \times 10^{-15}$ (b) Width $\approx 4 \times 10^{-11}$ keV;
(Width)/$E_o \approx 2.8 \times 10^{-13}$ (c) With many measurements, uncertainty in line center can be $\ll$ width. Special experimental tricks were also used (see their paper).

8-23 (a) About 10^{-33} eV and 2×10^{-9} eV (b) $\Delta E/E$ far too small to be measurable

CHAPTER 9

9-1 Values of R: (a) 0.11 (b) 0.88 (c) 0.029 (d) 7×10^{-4}
(e) 6×10^{-8}

9-3 (e) $0.22; 0; 0.22; 0.36$. R for single step ($V_o = 15E/16$) is 0.36

9-6 (a) $J(x, t) = (h/mL^2) \sin^3(\pi x/L) \sin(\omega_2 - \omega_1)t$

9-8 (a) Sufficient (b) Insufficient (d) Sufficient (e) Insufficient

9-9 (a) $(\hbar k_o/m)[f(x)]^2$ (b) Unit total probability moving at mean speed $\hbar k_o/m$ (c) $x \ll x_o - a$, no current at any time. $x \gg x_o + a$, pulse of current at some time $t > 0$ as packet passes by

9-13 (b) $T = 1/[1 + (kL/2)^2]$ (c) $L/\lambda = 1/\pi$

9-14 For $E = \frac{1}{2}V_o$, $\alpha = k = \sqrt{2mE}/\hbar$, and Eq. 9-15 leads to the simple, exact result $T = 4/(e^{\alpha L} + e^{-\alpha L})^2$. Here $\alpha L = 7.24$ and $T \approx 4e^{-2\alpha L} = 2 \times 10^{-6}$. Same result follows from direct use of the approximate equation (Eq. 9-17).

9-15 (a) Deuteron has only half as much K.E. per nucleon, so proton will get through more easily. (b) Using Eq. 9-17, $T_p = 0.8 \times 10^{-4}$, $T_d = 1.0 \times 10^{-6}$

9-16 (b) No periodicities for $E < V_o$, because no cancellations of reflected waves can occur

9-17 $16k_1k_0\alpha^2/\{(\alpha^2 + k_0{}^2)(\alpha^2 + k_1{}^2)(e^{2\alpha L} + e^{-2\alpha L})$
$- 2[(\alpha^2 - k_0{}^2)(\alpha^2 - k_1{}^2) - 4k_0k_1\alpha^2]$

9-20   (b) -1.64×10^6 volts

9-21   (a) $-4\pi(\hbar k/m)|A|^2(1 - |b|^2)$ (c) 100% scattering; some absorption; extra particles emerging from scatterer

9-22   (a) 1:4 (b) About 35%

9-23   (a) $\hbar k/m$ (b) $4\pi|A|^2(\hbar k/m)$ (c) $|A|^2$ is a measure of the target area (cross section) presented by the scattering center to the incident beam.

9-25   (a) F (b) Fairly sharp mean wavelength (c) Momentum increases in well (d) F (e) T (f) Results of scattering at both edges of well

CHAPTER 10

10-1   One rev. in about 10^{15} years (cf. age of universe $\approx 10^{10}$ years!)

10-2   (a) $T = 2\pi(\rho\pi R^4H/2K)^{1/2}$ (b) $L = (\rho\pi R^2HS_a)/(Am_o)$; $\phi_o = (2\pi\rho H/K)^{1/2}(S_a/Am_o)$ where $m_o = 1$ atomic mass unit (c) Use smallest available cylinder (d) $r = 6.2 \times 10^{-4}$ cm; $T \approx 50$ sec (e) 2.15×10^{-2} degrees (about 100 times larger than ϕ_{th})

10-3 2.9×10^{-21} ergs/gauss. (This is much lower than the true value, because the actual thermal distribution of atomic speeds causes the peaks to correspond to an energy of about $3kT$ rather than kT.)

10-4   (a) Assuming $B \approx 10^4$ gauss, the evB/c force would be about 30,000 times larger than $\mu(\partial B/\partial z)$ (b) The electric force would be about 10,000 times larger.

10-6   (a) No (c) $r \lesssim 2$ cm

10-7   (a) 2.1×10^{-20} erg/gauss (b) Underestimate—but corresponds to more than two lined-up electron spins per atom. Alignment must be close to 100%.

10-11   (b) 5/2; 5/3

10-13   (a) $(\psi_1 \pm i\psi_2)/\sqrt{2}$ (b) $\pm\hbar$

10-15 $(N + 1)(N + 2)/2$

10-16   (c) 16

10-17   (b) Non-degenerate (c) L_z not a good quantum number (except $L_z = 0$) (e) No (both questions)

CHAPTER 11

11-1   (b) Electron not localizable along orbit if L_z is sharply defined (single eigenvalue) (c) At least about three

11-2   (b) $\int_0^\infty [R(r)]^2 r^2 \, dr = 1$

11-3   (b) (i) T (ii) $\int_0^\infty R_1R_2 r^2 \, dr = 0$

11-4   (b) No (c) $2l + 1$

11-5   (b) Confirms that simultaneous eigenfunctions exist (c) Yes; No

11-7   1.131 A

11-8   (a) Ratio $= \sqrt{(l + 1)/l}$ (b) Assuming $L = \sqrt{l(l + 1)}\hbar$, $v_H \approx 700$ m/sec, $v_{Cl} \approx 20$ m/sec

11-9 Interatomic distance = 1.37 Å. Stretching is about 0.9 F between lowest and highest states.

11-12 (a) $r_o = 1.32 \times 10^{-8}$ cm

11-13 $v/c \approx \hbar c/e^2 = 1/$(Fine structure constant), $v \approx 100c$!

11-14 (a) 2.8×10^6 rev/sec   (b) 2.2×10^{12} rad/sec   (c) Ratio $\approx 5.3 \times 10^{-5}$

11-15 (b) 1.84×10^5 gauss

11-16 (b) Components L_z and S_z are no longer defined.

11-17 (b) $g_j = 1 \pm 1/(2l + 1)$; 4/5

CHAPTER 12

12-1 (a) $E_n = n^2 h^2/8mR^2$

12-2 (b) Two (c) $l < 5$ (conservative) (c) $\alpha < 0.6$ (conservative)

12-4 $0.74a_o$ and $5.24a_o$

12-6 Integrals (b) and (c) must vanish, since over-all orthogonality must hold even when $m' = m$ (as well as the given $l' = l$). Integrals (a) and (d) need not vanish since orthogonality is guaranteed by $l' \neq l$.

12-8 $\langle r \rangle = 3a_o/2$; $\langle r^2 \rangle = 3a_o^2$

12-9 (b) $n(n + \frac{1}{2})a_o$

12-10 (a) $\langle V \rangle = -e^2/a_o$

12-11 About 10^{-14}

12-13 (a) 1/3 (b) $-3E_R/4$ (= -10.2 eV); $2\hbar^2/3$; $\hbar/3$ (c) Depends on all. Repetition period = $2\pi/(E_2 - E_1)$

12-15 Possible: $^2S_{1/2}$; $^2P_{1/2}$; $^2F_{7/2}$; $^2D_{5/2}$; $^2P_{3/2}$; $^2D_{3/2}$

12-17 Transitions (i) and (iii) would have the same wavelength

12-20 6.8 eV; 2440 Å. The shortest wavelength in the positronium "Balmer series" would be 7291 Å. Thus positronium has no visible spectrum (4000–7000 Å); its longest-wavelength "Lyman line" is too short, and its shortest-wavelength "Balmer line" is too long.

CHAPTER 13

13-3 (a) 0.019 (b) 0.038 (c) 0.01

13-4 (b) Ground state, $\frac{1}{2}$ only. Excited state, $\frac{1}{2}$ or $\frac{3}{2}$

13-6 (a) Ratio 11:1

13-8 (a) 28.7 MeV (b) 19.6 MeV

13-9 (a) $4a_o/7 = 0.30$ Å (b) -83.3 eV; 28.9 eV

13-11 (b) Actual ground-states: N($^4S_{3/2}$); Si(^3P_o); As($^4S_{3/2}$); Ag($^2S_{1/2}$)

13-12 (a) $^2P_{1/2}$ (but it is actually $^2P_{3/2}$); $^2S_{1/2}$ (b) Ultraviolet would be a good guess (c) Fluorine; Ne$^+$ wavelengths will be shorter (quantitative estimates difficult).

13-13 (a) In $3S$ state, $Q_{eff} = 1.84e$; in $3P$ state, $Q_{eff} = 1.45e$ (b) the next electron removed sees a charge $\approx 3.7e$.

CHAPTER 14

14-1 (a) About 3×10^{11} eV/sec (c) About -1 MeV! (d) About $3 \, \mu$sec (e) 1.65×10^{-11} sec

14-2 (a) $t_r = (3m^2c^3a_o^3/2e^4)n^5$; about $\frac{1}{20}$ sec, 2 sec, $1\frac{1}{2}$ hr
(b) $(3\pi/2)(\hbar c/e^2)(Na_o^3)(v_{th}/c)n^9 < 1$ (more restrictive than $nr_n^3 < 1$). Numerically, this gives $n^9N\sqrt{T} \lesssim 10^{22}$. (c) Just marginal for $n = 100$ ($T = 3$K gives $n \approx 120$; $T = 1000$ K gives $n \approx 90$).

14-3 (a) $E2:E1 \approx 10^{-4}$; $M1:E1 \approx 2 \times 10^{-5}$ (b) $E2:E1 \approx 1$; $M1:E1 \approx 0.03$

14-4 (a) $D = (48qL/25\pi^2)e^{-i\omega t}$, with $\omega = 5\pi h/4mL^2$
(b) 0.93×10^{-22} cgs; 1.2×10^{-11} cm; 2.6×10^{-18} sec

14-5 (a) $D = qa/\sqrt{2}$ (b) 0.047 Å

14-6 For $E1$, about 10^{-10} sec; for $E2$, about 3×10^{-4} sec; for $M1$, about 2×10^{-8} sec. (Thus the identification through these estimates works quite well.)

14-7 About one in a million

14-8 (a) and (b) Δn must be odd

14-9 (a) Transitions to $n = 3$, $n = 2$, with $l = 1$, $m = 0, \pm 1$ (b) Ignoring fine structure, two lines only (c) A transition to states with $n = 4$, $l = 1$ would now be significant.

14-10 (a) 5.8×10^{-5} eV

14-12 (a) Of the order of 10^{-24} (b) About 10^{-6}

Answers to exercises

Bibliography

The literature of atomic and quantum physics is very large and very diverse. We present here a selected bibliography, subdivided into categories (although the divisions are not always clear-cut). In particular, a separation has been made between books about quantum theory *per se* and books on "modern physics" (the category entitled "Surveys of Atomic Physics," below) which provide descriptive accounts of atomic and nuclear physics and relate the phenomena in greater or less detail to their quantum-theoretical bases.

The bibliography makes no claim to completeness, but the authors apologize in advance if, as is all too probable, there are glaring omissions.

SCIENTIFIC AND HISTORICAL BACKGROUND

Anderson, D. L., *The Discovery of the Electron*, Van Nostrand, Princeton, New Jersey, 1964.
> A good brief monograph about the atomicity of electric charge.

Andrade, E. N. da C., *Rutherford and the Nature of the Atom,* Doubleday Anchor, New York, 1964
> An excellent, brief scientific biography by a man who worked under Rutherford during the great days at Manchester when the nucleus was being discovered.

Boorse, H. A. and Motz, L., *The World of the Atom* (2 vols.), Basic Books, New York, 1966.
> A treasurehouse of extended selections from the original litera-

ture, with commentaries, covering the whole history of atomic and quantum physics from Lucretius to elementary-particle research in the 1960s.

Cropper, W. H., *The Quantum Physicists,* Oxford University Press, London, 1970.
A lively book about the founders of quantum physics, with substantial details about the development of quantum theory.

De Broglie, L., *The Revolution in Physics* (trans. R. W. Niemeyer), Noonday Press, New York, 1953.
Described in its subtitle as "a non-mathematical survey of quanta."

————, *Physics and Microphysics* (trans. M. Davidson), Hutchinson, London, 1955.
Contains de Broglie's own account of the beginnings of wave mechanics.

Fierz, M. and Weisskopf, V. F. (eds.), *Theoretical Physics in the Twentieth Century,* Interscience, New York, 1960.
A memorial volume to Wolfgang Pauli, with some interesting essays about the historical development of quantum theory.

Friedman, F. L. and Sartori, L., *The Classical Atom,* Addison-Wesley, Reading, Massachusetts, 1965.
A good survey of atomic physics up to and including Rutherford's discovery of the nucleus.

Gamow, G., *Thirty Years That Shook Physics,* Doubleday Anchor, New York, 1966.
A highly entertaining, nontechnical, and personal account of the birth of quantum physics.

Heisenberg, W., *The Physical Principles of the Quantum Theory* (trans. C. Eckart and F. C. Hoyt), Dover, New York, 1930.
A classic in the literature of quantum physics, based on lectures given by Heisenberg in 1929.

Hermann, A., *The Genesis of Quantum Theory (1899 – 1913)* (trans. C. W. Nash), MIT Press, Cambridge, Massachusetts, 1971.
A careful account of the early history, up to Bohr.

Hoffmann, B., *The Strange Story of the Quantum,* Dover, New York, 1959.
A lively and literate semipopular account.

Jaffe, B., *Moseley and the Numbering of the Elements,* Doubleday Anchor, New York, 1971.
A very nice mixture of biography and physics, describing Moseley's brilliant and tragically brief career.

Jammer, M., *The Conceptual Development of Quantum Mechanics,* McGraw-Hill, New York, 1966.
A thorough and lavishly referenced history, embodying both biographical and technical details.

Ludwig, G., *Wave Mechanics,* Pergamon, London, 1968.
A brief history with excerpts from some of the early papers.

Millikan, R. A., *The Electron,* (1917), Univ. of Chicago Press, Chicago, 1963.
A reprint of Millikan's book in which he describes in great detail the work that culminated in his precise measurement of e. With a biographical/historical introduction by J. W. M. Du-Mond.

Rozental, S. (ed.), *Niels Bohr,* North-Holland, Amsterdam, 1967.
A charming collection of biographical essays about Bohr by his friends, students and co-workers.

Schrödinger, E., *Collected Papers on Wave Mechanics* (trans. J. F. Shearer and W. M. Deans), Blackie, London and Glasgow, 1928.
All of Schrödinger's pioneering papers.

———, *Letters on Wave Mechanics* (ed. K. Przibram), Philosophical Library, New York, 1967.
A fascinating short collection of correspondence that Schrödinger had with Planck, Einstein, and Lorentz, translated and with an introduction by Martin J. Klein.

Ter Haar, D., *The Old Quantum Theory,* Pergamon, London, 1967.
A survey, with excerpts from original papers, of quantum theory before wave mechanics.

Trigg, G. L., *Crucial Experiments in Modern Physics,* Van Nostrand Reinhold, New York, 1971.
A nicely documented account of some now classic experiments.

Van der Waerden, B. L. (ed.), *Sources of Quantum Mechanics,* Horth-Holland, Amsterdam, 1967.
A collection of important papers up to 1926, with a lengthy introduction by the editor. The book is concerned with the matrix mechanics of Heisenberg *et al.*; Schrödinger's wave mechanics is not represented.

Weisskopf, V. F., *Physics in the Twentieth Century,* MIT Press, Cambridge, Massachusetts, 1972.
A collection of essays about quantum theory and atomic and nuclear physics by a distinguished physicist and master expositor who came to professional maturity just as modern quantum theory was being born.

Whittaker, E. T., *A History of the Theories of Aether and Electricity* (Vol. II, *The Modern Theories*), Harper and Row, New York, 1960.

A detailed historical and technical survey of "modern" physics from 1900 to 1926.

SURVEYS OF ATOMIC PHYSICS

Bitter, F. and Medicus, H. A., *Fields and Particles,* American Elsevier, New York, 1973.

Mainly a survey of atomic, nuclear, and particle physics, presented in a way that brings the physics alive.

Blanpied, W. A., *Modern Physics,* Holt, Rinehart and Winston, New York, 1971.

A primarily theoretical survey.

Born, M., *The Restless Universe* (trans. W. M. Deans), Dover, New York, 1951.

A fascinating elementary (but by no means trivial) account of atomic and quantum physics by one of the creators of quantum theory.

————. (rev. J. Dougal), *Atomic Physics* (8th ed.), Blackie, London and Glasgow, 1969.

A classic textbook. To avoid interrupting the narrative, the theoretical material is largely relegated to a copious set of discursive appendices.

Cagnac, B. and Pebay-Peyroula, J. -C., *Modern Atomic Physics* (2 vols.), Macmillan, London, 1975.

A well-balanced combination of experimental and theoretical material.

Enge, H. A., Wehr, M. R., and Richards, J. A., *Introduction to Atomic Physics,* Addison-Wesley, Reading, Massachusetts, 1972.

Harnwell, G. P. and Livingood, J. J., *Experimental Atomic Physics,* McGraw-Hill, New York, 1933.

Written to provide background for a course in experimental physics, but really a fine (though now somewhat dated) general text on atomic physics.

Herzberg, G., *Atomic Spectra and Atomic Structure,* Dover, New York, 1944.

An excellent short introduction to atomic and quantum physics via spectroscopy.

Leighton, R. B., *Principles of Modern Physics,* McGraw-Hill, New York, 1959.
A very fine general text at the advanced undergraduate level.

Livesey, D. L., *Atomic and Nuclear Physics,* Blaisdell, Waltham, Massachusetts, 1966.

McGervey, J. D., *Introduction to Modern Physics,* Academic Press, New York, 1971.

Norwood, J., Jr., *Twentieth Century Physics,* Prentice-Hall, Englewood Cliffs, New Jersey, 1976.

Richtmeyer, F. K., Kennard, E. H., and Cooper, J. N., *Introduction to Modern Physics* (6th ed.), McGraw-Hill, New York, 1969.
A deservedly long-lived general survey.

Semat, H. and Albright, J. R., *Introduction to Atomic and Nuclear Physics* (5th ed.), Holt, Rinehart and Winston, New York, 1972.
Another durable text.

Shankland, R. S., *Atomic and Nuclear Physics* (2nd ed.), Macmillan, New York, 1960.
Particularly good for its copious references to the original literature.

Sproull, R. L., *Modern Physics* (2nd ed.), Wiley, New York, 1963.

Tipler, P. A., *Foundations of Modern Physics,* Worth, New York, 1969.

Weidner, R. T. and Sells, R. L., *Elementary Modern Physics,* Allyn and Bacon, Boston, 1973.

Willmott, J. C., *Atomic Physics,* Wiley, London, 1970

QUANTUM THEORY

Bohm, D., *Quantum Theory,* Prentice-Hall, New York, 1951.
A very discursive development, pervaded with Bohm's concern for the physical meaning underlying the mathematics (a concern that later led him to his investigations into "hidden variables" that would remove the indeterminism from quantum theory).

Dicke, R. H. and Wittke, J. P., *Introduction to Quantum Mechanics,* Addison-Wesley, Reading, Massachusetts, 1960.
A stimulating but rather terse presentation of nonrelativistic quantum theory.

Eisberg, R. M., *Fundamentals of Modern Physics*, Wiley, New York, 1961.
A very thorough development of Schrödinger's wave mechanics, with applications to atomic and nuclear problems.

————., and Resnick, R., *Quantum Physics of Atoms, Molecules, Solids, Nuclei and Particles*, Wiley, New York, 1974.
A simplified but modernized presentation with its roots in the preceding reference.

Feynman, R. P., Leighton, R. B., and Sands, M., *The Feynman Lectures on Physics* (Vol. III, *Quantum Mechanics*), Addison-Wesley, Reading, Massachusetts, 1965.
An introductory but sophisticated treatment based on matrix mechanics rather than the Schrödinger approach.

Flint, H. T., *Wave Mechanics* (9th ed.), Methuen, London, 1967.
A compact monograph emphasizing the mathematical aspects of the Schrödinger method.

Holden, A. N., *The Nature of Atoms, Stationary States*, and *Bonds between Atoms*, Clarendon Press, Oxford, 1971.
A set of three short monographs that collectively provide an excellent presentation of the essentials of wave mechanics and its application to atomic and solid-state physics.

Houston, W. V. and Phillips, G. C., *Principles of Quantum Mechanics*, North-Holland, Amsterdam, 1973.

Mandl, F., *Quantum Mechanics*, Butterworths, London, 1954.
A clear and compact presentation of the mathematical principles.

Matthews, P. T., *Introduction to Quantum Mechanics* (3rd ed.), McGraw-Hill, London, 1974.
A mathematically oriented introductory development of wave and matrix mechanics, based on the postulational approach.

Mott, N. F., *Elements of Wave Mechanics*, Cambridge Univ. Press, London, 1952.
An elementary theoretical presentation by one of the pioneers.

Park, D., *Introduction to the Quantum Theory* (2nd ed.), McGraw-Hill, New York, 1974.
An excellent presentation by an outstanding pedagogue.

Pauling, L. and Wilson, E. B., *Introduction to Quantum Mechanics*, McGraw-Hill, New York, 1935.
One of the older texts, but still worth studying.

Powell, J. L. and Craseman, B., *Quantum Mechanics,* Addison-Wesley, Reading, Massachusetts, 1961.
> An intermediate-level presentation of wave and matrix mechanics.

Rojansky, V., *Introductory Quantum Mechanics,* Prentice-Hall, Englewood Cliffs, New Jersey, 1938.
> A classic text, particularly valuable for its detailed exposition of the basic principles of wave and matrix mechanics.

Saxon, D. S., *Elementary Quantum Mechanics,* Holden-Day, San Francisco, 1968.
> A primarily mathematical development of both wave and matrix mechanics.

Sherwin, C. W., *Introduction to Quantum Mechanics,* Holt, Rinehart and Winston, New York, 1959.
> A highly original and stimulating presentation of key ideas, discussed with much physical insight.

Sillitto, R. M., *Non-relativistic Quantum Mechanics* (2nd ed.), University Press, Edinburgh, 1967.
> A good and scholarly text.

Wichmann, E. E., *Quantum Physics* (Berkeley Physics Course, Vol. 4), McGraw-Hill, New York, 1971.
> A very thorough discussion of the physical results and ideas underlying quantum mechanics.

Wieder, S., *The Foundations of Quantum Theory,* Academic Press, New York, 1973.
> A text based, à la Dirac, on the "canonical" formulation of classical mechanics, with a strongly mathematical flavor.

SOME MORE ADVANCED TEXTS

Dirac, P. A. M., *Quantum Mechanics* (4th ed.), Oxford Univ. Press, London, 1958.
> A classic by one of the founders of quantum theory. The whole development is, however, in terms of matrices and matrix operators, rather than the particle-wave duality.

Feenberg, E. and Pake, G. E., *Notes on the Quantum Theory of Angular Momentum,* Stanford Univ. Press, Stanford, California, 1959.

Landau, L. and Lifshitz, E. M., *Quantum Mechanics, Non-relativistic Theory* (2nd ed.) (trans. J. B. Sykes and J. S. Bell), Addison-Wesley, Reading, Massachusetts, 1965.

Merzbacher, E., *Quantum Mechanics* (2nd ed.), Wiley, New York, 1970.

Messiah, A., *Quantum Mechanics* (2 vols.) (trans. G. M. Temmer), Wiley, New York, 1959.

Schiff, L. I., *Quantum Mechanics* (3rd ed.), McGraw-Hill, New York, 1968.

SPECIAL TOPICS

American Association of Physics Teachers, *Quantum and Statistical Aspects of Light,* AAPT, Graduate Physics Building, SUNY, Stony Brook, New York, 1963.
A fine collection of selected reprints, plus an annotated bibliography by P. Carruthers.

Cohen, E. R., Crowe, K. M., and DuMond, J. W. M., *The Fundamental Constants of Physics,* Interscience, New York, 1957.
A fascinating account of how our knowledge of the atomic constants was developed.

Frauenfelder, H. and Henley, E. M., *Subatomic Physics,* Prentice-Hall, Englewood Cliffs, New Jersey, 1974.
An excellent introduction to the theoretical and experimental aspects of the physics of elementary particles.

Harnwell, G. P. and Stephens, W. E., *Atomic Physics,* McGraw-Hill, New York, 1955.
Detailed discussion of topics in atomic (nonnuclear) physics.

Herzberg, G., *Spectra of Diatomic Molecules* (2nd ed.), Van Nostrand-Reinhold, New York, 1950.

Kuhn, H. G., *Atomic Spectra,* Academic Press, New York, 1962.

Morrison, M. A., Estle, T. L., and Lane, N. F., *Quantum States of Atoms, Molecules, and Solids,* Prentice-Hall, Englewood Cliffs, New Jersey, 1976.
"This book will show you some of what you can do with quantum mechanics" (from the Preface).

Pauling, L. and Goudsmit, S. A., *The Structure of Line Spectra,* McGraw-Hill, New York, 1930.

Sanders, J. H., *The Fundamental Atomic Constants,* Oxford Univ. Press, London, 1961.
A briefer but more up-to-date account than the book by Cohen, Crowe, and DuMond (v. sup.).

Series, G. W., *The Spectrum of Atomic Hydrogen,* Oxford Univ. Press, London, 1957.

Slater, J. C., *Quantum Theory of Matter* (2nd ed.), McGraw-Hill, New York, 1968.

Smith, K. F., *Molecular Beams,* Methuen, London, 1955.
Provides valuable background to particle-beam experiments (e.g., Stern-Gerlach, diffraction of atomic beams).

Whiffen, D. H., *Spectroscopy,* Longman, London, 1972.

White, H. E., *Introduction to Atomic Spectra,* McGraw-Hill, New York, 1934.

PHILOSOPHICAL AND AUTOBIOGRAPHICAL

Bohr, N., *Atomic Physics and Human Knowledge,* Wiley, New York, 1958.
A collection of essays; of special interest is Bohr's discussion with Einstein (1949) entitled "Epistemological Problems in Atomic Physics."

Born, M., *Natural Philosophy of Cause and Chance,* Dover, New York, 1964.
Discusses many of the questions raised by the generally accepted probabilistic (indeterministic) character of quantum theory.

De Broglie, L., *New Perspectives in Physics* (trans. A. J. Pomerans), Basic Books, New York, 1962.
A collection of essays, including several on the interpretation of wave mechanics.

Heisenberg, W., *Physics and Philosophy,* Harper, New York, 1958.
An extended, nontechnical essay.

Jammer, M., *The Philosophy of Quantum Mechanics,* Wiley, New York, 1974.
A full account of the various interpretations of quantum theory from 1926 to the present.

Jauch, J. M., *Are Quanta Real?,* Indiana Univ. Press, Bloomington, Indiana, 1973.
An entertaining and instructive fictional debate in the style of Galileo's "Two New Sciences."

Wigner, E. P., *Symmetries and Reflections,* MIT Press, Cambridge, Massachusetts, 1967.
A collection of scientific essays by one of the most distinguished theoretical physicists of the 20th century.

Selected physical constants and conversion factors

PHYSICAL CONSTANTS

Quantity	Symbol	Value	Units	
Atomic mass unit	amu	$1.661 \times$	$\begin{cases} 10^{-27} \\ 10^{-24} \end{cases}$	kg g
Avogadro's number	N_A	$6.022 \times$	$\begin{cases} 10^{26} \\ 10^{23} \end{cases}$	$(\text{kg-mole})^{-1}$ $(\text{g-mole})^{-1}$
Bohr magneton	μ_B	$9.274 \times$	$\begin{cases} 10^{-24} \\ 10^{-21} \end{cases}$	J/T erg/gauss
Bohr radius	a_0	$5.292 \times$	$\begin{cases} 10^{-11} \\ 10^{-9} \end{cases}$	m cm
Boltzmann's constant	k	$1.381 \times$	$\begin{cases} 10^{-23} \\ 10^{-16} \end{cases}$	J/°K erg/°K
Electron charge/mass	e/m_e	$\begin{cases} 1.759 \times \\ 5.273 \times \end{cases}$	10^{11} 10^{17}	C/kg esu/g
Electron mass	m_e	$9.109 \times$	$\begin{cases} 10^{-31} \\ 10^{-28} \end{cases}$	kg g
Elementary charge	e	$\begin{cases} 1.602 \times \\ 4.803 \times \end{cases}$	10^{-19} 10^{-10}	C esu
Fine structure constant	α $(=e^2/\hbar c)$	$7.297 \times$	10^{-3}	$(\approx 1/137)$
Ionization energy of H atom		13.6		eV
Planck's constant	h	$\begin{cases} 6.626 \times \\ 4.136 \times \end{cases}$	$\begin{cases} 10^{-34} \\ 10^{-27} \end{cases}$ 10^{-15}	J-sec erg-sec eV-sec

Quantity	Symbol	Value	Units
Quantum of angular momentum	$\hbar$	$1.054 \times$ $\begin{cases}10^{-34} \\ 10^{-27}\end{cases}$	kg-m²/sec g-cm²/sec
Rydberg constant	R_∞	$1.097 \times$ $\begin{cases}10^{7} \\ 10^{5}\end{cases}$	m⁻¹ cm⁻¹
Speed of light in vacuum	c	$2.998 \times$ $\begin{cases}10^{8} \\ 10^{10}\end{cases}$	m/sec cm/sec

CONVERSION FACTORS

$1 \text{ eV} = 1.602 \times 10^{-19} \text{ J} = 1.602 \times 10^{-12} \text{ erg}$

$1 \text{ amu} \equiv 1.492 \times 10^{-10} \text{ J} = 1.492 \times 10^{-3} \text{ erg} = 931.5 \text{ MeV}$

$m_e \text{ (electron mass)} \equiv 8.19 \times 10^{-14} \text{ J} = 8.19 \times 10^{-7} \text{ erg}$
$= 0.511 \text{ MeV}$

$1 \text{ Å (angstrom)} = 10^{-10} \text{m} = 10^{-8} \text{ cm}$

$1 \text{ F (fermi)} \quad = 10^{-15} \text{ m} = 10^{-13} \text{ cm}$

$1 \text{ T (tesla)} \quad = 10^{4} \text{ gauss}$

Selected physical constants and conversion factors

Index

The parenthetical symbol (ex) following an entry|indicates treatment of the subject in an exercise.

664   Index

666 Index

668   Index

669   Index